Nutrition and Metabolism

Nutrition and Metabolism

Edited by **Dave Stewart**

R CALLISTO
REFERENCE

New York

Published by Callisto Reference,
106 Park Avenue, Suite 200,
New York, NY 10016, USA
www.callistoreference.com

Nutrition and Metabolism
Edited by Dave Stewart

International Standard Book Number: 978-1-63239-655-6 (Hardback)

Printed in the United States of America.

Contents

Preface   IX

Chapter 1   **Nutrition before and during Surgery and the Inflammatory Response of the Heart: A Randomized Controlled Trial**   **1**
Marlieke Visser, Hans W. M. Niessen, Wouter E. M. Kok, Riccardo Cocchieri, Willem Wisselink, Paul A. M. van Leeuwen and Bas A. J. M. de Mol

Chapter 2   **Prevalence of Anemia and Associated Factors among Pregnant Women in North Western Zone of Tigray, Northern Ethiopia: A Cross-Sectional Study**   **9**
Abel Gebre and Afework Mulugeta

Chapter 3   **Dietary Pattern and Metabolic Syndrome in Thai Adults**   **16**
W. Aekplakorn, W. Satheannoppakao, P. Putwatana, S. Taneepanichskul, P. Kessomboon, V. Chongsuvivatwong and S. Chariyalertsak

Chapter 4   **Maternal Fructose Intake Induces Insulin Resistance and Oxidative Stress in Male, but Not Female, Offspring**   **26**
Lourdes Rodríguez, Paola Otero, María I. Panadero, Silvia Rodrigo, Juan J. Álvarez-Millán and Carlos Bocos

Chapter 5   **Performance Enhancing Diets and the PRISE Protocol to Optimize Athletic Performance**   **34**
Paul J. Arciero, Vincent J. Miller and Emery Ward

Chapter 6   **Nutrition and Reproductive Health: Sperm versus Erythrocyte Lipidomic Profile and ω-3 Intake**   **73**
Gabriela Ruth Mendeluk, Mariano Isaac Cohen, Carla Ferreri and Chryssostomos Chatgilialoglu

Chapter 7   **Plasma Fatty Acids in Zambian Adults with HIV/AIDS: Relation to Dietary Intake and Cardiovascular Risk Factors**   **81**
Christopher K. Nyirenda, Edmond K. Kabagambe, John R. Koethe, James N. Kiage, Benjamin H. Chi, Patrick Musonda, Meridith Blevins, Claire N. Bosire, Michael Y. Tsai and Douglas C. Heimburger

Chapter 8   **Prevalence of Overweight and Obesity among Students in the Kumasi Metropolis**   **89**
D. B. Kumah, K. O. Akuffo, J. E. Abaka-Cann, D. E. Affram and E. A. Osae

Chapter 9   **A Comparative Study of the Metabolic and Skeletal Response of C57BL/6J and C57BL/6N Mice in a Diet-Induced Model of Type 2 Diabetes**   **93**
Elizabeth Rendina-Ruedy, Kelsey D. Hembree, Angela Sasaki, McKale R. Davis, Stan A. Lightfoot, Stephen L. Clarke, Edralin A. Lucas and Brenda J. Smith

Chapter 10   **Efficacy of Multiple Micronutrients Fortified Milk Consumption on Iron Nutritional Status in Moroccan Schoolchildren**   106
Imane El Menchawy, Asmaa El Hamdouchi, Khalid El Kari, Naima Saeid, Fatima Ezzahra Zahrou, Nada Benajiba, Imane El Harchaoui, Mohamed El Mzibri, Noureddine El Haloui and Hassan Aguenaou

Chapter 11   **Fasting Blood Glucose Profile among Secondary School Adolescents in Ado-Ekiti, Nigeria**   114
I. O. Oluwayemi, S. J. Brink, E. E. Oyenusi, O. A. Oduwole and M. A. Oluwayemi

Chapter 12   **Effects of 8-Prenylnaringenin and Whole-Body Vibration Therapy on a Rat Model of Osteopenia**   118
Daniel B. Hoffmann, Markus H. Griesel, Bastian Brockhusen, Mohammad Tezval, Marina Komrakova, Bjoern Menger, Marco Wassmann, Klaus Michael Stuermer and Stephan Sehmisch

Chapter 13   **Factors Associated with Anemia among Children Aged 6–23 Months Attending Growth Monitoring at Tsitsika Health Center, Wag-Himra Zone, Northeast Ethiopia**   127
Haile Woldie, Yigzaw Kebede and Amare Tariku

Chapter 14   **Fructose Metabolism and Relation to Atherosclerosis, Type 2 Diabetes, and Obesity**   136
Astrid Kolderup and Birger Svihus

Chapter 15   **Breakfast Protein Source Does Not Influence Postprandial Appetite Response and Food Intake in Normal Weight and Overweight Young Women**   148
Christina M. Crowder, Brianna L. Neumann and Jamie I. Baum

Chapter 16   **Short-Term High Fat Intake Does Not Significantly Alter Markers of Renal Function or Inflammation in Young Male Sprague-Dawley Rats**   156
Catherine Crinigan, Matthew Calhoun and Karen L. Sweazea

Chapter 17   **Awareness and Perception of Plant-Based Diets for the Treatment and Management of Type 2 Diabetes in a Community Education Clinic: A Pilot Study**   165
Vincent Lee, Taylor McKay and Chris I. Ardern

Chapter 18   **Evaluation of Antioxidant Capacity of *Solanum sessiliflorum* (Cubiu) Extract: An *In Vitro* Assay**   171
Diego Rocha de Lucena Herrera Mascato, Janice B. Monteiro, Michele M. Passarinho, Denise Morais Lopes Galeno, Rubén J. Cruz, Carmen Ortiz, Luisa Morales, Emerson Silva Lima and Rosany Piccolotto Carvalho

Chapter 19   **Absorption, Metabolism, and Excretion by Freely Moving Rats of 3,4-DHPEA-EDA and Related Polyphenols from Olive Fruits (*Olea europaea*)**   179
Shunsuke Kano, Haruna Komada, Lina Yonekura, Akihiko Sato, Hisashi Nishiwaki and Hirotoshi Tamura

Chapter 20   **Alanine with the Precipitate of Tomato Juice Administered to Rats Enhances the Reduction in Blood Ethanol Levels**   189
Shunji Oshima, Sachie Shiiya, Yoshimi Tokumaru and Tomomasa Kanda

Chapter 21   **Associations among Physical Activity, Diet, and Obesity Measures Change during Adolescence**   **197**
Janne H. Maier and Ronald Barry

Chapter 22   **Modulation of Metabolic Detoxification Pathways Using Foods and Food-Derived Components: A Scientific Review with Clinical Application**   **205**
Romilly E. Hodges and Deanna M. Minich

Chapter 23   **Treatment Outcome of Severe Acute Malnutrition Cases at the Tamale Teaching Hospital**   **228**
Mahama Saaka, Shaibu Mohammed Osman, Anthony Amponsem,
Juventus B. Ziem, Alhassan Abdul-Mumin, Prosper Akanbong, Ernestina Yirkyio,
Eliasu Yakubu and Sean Ervin

            Permissions

            List of Contributors

Preface

Over the recent decade, advancements and applications have progressed exponentially. This has led to the increased interest in this field and projects are being conducted to enhance knowledge. The main objective of this book is to present some of the critical challenges and provide insights into possible solutions. This book will answer the varied questions that arise in the field and also provide an increased scope for furthering studies.

Nutrition studies the interaction of nutrients with other substances present in food while metabolism deals with biochemical transformations occurring inside body cells. They together combine the holistic study of diabetes, lipidemias, obesity, exercise physiology and metabolic syndrome. Both of these work in tandem to provide human beings the knowledge of the effects of various foods and physical activities. From theories to research to practical applications, case studies related to all contemporary topics of relevance to this field have been included in this book. It has been compiled in such a manner, that it will provide in-depth knowledge about the theory and practice of these disciplines. This book will prove immensely beneficial to students and researchers in these fields.

I hope that this book, with its visionary approach, will be a valuable addition and will promote interest among readers. Each of the authors has provided their extraordinary competence in their specific fields by providing different perspectives as they come from diverse nations and regions. I thank them for their contributions.

Editor

Nutrition before and during Surgery and the Inflammatory Response of the Heart: A Randomized Controlled Trial

Marlieke Visser,[1,2] **Hans W. M. Niessen,**[3] **Wouter E. M. Kok,**[4] **Riccardo Cocchieri,**[1] **Willem Wisselink,**[2] **Paul A. M. van Leeuwen,**[2] **and Bas A. J. M. de Mol**[1]

[1]*Department of Cardiothoracic Surgery, Academic Medical Center, University of Amsterdam,*
P.O. Box 22700, 1100 DE Amsterdam, Netherlands
[2]*Department of Surgery, VU University Medical Center, P.O. Box 7057, 1007 MB Amsterdam, Netherlands*
[3]*Department of Pathology and Cardiac Surgery, ICaR-VU, VU University Medical Center,*
P.O. Box 7057, 1007 MB Amsterdam, Netherlands
[4]*Department of Cardiology, Academic Medical Center, University of Amsterdam, P.O. Box 22700, 1100 DE Amsterdam, Netherlands*

Correspondence should be addressed to Bas A. J. M. de Mol; b.a.demol@amc.uva.nl

Academic Editor: Maurizio Muscaritoli

Major surgery induces a long fasting time and provokes an inflammatory response which increases the risk of infections. Nutrition given before and during surgery can avoid fasting and has been shown to increase the arginine/asymmetric dimetlhylarginine ratio, a marker of nitric oxide availability, in cardiac tissue and increased concentrations of branched chain amino acids in blood plasma. However, the effect of this new nutritional strategy on organ inflammatory response is unknown. Therefore, we studied the effect of nutrition before and during cardiac surgery on myocardial inflammatory response. In this trial, 32 patients were randomised between enteral, parenteral, and no nutrition supplementation (control) from 2 days before, during, up to 2 days after coronary artery bypass grafting. Both solutions included proteins or amino acids, glucose, vitamins, and minerals. Myocardial atrial tissue was sampled before and after revascularization and was analysed immunohistochemically, subdivided into cardiomyocytic, fatty, and fibrotic areas. Inflammatory cells, especially leukocytes, were present in cardiac tissue in all study groups. No significant differences were found in the myocardial inflammatory response between the enteral, parenteral, and control groups. In conclusion, nutrition given before and during surgery neither stimulates nor diminishes the myocardial inflammatory response in patients undergoing coronary artery bypass grafting. The trial was registered in Netherlands Trial Register (NTR): NTR2183.

1. Introduction

Surgery provokes an inflammatory response in order to heal tissue damage but the same inflammatory response predisposes patients to the development of infections [1]. Surgical patients commonly receive only clear fluids during the period prior to surgery and the day after surgery leading to starvation of the patient over a longer period of time. Fasting can induce thirst, stress, insulin resistance [2], and nutrient deficiencies which can impair immune defence [3]. As most surgical patients are already in a catabolic state [4], prolonged fasting will impair recovery after surgery [2]. A previous study of patients undergoing cardiac surgery found that preoperative nutritional supplements decreased plasma concentrations of interleukin 6 (IL-6) and reduced the number of postoperative infections [5]. In other surgical populations too, pre- and postoperative nutritional supplementation has been found to reduce the number of postoperative infections [6]. Recently, for the first time, nutrition was given before, during, and after cardiac surgery in order to avoid a long fasting time [7]. It was found that this new nutritional strategy increased the arginine/asymmetric dimethylarginine (ADMA) ratio, a marker of nitric oxide (NO) availability, in cardiac tissue. Furthermore, nutrition increased the arginine/ADMA ratio and concentrations of branched chain amino acids (BCAA, i.e., leucine, isoleucine, and valine) in blood plasma, and an increase in plasma arginine/ADMA ratio correlated with improved myocardial viability. Both NO [8] and amino acids [9] are known to play an essential role in the inflammatory response. NO enhances immunity at a cellular level by increasing proliferation of lymphocytes and monocytes, enhancing T-helper cell formation, activating macrophage cytotoxicity, reinforcing natural killer cells, increasing phagocytosis, and enhancing cytokine production [8].

The amino acid arginine is an immune enhancing nutrient and plays an important role in immunity by many pathways from which its roles as substrate of the enzymes arginase and nitric oxide synthase (NOS) are most familiar [8]. However, while arginine is essential in the inflammatory response, high levels of the amino acids can result in negative outcome [10, 11]. Here, arginine might have induced excessive NO production by inducible nitric oxide synthase (iNOS) which in turn could have led to detrimental systemic vasodilation [8]. On the other hand, arginine is needed as precursor of NO mediated microvascular vasodilatation facilitated by endothelial NOS (eNOS) which is crucial for organ perfusion and coronary patency. Likely, NO availability needs to be perfectly balanced [12]. Therefore, nutritional formulas must be used that contain arginine but not in elevated levels. Furthermore, inadequate access to amino acids during immune cell activation may diminish immune response by inhibiting immune cell division, differentiation, and migration [9]. Therefore, by increasing myocardial and plasma arginine/ADMA ratio and amino acids, supplementation of nutrition before and during surgery without elevated concentrations of the immune enhancing nutrients might be a safe way that may influence myocardial inflammatory response.

Patients undergoing coronary artery bypass grafting (CABG) offer the possibility to study the effects of nutrition on the inflammatory response of the heart as cardiac tissue biopsies can be taken during this procedure. In order to prevent cardioplegic effects on cardiac cells, patients that can undergo an off-pump (i.e., without cardiopulmonary bypass) CABG procedure should be selected. Therefore, in this proof-of-concept trial, we investigated the effect of nutrition before and during surgery on the myocardial inflammatory response in patients undergoing off-pump CABG.

2. Materials and Methods

2.1. Study Design. This randomised controlled intervention study was carried out at the Department of Cardiothoracic Surgery at the Academic Medical Center of the University of Amsterdam (AMC) between July 2010 and August 2012. The study protocol [14] was approved by the Medical Ethics Committee of the AMC and the competent authority in Netherlands. A monitor verified that the trial was carried out in accordance with the protocol as described in the European Medicine Agency's "Note for guidance on good clinical practice CPMP/ICH/135/95" as well as the Declaration of Helsinki. Monitoring was performed and reported following the sponsor's standing operating procedures.

The primary end points of the study were cardiomyocytes structure (i.e., immunohistochemical analysis) and the concentrations of amino acids in myocardial tissue.

2.2. Nutrition Protocol. Patients were randomised to the enteral ($n = 12$), parenteral ($n = 9$), or control ($n = 11$) group. Randomization was performed online via a secure internet facility in a $1 : 1 : 1$ ratio by the TENALEA Clinical Trial Data Management System using randomly permuted blocks of sizes three and six.

2.2.1. Enteral Nutrition. During the four days before hospital admission, the enteral group received 125 mL per day of a nutrient drink (Nutridrink Compact, Nutricia, Zoetermeer, Netherlands) consisting of proteins, carbohydrates, fats, vitamins, and minerals (Table 1). When admitted to the hospital, patients in the enteral group received a solution containing amino acids (PeptoPro, DSM, Delft, Netherlands), carbohydrates (Fantomalt, Nutricia, Zoetermeer, Netherlands), and vitamins and minerals (Phlexy-Vits, SHS International Ltd., Liverpool, UK) which was prepared at the hospital each day. An amount of 1050 mL of the enteral nutrition was given during 24 hours. This nutrition was given two days before, during, and two days after CABG by a computerized guidance system-placed nasoduodenal tube (Cortrak, Viasys Healthcare, Wheeling, IL, USA). On the morning of surgery, the position of the duodenal tube was verified. Patients were permitted to eat and drink in addition to their supplemental nutrition.

2.2.2. Parenteral Nutrition. Patients in the parenteral group received 1250 mL of nutrition (Nutriflex Lipid peri, B. Braun, Oss, Netherlands) containing amino acids, lipids, and glucose. An amount of 1250 mL of the amino acid infusion (840 mOsm/L) was given in 24 hours for 5 days (Table 3). In addition, vitamins (Cernevit, Baxter, Utrecht, Netherlands) and trace elements (Nutritrace, B. Braun, Oss, Netherlands) were added to the parenteral nutrition. This nutrition was given two days before, during, and two days after CABG. Patients were permitted to eat and drink in addition to their supplemental nutrition.

2.2.3. Controls. The control group followed the standard protocol of the department of cardiothoracic surgery of the AMC allowing patients to eat and drink until six hours before surgery. The day after surgery, this standard protocol prescribes a (clear) liquid diet. On the second day after surgery, a normal diet is recommended for patients.

2.3. Patients. All 32 patients were due to undergo off-pump cardiac surgery for CABG and were aged between 55 and 76 years. Exclusion criteria were a combined valve and CABG procedure, pregnancy, renal insufficiency (defined as creatinine >95 μmol/L for women and >110 μmol/L for men), and liver insufficiency (defined as alanine aminotransferase >34 U/L for women and >45 U/L for men). All patients gave informed consent.

2.4. Surgical Procedures. Standard anesthestetic and off-pump CABG surgical procedures were used in this study.

2.5. Myocardial Tissue Sampling. During surgery, two tissue samples of the appendix of the right atrium were taken by

TABLE 1: Composition of enteral and parenteral nutrition.

	Enteral group		Parenteral group
	Drink (at home) per day	Nutrition (at hospital) per day	Nutrition (at hospital) per day
Volume (mL)	125	1050	1250
Amino acids (g)	12	80.5	40
Carbohydrates (g)	37.1	95	80
Fat (g)	11.6	1.5	50
Energy (kcal)	300	745	955
Vitamins and minerals	Yes	Yes	Yes

the surgeon. One sample was taken after the harvesting of the left mammary artery and before performing the distal anastomosis and one sample at the end of the procedure before closing the pericardium. Each sample was immediately put in formalin prior to immunohistochemical analysis.

2.6. Immunohistochemical Analyses. In myocardial samples, inflammatory cells were analysed and quantified using antibodies against CD45 (lymphocytes), myeloperoxidase (MPO: neutrophilic granulocytes), and CD68 (macrophages). In addition, cytokines (interleukin-6 (IL-6), IL-1β, and tumour necrosis factor-α (TNF-α)) were determined in endothelial cells. Endothelial cell activation was also analysed using antibodies against complement (C3d), P-selectin, E-selectin, and the advanced glycation end product (AGE) carboxymethyl lysine (CML). Complement was further analysed in erythrocytes and fibrotic tissue.

Tissue samples fixed in formalin were embedded in paraffin and sliced in 4 μm sections. Slides were deparaffinised and hydrated, and endogenous peroxidase activity was blocked by 0.03% hydrogen peroxide in methanol for 30 minutes. Enzymatic CD68, CML, E-selectin, and P-selectin antigen retrieval was done by incubating the tissue samples with 0.1% pepsin (activated with hydrochloric acid 37%, 1 : 600) for 30 minutes at 37°C. MPO, CD45, and C3d heat antigen retrieval was performed by heating the slides for 15 minutes in citrate (MPO; pH 6.0) and in Tris-EDTA (CD45; pH 9.0) at 100°C. After washing the sections in demineralised water and in PBS (pH 7.4), the slides were incubated with specific antibody solutions (diluted in PBS-BSA) for 60 minutes (anti-CD68 1 : 100, anti-MPO 1 : 500, anti-CD45 1 : 50, anti-CML 1 : 500, anti-C3d 1 : 1000, anti-E-selectin, and anti-P-selectin 1 : 50). Enzymatic IL-6, IL-1β, and TNF-α antigen retrieval was performed by boiling slides in a citrate pH 6.0 (IL-6 and IL-1β) or a Tris-EDTA pH 9.0 (TNF-α) solution in a microwave for 10 minutes. Next, sections were incubated with 1 : 500 monoclonal rabbit anti-human IL-1β (Abcam, UK) and 1 : 1000 monoclonal mouse anti-human TNF-α (Abcam, UK) for 1 hour at room temperature. For IL-6 staining, sections were incubated overnight with 1 : 100 monoclonal rabbit anti-human IL-6 (Santa Cruz, USA) at room temperature.

The slides were again washed in PBS, followed by 30-minute incubation with anti-rabbit and anti-mouse EnVision-HRP (DakoCytomation, Denmark). Staining was

visualised using 3,3-diaminobenzidine (DAB, 0.1 mg/mL, 0,02% H_2O_2). Sections were counterstained with haematoxylin, dehydrated, and covered. The same staining procedures were used as a control, but instead of the primary monoclonal or polyclonal antibody, PBS was used; these heart tissue slides were found to be negative.

2.7. Morphometrical Analyses. In atrial tissue, fatty tissue and fibrotic areas are found next to cardiomyocytic areas in amounts that vary not only between patients but also within different areas of the atrium of one patient. In all these three areas of the atria (as seen by histological view) within each myocardial tissue slide, the number of CD45-, CD68-, MPO-, and IL-6-positive inflammatory cells and E-selectin-, P-selectin-, IL-1β-, TNF-α-, and C3d-positive endothelial vessels was measured separately as the number of positive cells or vessels/mm^2 myocardium using Q-PRODIT (Leica, Cambridge, UK). Due to the high number of CD45-positive cells in fibrotic tissue, separate positive cells could not be identified and quantified as such. Therefore, the surface of the CD45-positive area was calculated and presented per μm^2 myocardium. Additionally, C3d-positive erythrocytes and fibrotic areas were graded based on an intensity score; namely, 1 = no or weak intensity, 2 = moderate intensity, and 3 = strong intensity. CML-positive endothelial cells were semiquantified based on an intensity score for each positive vessel as follows: 1 = weak positivity, 2 = moderate positivity, and 3 = strong positivity [15]. Each CML intensity score was multiplied by the number of vessels that scored positively. The multiplication scores were then added and the sum was divided by the area of the slide, resulting in an immunohistochemical score per mm^2.

2.8. Statistical Analysis. Data were expressed as mean ± standard deviation (SD) where data were normally distributed and as median with interquartile range (IQR) where data were not normally distributed. To investigate the effects of administration of enteral or parenteral nutrition during surgery, differences between the study groups in the first and second tissue samples were analysed using the Kruskal-Wallis test. A p value of < 0.05 (2-tailed) was considered statistically significant. All statistical analyses were performed with SPSS 20.0 for Windows (SPSS Inc., Chicago, IL, USA).

TABLE 2: Patient characteristics and postoperative outcome.

	Enteral group ($n = 12$)	Parenteral group ($n = 9$)	Control group ($n = 11$)
Patients			
Age (years)	66.1 ± 6.1	66.6 ± 7.1	63.3 ± 6.7
Gender (% male)	12 (100)	10 (100)	11 (100)
BMI (kg/m^2)	27.8 (27.0–30.9)	27.4 (26.2–30.3)	27.7 (24.2–30.0)
Fat-free mass index (kg/m^2)	19.8 (18.3–21.0)	19.5 (18.2–21.1)	20.9 (18.4–21.7)
Diabetes mellitus (%)	3 (25.0)	4 (44.4)	3 (27.3)
EuroSCORE*	2.0 (1.3–3.0)	3.0 (3.0–4.0)	2.0 (0.0–3.0)
Previous acute myocardial infarction (%)	5 (41.7)	6 (66.7)	2 (18.2)
Preoperative laboratory tests			
Plasma CRP (mg/L)	0.8 (0.7–1.6)	2.7 (0.9–5.9)	0.6 (0.6–3.2)
Plasma albumin (g/L)	46.5 (45.3–48.8)	43.0 (41.0–47.0)	46.0 (45.0–48.0)
Plasma NT-proBNP (ng/L)	234 (54–522)	204 (121–439)	99 (71–431)
Intraoperative period			
Propofol use (%)	8 (66.7)	4 (44.4)	8 (72.7)
Surgery duration (min)	268 ± 39	279 ± 27	294 ± 40
Postoperative period			
Plasma CK-MB (μg/L)	9.0 (6.4–15.0)	8.2 (6.2–8.9)	10.3 (7.6–15.5)
Intensive care stay (hours)	21.5 (18.0–23.8)	24.0 (22.0–34.5)	21.0 (20.0–22.0)
Stent (%)	0 (0)	0 (0)	0 (0)
Catheterisation < 3 months (%)	0 (0)	0 (0)	1 (9.1)
Revascularisation < 3 months (%)	0 (0)	0 (0)	0 (0)
Infections < 3 months (%)	0 (0)	0 (0)	1 (9.1)
Mortality (%)	0 (0)	0 (0)	0 (0)

Values are expressed as median (IQR) or mean ± SD. *EuroSCORE, European System for Cardiac Operation Risk Evaluation score. The EuroSCORE is a validated risk stratification system to determine the risk profile for mortality of cardiothoracic surgery patients [13].

3. Results and Discussion

3.1. Patient Characteristics and Myocardial Tissue Samples. Although 38 patients were enrolled, only 32 patients were ultimately included. In two patients, the CABG procedure was switched from off-pump to on-pump (which was an exclusion criterion). In one patient, the appendix of the right atrium could not be reached, the atrial tissue of one patient was lost during immunohistochemical analysis, and in two patients the peripheral line was removed before surgery. Patient, laboratory and operation characteristics, and data about postoperative outcome are presented in Table 2. Concerning these characteristics, no statistically significant differences between the groups were observed.

3.2. The Effect of Nutrition during Surgery on Myocardial Inflammatory Response. Table 3 shows the number of inflammatory cells and cytokines and endothelial cell activation in the first and second myocardial tissue samples in each study group. Mean time between sampling of the first and second tissue samples was 97 ± 28 min.

The first tissue samples comprised (median (IQR)) 56.6% (27.8–74.7) cardiomyocyte areas, 29.5% (22.7–45.0) fibrotic areas, and 0.98% (0.00–26.3) fatty areas. In the second tissue samples, these values were 56.5% (14.0–73.6), 31.1% (15.6–42.6), and 0.82% (0.00–33.30), respectively. These percentages had no statistically significant differences between the first and second tissue samples.

Lymphocytes were present in all three areas in both the first and the second tissue samples. Neutrophilic granulocytes in cardiomyocytic and fibrotic tissue were present in the second tissue sample but were not present in most samples of all areas taken at the start of surgery and of fibrotic and fatty tissue taken at the end of surgery. Macrophages were not present in most tissue samples of all three areas, both at the start and at the end of surgery. IL-6 was frequently present in cardiomyocytic areas of both the first and the second tissue samples. In fibrotic and fatty tissue areas, IL-6 was not found

TABLE 3: Inflammatory cells in the 1st and 2nd myocardial tissue samples (number/mm^2) in study groups.

	Start of surgery (1st tissue sample)			End of surgery (2nd tissue sample)		
	Enteral group ($n = 12$)	Parenteral group ($n = 9$)	Control group ($n = 11$)	Enteral group ($n = 12$)	Parenteral group ($n = 9$)	Control group ($n = 11$)
Lymphocytes (CD45)						
Cardiomyocytic area	5.03 (1.58–15.17)	0.00 (0.00–8.64)	10.23 (3.44–15.51)	5.75 (0.81–6.88)	3.92 (0.00–6.53)	1.91 (0.00–10.10)
Fibrotic area (μm^2)	154.10 (35.43–1215.60)	93.90 (36.17–2080.91)	241.18 (120.80–481.83)	178.09 (31.01–5449.95)	47.39 (9.23–147.22)	60.25 (0.08–138.74)
Fatty area	8.54 (0.89–12.54)	4.06 (0.16–10.43)	6.22 (0.00–120.05)	3.22 (0.00–5.08)	2.56 (1.19–93.43)	7.24 (3.22–22.85)
Neutrophil granulocytes (MPO)						
Cardiomyocytic area	0.00 (0.00–14.23)	0.00 (0.00–0.31)	1.47 (0.00–3.62)	4.10 (0.00–15.86)	3.07 (0.32–6.53)	0.65 (0.00–1.94)
Fibrotic area	0.00 (0.00–0.00)	0.00 (0.00–11.87)	0.00 (0.00–1.72)	0.00 (0.00–32.44)	6.72 (0.00–12.92)	1.64 (0.00–15.56)
Fatty area	0.00 (0.00–0.00)	0.00 (0.00–0.16)	0.00 (0.00–0.00)	0.00 (0.00–1.81)	0.00 (0.00–2.49)	0.00 (0.00–8.52)
Macrophages (CD68)						
Cardiomyocytic area	0.00 (0.00–0.00)	0.00 (0.00–0.00)	0.00 (0.00–1.03)	0.00 (0.00–0.00)	0.00 (0.00–1.12)	0.00 (0.00–0.38)
Fibrotic area	0.00 (0.00–0.95)	0.00 (0.00–9.86)	1.72 (0.00–8.24)	0.00 (0.00–4.32)	0.72 (0.00–13.49)	2.11 (0.00–6.62)
Fatty area	0.00 (0.00–0.49)	0.00 (0.00–1.28)	0.00 (0.00–1.05)	0.00 (0.00–1.21)	0.00 (0.00–2.32)	0.00 (0.00–0.66)
IL-6						
Cardiomyocytic area	28.72 (0.48–77.38)	0.00 (0.00–18.91)	19.20 (1.20–88.94)	36.03 (0.00–80.77)	20.53 (0.79–43.25)	14.53 (0.00–55.17)
Fibrotic area	0.00 (0.00–7.04)	0.00 (0.00–0.000)	2.47 (0.00–6.39)	0.00 (0.00–2.26)	0.00 (0.00–5.90)	0.00 (0.00–3.37)
Fatty area	0.00 (0.00–0.00)	0.00 (0.00–2.50)	0.00 (0.00–0.00)	0.00 (0.00–0.00)	0.00 (0.00–1.67)	0.00 (0.00–0.00)
IL-1β endothelium	0.00 (0.00–0.60)	0.00 (0.00–0.50)	0.00 (0.00–1.72)	0.00 (0.00–1.86)	0.30 (0.00–0.79)	0.00 (0.00–0.00)
TNF-α endothelium	0.00 (0.00–0.00)	0.00 (0.00–0.00)	0.00 (0.00–0.00)	0.00 (0.00–0.00)	0.00 (0.00–0.00)	0.00 (0.00–0.00)
Activated endothelium						
P-selectin	2.17 (0.17–3.76)	1.41 (0.87–5.43)	3.37 (0.69–6.29)	2.29 (0.57–6.93)	3.33 (1.27–5.06)	1.70 (0.32–6.46)
E-selectin	0.00 (0.00–0.00)	0.00 (0.00–0.30)	0.00 (0.00–0.00)	0.17 (0.00–0.48)	0.14 (0.00–0.46)	0.00 (0.00–0.42)
Proinflammatory vessel damage (CML)	26.00 (7.58–77.44)	20.78 (5.23–68.66)	46.31 (4.12–134.68)	28.78 (5.29–85.01)	21.18 (12.12–45.80)	21.21 (9.28–81.26)
C3d						
Endothelium	0.00 (0.00–0.36)	0.09 (0.00–0.25)	0.00 (0.00–0.36)	0.00 (0.00–0.32)	0.00 (0.00–0.27)	0.00 (0.00–0.14)
Erythrocytes intensity score > 1	41.7%	22.2%	9.1%	58.3%	44.4%	27.3%
Fibrotic area intensity score > 1	8.3%	11.1%	9.1%	75.0%	55.6%	54.5%

Values are median (IQR).
MPO, myeloperoxidase; CML, carboxymethyl lysine.

in most samples taken at the start and the end of surgery. IL-1β was not shown in most endothelial cells in all areas of the atria of the first and second tissue samples. TNF-α was not found in endothelial vessels of the first or second tissue samples. In all areas of the atria, P-selectin was present in both the first and the second tissue samples while E-selectin was not found in the first tissue sample and in very low numbers in the second tissue sample. CML was present in both the first and the second tissue samples of all areas of the atria. C3d was not present in most endothelium of all areas of the atria of the first and second tissue samples. C3d-positive erythrocytes and fibrotic areas were found in both the first and the second tissue samples.

In both the first and the second tissue samples, the number of inflammatory cells and cytokines and endothelial cell activation did not differ significantly between the enteral, parenteral, and control groups.

3.3. Discussion. In this study, the effect of nutritional supplementation before and during surgery on the myocardial inflammatory response in cardiac surgery was investigated. While inflammatory cells were demonstrated in cardiac tissue at the start and end of surgery, nutrition before and during cardiac surgery did not affect myocardial inflammatory response in these patients.

Both enteral and parenteral nutrition included amino acids. Amino acids are essential for a proper immune cell response [9]. When immune cells are activated by inflammatory signals, their demand for amino acids increases rapidly which may explain the low plasma levels of amino acids in surgical patients [16]. In a previous analysis of this study data, nutrition before and during surgery increased the myocardial and plasma arginine/ADMA ratio which may have beneficial effects on myocardial glucose metabolism [7]. Furthermore, plasma concentrations of BCAA were higher in the enteral and parenteral groups than in the control group. Increasing BCAA might be beneficial as they function as precursors for myocardial protein synthesis [17, 18] while inflammation is known to increase protein degradation and to attenuate protein synthesis [19]. By being precursor of NO, the amino acid arginine is important for regulation of inflammation and immunity [8]. However, production of NO can be reduced by the NO synthase (NOS) inhibitor ADMA which is considered a risk factor for cardiovascular disease and an indicator of worse outcome in patients with cardiac dysfunction [20, 21]. The net production of NO probably depends on the ratio between substrate and inhibitor: the arginine/ADMA ratio. While previous data analysis showed that nutrition before and during surgery increased the arginine/ADMA ratio and BCAA [7], nutrition during surgery did not affect myocardial inflammatory response.

A lack of any effect of our nutritional intervention may be found in the relatively low amount of immune-modulating nutrients and low amount of calories in the (par)enteral nutrition. However, as this was the first study investigating the effect of nutrition during surgery, we supplied hypocaloric nutrition which included the immunomodulator arginine but not in elevated concentrations [14]. These characteristics of our nutrition are supported by results from recent studies showing that high levels of immunomodulating nutrients [22, 23] and overfeeding [24–27] can have negative effects on clinical outcome. The negative effect of the immunomodulating nutrients is ascribed to their stimulating effects on proliferation of immune cells [28] and production of excessive NO [21] and thereby augmenting the inflammatory response which negatively influences outcome. In inflammatory states, supplementation of high amounts of arginine or glutamine may induce an excess in NO production by iNOS which can be deleterious because it may lead to detrimental vasodilatation [8] and to increased formation of ROS leading to cellular damage [29]. Furthermore, glutamine might stimulate lymphocyte proliferation and cytokine production and may enhance the immune response [28]. Therefore, immunomodulating nutrition is not recommended in inflammatory states [21, 28, 30]. On the other hand, an increase in NO facilitated by eNOS is of vital importance as it mediates microvascular vasodilatation. Likely, NO availability needs to be perfectly balanced [12, 21]. The negative findings of overfeeding in previous studies are ascribed to the increase in fat mass and hyperglycemia [24–27]. For example, early parenteral nutrition in addition to enteral nutrition has been shown to induce fat incorporation in muscle tissue and to negatively affect clinical outcome [24, 25]. Results of previous studies investigating the effect of hypocaloric nutrition show contrary results. In the ICU, hypocaloric nutrition has been shown to have similar [31] and negative effects [32] compared to normocaloric nutrition. However, both studies included patients requiring artificial nutrition who might not have received enough protein and/or calories while in our study patients received nutrition in order to avoid fasting. Additionally, we hypothesize that a lack of physical activity plays an important role in the lack of any (beneficial) effect in our study and previous studies as nutritional interventions may be most effective together with anabolic stimuli [33–37]. Probably, nutritional supplementation with an anabolic stimulus like physical activity may result in an increase in muscle mass and strength (instead of fat mass) which is related to better clinical outcome [38, 39]. Unfortunately, studies combining the effects of nutrition and physical activity are very scarce. Therefore, future studies investigating nutritional supplementation should incorporate an anabolic stimulus, like exercise training, in their study protocol.

As a first in line, this study had limitations. First, it was a proof-of-concept trial that tested the novel strategy of nutrition during surgery as a form of perioperative treatment. The sample size was powered for the primary outcome (i.e., amino acids and immunohistochemistry in myocardial tissue) and a significant difference between the (par)enteral and control groups. Unfortunately, the intended sample size could not be reached because of the low consent numbers. Nevertheless, in the current study population, our results show that it is feasible to supply nutrition during surgery which did not affect myocardial inflammation. No adverse events were seen in our study. Second, it is not known if the inflammatory cells are evenly distributed throughout the right atrial appendix and other parts of the myocardium. However, the availability of small samples of atrial appendix obtained in the course of off-pump CABG provided a sound model to study inflammatory response effects. Finally, true baseline values of the inflammatory response in myocardial tissue could not be measured as it was not possible to take myocardial biopsies before surgery.

4. Conclusions

Our study demonstrated the presence of inflammatory cells in the heart of patients undergoing off-pump CABG. Nutrition before and during surgery neither stimulated nor diminished the inflammatory response of the myocardium. Future studies should focus on the effects of nutritional supplementation in combination with physical activity and investigate whether continuing of nutritional interventions during surgery is beneficial for immune response and clinical outcome.

Conflict of Interests

The authors declare that there is no conflict of interests regarding the publication of this paper.

Acknowledgments

The authors would like to thank the attending nurses, physicians, residents, and nonmedical staff of the Department of Cardiothoracic Surgery, Anaesthesiology, Intensive Care Unit, and Pharmacy at the AMC for their help in patient care and protocol compliance, Max Ramali from the Department of Gastroenterology at the AMC for the placement and verification of the duodenal tubes, and Ibrahim Korkmaz from the Department of Pathology of the VUmc for expert assistance in immunohistochemical analysis of tissue. Last, the authors would like to thank all participants who were willing to participate in the study.

References

[1] B. A. Kohl and C. S. Deutschman, "The inflammatory response to surgery and trauma," *Current Opinion in Critical Care*, vol. 12, no. 4, pp. 325–332, 2006.

[2] O. Ljungqvist and E. Søreide, "Preoperative fasting," *British Journal of Surgery*, vol. 90, no. 4, pp. 400–406, 2003.

[3] S. Bengmark, R. Andersson, and G. Mangiante, "Uninterrupted perioperative enteral nutrition," *Clinical Nutrition*, vol. 20, no. 1, pp. 11–19, 2001.

[4] S. M. Jakob and Z. Stanga, "Perioperative metabolic changes in patients undergoing cardiac surgery," *Nutrition*, vol. 26, no. 4, pp. 349–353, 2010.

[5] R. Tepaske, H. Te Velthuis, H. M. Oudemans-van Straaten et al., "Effect of preoperative oral immune-enhancing nutritional supplement on patients at high risk of infection after cardiac surgery: a randomised placebo-controlled trial," *The Lancet*, vol. 358, no. 9283, pp. 696–701, 2001.

[6] P. E. Marik and M. Flemmer, "Immunonutrition in the surgical patient," *Minerva Anestesiologica*, vol. 78, no. 3, pp. 336–341, 2012.

[7] M. Visser, M. Davids, H. J. Verberne et al., "Nutrition before, during, and after surgery increases the arginine: asymmetric

dimethylarginine ratio and relates to improved myocardial glucose metabolism: a randomized controlled trial," *The American Journal of Clinical Nutrition*, vol. 99, no. 6, pp. 1440–1449, 2014.

[8] U. Suchner, D. K. Heyland, and K. Peter, "Immune-modulatory actions of arginine in the critically ill," *British Journal of Nutrition*, vol. 87, no. 1, pp. S121–S132, 2002.

[9] T. L. Mcgaha, L. Huang, H. Lemos et al., "Amino acid catabolism: a pivotal regulator of innate and adaptive immunity," *Immunological Reviews*, vol. 249, no. 1, pp. 135–157, 2012.

[10] G. Bertolini, G. Iapichino, D. Radrizzani et al., "Early enteral immunonutrition in patients with severe sepsis: results of an interim analysis of a randomized multicentre clinical trial," *Intensive Care Medicine*, vol. 29, no. 5, pp. 834–840, 2003.

[11] D. L. Dent, "Immunonutrition may increase mortality in critically ill patients with pneumonia: results of a randomized trial," *Critical Care Medicine*, vol. 30, no. 12, article A17, 2002.

[12] J.-C. Preiser, Y. Luiking, and N. Deutz, "Arginine and sepsis: a question of the right balance?" *Critical Care Medicine*, vol. 39, no. 6, pp. 1569–1570, 2011.

[13] S. A. M. Nashef, F. Roques, P. Michel, E. Gauducheau, S. Lemeshow, and R. Salamon, "European system for cardiac operative risk evaluation (EuroSCORE)," *European Journal of Cardiothoracic Surgery*, vol. 16, no. 1, pp. 9–13, 1999.

[14] M. Visser, M. Davids, H. J. Verberne et al., "Rationale and design of a proof-of-concept trial investigating the effect of uninterrupted perioperative (par)enteral nutrition on amino acid profile, cardiomyocytes structure, and cardiac perfusion and metabolism of patients undergoing coronary artery bypass grafting," *Journal of Cardiothoracic Surgery*, vol. 6, no. 1, article 36, 2011.

[15] A. Baidoshvili, H. W. M. Niessen, W. Stooker et al., "N^{ε}-(carboxymethyl)lysine depositions in human aortic heart valves: similarities with atherosclerotic blood vessels," *Atherosclerosis*, vol. 174, no. 2, pp. 287–292, 2004.

[16] M.-S. Suleiman, H. C. Fernando, W. C. Dihmis, J. A. Hutter, and R. A. Chapman, "A loss of taurine and other amino acids from ventricles of patients undergoing bypass surgery," *Heart*, vol. 69, no. 3, pp. 241–245, 1993.

[17] H. E. Morgan, D. C. Earl, A. Broadus, E. B. Wolpert, K. E. Giger, and L. S. Jefferson, "Regulation of protein synthesis in heart muscle. I. Effect of amino acid levels on protein synthesis," *The Journal of Biological Chemistry*, vol. 246, no. 7, pp. 2152–2162, 1971.

[18] L. H. Young, P. H. McNulty, C. Morgan, L. I. Deckelbaum, B. L. Zaret, and E. J. Barrett, "Myocardial protein turnover in patients with coronary artery disease: effect of branched chain amino acid infusion," *Journal of Clinical Investigation*, vol. 87, no. 2, pp. 554–560, 1991.

[19] H. Nicastro, C. R. da Luz, D. F. S. Chaves et al., "Does branched-chain amino acids supplementation modulate skeletal muscle remodeling through inflammation modulation? Possible mechanisms of action," *Journal of Nutrition and Metabolism*, vol. 2012, Article ID 136937, 10 pages, 2012.

[20] M. Visser, W. J. Paulus, M. A. R. Vermeulen et al., "The role of asymmetric dimethylarginine and arginine in the failing heart and its vasculature," *European Journal of Heart Failure*, vol. 12, no. 12, pp. 1274–1281, 2010.

[21] M. Visser, M. A. R. Vermeulen, M. C. Richir et al., "Imbalance of arginine and asymmetric dimethylarginine is associated with markers of circulatory failure, organ failure and mortality in shock patients," *The British Journal of Nutrition*, vol. 107, no. 10, pp. 1458–1465, 2012.

[22] D. Heyland, J. Muscedere, P. E. Wischmeyer et al., "A randomized trial of glutamine and antioxidants in critically Ill patients," *The New England Journal of Medicine*, vol. 368, no. 16, pp. 1489–1497, 2013.

[23] A. R. H. van Zanten, F. Sztark, U. X. Kaisers et al., "High-protein enteral nutrition enriched with immune-modulating nutrients vs standard high-protein enteral nutrition and nosocomial infections in the ICU: a randomized clinical trial," *The Journal of the American Medical Association*, vol. 312, no. 5, pp. 514–524, 2014.

[24] M. P. Casaer, D. Mesotten, G. Hermans et al., "Early versus late parenteral nutrition in critically ill adults," *The New England Journal of Medicine*, vol. 365, no. 6, pp. 506–517, 2011.

[25] M. P. Casaer, L. Langouche, W. Coudyzer et al., "Impact of early parenteral nutrition on muscle and adipose tissue compartments during critical illness," *Critical Care Medicine*, vol. 41, no. 10, pp. 2298–2309, 2013.

[26] P. Singer, M. M. Berger, G. van den Berghe et al., "ESPEN guidelines on parenteral nutrition: intensive care," *Clinical Nutrition*, vol. 28, no. 4, pp. 387–400, 2009.

[27] L. M. W. van Venrooij, P. A. M. van Leeuwen, R. de Vos, M. M. M. J. Borgmeijer-Hoelen, and B. A. J. M. de Mol, "Preoperative protein and energy intake and postoperative complications in well-nourished, non-hospitalized elderly cardiac surgery patients," *Clinical Nutrition*, vol. 28, no. 2, pp. 117–121, 2009.

[28] H. M. Oudemans-van Straaten and A. R. H. van Zanten, "Glutamine supplementation in the critically ill: friend or foe?" *Critical Care*, vol. 18, no. 3, article 143, 2014.

[29] Y. Xia, V. L. Dawson, T. M. Dawson, S. H. Snyder, and J. L. Zweier, "Nitric oxide synthase generates Superoxide and nitric oxide in arginine-depleted cells leading to peroxynitrite-mediated cellular injury," *Proceedings of the National Academy of Sciences of the United States of America*, vol. 93, no. 13, pp. 6770–6774, 1996.

[30] G. van den Berghe, "Low glutamine levels during critical illness—adaptive or maladaptive?" *The New England Journal of Medicine*, vol. 368, no. 16, pp. 1549–1550, 2013.

[31] E. J. Charles, R. T. Petroze, R. Metzger et al., "Hypocaloric compared with eucaloric nutritional support and its effect on infection rates in a surgical intensive care unit: a randomized controlled trial," *American Journal of Clinical Nutrition*, vol. 100, no. 5, pp. 1337–1343, 2014.

[32] S. Petros, M. Horbach, F. Seidel, and L. Weidhase, "Hypocaloric vs normocaloric nutrition in critically ill patients: a prospective randomized pilot trial," *Journal of Parenteral & Enteral Nutrition*, 2014.

[33] R. R. Wolfe, "Protein supplements and exercise," *The American Journal of Clinical Nutrition*, vol. 72, pp. 551–557, 2000.

[34] M. Beelen, L. M. Burke, M. J. Gibala, and L. J. C. van Loon, "Nutritional strategies to promote postexercise recovery," *International Journal of Sport Nutrition and Exercise Metabolism*, vol. 20, no. 6, pp. 515–532, 2010.

[35] G. Biolo, B. Ciocchi, M. Lebenstedt et al., "Short-term bed rest impairs amino acid-induced protein anabolism in humans," *Journal of Physiology*, vol. 558, no. 2, pp. 381–388, 2004.

[36] S. Baldi, R. Aquilani, G. D. Pinna, P. Poggi, A. De Martini, and C. Bruschi, "Fat-free mass change after nutritional rehabilitation in weight losing COPD: role of insulin, C-reactive protein and tissue hypoxia," *International Journal of Chronic Obstructive Pulmonary Disease*, vol. 5, no. 1, pp. 29–39, 2010.

[37] K. N. Jeejeebhoy, "Malnutrition, fatigue, frailty, vulnerability, sarcopenia and cachexia: overlap of clinical features," *Current Opinion in Clinical Nutrition and Metabolic Care*, vol. 15, no. 3, pp. 213–219, 2012.

[38] L. M. W. van Venrooij, H. J. Verberne, R. de Vos, M. M. M. J. Borgmeijer-Hoelen, P. A. M. van Leeuwen, and B. A. J. M. de Mol, "Postoperative loss of skeletal muscle mass, complications and quality of life in patients undergoing cardiac surgery," *Nutrition*, vol. 28, no. 1, pp. 40–45, 2012.

[39] M. Visser, L. M. W. van Venrooij, L. Vulperhorst et al., "Sarcopenic obesity is associated with adverse clinical outcome after cardiac surgery," *Nutrition, Metabolism and Cardiovascular Diseases*, vol. 23, no. 6, pp. 511–518, 2013.

Prevalence of Anemia and Associated Factors among Pregnant Women in North Western Zone of Tigray, Northern Ethiopia: A Cross-Sectional Study

Abel Gebre[1] and Afework Mulugeta[2]

[1]Department of Public Health, College of Health Sciences, Samara University, Samara, Ethiopia
[2]Department of Public Health, College of Health Sciences, Mekelle University, Mekelle, Ethiopia

Correspondence should be addressed to Abel Gebre; abelgebre21@gmail.com

Academic Editor: Simin Liu

Background. Anemia affects the lives of more than 2 billion people globally, accounting for over 30% of the world's population. Anemia is a global public health problem occurring at all stages of the life cycle but the burden of the problem is higher in pregnant women particularly in developing countries. The aim of this study was to determine the prevalence of anemia and associated factors among pregnant women attending antenatal clinics in north western zone of Tigray, northern Ethiopia. *Methods.* A facility based cross-sectional study was employed. A systematic random sampling procedure was employed to select 714 pregnant women who were attending antenatal clinics in health facilities found in the study area from April to May 2014. The data was entered and analyzed using Epi-info version 3.5.1 and SPSS version 20.0 statistical software, respectively. Logistic regression analysis was used to identify factors associated with anemia among the study participants. All tests were two-sided and p value < 0.05 was considered statistically significant. *Results.* The overall prevalence of anemia (hemoglobin < 11 g/dL) among the pregnant women was 36.1% (95% CI = 32.7%–39.7%) of which 58.5% were mildly, 35.7% moderately, and 5.8% severely anemic. In pregnant women, rural residence (AOR = 1.75, 95% CI = 1.01–3.04), no education/being illiterate (AOR = 1.56, 95% CI = 1.03–2.37), absence of iron supplementation during pregnancy (AOR = 2.76, 95% CI = 1.92–5.37), and meal frequency of less than two times per day (AOR = 2.28, 95% CI = 1.06–4.91) were the independent predictors for increased anemia among the pregnant women. *Conclusions.* Anemia was found to be moderate public health problem in the study area. Residence, educational status, iron supplementation during pregnancy, and meal frequency per day were statistically associated with anemia among the pregnant women. Awareness creation and nutrition education on the importance of taking iron supplementation and nutritional counseling on consumption of extra meal and iron-rich foods during pregnancy are recommended to prevent anemia in the pregnant women.

1. Background

Anemia affects the lives of more than 2 billion people globally, accounting for over 30% of the world's population which is the most common public health problem particularly in developing countries occurring at all stages of the life cycle [1, 2]. The prevalence of anemia in developing and developed countries is estimated to be 43% and 9%, respectively [3].

Anemia in pregnant women remains one of the most intractable public health problems in developing countries because of various sociocultural problems like illiteracy, poverty, lack of awareness, cultural and religious taboos, poor dietary habits, and high prevalence of parasitic infestation. Current estimates from the World Health Organization (WHO) put prevalence of anemia at 41.8% among pregnant women, with the highest prevalence rate (61.3%) found among pregnant women in Africa and 52.5% among South East Asia [4–6]. Sub-Saharan Africa is the most affected region, with anemia prevalence among pregnant women estimated to be 17.2 million, which corresponds to approximately 30% of total global cases [7]. Globally, anemia contributes to 20% of all maternal deaths. Anemia in pregnancy may also lead to premature births, low birth weight, fetal impairment, and infant deaths [4, 8, 9].

Anemia during pregnancy is defined by the Centers of Disease Control and Prevention (CDC) and World Health Organization as a hemoglobin concentration less than 11 g/dL. It is considered severe when hemoglobin concentration is less than 7.0 g/dL, moderate when hemoglobin falls between 7.0 and 9.9 g/dL, and mild when hemoglobin is from 10.0 to 10.9 g/dL [10–12]. According to the Ethiopian Demographic and Health Survey (EDHS) report, 17% of reproductive age women are estimated to be anemic and 22% of the pregnant women are anemic [13]. The contextual factors contributing for anemia among pregnant women are different. Interaction of multiple factors like women's sociodemographic, economic, nutritional, and health related factors causes anemia in pregnant women. The availability of local information on the magnitude and related risk factors has a major role in the management and control of anemia in pregnancy. However, there is no adequate and reputable information on the prevalence and factors leading to anemia in pregnant women in Ethiopia and the study area in particular. Therefore, the aim of this study was to determine the prevalence of anemia and associated factors among pregnant women attending antenatal clinics in north western zone of Tigray, northern Ethiopia.

2. Methods

2.1. Study Setting and Design. A facility based cross-sectional study was conducted from April to May 2014 in randomly selected health facilities found in north western zone of Tigray which is located at 1087 Kilometer from Addis Ababa and 304 Kilometer from Mekelle, the regional capital city. According to the 2007 national census, the total estimated pregnant women in study area are 25,052. There are two governmental hospitals and twenty-eight public health centers which provides routine antenatal care service to the community [14].

2.2. Study Population. All pregnant women attending antenatal care in the governmental health facilities in the study area were target for the study. The study population consisted of a sample of pregnant women who were residing in the study area during the study period and attending heath facilities found in the study area. Those pregnant women who were not long-term residents of the study area (less than 6 months) were excluded. All pregnant women were excluded from the study if they have any of the following disorders including being seriously ill, mental disorder, and women who are unable to hear and/or speak during data collection period.

2.3. Sample Size and Sampling Procedures. Sample size was determined based on the single population proportion formula using $Z^2 \times p \times q/d^2$ with a 95% CI, 5% margin of error, and an assumption that 31.6% of pregnant women are anemic in the study area [15]. Assuming a 10% nonresponse rate and a design effect of 2, a total sample size of 731 pregnant women was required. Multistage sampling technique was used to select the study participants. One hospital and twelve health centers which provide routine antenatal care services for the pregnant women were selected using a lottery method.

A proportional allocation was employed to obtain the sample size from the selected health facilities and a systematic random sampling method was used to select the study participants from each antenatal clinic in the respective health facilities during the data collection period.

2.4. Data Collection. Data was collected using pretested interviewer administered questionnaire, which contains sociodemographic characteristics (age, education, occupation, marital status, and others), obstetric and gynecological history (trimester, gravidity, parity, and others), and dietary factors (iron intake, meal frequency, intake of coffee or tea, etc.). Blood hemoglobin concentration was measured using a HemoCue Hb 301 analyzer (manufactured by Ängelholm, Sweden), a precalibrated instrument designed for the measurement of hemoglobin concentration. Venous blood was drawn, through microcuvettes, and inserted into the HemoCue Hb analyzer and the result was recorded.

2.5. Statistical Analysis. Data were analyzed using SPSS version 20 after the data were entered to Epi-info version 3.5.1 and exported to it. Categorical variables were summarized as numbers and percentages, whereas normally distributed continuous variables were presented as means and standard deviations. To identify factors associated with the outcome variable (anemia), first a bivariate logistic regression analysis was performed for each independent variable and crude odds ratio (COR) with 95% confidence intervals was obtained. Then, significant variables observed in the bivariate logistic regression analysis (p value < 0.2) were subsequently included in the multivariable logistic regression model to determine independent predictors for the outcome variable among the pregnant women. The strength of statistical association was measured by adjusted odds ratios (AOR) and 95% confidence intervals. All tests were two-sided and p value < 0.05 was considered statistically significant. The goodness of fit of the final logistic model was tested using Hosmer and Lemeshow test at a p value > 0.05.

2.6. Ethical Considerations. The study was conducted after getting ethical clearance from Mekelle University, College of Health Sciences, Institutional Review Board (IRB). Support letter was obtained from Tigray Regional Health Bureau and concerned health departments. Written informed consent was secured from study participants after explaining about the objective and purpose of the study to each study participants. The participants were also assured about the confidentiality of the data. While assessing anemia status, the result of the test was communicated immediately to each participant and if the pregnant woman was anemic, she was referred to the health personnel for treatment and follow-up.

3. Results

3.1. Socioeconomic and Demographic Characteristics of the Pregnant Women. A total of 714 study participants were included in the study making a response rate of 97.7% (Table 1). The mean age (±SD) of the study participants at present

TABLE 1: Socioeconomic and demographic characteristics of the pregnant women attending antenatal clinics in north western zone of Tigray, northern Ethiopia, 2014 ($n = 714$).

Variables	n (%)	Mean ± SD
Age (years)		
<18	23 (3.2)	
18–24	288 (40.3)	
25–29	223 (31.2)	25.8 ± 5.84
30–34	116 (16.2)	
≥35	64 (9.0)	
Residence		
Urban	401 (56.2)	
Rural	315 (43.8)	
Ethnicity		
Tigray	655 (91.7)	
Amhara	45 (6.4)	
Others	14 (1.9)	
Religion		
Orthodox	643 (90.0)	
Muslim	64 (9.0)	
Others	7 (1.0)	
Marital status		
Married	695 (97.3)	
Divorced	14 (2.0)	
Widowed	3 (0.4)	
Single	2 (0.3)	
Educational status		
Cannot read and write	266 (37.3)	
Can read and write	94 (13.2)	
Primary (grades 1–8)	168 (23.5)	
Secondary (grades 9–12)	137 (19.2)	
Above secondary (above grade 12)	49 (6.9)	
Occupational status		
Housewife	453 (63.4)	
Government employee	219 (30.7)	
Private	30 (4.2)	
Others	12 (1.7)	
Family size		
≤4	511 (71.6)	
5–7	182 (25.5)	
≥8	21 (2.9)	
Age at first marriage (years)		
<18	191 (26.8)	18.2 ± 3.41
≥18	523 (73.2)	
Family income (ETB)		
<500	61 (8.5)	
500–1000	260 (36.4)	712.4 ± 289.90
≥1000	393 (55.0)	

ETB: Ethiopian birr; 1 Ethiopian birr equals 20 USD.

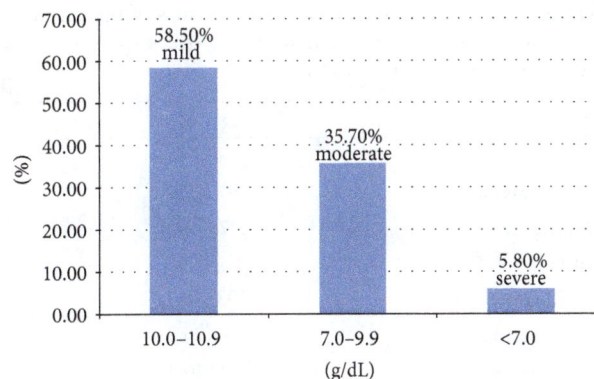

FIGURE 1: Percentage of anemia by severity among anemic pregnant women ($n = 258$).

occupations were housewife. The majority (91.7%) of the participants were Tigray in ethnicity followed by Amhara (6.4%). 47.4% of the study participants were in the age range of 25 to 34 years. The majority (90.0%) of the study participants were Orthodox Christian followers.

3.2. Obstetric and Nutrition Related Characteristics of the Pregnant Women. The mean current gestational age (±SD) of the study participants was 26.7 ± 8.05 weeks (Table 2). Above two-thirds of the participants were multigravida. More than half (52.2%) of the study participants were in their third trimester. Two-thirds of the participants did not have iron supplementation during pregnancy.

3.3. Prevalence of Anemia among the Pregnant Women. The overall prevalence of anemia in this study was 36.1% (95% CI = 32.7%–39.7%). The mean ± SD hemoglobin concentration among the study participants was 11.21 ± 1.18 g/dL. Of the anemic pregnant women, 151 (58.5%), 92 (35.7%), and 15 (5.8%) had mild anemia (Hb ranges 10.0–10.9 g/dL), moderate anemia (Hb ranges 7.0–9.9 g/dL), and severe anemia (Hb < 7.0 g/dL), respectively (Figure 1).

3.4. Factors Associated with Anemia among the Pregnant Women. The comparison between the profiles of the pregnant women who had anemia and who did not from the bivariate logistic regression analysis revealed that marital status, residence, educational status, family monthly income, number of visits, age of the women at first marriage, body mass index, iron supplementation, meal frequency per day, and nutrition education were significantly associated with maternal anemia (Table 3). However, in the multivariable logistic regression analysis level after controlling the effect of confounders revealed that variables that were independent predictors for maternal anemia among the pregnant women were maternal residence (AOR = 1.75, 95% CI = 1.01–3.04), educational status (AOR = 1.56, 95% CI = 1.03–2.37), iron supplementation during pregnancy (AOR = 3.76, 95% CI = 1.92–8.37), and meal frequency per day (AOR = 2.18, 95% CI = 1.06–4.91) (Table 3).

and at first marriage was 25.8 ± 5.84 and 18.2 ± 3.41 years, respectively. Majority (97.3%) of the study participants were currently married. Two-thirds of the study participants'

TABLE 2: Obstetric and nutritional characteristics of the pregnant women attending antenatal clinics in north western zone of Tigray, northern Ethiopia, 2014 ($n = 714$).

Variables	n (%)	Mean ± SD
Gravidity		
Primigravida	197 (27.6)	
Multigravida	517 (72.4)	
Parity		
Nulliparous	203 (28.4)	
Primiparous	180 (25.2)	
Multiparous	331 (46.4)	
Birth interval ($n = 511$)		
<2 years	232 (45.4)	
≥2 years	279 (54.6)	
Trimester		
First	84 (11.8)	
Second	207 (29.0)	26.7 ± 8.05 weeks
Third	423 (59.2)	
Number of visits		
1 time	169 (23.7)	
2-3 times	434 (60.8)	
≥4 times	111 (15.5)	
Meal frequency per day		
≤2 times	156 (21.8)	
3 times	409 (57.3)	
>3 times	14967 (20.1)	
Iron supplementation		
Yes	236 (33.1)	
No	478 (66.9)	
Malaria infection in the previous one year		
Yes	104 (14.6)	
No	610 (85.4)	
Nutrition education		
Yes	450 (63.0)	
No	264 (37.0)	
Taking tea/coffee immediately after meal		
Yes	420 (58.8)	
No	294 (41.2)	
Body mass index (BMI)		
Low (BMI ≤ 20 kg/m^2)	156 (21.8)	
Normal (BMI: 20–24.9 kg/m^2)	149 (20.9)	
High (BMI ≥ 25 kg/m^2)	409 (57.3)	

4. Discussion

Pregnant women are one of the vulnerable groups of a population to develop anemia particularly in developing countries [16]. Therefore, the aim of this study was to determine the prevalence of anemia and associated factors among pregnant women attending antenatal clinics in north western zone of Tigray, Ethiopia.

This study found that 36.1% (95% CI = 32.7%–39.7%) of the pregnant women in the study area were anemic. According to the World Health Organization classification of the public health importance of anemia, the magnitude indicates that there is moderate public health significance of anemia among the pregnant women in the study area [11, 12].

The prevalence of anemia obtained in this study is almost consistent with other studies conducted among pregnant women attending antenatal clinics in Sidama [15], West Arsi [17], and northern Nigeria [18], with the prevalence of 31.6%, 36.6%, and 30%, respectively. The result of this study was much lower than the previous studies conducted among pregnant women attending antenatal clinics in Gode town [19], north Bengal [20], Udupi district [21], Pakistan [22], Bangladesh [23], and West Bengal [24] with the prevalence of 56.8%, 82%, 50.1%, 90.5%, 73%, and 67.8%, respectively, but higher than a study conducted in Mekelle [3] and Addis Ababa [25] and the national prevalence of anemia noted in 2011, Ethiopian Demographic and Health Survey report [13], where the prevalence anemia among the pregnant women was found to be 11%, 21.3%, and 22%, respectively. Socioeconomic and geographical variations might be the reasons for the different prevalence of anemia among pregnant women across countries and regions. Using different cutoff points and hemoglobin measurement for anemia and study areas may also result in varied prevalence of anemia in pregnant women.

In this study, the multivariable logistic regression analysis revealed that maternal residence, educational status, iron supplementation during pregnancy, and meal frequency per day were significantly associated with anemia among the pregnant women at p value ≤ 0.05 (Table 3). However, maternal marital status, family monthly income, number of visits, age of the women at first marriage, body mass index, and nutrition education showed significant association with bivariate analysis but not with the multivariable analysis (Table 3).

In the present study, the prevalence of anemia was higher among pregnant women who were from rural areas as compared to pregnant women residents of urban areas in which the risk of developing anemia among rural pregnant women was 1.75 times higher to be anemic as compared to those pregnant women living in urban areas (AOR = 1.75, 95% CI = 1.01–3.04). This could be due to the reason that pregnant women from rural areas might have lack of information about adequate nutrition during pregnancy, economic factors, and inaccessibility to health care facilities. Similar results were reported by other studies conducted in south eastern Ethiopia (AOR = 3.3, 95% CI = 1.5–7.4) [9], Gondar (AOR = 2.14, 95% CI = 1.51–3.38) [26], and southwest Ethiopia (AOR = 1.62, 95% C.I = 1.02–2.62) [27].

In the present study, the prevalence of anemia was higher among pregnant women who are not educated as compared to those pregnant women who are educated in which pregnant women who were not educated were at 1.56 times higher risk to be anemic as compared to pregnant women who had formal education (AOR = 1.56, 95% CI = 1.03–2.37). The reason for this might be the fact that pregnant women who

TABLE 3: Factors associated with anemia among pregnant women attending antenatal clinics in north western zone of Tigray, northern Ethiopia, 2014 ($n = 714$).

Variables	Anemia		COR (95% CI)	AOR (95% CI)
	Yes (258)	No (456)		
Marital status				
In marital union	247 (35.5%)	448 (64.2%)	1	1
Not in marital union	11 (57.9%)	8 (42.1%)	2.49 (0.99–6.28)	1.23 (0.74–2.00)
Residence				
Urban	126 (31.6%)	275 (68.4%)	1	1
Rural	132 (41.9%)	181 (58.1%)	1.59 (1.15–2.13)	**1.75 (1.01–3.04)**[*]
Educational status				
Not educated	160 (44.4%)	200 (55.6%)	2.09 (1.53–2.86)	**1.56 (1.03–2.37)**[*]
Educated	98 (27.7%)	256 (72.3%)	1	**1**
Family income (ETB)				
<500	26 (42.6%)	35 (57.4%)	1.82 (0.05–3.16)	0.82 (0.46–1.48)
500–1000	118 (45.4%)	142 (54.6%)	2.03 (2.51–5.57)	1.30 (0.72–2.39)
≥1000	114 (29.0%)	279 (71.0%)	1	1
Age at first marriage (years)				
<18	61 (31.9%)	130 (68.1%)	0.78 (0.55–1.10)	1.03 (0.67–1.58)
≥18	197 (37.7%)	326 (76.8%)	1	1
Number of visits				
1 time	49 (29.0%)	120 (71.0%)	1.55 (0.93–2.57)	1.13 (0.60–2.11)
2-3 times	166 (38.2%)	268 (61.8%)	1.02 (0.67–1.57)	0.93 (0.54–1.59)
≥4 times	43 (38.7%)	68 (61.3%)	1	1
Iron supplementation				
Yes	63 (26.7%)	173 (73.3%)	1	**1**
No	195 (40.8%)	283 (59.2%)	1.89 (1.35–2.66)	**3.76 (1.92–8.37)**[*]
Nutrition education				
Yes	146 (32.4%)	304 (67.6%)	1	1
No	112 (42.4%)	157 (57.6%)	1.56 (1.14–2.13)	0.50 (0.22–1.15)
Meal frequency per day				
≤2 times	60 (42.6%)	81 (57.4%)	1.36 (0.79–1.08)	**2.18 (1.06–4.91)**[*]
3 times	52 (32.9%)	106 (67.1%)	0.90 (0.75–1.63)	2.31 (0.86–3.39)
>3 times	146 (35.2%)	269 (64.8%)	1	1
Body mass index				
Low (BMI < 20 kg/m^2)	56 (35.9%)	100 (64.1%)	1.43 (0.88–2.32)	1.59 (0.84–2.98)
Normal (BMI; 20–24.9 kg/m^2)	42 (28.2%)	107 (71.8%)	1	1
High (BMI ≥ 25 kg/m^2)	160 (39.1%)	249 (60.9%)	0.87 (0.59–1.28)	0.74 (0.46–1.20)

[*]Statistically significant ($p < 0.05$) 1: reference group; ETB: Ethiopian birr; 1 Ethiopian birr equals 20 USD.

have some level of formal education can be aware of anemia during pregnancy and take some preventive measures like eating iron-rich food and taking iron tables. The result of this study is consistent with other studies conducted in Addis Ababa (AOR = 2.12, 95% CI = 2.47-6.80) [25], West Bengal (AOR = 17.50, 95% CI = 3.77–90.68) [24], and West Algeria (AOR = 0.79, 95% CI = 0.07–0.49) [28].

In the present study, the prevalence of anemia was higher among pregnant women having a meal frequency of less than 3 times per day as compared to pregnant women who had a meal frequency of more than 2 times per day in which the pregnant women having a meal frequency of less than 3 times per day were at 2 times higher risk of developing anemia as compared to pregnant women who had a meal frequency of more than 3 times per day (AOR = 2.18, 95% CI = 1.06-4.91). This might be due to the reason that pregnancy is a critical period with increased energy and nutrient demand for the mother which should be fulfilled with increased meal frequency per day. This result is consistent with other studies conducted in Mekelle (AOR = 3.88, 95% CI = 1.93, 7.79) [3] and West Arsi (AOR = 4.66, 95% CI = 2.94, 7.38) [17].

In the present study, the prevalence of anemia was significantly higher among pregnant women who did not take iron supplementation during pregnancy as compared to those pregnant women who took their iron supplementation in which pregnant women who did not take iron supplementation were at 3.76 times higher risk to be anemic as compared to pregnant women who took their iron supplementation

(AOR = 3.76, 95% CI = 1.92–8.37). The reason for this might be pregnant women who take their iron tablets which can help them to increase their hemoglobin level and prevent anemia during pregnancy time. This result was consistent with other studies conducted in Sidama (AOR = 1.90, 95% CI = 1.14–3.19) [15], Gode town (AOR = 1.54, 95% CI = 1.04–2.27) [19], West Bengal, (AOR = 5.65, 95% CI = 1.78–18.54) [24], and West Algeria (AOR = 0.71, 95% CI = 0.26–0.99) [28].

5. Limitation of the Study

There may be a social desirability bias for dietary information and monthly income. This may overestimate the association between the variables and anemia among the pregnant women.

6. Conclusion and Recommendations

Anemia was found to be moderate public health problem in the study area. Residence, educational status, iron supplementation during pregnancy, and meal frequency of the woman per day were statistically significant independent predictors for maternal anemia among the pregnant women in the study area.

Awareness creation and nutrition education on the importance of taking iron supplementation and nutritional counseling on consumption of extra meal and iron-rich foods during pregnancy are recommended to prevent anemia in the pregnant women.

Abbreviations

AOR: Adjusted odds ratio
CDC: Center of Disease Control and Prevention
COR: Crude odds ratio
CI: Confidence interval
EDHS: Ethiopian Demographic and Health Survey
g/dL: Gram/deciliter
Hb: Hemoglobin
SD: Standard deviation
WHO: World Health Organization.

Conflict of Interests

The authors declared that they have no conflict of interests.

Authors' Contribution

All the authors participated in the designing, data collection, analysis, and writing of the study. The authors have read and approved the final paper.

Acknowledgments

The authors acknowledge Save the Children, UK, for its financial support to make their work easy in the research process. The authors would also like to thank health departments for giving permission to carry out the study at the respective health facilities. They are indebted to all pregnant women involved in this study.

References

[1] World Health Organization, *Micronutrient Deficiencies: Prevention and Control Guidelines*, World Health Organization, Geneva, Switzerland, 2015, http://www.who.int/nutrition/topics/ida/en/index.html.

[2] World Health Organization/United Nations University/UNICEF, *Iron Deficiency Anemia, Assessment, Prevention and Control: A Guide for Programme Managers*, WHO, Geneva, Switzerland, 2001.

[3] A. Abriha, M. E. Yesuf, and M. M. Wassie, "Prevalence and associated factors of anemia among pregnant women of Mekelle town: a cross sectional study," *BMC Research Notes*, vol. 7, article 888, 2014.

[4] A. Levy, D. Fraser, M. Katz, M. Mazor, and E. Sheiner, "Maternal anemia during pregnancy is an independent risk factor for low birth weight and preterm delivery," *European Journal of Obstetrics Gynecology and Reproductive Biology*, vol. 122, no. 2, pp. 182–186, 2005.

[5] P. Lacerte, M. Pradipasen, P. Temcharoen, N. Imamee, and T. Vorapongsathorn, "Determinants of adherence to iron/folate supplementation during pregnancy in two provinces in Cambodia," *Asia-Pacific Journal of Public Health*, vol. 23, no. 3, pp. 315–323, 2011.

[6] E. Ejeta, B. Alemnew, A. Fikadu, M. Fikadu, L. Tesfaye, and T. Birhanu, "Prevalence of anaemia in pregnant womens and associated risk factors in Western Ethiopia," *Food Science and Quality Management*, vol. 31, 2014.

[7] A. Bilimale, J. Anjum, H. N. Sangolli, and M. Mallapur, "Improving adherence to oral iron supplementation during pregnancy," *Australasian Medical Journal*, vol. 3, no. 5, pp. 281–290, 2010.

[8] F. Bánhidy, N. Acs, E. H. Puhó, and A. E. Czeizel, "Iron deficiency anemia: pregnancy outcomes with or without iron supplementation," *Nutrition*, vol. 27, no. 1, pp. 65–72, 2011.

[9] F. Kefiyalew, E. Zemene, Y. Asres, and L. Gedefaw, "Anemia among pregnant women in Southeast Ethiopia: prevalence, severity and associated risk factors," *BMC Research Notes*, vol. 7, no. 1, article 771, 2014.

[10] S. Salhan, V. Tripathi, R. Singh, and H. S. Gaikwad, "Evaluation of hematological parameters in partial exchange and packed cell transfusion in treatment of severe anemia in pregnancy," *Anemia*, vol. 2012, Article ID 608658, 7 pages, 2012.

[11] WHO/CDC, *Worldwide Prevalence of Anemia 1993–2005: WHO Global Database on Anemia*, WHO Press, Geneva, Switzerland, 2008.

[12] WHO, *Hemoglobin Concentrations for the Diagnosis of Anemia and Assessment of Severity. Vitamin and Mineral Nutrition Information System*, World Health Organization, Geneva, Switzerland, 2011, http://www.who.int/vmnis/indicators/haemoglobin.pdf.

[13] EDHS, Ethiopian Central Statistical Agency, Addis Ababa, Ethiopia, 2011, http://www.measuredhs.com.

[14] The 2007 Population and Housing Census of Ethiopia; CSA, Addis Ababa, http://www.CSA.com.

[15] S. Gebremedhin, F. Enquselassie, and M. Umeta, "Prevalence and correlates of maternal anemia in rural Sidama, Southern Ethiopia," *African Journal of Reproductive Health*, vol. 18, no. 1, pp. 44–53, 2014.

[16] T. A. Zerfu and H. T. Ayele, "Micronutrients and pregnancy; Effect of supplementation on pregnancy and pregnancy outcomes: a systematic review," *Nutrition Journal*, vol. 12, article 20, 2013.

[17] N. Obse, A. Mossie, and T. Gobena, "Magnitude of anemia and associated risk factors among pregnant women attending antenatal care in Shalla Woreda, West Arsi Zone, Oromia Region, Ethiopia," *Ethiopian Journal of Health Sciences*, vol. 23, no. 2, pp. 165–173, 2013.

[18] D. J. VanderJagt, H. S. Brock, G. S. Melah, A. U. El-Nafaty, M. J. Crossey, and R. H. Glew, "Nutritional factors associated with anaemia in pregnant women in northern Nigeria," *Journal of Health, Population and Nutrition*, vol. 25, no. 1, pp. 75–81, 2007.

[19] K. Addis Alene and A. Mohamed Dohe, "Prevalence of anemia and associated factors among pregnant women in an urban area of eastern Ethiopia," *Anemia*, vol. 2014, Article ID 561567, 7 pages, 2014.

[20] S. Dutta, S. Chatterjee, D. Sinha, B. Pal, M. Basu, and A. Dasgupta, "Correlates of anaemia and worm infestation among rural pregnant women: a cross sectional study from Bengal," *National Journal of Community Medicine*, vol. 4, no. 4, pp. 603–607, 2013.

[21] J. A. Noronha, A. Bhaduri, and H. V. Bhat, "Prevalence of anaemia among pregnant women: a community-based study in Udupi district," *Health and Population: Perspectives and Issues*, vol. 31, no. 1, pp. 31–40, 2008.

[22] N. Baig-Ansari, S. H. Badruddin, R. Karmaliani et al., "Anemia prevalence and risk factors in pregnant women in an urban area of Pakistan," *Food and Nutrition Bulletin*, vol. 29, no. 2, pp. 132–139, 2008.

[23] S. M. Ziauddin Hyder, L.-Å. Persson, A. M. R. Chowdhury, and E.-C. Ekström, "Anaemia among non-pregnant women in rural Bangladesh," *Public Health Nutrition*, vol. 4, no. 1, pp. 79–83, 2001.

[24] S. Bisoi, D. Haldar, T. K. Majumdar, N. Bhattacharya, G. N. Sarkar, and S. K. Ray, "Correlates of anemia among pregnant women in a rural area of West Bengal," *The Journal of Family Welfare*, vol. 57, no. 1, 2011.

[25] A. H. Jufar and T. Zewde, "Prevalence of anemia among pregnant women attending antenatal care at Tikur Anbessa specialized Hospital, Addis Ababa Ethiopia," *Journal of Hematology & Thromboembolic Diseases*, vol. 2, article 125, 2014.

[26] M. Alem, B. Enawgaw, A. Gelaw, T. Kenaw, M. Seid, and Y. Olkeba, "Prevalence of anemia and associated risk factors among pregnant women attending antenatal care in Azezo Health Center Gondar town, Northwest Ethiopia," *Journal of Interdisciplinary Histopatholog*, vol. 1, no. 3, pp. 137–144, 2013.

[27] M. Getachew, D. Yewhalaw, K. Tafess, Y. Getachew, and A. Zeynudin, "Anaemia and associated risk factors among pregnant women in Gilgel Gibe dam area, Southwest Ethiopia," *Parasites and Vectors*, vol. 5, no. 1, article no. 296, 2012.

[28] A. Demmouche, S. Khelil, and S. Moulessehoul, "Anemia among pregnant women in the Sidi Bel Abbes Region (West Alegria): an epidemiologic study," *Journal of Blood Disorders & Transfusion*, vol. 2, article 113, 2011.

3

Dietary Pattern and Metabolic Syndrome in Thai Adults

W. Aekplakorn,[1,2] **W. Satheannoppakao,**[3] **P. Putwatana,**[4] **S. Taneepanichskul,**[5]
P. Kessomboon,[6] **V. Chongsuvivatwong,**[7] **and S. Chariyalertsak**[8]

[1]*Department of Community Medicine, Faculty of Medicine, Ramathibodi Hospital, Mahidol University,*
 Rama VI Road, Ratchathewi, Bangkok 10400, Thailand
[2]*National Health Examination Survey Office, Nonthaburi 11000, Thailand*
[3]*Department of Nutrition, Faculty of Public Health, Mahidol University, Bangkok 10400, Thailand*
[4]*Ramathibodi School of Nursing, Faculty of Medicine, Ramathibodi Hospital, Mahidol University, Bangkok 10400, Thailand*
[5]*College of Public Health Sciences, Chulalongkorn University, Bangkok 10330, Thailand*
[6]*Faculty of Medicine, Khon Kaen University, Khon Kaen 40002, Thailand*
[7]*Epidemiology Unit, Faculty of Medicine, Prince of Songkla University, Songkhla 90110, Thailand*
[8]*Faculty of Medicine, Chiang Mai University, Chiang Mai 50002, Thailand*

Correspondence should be addressed to W. Aekplakorn; wichai.aek@mahidol.ac.th

Academic Editor: Michael B. Zemel

Objectives. To determine the dietary patterns of middle-aged Thais and their association with metabolic syndrome (MetS). *Methods.* The Thai National Health Examination Survey IV data of 5,872 participants aged ≥30–59 years were used. Dietary patterns were obtained by factor analysis and their associations with Mets were examined using multiple logistic regression. *Results.* Three major dietary patterns were identified. The first, meat pattern, was characterized by a high intake of red meat, processed meat, and fried food. The second, healthy pattern, equated to a high intake of beans, vegetables, wheat, and dairy products. The third, high carbohydrate pattern, had a high intake of glutinous rice, fermented fish, chili paste, and bamboo shoots. Respondents with a healthy pattern were more likely to be female, higher educated, and urban residents. The carbohydrate pattern was more common in the northeast and rural areas. Compared with the lowest quartile, the highest quartile of carbohydrate pattern was associated with MetS (adjusted odds ratio: 1.82; 95% CI 1.31, 2.55 in men and 1.60; 95% CI 1.24, 2.08 in women), particularly among those with a low level of leisure time physical activity (LTPA). *Conclusion.* The carbohydrate pattern with low level of LTPA increased the odds of MetS.

1. Introduction

Metabolic syndrome (MetS), a cluster of metabolic risk factors, is considered to be an intermediate outcome preceding disability and death from various related diseases such as diabetes mellitus and cardiovascular diseases (CVD) [1]. Dietary pattern is an important factor associated with components of MetS such as obesity, diabetes, dyslipidemia, hypertension, and subsequently CVD [2, 3]. Several studies have demonstrated that an unhealthy dietary pattern is associated with increased CVD risk factors, and a healthy diet is linked to decreased risk [2, 4]. Dietary patterns with red meat and processed meat have been reported to be associated

with metabolic factors and CVD, whereas a Mediterranean diet is beneficial to metabolic factors [2, 3].

Similar to other countries in Asia, rice, a carbohydrate-rich source, is a staple food among Thais, and glutinous rice, in particular, is common in the northeast region of Thailand. In Asian populations, studies have reported the association between a carbohydrate-rich dietary pattern and metabolic risk, for instance, in Korea and Japan [5, 6]. A diet high in carbohydrate was reported to be associated with dyslipidemia, diabetes, and MetS in Korean adults [7, 8], whereas a western dietary pattern has been shown to be associated with dyslipidemia in Japanese adults [9]. In addition, a healthy dietary pattern combined with a physically active lifestyle has been reported to reduce the risk of CVD and related risk factors [10]. A study in Korea also reported that the risk of MetS associated with carbohydrate intake was dependent on BMI [5].

The prevalence of metabolic risk factors in developing countries, including Thailand, has increased in the past two decades. Previous studies have shown variation in the distribution of metabolic risk factors by geographic region and area of residence in the Thai population [11–14]. However, studies describing the dietary patterns of Thais are scarce. It is less clear whether dietary patterns play a role in the differences in metabolic profile and whether the effect is influenced by physical activity level or BMI. Knowledge of dietary patterns in the population is important for the prevention and control of CVD and related diseases. Therefore, the aim of this study was to examine whether there were variations in dietary patterns by geographic region, area of residence, and socioeconomic status (SES). We examined the associations of dietary patterns with metabolic risk factors in middle-aged Thais and evaluated whether the associations were modified by physical activity or obesity status.

2. Methods

2.1. Study Population.
Data from the Thai National Health Examination Survey IV (NHES IV) were used. NHES IV was a nationally representative cross-sectional study of the Thai population conducted in 2009. Details of the study design have been described elsewhere [12]. Briefly, a total of 20,426 individuals aged 15 and over, of which 8,582 individuals were aged 30–59 years, were randomly selected using multistage cluster sampling. To rule out the effect of behavioral change due to existing chronic diseases in the present study, we excluded those who were previously diagnosed with diabetes, hypertension, ischemic heart disease, and stroke and those taking lipid lowering medication. Thus, a total of 5,872 adults (2,693 men and 3,179 women) aged 30–59 years were included in the analysis. This study was approved by the Ethical Clearance Committee on Human Rights Related to Research involving Human Subjects, Faculty of Medicine, Ramathibodi Hospital, Mahidol University.

2.2. Life Style and Health Data.
Data included in the survey were comprised of information about age, education level, area of residence, geographic region, food consumption frequency, smoking status, medical history of diabetes, hypertension, and dyslipidemia, and medication use. Anthropometric measurements including height, weight, and waist circumference were performed using standard procedures. Weight was measured by using a calibrated digital scale, TANITA model HD316, and height was measured by using a stadiometer while standing barefoot with shoulders in a normal position. BMI was calculated as weight in kilograms divided by height squared in meters. Waist circumference was measured at a horizontal plane midway between the iliac crest and lower rib margin in centimeters to the nearest 0.1 cm. Venous blood samples were obtained from participants who had fasted for 12 hours overnight. Blood samples were analyzed to determine fasting plasma glucose, high density lipoprotein cholesterol (HDL-C), and triglyceride (TG) at the central laboratory center in the Faculty of Medicine, Ramathibodi Hospital. Plasma glucose was analyzed by a hexokinase enzyme method. TG was measured by enzymatic colorimetric methods and HDL-C was analyzed by homogeneous enzymatic colorimetric methods using the Hitachi 917 model. Physical activity status was assessed by using the Global Physical Activity Questionnaire (GPAQ) [15]. Leisure time physical inactivity (LTPA) level was assessed and categorized into two groups as high LTPA and low LTPA. High LTPA was defined as having 20 minutes of vigorous intensity activity per day on at least 3 days in a typical week or 30 minutes of moderate-intensity leisure-time activity or 5 days of combination of moderate and vigorous intensity activities at least 600 MET-minutes per week [16]; otherwise LTPA was classified as low. Alcohol consumption was assessed and average daily alcohol intake (gm/day) was calculated according to the WHO guide [17] and then classified into two groups as abstinence to low-risk drinking versus moderate to high risk drinking (cut-off point at >41 gm/day for men and >21 gm/day for women).

2.2.1. Dietary Assessment.
Trained interviewers collected participants' food consumption data using a food frequency questionnaire (FFQ) which featured commonly consumed food items. The FFQ was developed and validated during a pilot test [18]. All food items were later categorized into 22 key food groups, formed according to key nutrient component, main food group, culinary use, and risk to chronic diseases in particular CVD (low fat, high fat, fiber, etc.), as shown in Table 6. Trained nurses and interviewers performed a face to face interview using pictures of common food items and a frequency card to facilitate answers. The food groups were as follows: meat, fatty meat, processed meat with high fat, processed meat with high salt, fish, shellfish and squid, animal organ, egg, beans, rice, wheat, glutinous rice, fried food, food with coconut milk, fermented fish/soybean, chili sauce/dip, fruit, milk, soymilk, beverage, bamboo shoot, and vegetables. A pilot test was done in order to test reliability and Cronbach's alpha coefficient of 0.80 was obtained, indicating a relatively acceptable level of interitem reliability for the FFQ.

2.2.2. Dietary Pattern.
Dietary patterns were derived from factor analysis using principal component analysis based on 22 food groups from the FFQ. A factor score was created for each individual based on the factor analysis. The factors were analyzed using orthogonal transformed with Varimax options. An eigenvalue of >1.5 and the point in the scree plot where the slope of the curve clearly leveled off indicated the number of factors to be retained. The factor score for each pattern was calculated and assigned to each participant. Three main dietary patterns were labeled based on the food groups for each factor. The three factors with factors loading are shown in Table 1. Pearson's correlation coefficients between scores for the three patterns were very small ($r < 0.001$, all $P = 0.99$), indicating that they had no linear correlation. The scores of each pattern were categorized into quartiles. For SES, a questionnaire regarding household assets was used during the field visit. Consumer items in each participant's

TABLE 1: Factor loading for the first three factors from principal components analysis of food frequency questionnaire interviews, among Thai adults aged 30–59 years, NHES IV, 2009.

Food	Men			Women		
	Meat pattern	Carbohydrate pattern	Healthy pattern	Meat pattern	Carbohydrate pattern	Healthy pattern
Meat	0.60			0.52		
Fatty meat	0.66			0.58		
Processed meat with high fat	0.44			0.48		
Processed meat with high salt	0.45			0.45		
Fish			0.32			0.36
Shellfish and squid	0.40			0.39		
Animal organ	0.47			0.46		
Egg	0.37			0.49		
Beans			0.58			0.59
Rice	0.18			0.25		
Wheat			0.42			0.48
Glutinous rice		0.81			0.79	
Fried food	0.56			0.60		
Food with coconut milk	0.51			0.51		
Fermented fish/soybean		0.81			0.82	
Chili sauce/dip		0.50			0.46	
Fruit			0.61			0.58
Milk			0.56			0.56
Soy milk			0.60			0.61
Beverage	0.36			0.46		
Bamboo shoot		0.48			0.56	
Vegetables			0.29			0.30
Variance explained (%)	11.93	10.81	10.0	12.6	10.6	9.9

Variance in men 32.74; women 33.1.

household were observed and recorded. The household assets included bed, air conditioner, electric water boiler, washing machine, microwave, personal computer, house telephone, car, and flushing toilet. A SES variable, wealth index score, was calculated based on ownership of household items using factor analysis to assign the indicator weights [19]. The standardized score was assigned to individuals. Wealth quintile was created according to quintile of the scores. The lowest quintile indicated the poorest group and the highest indicated the richest group. To describe smoking habit, each participant was classified as one of the following: nonsmoker, ex-smoker, or current smoker.

2.3. Outcome. Metabolic syndrome (MetS) was defined according to the harmonizing criteria [1] as having three or more of the following factors: abdominal obesity (waist circumference ≥ 80 cm in women and ≥ 90 cm in men); high blood pressure, systolic blood pressure (SBP) ≥ 130 mmHg, or diastolic blood pressure (DBP) ≥ 85 mmHg; low HDL-C < 50 mg/dL (1.29 mmol/L) in women and < 40 mg/dL (1.04 mmol/L) in men, high triglycerides (TG ≥ 150 mg/dL (1.69 mmol/L)); hyperglycemia (fasting plasma glucose ≥ 100 mg/dL (5.6 mmol/L)).

2.4. Statistical Analysis. All statistical analyses were weighted to account for the complex survey design. Since dietary pattern is likely to differ by sex, all the statistical tests were performed separately by sex. Means and proportions of participants' metabolic and demographic characteristics, for example, age, sex, urban/rural area, geographic region, and wealth index, were calculated according to dietary pattern quartiles. The Chi-square test was used to compare categorical variables among quartiles. Continuous variables were compared across quartiles using ANOVA. Age-adjusted means of metabolic variables by quartile for each dietary pattern were calculated by using ANCOVA. Multiple logistic regression was used to obtain age-adjusted percentage of each metabolic component (coded as 0 = no, 1 = yes) and MetS prevalence across quartiles. For linear trend analysis, linear regression was used to examine the linear trend of continuous variable of characteristics of subjects across quartile categories for each dietary pattern with each category coded on an ordinal scale. For categorical variables, multivariate logistic regression was used. Multiple logistic regression was used to examine the association of dietary patterns with each metabolic component and MetS, controlling for potential confounding factors including age, family history of diabetes,

TABLE 2: Socio-demographic factors by quartile of factor scores of dietary patterns in Thai adults aged 30–59 years, NHES IV, 2009.

	Meat				P value	Carbohydrate				P value	Healthy				P value
	Q1	Q2	Q3	Q4		Q1	Q2	Q3	Q4		Q1	Q2	Q3	Q4	
Men (n = 2693)															
Age (mean, yrs)	45.3	44.3	44.1	42.5	<0.001	43.6	44.0	44.0	44.5	0.01	43.5	44.4	43.8	44.6	0.08
Area of residence															
Urban (%)	25.9	23.9	22.4	27.9	0.02	33.2	25.5	24.7	16.5	<0.001	23.1	21.0	24.7	31.2	<0.001
Rural (%)	27.3	25.8	24.7	22.3		13.6	19.7	24.6	42.0		27.4	29.9	22.9	19.8	
Region															
North (%)	28.0	28.6	26.7	16.6		5.7	17.8	34.5	41.9		24.4	30.4	25.0	20.1	
Central (%)	20.6	21.6	26.0	31.9		31.9	38.2	24.8	5.1		25.5	25.5	22.9	26.1	
Northeast (%)	33.3	26.1	21.2	19.4	<0.001	2.4	8.5	23.7	65.4	<0.001	28.6	30.9	21.7	18.9	<0.001
South (%)	16.3	25.6	27.3	30.8		52.3	33.3	12.4	2.0		26.2	23.1	25.6	25.0	
Bangkok (%)	25.4	23.3	20.9	30.3		48.9	26.9	21.1	3.1		22.1	15.5	25.7	36.7	
Wealth quintile															
Q1 (%)	31.8	25.3	23.8	19.1		10.4	17.3	24.2	48.1		32.4	32.6	19.8	15.1	
Q2 (%)	26.1	28.2	24.5	21.1		12.0	18.8	26.7	42.5		26.7	32.6	24.0	16.7	
Q3 (%)	24.3	23.1	24.9	27.6	0.01	20.5	19.0	25.5	34.9	<0.001	26.1	26.5	25.1	21.9	<0.001
Q4 (%)	24.0	23.3	25.2	27.6		22.4	27.6	24.0	26.0		24.5	22.2	23.1	30.2	
Q5 (%)	27.2	26.1	21.6	25.1		34.0	26.7	22.0	17.2		18.8	19.9	21.1	36.1	
Women (n = 3179)															
Age (yr)	44.7	44.3	43.6	41.9	<0.001	43.4	43.6	43.8	43.7	0.51	43.8	43.2	43.9	43.6	0.95
Area of residence															
Urban (%)	26.8	23.2	22.3	27.8	0.08	31.7	26.7	23.7	17.8	<0.001	23.5	21.3	26.0	29.1	<0.001
Rural (%)	26.4	25.8	24.3	23.4		14.5	17.7	25.3	42.6		28.6	27.8	24.3	19.3	
Region															
North (%)	29.8	25.2	25.6	19.4		6.5	15.0	29.1	49.4		25.5	29.8	27.5	17.2	
Central (%)	20.7	23.4	25.7	30.2		34.2	31.2	29.4	5.2		31.4	21.1	23.2	24.3	
Northeast (%)	33.6	27.2	19.8	19.3	<0.001	2.6	6.6	23.8	67.0	<0.001	26.4	29.8	23.9	19.9	<0.001
South (%)	15.1	24.5	31.6	28.8		44.9	39.4	13.2	2.5		24.9	23.6	23.8	27.6	
Bangkok (%)	22.2	21.4	19.0	37.4		41.6	31.7	22.3	4.4		24.8	17.7	27.9	29.6	
Wealth quintile															
Q1 (%)	32.4	26.5	20.2	20.9		8.9	12.6	25.8	52.6		33.3	34.2	23.1	9.5	
Q2 (%)	23.5	26.3	25.4	24.9		12.3	17.9	26.7	43.1		33.3	26.8	24.1	15.8	
Q3 (%)	21.8	26.9	26.1	25.3	<0.001	17.3	24.2	24.6	33.9	<0.001	28.4	28.7	22.5	20.3	<0.001
Q4 (%)	23.2	22.5	26.2	28.1		23.8	24.4	23.2	28.5		26.6	19.6	26.4	27.4	
Q5 (%)	31.0	22.9	21.3	24.7		34.0	22.3	23.9	19.8		15.2	19.4	28.1	37.2	

Data are shown in means and percent.

BMI, and leisure time physical activity. We included BMI as continuous scale variable in the model to further assess whether the effect of dietary pattern was mediated by general obesity. To explore whether the effect of dietary pattern was modified by general obesity or LTPA status, we performed stratified analysis by obesity (at BMI ≥25 and BMI <25 kg/m^2) or by LTPA status. For obesity status, the results of both obesity strata were similar. We also tested the interaction of dietary patterns with BMI and with LTPA and found no interaction for dietary pattern with BMI, but possible interaction with LTPA ($P < 0.1$ was considered as potential interaction). Consequently, we assessed the association of dietary pattern and MetS stratified by LTPA status. We did not include wealth quintile and geographic region in the multivariable regression, since they were highly correlated with dietary pattern. For women, we also included a variable to further control for menopausal status. Statistical significance level was considered at $P < 0.05$. All the statistical tests were performed using STATA version 10 (Texas, USA).

3. Results

Table 1 presents the food groups and factor loadings of 3 dietary patterns labeled as meat, carbohydrate, and healthy patterns. The 3 dietary patterns represent 32.7% of variance explained. The meat pattern explained 11.9% of the variance

TABLE 3: Age-adjusted means and percentage of metabolic risk factors by dietary pattern in Thai adults aged 30–59 years, NHES IV, 2009.

	Meat				P for trend	Carbohydrate				P for trend	Healthy				P for trend
	Q1	Q2	Q3	Q4		Q1	Q2	Q3	Q4		Q1	Q2	Q3	Q4	
Men ($n = 2693$)															
BMI (kg/m^2)	23.1	23.1	23.2	23.5	0.08	23.6	23.5	23.0	23.0	<0.01	23.2	22.9	23.3	23.6	0.01
WC (cm)	79.5	79.1	80.3	80.9	<0.01	81.2	81.4	79.4	78.8	<0.001	80.1	78.9	80.0	81.1	0.02
SBP (mm Hg)	120.6	119.8	121.9	124.0	0.001	123.4	122.0	121.0	120.7	0.01	121.1	120.3	122.5	122.5	0.20
HDL-C (mg/dL)	43.8	45.3	45.5	45.4	0.03	46.7	46.2	46.0	42.7	<0.001	45.3	44.5	44.3	46.0	0.39
Triglyceride (mg/dL)	194.3	169.6	172.3	173.1	0.005	154.9	180.5	170.8	193.8	0.001	176.1	179.5	186.9	168.3	0.42
FPG (mg/dL)	85.7	87.6	88.0	88.8	<0.01	90.5	91.5	88.4	82.8	<0.001	86.4	87.8	88.0	87.8	0.04
MetS (%)	18.8	16.7	18.5	21.5	0.10	17.3	20.7	17.5	19.4	0.65	18.2	17.8	20.5	19.0	0.37
Abdominal obesity (%)	15.0	13.5	19.0	18.7	<0.01	21.0	19.3	16.4	12.4	<0.001	17.6	13.7	15.5	19.4	0.25
High TG (%)	49.7	43.2	46.1	47.2	0.57	35.4	42.4	46.2	55.2	<0.001	43.0	48.9	52.2	42.2	0.78
Low HDL-C (%)	39.9	35.6	33.6	33.5	0.02	29.8	31.8	28.7	46.2	<0.001	34.3	37.5	39.1	32.0	0.46
High BP (%)	29.3	25..5	28.8	39.2	0.001	35.5	33.0	27.3	28.7	0.01	30.3	27.4	34.8	30.3	0.35
Hyperglycemia (%)	10.4	14.7	16.4	17.2	0.001	17.7	17.2	16.8	9.7	<0.001	13.5	14.4	15.1	15.0	0.36
Alcohol drinker	10.3	13.0	14.2	16.8	<0.01	13.9	14.3	16.4	10.7	0.16	15.1	13.3	14.2	11.1	0.04
Low LTPA (%)	75.7	82.4	76.2	80.0	0.47	79.1	75.0	77.9	80.8	0.16	84.9	81.7	77.0	69.0	<0.001
Women ($n = 3179$)															
BMI (kg/m^2)	24.6	25.1	24.7	25.0	0.20	24.7	25.2	24.9	24.7	0.39	25.1	25.1	24.6	24.5	0.001
WC (cm)	78.4	79.5	79.0	79.7	0.04	78.8	79.4	79.6	78.9	0.96	79.9	79.3	78.9	78.3	<0.01
SBP (mm Hg)	117.9	118.4	118.4	117.7	0.79	117.2	118.6	118.3	118.2	0.32	119.1	118.5	117.4	117.3	0.001
HDL-C (mg/dL)	48.5	48.4	49.9	51.1	<0.001	51.3	51.3	50.4	46.7	<0.001	49.2	49.6	48.6	50.4	0.10
Triglyceride (mg/dL)	134.9	133.8	134.5	124.0	0.05	119.9	122.0	126.4	148.0	0.001	133.1	131..9	133.9	128.0	0.19
FPG (mg/dL)	83.6	86.4	85.5	85.9	0.04	88.3	86.8	85.1	83.0	<0.01	85.4	85.4	85.3	85.2	0.74
MetS (%)	23.4	25.6	24.1	22.5	0.62	22.1	21.4	22.3	27.5	0.02	28.0	22.8	22.5	21.8	0.01
Abdominal obesity (%)	43.4	44.0	42.5	46.4	0.36	43.5	43.5	45.6	44.2	0.51	50.0	43.7	43.1	39.6	<0.001
High TG (%)	27.8	31.8	29.1	28.1	0.87	22.8	23.2	27.2	37.6	<0.001	30.6	29.2	31.4	25.1	0.03
Low HDL-C (%)	57.0	61.5	55.0	50.7	<0.01	50.1	51.2	50.3	66.3	<0.001	56.5	55.6	58.1	54.1	0.49
High BP (%)	22.1	23.1	24.8	22.8	0.61	23.7	25.4	24.9	20.4	0.09	25.5	23.4	19.8	23.8	0.19
Hyperglycemia (%)	6.6	9.9	10.2	12.2	0.001	13.3	9.5	8.7	8.1	0.021	11.4	8.8	8.2	9.6	0.29
Alcohol drinker (%)	0.5	1.1	1.7	3.6	<0.001	0.9	0.7	3.0	1.8	0.02	2.3	2.2	1.3	0.9	<0.01
Low LTPA (%)	79.9	84.7	82.0	83.3	0.21	83.3	83.7	82.1	81.5	0.34	90.5	82.9	81.1	73.6	<0.001
Menopause (%)	23.3	22.0	22.6	23.0	0.88	23.6	22.3	23.9	21.6	0.27	21.7	23.1	23.1	23.1	0.38

BMI: body mass index. WC: waist circumference. High TG: high triglycerides. HDL-C: high density-lipoprotein cholesterol. FPG: fasting plasma glucose. SBP: systolic blood pressure. LTPA: leisure time physical activity. Alcohol drinker: consumed alcohol >41 gm/day for men and >21 gm/day for women.

in men and 12.6% in women; this pattern was characterized by the high factor loadings of meat, fatty meat, processed meat with high fat, processed meat with high salt, shellfish and squid, animal organ, egg, fried food, food with coconut milk, and beverage. The carbohydrate pattern (explained 10.8% and 10.6% of variance in men and women, resp.) included glutinous rice, chili sauce/dip, fermented fish/soybean, and bamboo shoot. The healthy pattern was characterized by a high intake of fish, bean, wheat, fruit, milk, soy milk, and vegetables and explained 10% of the variance in men and 9.9% in women.

Table 2 shows the sociodemographic characteristics of participants in each dietary pattern. Individuals in the highest quartile of intake for the meat pattern were more likely to be younger and resided in the central and southern regions and in Bangkok. The percentage of the meat pattern was different for urban versus rural areas in men, but not in women, as

men in urban areas were more likely to be in the fourth quartile compared with men in rural areas. Urban residents were more likely to be in the fourth quartile of the healthy pattern, but less likely to be in the highest quartile of the carbohydrate pattern, in contrast to rural residents. Those in the highest quartile of intake for the carbohydrate pattern were more likely to be residents in rural areas, in the north and northeast regions, whereas those residing in Bangkok were more likely to have a healthy pattern. For socioeconomic status (SES), those in the highest quintile of wealth index were more likely to have the healthy pattern, whereas those in the lower quintile were more likely to have a high carbohydrate intake. Individuals in the highest quartile of the carbohydrate pattern were more likely to be among those with low SES.

Table 3 shows the distribution of metabolic risk factors by dietary patterns. The mean BMI by quartile of meat pattern was relatively similar in both sexes. For the carbohydrate

TABLE 4: Adjusted odds ratio (95% CI) for metabolic syndrome and its components according to quartile of factor scores of dietary patterns in Thai adults aged 30–59 years, NHES IV, 2009.

	Abdominal obesity	High TG	Low HDL-C	High BP	Hyperglycemia	MetS
Men (2693)						
Age	1.02 (1.01, 1.03)	1.00 (0.99, 1.01)	1.01 (1.00, 1.02)	1.04 (1.03, 1.05)	1.05 (1.04, 1.06)	1.05 (1.04, 1.07)
Meat						
Q1	1	1	1	1	1	1
Q2	0.93 (0.70, 1.23)	0.75 (0.62, 0.91)	0.83 (0.66, 1.03)	0.81 (0.63, 1.04)	1.55 (1.19, 2.01)	0.82 (0.61, 1.10)
Q3	1.45 (1.10, 1.92)	0.87 (0.69, 1.11)	0.78 (0.64, 0.95)	0.99 (0.77, 1.25)	1.74 (1.33, 2.27)	0.89 (0.63, 1.25)
Q4	1.40 (1.10, 1.83)	0.92 (0.74, 1.14)	0.80 (0.64, 0.99)	1.56 (1.24, 1.97)	1.78 (1.32, 2.41)	1.01 (0.82, 1.23)
Carbohydrate						
Q1	1	1	1	1	1	1
Q2	0.94 (0.69, 1.28)	1.32 (1.08, 1.60)	1.08 (0.87, 1.35)	0.90 (0.72, 1.12)	0.93 (0.690, 1.25)	1.34 (1.00, 1.80)
Q3	0.79 (0.57, 1.07)	1.58 (1.26, 1.98)	0.94 (0.75, 1.19)	0.68 (0.54, 0.85)	0.94 (0.66, 1.35)	1.40 (1.07, 1.84)
Q4	0.60 (0.46, 0.79)	2.26 (1.69, 3.03)	1.93 (1.50, 2.47)	0.79 (0.62, 1.00)	0.52 (0.36, 0.75)	1.82 (1.31, 2.55)
Healthy						
Q1	1	1	1	1	1	1
Q2	0.79 (0.59, 1.07)	1.18 (0.94, 1.48)	1.05 (0.87, 1.26)	0.90 (0.69, 1.18)	1.13 (0.82, 1.54)	1.09 (0.79, 1.51)
Q3	0.83 (0.63, 1.10)	1.40 (1.16, 1.69)	1.20 (1.00, 1.43)	1.19 (0.93, 1.52)	1.05 (0.77, 1.42)	1.11 (0.76, 1.61)
Q4	1.03 (0.79, 1.35)	0.92 (0.74, 1.15)	0.88 (0.75, 1.02)	0.91 (0.71, 1.17)	1.03 (0.76, 1.39)	0.91 (0.67, 1.23)
Women (3179)						
Age	1.03 (1.02, 1.04)	1.04 (1.03, 1.05)	0.99 (0.98, 1.00)	1.08 (1.07, 1.09)	1.08 (1.06, 1.10)	1.07 (1.05, 1.08)
Meat						
Q1	1	1	1	1	1	1
Q2	1.05 (0.82, 1.34)	1.31 (1.03, 1.66)	1.28 (1.04, 1.58)	1.06 (0.82, 1.39)	1.65 (1.07, 2.53)	1.07 (0.84, 1.36)
Q3	0.99 (0.83, 1.19)	1.21 (0.94, 1.56)	1.02 (0.82, 1.26)	1.17 (0.86, 1.61)	1.62 (1.19, 2.20)	1.13 (0.84, 1.52)
Q4	1.17 (0.98, 1.41)	1.01 (0.88, 1.49)	0.86 (0.70, 1.05)	1.03 (0.79, 1.35)	1.96 (1.30, 2.96)	0.94 (0.72, 1.21)
Carbohydrate						
Q1	1	1	1	1	1	1
Q2	1.03 (0.85, 1.27)	1.01 (0.81, 1.26)	1.03 (0.84, 1.27)	1.09 (0.90, 1.32)	0.66 (0.48, 0.92)	0.86 (0.65, 1.14)
Q3	1.20 (1.00, 1.45)	1.31 (1.07, 2.77)	1.04 (0.83, 1.30)	1.11 (0.89, 1.40)	0.65 (0.48, 0.89)	1.02 (0.82, 1.28)
Q4	1.11 (0.88, 1.40)	2.12 (1.62, 2.77)	2.03 (1.63, 2.54)	0.85 (0.66, 1.09)	0.63 (0.39, 1.02)	1.60 (1.24, 2.08)
Healthy						
Q1	1	1	1	1	1	1
Q2	0.81 (0.65, 1.01)	0.84 (0.56, 1.09)	0.92 (0.78, 1.09)	0.89 (0.74, 1.09)	0.83 (0.55, 1.24)	0.64 (0.50, 0.81)
Q3	0.79 (0.70, 0.94)	0.93 (0.73, 1.19)	1.02 (0.88, 1.18)	0.72 (0.57, 0.90)	0.73 (0.46, 1.17)	0.69 (0.523, 0.91)
Q4	0.67 (0.57, 0.79)	0.74 (0.59, 0.92)	0.97 (0.82, 1.15)	0.89 (0.67, 1.18)	0.86 (0.54, 1.37)	0.72 (0.52, 0.99)

High TG: high triglycerides. Low HDL-C: low high density-lipoprotein cholesterol. High BP: high blood pressure. MetS: metabolic syndrome.
All models were controlled for age, alcohol drinking, family history of diabetes and smoking, leisure time physical activity, and BMI.

pattern, BMI decreased significantly in the highest quartile in men but was not different in women. Compared with participants in the lowest quartile, both men and women in the highest quartile of the carbohydrate pattern had lower HDL-C levels, higher TG levels, and a higher prevalence of MetS. Both men and women in the highest quartile of the meat pattern had higher levels of HDL-C. Only men in the highest quartile of meat pattern had a higher level of SBP but a lower level of TG compared with those in the lowest quartile. Menopausal status was relatively equally distributed across each dietary pattern.

Table 4 shows the results of multiple logistic regression models for the associations of dietary patterns with each metabolic component and MetS controlling for other covariates. In men, compared with the first quartile, the highest quartile of carbohydrate pattern was associated with increased odds of high TG and low HDL-C, but lower odds of abdominal obesity and hyperglycemia. Meat pattern was associated with increased odds of abdominal obesity, high blood pressure, and hyperglycemia. In women, the meat pattern was associated with increased odds of hyperglycemia. Carbohydrate pattern was likely to be associated with hypertriglyceridemia and low HDL-C, while the healthy pattern was associated with decreased odds of abdominal obesity and hypertriglyceridemia.

The carbohydrate pattern was the only dietary pattern associated with increased odds of MetS in the fourth quartile compared with the first quartile in men (adjusted OR of 1.82,

TABLE 5: Adjusted odds ratio (95% CI) for metabolic syndrome according to quartile of factor scores of dietary patterns by leisure time physical activity status in Thai adults aged 30–59 years, NHES IV, 2009.

	Men		Women	
	Low LTPA (n = 2075)	High LTPA (n = 618)	Low LTPA (n = 2600)	High LTPA (n = 579)
Age	1.06 (1.04, 1.08)	1.04 (1.01, 1.07)	1.06 (1.04, 1.08)	1.11 (1.08, 1.14)
Meat				
Q1	1	1	1	1
Q2	0.81 (0.54, 1.22)	0.78 (0.32, 1.91)	1.07 (0.84, 1.35)	1.00 (0.58, 1.72)
Q3	0.75 (0.52, 1.09)	1.57 (0.83, 2.98)	0.99 (0.70, 1.42)	1.40 (0.87, 2.27)
Q4	1.02 (0.75, 1.39)	0.84 (0.33, 2.09)	0.84 (0.61, 1.16)	0.90 (0.38, 2.17)
Carbohydrate				
Q1	1	1	1	1
Q2	1.37 (0.99, 1.90)	1.38 (0.64, 2.98)	0.84 (0.64, 1.10)	0.90 (0.38, 2.17)
Q3	1.37 (1.00, 1.87)	1.58 (0.72, 3.44)	0.98 (0.72, 1.32)	1.08 (0.49, 2.35)
Q4	1.83 (1.27, 2.63)	1.88 (0.96, 3.68)	1.70 (1.28, 2.27)	1.08 (0.53, 2.22)
Healthy				
Q1	1	1	1	1
Q2	1.06 (0.80, 1.41)	1.17 (0.41, 3.30)	0.58 (0.44, 0.76)	0.97 (0.42, 2.24)
Q3	1.13 (0.74, 1.72)	0.82 (0.32, 2.14)	0.62 (0.45, 0.87)	1.28 (0.56, 2.91)
Q4	0.87 (0.64, 1.19)	0.89 (0.42, 1.90)	0.71 (0.49, 1.05)	0.84 (0.37, 1.93)

LTPA: leisure-time physical activity.
Model included age, family history of diabetes, smoking, alcohol drinking, and BMI.

95% CI 1.31, 2.55) and women (adjusted OR 1.60, 95% CI 1.24, 2.08).

The associations between dietary patterns and MetS as stratified by leisure time physical activity status are shown in Table 5. The odds of metabolic syndrome for the fourth quartile of the carbohydrate pattern remained significantly increased among men and women who were physically inactive (adjusted OR 1.83, 95% CI 1.27, 2.63, 1.70, 95% CI 1.28, 2.27, resp.), but not among those who were physically active, particularly in women.

4. Discussion

The important findings in the present study were that the carbohydrate pattern was more popular in the rural areas and in the northeastern and northern region of Thailand, whereas the meat pattern was more common in urban areas, Bangkok, and the central and southern regions. The healthy dietary pattern was also more prevalent in the residents living in urban areas and among those with higher socioeconomic status. Those with the healthy dietary pattern were more likely to have a better metabolic profile. To our knowledge, the present study is the first to report the association of high dietary intake of carbohydrate with glutinous rice with an increased risk of MetS due to high TG and low HDL-C. On the other hand, individuals with a high intake of meat and its product were more likely to have high BP, high FPG, and abdominal obesity. However, the effect of carbohydrate pattern decreased among those with high LTPA and increased among those with low LTPA.

The findings of an association between high carbohydrate intake and increased risk of metabolic risk factors and MetS

in the present study were consistent with other studies [5, 8]. A study in Korea reported that risk of MetS increased among people eating white rice compared with those eating rice with beans and rice with multigrains [8]. Kim et al. [5] reported a higher risk of MetS among those in the highest quartile of carbohydrate intake, particularly among those with BMI ≥25 kg/m². However, the present study did not reveal such interaction between BMI and carbohydrate; the effect of the dietary pattern was not different by BMI status. Although it is not clear whether total energy intake is different by pattern, it is very likely that the proportion of energy derived from carbohydrate among those favoring a carbohydrate dietary pattern is greater than in those who favor the other two patterns. A survey of national food insecurity in Thailand reported that people in the northeast had the highest daily consumption of carbohydrate (338.7 gm/person), followed by those in the north at 322.4 gm/person. Daily carbohydrate intake was lowest in Bangkok (268.8 gm/person). But the northeast had the lowest daily intake of energy from fats (51.1 gm/person) compared with a national average of 60.3 gm/person [20]. We found that the healthy pattern including fruit and vegetables, and beans decreased the risk of MetS. MetS has been negatively associated with a healthy diet comprising of fruit and vegetables in several populations [21–23]. A high intake of fruit, vegetables, and fish as part of a healthy diet is associated with a decreased risk of coronary heart disease (CHD) [4, 24] and negatively associated with MetS [25]. The lower risk of MetS among individuals with a healthy diet might be due to higher intakes of dietary fiber, minerals, phytoestrogen, and whole grains [26]. A healthy diet which is high in soy-derived isoflavones decreases the risk of coronary heart diseases and CVD by protecting against

TABLE 6: Food groups and food items from the food frequency questionnaire.

Food groups	Food items
Meat	Lean pork/beef, chicken/duck without skin
Fatty meat	Meat with fat, chicken/duck with skin, streaky pork
Processed meat with high fat	Processed meat high in fat, that is, sausage, Thai style sausage, Chinese sausage, sour sausage, bacon, ham, Vietnamese ham
Processed meat with high salt	Processed meat high in salt, that is, meat floss, salted fish, salted sun-dried beef/pork/fish
Fish	Fresh-water fish, salt-water fish, and so forth
Shellfish and squid	Crustacean and mollusk seafood, that is, shrimp, crab, squid, clam, and so forth
Animal organ	Liver, blood jelly, intestine, gizzard, and so forth
Egg	Egg
Beans	Beans and its products, that is, mung bean, soybean, peanut, tofu, Kaset protein
Rice	Polished rice
Wheat	Whole wheat, whole grain bread
Glutinous rice	Sticky rice
Fried food	Fried pork, fried chicken, friend banana, friend potato, fried meat ball, fish cake, and so forth
Food with coconut milk	Any dishes cooked with coconut milk, that is, spicy curry with coconut milk, green beef curry with coconut milk, and so forth
Fermented fish/soybean	Condiment of fermented fish, southern style fish sauce, Thai style fermented soybean, and so forth
Chili sauce/dip	Thai dipping sauce, that is, roasted chili paste, shrimp paste sauce, green chili dip, fermented fish spicy dip
Fruit	Fresh fruits, that is, orange, banana, guava, ripe papaya, pineapple, longan, watermelon, sugar apple, grapes, rambutan
Milk	Milk
Soy milk	Soy milk
Beverage	Soda beverage
Bamboo shoots	Bamboo shoots
Vegetables	Chinese kale, cauliflower, cabbage, ivy gourd, sponge gourd, Thai water morning glory, green onion

dyslipidemia. Clinical research suggests that soy reduces low-density lipoprotein cholesterol (LDL-C) and improves HDL-C and glycemic control [27, 28].

The association of a high meat dietary pattern intake with MetS has been reported in several studies [3, 21, 25]. The Malmo study [20] found an increased risk of hyperglycemia with fatty meat intake in men. Moreover, a significant body of knowledge suggests that high meat intake, especially red meat, is associated with increased risk of CHD and diabetes. The present study found a positive association of meat and its product intake with odds of abdominal obesity, high blood pressure, and hyperglycemia in men, but only with hyperglycemia in women. We did not observe an association between meat pattern and MetS. The difference in the effect of meat pattern from other studies might be due to difference in the data collection tools, amount of meat consumed, and the fat content of the meat consumed. With regard to the benefit of physical activity, the decrease in association of carbohydrate pattern with MetS among physically active individuals found in the present study is consistent with other observational and experimental studies [10, 29, 30]. A population-based study in China reported a lower likelihood of CVD risk factors among those who had both physically active lifestyles and healthy dietary patterns [10]. An increase from moderate to vigorous LTPA was associated with

a decreased probability of developing MetS and diabetes in a Finnish randomized controlled study [29]. Menopause might be a risk for MetS; however, in the present study menopause was relatively equally distributed across each dietary quartile and was not significantly associated with MetS. Hence, it was not included in the final model. We included BMI in the model to assess the direct effect of dietary pattern on MetS, not mediated through obesity. However, general obesity was associated with MetS; it is possible that controlling for BMI in the model might have underestimated the effect of diet.

Some limitations in the present study include the shortcoming of factor analysis, as it is not highly reproducible due to some degree of subjective decision. The FFQ used in the present study was not semiquantitative, so only frequency of intakes and not quantities of nutrients consumed could be estimated and quantified as energy intake. The cross-sectional study is a useful design to assess the dietary patterns of people with particular characteristics, but it cannot establish a causal relationship between dietary patterns and health outcomes. Also, in our analysis we did not take into account hormone and supplement use among women; whether this might have confounded the results is not clear and needs further study. Nevertheless, a major strength of the present study is that it involved a nationally representative sample of the Thai population. The implication of this study is that knowledge of regional preferences in dietary patterns allows the design of more specific dietary recommendations for specific groups, such as advice about healthy dietary patterns and more LTPA among those with high consumption of carbohydrate pattern.

5. Conclusion

The difference in dietary patterns by demographic characteristics and geographic areas might contribute to the variation in metabolic profile. The carbohydrate dietary pattern increased the odds of MetS; however, the risk decreased among those with high level of LTPA.

Conflict of Interests

The authors declare that there is no conflict of interests regarding the publication of this paper.

Acknowledgments

NHES IV was conducted by the National Health Examination Survey Office, Health Systems Research Institute, Thailand. The Thai National Health Examination Survey IV was supported by the Bureau of Policy and Strategy, Ministry of Public Health; the Thai Health Promotion Foundation; and the National Health Security Office, Thailand. The authors thank Professor Amnuay Thithapandha, Faculty of Medicine, and Dr. Carol Hutchinson, Faculty of Public Health, Mahidol University, for their help in editing the paper.

References

[1] K. G. Alberti, R. H. Eckel, S. M. Grundy et al., "Harmonizing the metabolic syndrome: a joint interim statement of the International Diabetes Federation Task Force on Epidemiology and Prevention; National Heart, Lung, and Blood Institute; American Heart Association; World Heart Federation; International Atherosclerosis Society; and International Association for the Study of Obesity," *Circulation*, vol. 120, no. 16, pp. 1640–1645, 2009.

[2] S. Liu, W. C. Willett, M. J. Stampfer et al., "A prospective study of dietary glycemic load, carbohydrate intake, and risk of coronary heart disease in US women," *The American Journal of Clinical Nutrition*, vol. 71, no. 6, pp. 1455–1461, 2000.

[3] A. J. Baxter, T. Coyne, and C. McClintock, "Dietary patterns and metabolic syndrome—a review of epidemiologic evidence," *Asia Pacific Journal of Clinical Nutrition*, vol. 15, no. 2, pp. 134–142, 2006.

[4] F. B. Hu and W. C. Willett, "Optimal diets for prevention of coronary heart disease," *The Journal of the American Medical Association*, vol. 288, no. 20, pp. 2569–2578, 2002.

[5] K. Kim, S. H. Yun, B. Y. Choi, and M. K. Kim, "Cross-sectional relationship between dietary carbohydrate, glycaemic index, glycaemic load and risk of the metabolic syndrome in a Korean population," *British Journal of Nutrition*, vol. 100, no. 3, pp. 576–584, 2008.

[6] K. Murakami, S. Sasaki, Y. Takahashi et al., "Dietary glycemic index and load in relation to metabolic risk factors in Japanese female farmers with traditional dietary habits," *The American Journal of Clinical Nutrition*, vol. 83, no. 5, pp. 1161–1169, 2006.

[7] S. J. Song, J. E. Lee, H.-Y. Paik, M. S. Park, and Y. J. Song, "Dietary patterns based on carbohydrate nutrition are associated with the risk for diabetes and dyslipidemia," *Nutrition Research and Practice*, vol. 6, no. 4, pp. 349–356, 2012.

[8] Y. Ahn, S.-J. Park, H.-K. Kwack, M. K. Kim, K.-P. Ko, and S. S. Kim, "Rice-eating pattern and the risk of metabolic syndrome especially waist circumference in Korean Genome and Epidemiology Study (KoGES)," *BMC Public Health*, vol. 13, no. 1, article 61, 2013.

[9] A. Sadakane, A. Tsutsumi, T. Gotoh et al., "Dietary patterns and levels of blood pressure and serum lipids in a Japanese population," *Journal of Epidemiology*, vol. 18, no. 2, pp. 58–67, 2008.

[10] D. Wang, Y. He, Y. Li et al., "Joint association of dietary pattern and physical activity level with cardiovascular disease risk factors among Chinese men: a cross-sectional study," *PLoS ONE*, vol. 8, no. 6, Article ID e66210, 2013.

[11] W. Aekplakorn, "Prevalence, treatment, and control of metabolic risk factors by BMI status in Thai Adults: national health examination survey III," *Asia-Pacific Journal of Public Health*, vol. 23, no. 3, pp. 298–306, 2011.

[12] W. Aekplakorn, P. Kessomboon, R. Sangthong et al., "Urban and rural variation in clustering of metabolic syndrome components in the Thai population: results from the fourth National Health Examination Survey 2009," *BMC Public Health*, vol. 11, article 854, 2011.

[13] W. Aekplakorn, R. Sangthong, P. Kessomboon et al., "Changes in prevalence, awareness, treatment and control of hypertension in Thai population, 2004–2009: Thai National Health Examination Survey III-IV," *Journal of Hypertension*, vol. 30, no. 9, pp. 1734–1742, 2012.

[14] W. Aekplakorn, S. Chariyalertsak, P. Kessomboon et al., "Prevalence and management of diabetes and metabolic risk factors in Thai adults: the Thai National Health Examination Survey IV, 2009," *Diabetes Care*, vol. 34, no. 9, pp. 1980–1985, 2011.

[15] F. C. Bull, T. S. Maslin, and T. Armstrong, "Global physical activity questionnaire (GPAQ): nine country reliability and validity study," *Journal of Physical Activity and Health*, vol. 6, no. 6, pp. 790–804, 2009.

[16] B. E. Ainsworth, W. L. Haskell, M. C. Whitt et al., "Compendium of physical activities: an update of activity codes and MET intensities," *Medicine and Science in Sports and Exercise*, vol. 32, no. 9, pp. S498–S504, 2000.

[17] WHO, *International Guide for Monitoring Alcohol Consumption and Related Harm*, Department of Mental Health and Substance Dependence, World Health Organization, Geneva, Switzerland, 2000.

[18] N. Boontaveeyuwat, *Validity of Food Consumption and Nutrition Survey Questionnaire for the National Health Examination Survey IV*, National Health Examination Survey Office, Bangkok, Thailand, 2008.

[19] S. O. Rutstein and K. Johnson, "The DHS wealth index," Contract 6, U.S. Agency for International Development, Calverton, Md, USA, 2004.

[20] *Food insecurity assessment at National and subnational levels in Thailand, 2011*, National Statistical Office and Office of Agriculture Economic of the Kingdom of Thailand, Bangkok, Thailand, 2012.

[21] E. Wirfält, B. Hedblad, B. Gullberg et al., "Food patterns and components of the metabolic syndrome in men and women: a cross-sectional study within the Malmo diet and cancer cohort,"

The American Journal of Epidemiology, vol. 154, no. 12, pp. 1150–1159, 2001.

[22] L. Sonnenberg, M. Pencina, R. Kimokoti et al., "Dietary patterns and the metabolic syndrome in obese and non-obese framingham women," *Obesity Research*, vol. 13, no. 1, pp. 153–162, 2005.

[23] A. Esmaillzadeh, M. Kimiagar, Y. Mehrabi, L. Azadbakht, F. B. Hu, and W. C. Willett, "Dietary patterns, insulin resistance, and prevalence of the metabolic syndrome in women," *The American Journal of Clinical Nutrition*, vol. 85, no. 3, pp. 910–918, 2007.

[24] G. S. Savige, "Candidate foods in the Asia-Pacific region for cardiovascular protection: fish, fruit and vegetables," *Asia Pacific Journal of Clinical Nutrition*, vol. 10, no. 2, pp. 134–137, 2001.

[25] M. A. Pereira, D. R. Jacobs Jr., L. van Horn, M. L. Slattery, A. I. Kartashov, and D. S. Ludwig, "Dairy consumption, obesity, and the insulin resistance syndrome in young adults: the CARDIA study," *The Journal of the American Medical Association*, vol. 287, no. 16, pp. 2081–2089, 2002.

[26] J. L. Slavin, M. C. Martini, D. R. Jacobs Jr., and L. Marquart, "Plausible mechanisms for the protectiveness of whole grains," *American Journal of Clinical Nutrition*, vol. 70, no. 3, pp. 459S–463S, 1999.

[27] L. Azadbakht, M. Kimiagar, Y. Mehrabi et al., "Soy inclusion in the diet improves features of the metabolic syndrome: a randomized crossover study in postmenopausal women," *The American Journal of Clinical Nutrition*, vol. 85, no. 3, pp. 735–741, 2007.

[28] D. J. A. Jenkins, C. W. C. Kendall, C.-J. C. Jackson et al., "Effects of high- and low-isoflavone soyfoods on blood lipids, oxidized LDL, homocysteine, and blood pressure in hyperlipidemic men and women," *The American Journal of Clinical Nutrition*, vol. 76, no. 2, pp. 365–372, 2002.

[29] P. Ilanne-Parikka, D. E. Laaksonen, J. G. Eriksson et al., "Leisure-time physical activity and the metabolic syndrome in the Finnish diabetes prevention study," *Diabetes Care*, vol. 33, no. 7, pp. 1610–1617, 2010.

[30] Y. He, Y. Li, J. Lai et al., "Dietary patterns as compared with physical activity in relation to metabolic syndrome among Chinese adults," *Nutrition, Metabolism and Cardiovascular Diseases*, vol. 23, no. 10, pp. 920–928, 2013.

Maternal Fructose Intake Induces Insulin Resistance and Oxidative Stress in Male, but Not Female, Offspring

Lourdes Rodríguez,[1] Paola Otero,[1] María I. Panadero,[1] Silvia Rodrigo,[1] Juan J. Álvarez-Millán,[1,2] and Carlos Bocos[1]

[1]*Facultad de Farmacia, Universidad CEU San Pablo, Urbanización Montepríncipe, Boadilla del Monte, 28668 Madrid, Spain*
[2]*CQS Laboratory, C/Artistas 1, 28020 Madrid, Spain*

Correspondence should be addressed to Carlos Bocos; carbocos@ceu.es

Academic Editor: Phillip B. Hylemon

Objective. Fructose intake from added sugars correlates with the epidemic rise in metabolic syndrome and cardiovascular diseases. However, consumption of beverages containing fructose is allowed during gestation. Recently, we found that an intake of fructose (10% wt/vol) throughout gestation produces an impaired fetal leptin signalling. Therefore, we have investigated whether maternal fructose intake produces subsequent changes in their progeny. *Methods*. Blood samples from fed and 24 h fasted female and male 90-day-old rats born from fructose-fed, glucose-fed, or control mothers were used. *Results*. After fasting, HOMA-IR and ISI (estimates of insulin sensitivity) were worse in male descendents from fructose-fed mothers in comparison to the other two groups, and these findings were also accompanied by a higher leptinemia. Interestingly, plasma AOPP and uricemia (oxidative stress markers) were augmented in male rats from fructose-fed mothers compared to the animals from control or glucose-fed mothers. In contrast, female rats did not show any differences in leptinemia between the three groups. Further, insulin sensitivity was significantly improved in fasted female rats from carbohydrate-fed mothers. In addition, plasma AOPP levels tended to be diminished in female rats from carbohydrate-fed mothers. *Conclusion*. Maternal fructose intake induces insulin resistance, hyperleptinemia, and plasma oxidative stress in male, but not female, progeny.

1. Introduction

In the last few decades, obesity, metabolic syndrome, and diabetes have escalated to epidemic proportions in many countries worldwide. These metabolic diseases are multifactorial resulting from genetic, physiological, behavioural, and environmental influences. Genetic influence alone does not suffice to explain the rate at which these diseases have increased [1]. In fact, several studies demonstrate that metabolic events during pre- and postnatal development modulate metabolic disease risks in later life [2]. Among them, feeding conditions likely constitute one of the most influential parameters on the health of the adult [3]. Thus, diet manipulation in mothers during critical developmental periods (such as gestation and/or the early postnatal) has been used to identify their contribution to obesity and diabetes development in offspring [4]. Given the current

worldwide shift toward a westernized lifestyle, there is an urgent need to address the relationship between the quality and quantity of nutrient intake during pregnancy and/or lactation and the metabolic fate of the offspring [5].

Fructose, present in added sugars such as sucrose and high-fructose corn syrup, has been linked to obesity and metabolic syndrome [6–8]. Experimental studies have shown that fructose can induce leptin resistance and features of metabolic syndrome in rats, whereas glucose intake does not [9–11]. Clinical studies also support fructose as a cause of metabolic syndrome [12–14].

Interestingly, sex-dependent differences in the influence of fructose for inducing metabolic diseases have been reported. Thus, women, but not men, exhibit an association between fructose consumption and an increased risk of type 2 diabetes mellitus. In relation to this, it has been shown that female rats subjected to fructose have a more detrimental response than their male counterparts. Fructose-fed male rats were resistant to the hepatic effects of leptin, whereas fructose-fed females had no signs of leptin resistance but had hyperinsulinemia and altered glucose tolerance test [15]. In contrast, in high-fructose-fed rodents, oxidative stress was observed in male, but not in female, rats [16].

Since it has been well-established that fructose intake modifies lipidemia in laboratory animals and humans [11, 17] and maternal diet manipulations can affect the progeny, it might be speculated that fructose administration during gestation and/or lactation could cause metabolic changes in the offspring. Unfortunately, studies investigating altered maternal nutrition have used quite different experimental designs to determine the role of fructose [18–24].

In our previous report, we investigated the effects of a low fructose intake throughout gestation in mothers and their fetuses [25], and we obtained intriguing results. Fructose-fed mothers presented a diminished leptin response to fasting and refeeding, their fetuses displaying an impaired transduction of the leptin signal, and these findings were not observed in glucose-fed rats. Therefore, the present study was designed to determine whether an intake of liquid fructose (10% wt/vol) during pregnancy had long-term consequences on the offspring. Furthermore, special attention has been given to determine whether any potential consequence of the treatment differed between female and male offspring.

2. Material and Methods

2.1. Animals and Experimental Design. Female Sprague-Dawley rats weighing 200–240 g were fed *ad libitum* a standard rat chow diet (B&K Universal, Barcelona, Spain) and housed under controlled light and temperature conditions (12 h light-dark cycle; 22 ± 1°C). The experimental protocol was approved by the Animal Research Committee of the University CEU San Pablo, Madrid, Spain. The experimental protocol to which pregnant rats were subjected was the same as previously reported [25]. Briefly, pregnant animals were randomly separated into a control group, a fructose-supplemented group (fructose), and a glucose-supplemented group (glucose). Fructose and glucose were supplied as a 10% (wt/vol) solution in drinking water throughout gestation. Control animals received no supplementary sugar. Pregnant rats were allowed to deliver and, on the day of birth, each suckling litter was reduced to nine pups per mother. After delivery, both mothers and their pups were maintained with water and food *ad libitum*. It is remarkable that these animals (mothers and pups) received no subsequent additive in the drinking water. On the 21st day after delivery, the lactating mothers were removed to stop the suckling period and pups were separated by gender. When the progeny were 90 days old, they were subjected to 24-hour fasting. After drawing a basal blood sample from the tail vein, pellets were removed from the cages. After 24 h fasting, a second blood sample from the tail vein was obtained. Blood samples were collected into EDTA (1 mg/mL) tubes and placed on ice. Samples were then centrifuged, and plasma was stored at −80°C until processed for glucose, insulin, leptin, and other determinations. Where

two or three pups from one litter were used, their data were averaged.

2.2. Determinations. Plasma aliquots were used to measure glucose by an enzymatic colorimetric test (Spinreact, Girona, Spain). NEFA (nonesterified ("free") fatty acids) (Wako, Neuss, Germany), glycerol (Sigma Chemical, St. Louis, MO), triglycerides, and uric acid (Spinreact) were measured using commercial kits. Ketone bodies were measured using a kinetic method (Randox Laboratories, United Kingdom). Insulin was determined in plasma samples using a specific ELISA kit for rats (Mercodia, Uppsala, Sweden). Leptin and adiponectin were assayed in plasma samples using a specific enzyme immunoassay (EIA) kit for rats (Biovendor, Brno, Czech Republic, and Millipore, Bedford, MA, resp.).

Estimates of insulin resistance were calculated as previously described [15, 25] by determination of the following indexes from the 24 h fasting plasma glucose and insulin values: homeostasis model assessment of insulin resistance (HOMA-IR) and insulin sensitivity index (ISI). The HOMA-IR was calculated as the product of the fasting plasma glucose (FPG) and insulin (FPI) divided by a constant, 22.5. FPI was expressed in microunits per milliliter and FPG as millimoles per liter [26]. The ISI was calculated as the ratio $2/[(FPI \times FPG) + 1]$, expressing FPI and FPG in micromoles per liter [15]. Additionally, estimates of insulin sensitivity were also calculated in fed-state as previously described [25] by determination of the ratio between glucose levels (in milligrams per deciliter) and insulin values (in microunits per milliliter).

Finally, plasma aliquots were also used to determine the oxidative stress state. The concentration of malondialdehyde (MDA) in plasma was measured as a marker of lipid peroxidation using the method previously described [27], by measuring the fluorescence of MDA-thiobarbituric acid (TBA) complexes at 515 nm/553 nm excitation/emission wavelengths. Further, the advanced oxidation protein products (AOPP) in plasma were determined as a protein oxidative stress biomarker using the spectrophotometric technique previously described [28, 29]. The AOPP concentrations were expressed as μmol/L of chloramine-T equivalents.

2.3. Statistical Analysis. Results were expressed as mean ± SE. Treatment effects were analyzed by one-way ANOVA. When treatment effects were significantly different ($P < 0.05$), means were tested by the Tukey multiple range test, using a computer program SSPS (version 15). When the variance was not homogeneous, a post hoc Tamhane test was performed.

3. Results

3.1. Ingestion of a 10% wt/vol Fructose Solution throughout Gestation Affects Leptinemia in Male Progeny. As shown in Table 1, neither fructose nor glucose intake throughout pregnancy produced alterations in the body weight of the male and female progeny.

TABLE 1: Body weight and plasma analytes in fed 90-day-old progeny from fructose- or glucose-supplemented mothers.

| | Male | | |
	Control	Fructose	Glucose
Body weight (g)	423.7 ± 12.3	403.4 ± 12.8	384.7 ± 21.3
Glucose (mg/dL)	133.4 ± 4.3	128.7 ± 3.2	138.0 ± 1.0
Insulin (μg/L)	0.53 ± 0.05	0.62 ± 0.06	0.47 ± 0.08
Glucose/insulin ratio	10.4 ± 0.4	8.1 ± 0.9	11.9 ± 1.4
Triglycerides (mg/dL)	95.6 ± 8.0	99.9 ± 2.2	82.5 ± 5.7
NEFA (mM)	0.45 ± 0.10	0.56 ± 0.15	0.46 ± 0.07
Glycerol (mg/dL)	2.87 ± 0.29	3.18 ± 0.22	2.55 ± 0.18
Adiponectin (μg/mL)	18.9 ± 1.1[a]	24.3 ± 0.9[b]	25.1 ± 1.8[b]
Leptin (ng/mL)	6.35 ± 0.57[a]	11.20 ± 1.44[b]	7.08 ± 0.92[a]
	Female		
	Control	Fructose	Glucose
Body weight (g)	260.4 ± 9.6	258.7 ± 6.4	253.7 ± 4.2
Glucose (mg/dL)	139.3 ± 7.0	134.1 ± 2.8	142.7 ± 1.8
Insulin (μg/L)	0.54 ± 0.05	0.49 ± 0.07	0.40 ± 0.05
Glucose/insulin ratio	10.7 ± 1.9	10.0 ± 0.7	14.7 ± 1.0
Triglycerides (mg/dL)	57.2 ± 2.9[a]	55.9 ± 8.1[a]	32.0 ± 4.3[b]
NEFA (mM)	0.48 ± 0.09	0.50 ± 0.09	0.37 ± 0.04
Glycerol (mg/dL)	2.81 ± 0.21	2.68 ± 0.29	2.10 ± 0.14
Adiponectin (μg/mL)	43.9 ± 4.0	48.7 ± 3.5	46.7 ± 3.8
Leptin (ng/mL)	5.27 ± 0.50	4.61 ± 0.24	4.41 ± 0.17

Data are means ± SE; n = 10–12 animals from four litters. Where two or three pups from one litter were studied, their data were averaged. Different letters indicate significant differences between the groups ($P < 0.05$).

TABLE 2: Plasma analytes in fasted 91-day-old progeny from fructose- or glucose-supplemented mothers.

| | Male | | |
	Control	Fructose	Glucose
Glucose (mg/dL)	89.3 ± 6.1	97.6 ± 4.8	96.8 ± 5.7
Insulin (μg/L)	0.044 ± 0.015[a]	0.144 ± 0.031[b]	0.031 ± 0.006[a]
Triglycerides (mg/dL)	52.1 ± 12.6	53.8 ± 6.8	35.6 ± 4.1
NEFA (mM)	1.66 ± 0.24	1.37 ± 0.13	1.42 ± 0.06
Glycerol (mg/dL)	5.04 ± 0.72	4.12 ± 0.34	4.25 ± 0.15
Adiponectin (μg/mL)	29.9 ± 6.8	27.7 ± 3.6	21.4 ± 1.0
Ketone bodies (mM)	0.64 ± 0.16	0.63 ± 0.11	0.73 ± 0.06
	Female		
	Control	Fructose	Glucose
Glucose (mg/dL)	104.1 ± 6.6	94.8 ± 2.8	102.9 ± 7.4
Insulin (μg/L)	0.101 ± 0.017[a]	0.042 ± 0.010[b]	0.026 ± 0.004[b]
Triglycerides (mg/dL)	26.4 ± 8.9	24.7 ± 4.4	23.6 ± 4.6
NEFA (mM)	1.49 ± 0.29	1.49 ± 0.21	1.29 ± 0.04
Glycerol (mg/dL)	4.62 ± 0.43	4.32 ± 0.48	4.94 ± 0.58
Adiponectin (μg/mL)	47.8 ± 8.6	48.2 ± 2.0	37.6 ± 5.3
Ketone bodies (mM)	0.63 ± 0.10	0.97 ± 0.12	1.00 ± 0.20

Data are means ± SE; n = 10–12 animals from four litters. Where two or three pups from one litter were studied, their data were averaged. Different letters indicate significant differences between the groups ($P < 0.05$).

Plasma NEFA, glycerol, and triglycerides concentrations were similar in the male rats from carbohydrate-fed mothers with respect to control values (Table 1). Male rats from carbohydrate-fed mothers showed higher levels of plasma adiponectin levels. Although glycemia and insulinemia showed no differences between the three groups (Table 1), glucose/insulin ratio tended to be lower in the male animals from fructose-fed mothers compared to the animals from control and glucose-supplemented rats (Table 1). Interestingly, male descendents from fructose-supplemented rats turned out to be clearly hyperleptinemic (Table 1).

On the other hand, since insulinemia and glycemia were similar in the female rats from carbohydrate-fed mothers with respect to control values, glucose/insulin ratio was found to be similar for the female progeny of control, fructose-fed, and glucose-fed pregnant rats (Table 1). Similar findings were recorded for plasma adiponectin levels. However, plasma NEFA, glycerol, and triglycerides concentrations were lower in female animals from glucose-fed pregnant rats in comparison to the other two groups (Table 1). In contrast to males, female descendents from fructose-supplemented rats showed similar values in their leptinemia to the animals from control and glucose-supplemented mothers (Table 1).

3.2. Ingestion of a 10% wt/vol Fructose Solution throughout Gestation Affects Insulin Sensitivity in Fasted Male Progeny. To confirm a possible disturbance in glucose homeostasis in

male, but not in female, rats born of mothers supplemented throughout pregnancy with liquid fructose, we subjected the progeny of the three experimental groups to 24 h fasting and measured plasma parameters related to insulin resistance.

After 24-hour fasting, plasma glucose, NEFA, glycerol, ketone bodies, and triglycerides concentrations did not show any differences in the male rats from carbohydrate-fed mothers with respect to control values (Table 2). Similar findings were recorded for plasma adiponectin levels. However, insulinemia was significantly higher in male animals from fructose-fed pregnant rats in comparison to the other two groups (Table 2). Thus, HOMA-IR and ISI ratios were significantly different in the male animals from fructose-fed mothers compared to the male animals from control and glucose-supplemented rats (Figures 1(a)-1(b)), and, therefore, these findings confirmed that an insulin resistant state exists in male progeny of fructose-fed mothers. In accordance with this result, male descendents from fructose-supplemented rats showed a clear increase in their plasma leptin concentrations (Figure 1(c)).

After 24-hour fasting, plasma glucose, NEFA, glycerol, ketone bodies, adiponectin, and triglycerides levels were similar in the female rats from carbohydrate-fed mothers with respect to the control values (Table 2). However, insulinemia was lower in female animals from carbohydrate-fed pregnant rats in comparison to the control group (Table 2). Thus, HOMA-IR and ISI ratios were significantly different in the female animals from carbohydrate-supplemented mothers compared to the female animals from control rats (Figures 1(a)-1(b)), these being more pronounced in the group from the glucose-fed mothers. Thus, not only did female rats from fructose-fed mothers not show an insulin

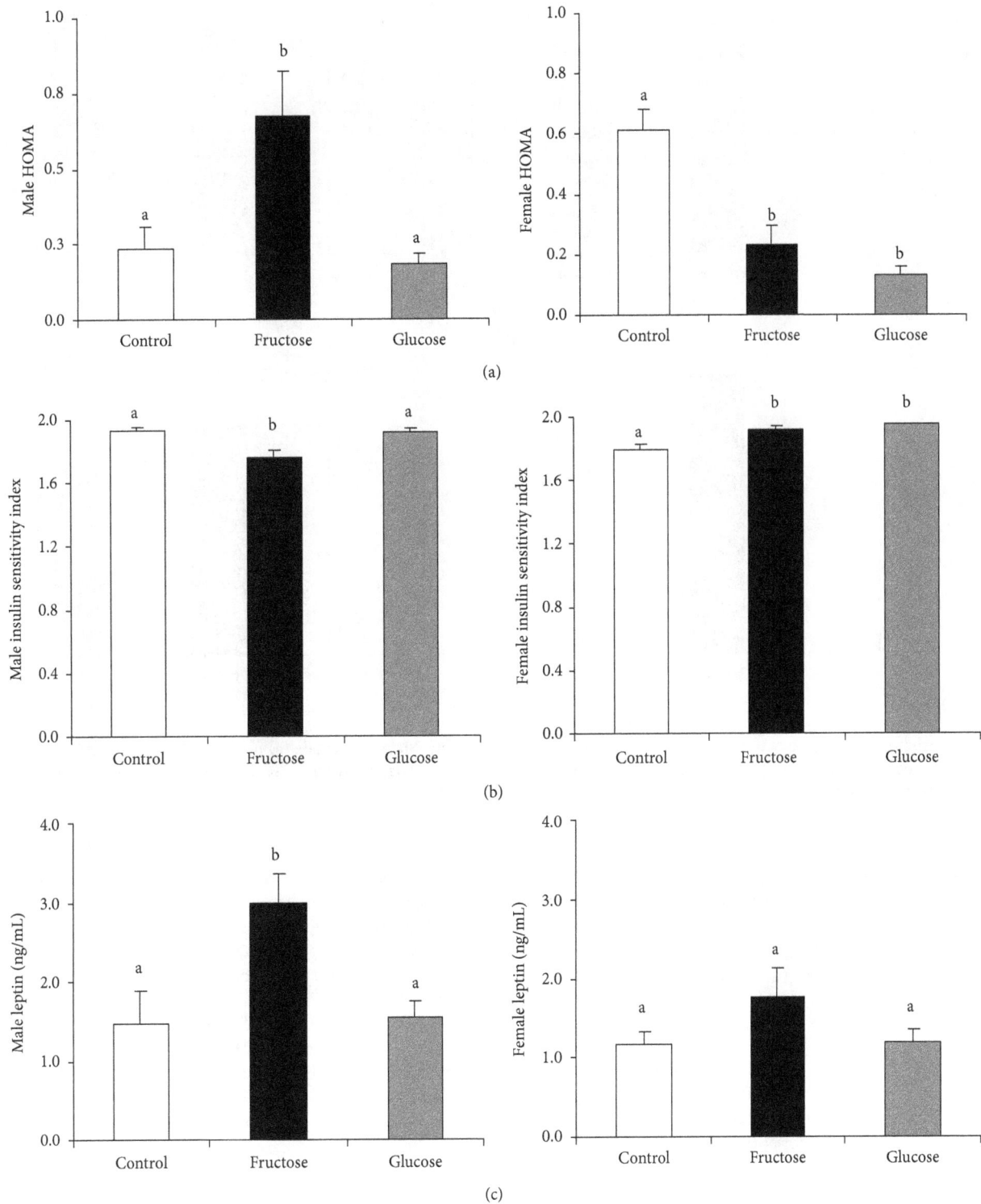

Figure 1: Fructose in pregnancy produces insulin resistance and hyperleptinemia in progeny. Male and female (a) HOMA-IR and (b) ISI ratios and (c) leptinemia of 24-hour fasted 90-day-old progeny from control, fructose-fed, and glucose-fed pregnant rats. Data are means ± SE; n = 10–12 animals from four litters. Different letters indicate significant differences between the groups ($P < 0.05$).

(a)

(b)

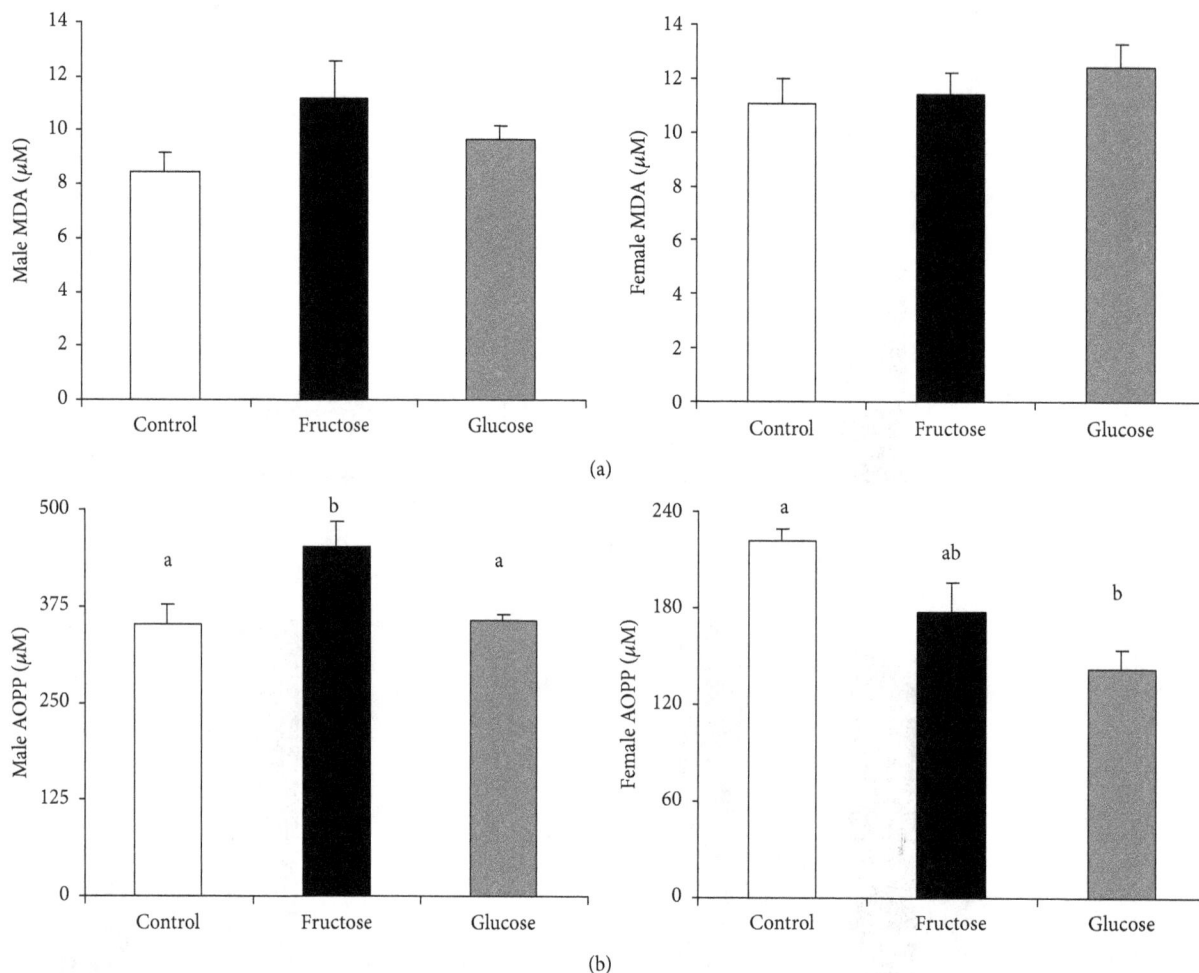

FIGURE 2: Fructose in pregnancy influences plasma oxidative stress in progeny. Male and female (a) plasma MDA (b) and AOPP values of 90-day-old progeny from control, fructose-fed, and glucose-fed pregnant rats. Data are means ± SE; n = 10–12 animals from four litters. Different letters indicate significant differences between the groups ($P < 0.05$).

resistant state as seen in male progeny, but female rats from carbohydrate-fed mothers showed improved insulin sensitivity. In fasted female descendents from fructose-supplemented rats, although plasma leptin concentrations tended to be higher than in the other two groups, the effect was not significant (Figure 1(c)).

3.3. Ingestion of a 10% wt/vol Fructose Solution throughout Gestation Affects Plasma Oxidative Stress in Male Progeny.

Since fructose intake during pregnancy seems to produce insulin resistance and hyperleptinemia in the male progeny, parameters which have been related to metabolic syndrome, we also checked whether other features related to that disturbance could be affected. Thus, as shown in Figure 2, plasma MDA and AOPP, which would indicate lipid and protein oxidation, respectively, tended to be augmented in the male rats from fructose-fed mothers in comparison to the animals from control and glucose-fed pregnant rats, becoming significant for AOPP levels. In contrast, plasma AOPP, but not MDA, were diminished in the female progeny of carbohydrate-fed

mothers (Figure 2), this effect being significantly different in the progeny from glucose-fed mothers.

Interestingly, male rats from fructose-supplemented mothers showed higher uricemia (3.4 ± 0.5; 5.4 ± 0.3; and 3.4 ± 0.4 mg/dL, for males born from control, fructose-fed, and glucose-fed mothers, resp., $P < 0.05$). In comparison, females from carbohydrate-fed mothers showed lower levels of plasma uric acid than the control group, this effect being significantly different in the progeny from glucose-fed mothers (4.1 ± 0.3; 3.4 ± 0.3; and 2.3 ± 0.2 mg/dL, for female progeny from control, fructose-fed, and glucose-fed mothers, resp., $P < 0.05$).

4. Discussion

In our previous study [25], we reported an impaired hepatic transduction of the leptin signal in the fetuses from fructose-fed, but not glucose-fed, pregnant rats and related it to a diminished leptin response to fasting and refeeding observed in their mothers. Since it has been proposed that a period of relative hypoleptinemia (or, in our case, leptin resistance) during development may induce some metabolic adaptations that underlie developmental programming [30], we speculated whether the findings found in our previous study [25] could be responsible for a developmental programming of those progeny and produce some metabolic disturbances, when adult.

Thus, male progeny born of fructose-fed rats presented higher leptinemia versus the other two groups, and that hyperleptinemia was observed in both fed and fasting conditions. In accordance with this finding, fasting insulinemia was higher in the male progeny from fructose-fed mothers, all these results being consistent with the insulin resistant state found in these animals (measured as HOMA and ISI ratios). The presence of hyperinsulinemia in fasted male fructose-fed progeny along with hyperleptinemia could also indicate leptin resistance at the level of pancreatic islets [31]. Since insulin stimulates adipogenesis and leptin production in adipocytes whereas leptin inhibits the production of insulin in pancreatic β-cells, a prolonged elevation of plasma leptin levels would result in dysregulation of the adipoinsular axis and a corresponding failure to suppress insulin secretion [32]. In fact, in the present study, although the male progeny from fructose-fed mothers was hyperleptinemic in both fed and fasted conditions, fasting produced a lower impact on insulin levels in comparison to the other two groups (12.1-, 4.3-, and 15.2-fold reduction, for control, fructose, and glucose groups, resp.). In relation to this, it has been reported that adipocytes from male offspring of lactating mothers consuming fructose spontaneously released more leptin than control rat-derived adipocytes and also displayed impaired response to insulin stimulation [4]. Since the progeny from fructose-fed mothers were already leptin resistant when they were fetuses [25], the male progeny of fructose-fed pregnant rats could present a vicious circle (leptin resistance, hypersecretion of insulin, and increasing insulin resistance) [23, 33].

On the other hand, it has been demonstrated that fructose is much more reactive than glucose with respect to participation in glycosylation reactions, which represent an important source of free radicals. In accordance with this fact, rats fed fructose (25% wt/vol) exhibited a significant increase in lipid peroxidation products, compared with rats fed the same amount of glucose or sucrose and those given pure water [34]. Moreover, it has been reported that a short-term administration of a fructose-rich diet to normal rats promotes an increase in several glycoxidative stress markers [16, 35]. Further, as a reducing sugar, fructose reacts with protein molecules to form toxic advanced glycation end-products (AGEs) and also induces protein oxidation possibly through the formation of hydroxyl radicals [36]. Both AGEs and AOPP appear to be involved in the pathogenesis of inflammation, diabetes complications, and cardiovascular diseases. In the present work, the male progeny from fructose-fed dams showed increased plasma levels of protein oxidation products, measured as AOPP. It is worthwhile to mention that AOPP levels have been associated with metabolic syndrome, since high levels of AOPP have been positively correlated with insulin and HOMA levels [37]. Given that it has been proposed that AGEs could accumulate indefinitely on long-lived molecules such as collagen and DNA [38], it might

be speculated that fructose intake during pregnancy could promote oxidation products formation in fetuses and these toxic compounds would accumulate, later appearing in the plasma of progeny.

Related to that increase in plasma oxidative stress of progeny from fructose-fed rats, we also determined uricemia and, surprisingly, male rats born to fructose-fed mothers turned out to be hyperuricemic. Some authors consider hyperuricemia in metabolic syndrome to be the consequence of elevated serum insulin levels, which have been shown to stimulate renal reabsorption of uric acid. In fact, thiazolidinediones which improve insulin sensitivity and lower insulin levels reduce the level of serum uric acid in diabetic patients. Conversely, Nakagawa et al. (2006) demonstrated that the reduction of uric acid levels with a xanthine oxidase inhibitor improved insulin sensitivity. Accordingly, in the present study, fructose-fed male progeny presented hyperuricemia along with insulin resistance. Interestingly, the xanthine oxidase pathway, which is related to uric acid production, has been shown to generate oxidants [39]. Our findings would agree with this result since the hyperuricemic male progeny also presents an increase in plasma oxidative stress. In fact, a recent study has shown a close relationship between fructose, hyperuricemia, oxidative stress, and diabetes [40].

The most prominent result found here is that the intake of just a small amount of fructose (10%) throughout gestation produces a clear impairment in the insulin action, hyperleptinemia, and other features of metabolic syndrome such as high values of oxidative stress biomarkers and uricemia in male progeny. However, female progeny born of fructose-fed mothers did not show any of these characteristics of metabolic syndrome. Nevertheless, as has been described, females born to mothers subjected to undernutrition expressed a programmed phenotype only in the presence of a high-fat diet, whereas the male progeny could manifest it independently of postnatal nutrition [30]. Therefore, it is possible that postnatal hypercaloric nutrition could amplify all these metabolic abnormalities induced by the fructose-fed fetal programming, and this deserves further investigation.

Finally, since the carbohydrate was administered only during gestation, we can assume that the effects of fructose intake would occur during intrauterine development. As previously reported [25], fetuses from fructose-fed mothers presented leptin resistance. Leptin has been implicated in the regulation of insulin secretion by islets and, in fact, leptin may influence the normal proliferation of pancreatic β-cells which occurs in the neonatal period [30]. If it is assumed that an interrelated endocrine insulin-leptin feedback system exists, we could speculate that the adipoinsular axis has been affected by fructose intake throughout pregnancy.

5. Conclusion

Maternal fructose intake seems to provoke features of metabolic syndrome, namely, insulin resistance, hyperleptinemia, hyperuricemia, and plasma oxidative stress in male, but not female, progeny.

Conflict of Interests

The authors declare that there is no conflict of interests regarding the publication of this paper.

Acknowledgments

The authors thank Jose M. Garrido for his help in handling the rats and Brian Crilly for his editorial help. This work was supported by a grant from Plan Nacional de Investigación Científica, Desarrollo e Innovación Tecnológica (I+D+i), Instituto de Salud Carlos III-Subdirección General de Evaluación y Fomento de la Investigación (PI-09/02192), European Community FEDER, and Fundación Universitaria San Pablo CEU (PC 09/2012). Silvia Rodrigo is a FUSP-CEU fellowship.

References

[1] G. J. Howie, D. M. Sloboda, T. Kamal, and M. H. Vickers, "Maternal nutritional history predicts obesity in adult offspring independent of postnatal diet," *The Journal of Physiology*, vol. 587, no. 4, pp. 905–915, 2009.

[2] B. Koletzko, I. Broekaert, H. Demmelmair et al., "Protein intake in the first year of life: a risk factor for later obesity? The E.U. childhood obesity project," *Advances in Experimental Medicine and Biology*, vol. 569, pp. 69–79, 2005.

[3] B. Beck, S. Richy, Z. A. Archer, and J. G. Mercer, "Ingestion of carbohydrate-rich supplements during gestation programs insulin and leptin resistance but not body weight gain in adult rat offspring," *Frontiers in Physiology*, vol. 3, article 224, 2012.

[4] A. Alzamendi, D. Castrogiovanni, R. C. Gaillard, E. Spinedi, and A. Giovambattista, "Increased male offspring's risk of metabolic-neuroendocrine dysfunction and overweight after fructose-rich diet intake by the lactating mother," *Endocrinology*, vol. 151, no. 9, pp. 4214–4223, 2010.

[5] L. Šedová, O. Šeda, L. Kazdová et al., "Sucrose feeding during pregnancy and lactation elicits distinct metabolic response in offspring of an inbred genetic model of metabolic syndrome," *The American Journal of Physiology—Endocrinology and Metabolism*, vol. 292, no. 5, pp. E1318–E1324, 2007.

[6] P. J. Havel, "Dietary fructose: implications for dysregulation of energy homeostasis and lipid/carbohydrate metabolism," *Nutrition Reviews*, vol. 63, no. 5, pp. 133–137, 2005.

[7] L. Tappy and K.-A. Lê, "Metabolic effects of fructose and the worldwide increase in obesity," *Physiological Reviews*, vol. 90, no. 1, pp. 23–46, 2010.

[8] R. J. Johnson, M. S. Segal, Y. Sautin et al., "Potential role of sugar (fructose) in the epidemic of hypertension, obesity and the metabolic syndrome, diabetes, kidney disease, and cardiovascular disease," *The American Journal of Clinical Nutrition*, vol. 86, no. 4, pp. 899–906, 2007.

[9] R. J. Johnson, S. E. Perez-Pozo, Y. Y. Sautin et al., "Hypothesis: could excessive fructose intake and uric acid cause type 2 diabetes?" *Endocrine Reviews*, vol. 30, no. 1, pp. 96–116, 2009.

[10] C. H. Taghibiglou, A. Carpentier, S. C. van Iderstine et al., "Mechanisms of hepatic very low-density lipoprotein overproduction in insulin resistance," *The Journal of Biological Chemistry*, vol. 275, no. 12, pp. 8416–8425, 2000.

[11] N. Roglans, L. Vilà, M. Farré et al., "Impairment of hepatic STAT-3 activation and reduction of PPARα activity in fructose-fed rats," *Hepatology*, vol. 45, no. 3, pp. 778–788, 2007.

[12] K. L. Stanhope, J. M. Schwarz, N. L. Keim et al., "Consuming fructose-sweetened, not glucose-sweetened, beverages increases visceral adiposity and lipids and decreases insulin sensitivity in overweight/obese humans," *Journal of Clinical Investigation*, vol. 119, no. 5, pp. 1322–1334, 2009.

[13] K. L. Stanhope, A. A. Bremer, V. Medici et al., "Consumption of fructose and high fructose corn syrup increase postprandial triglycerides, LDL-cholesterol, and apolipoprotein-B in young men and women," *The Journal of Clinical Endocrinology and Metabolism*, vol. 96, no. 10, pp. E1596–E1605, 2011.

[14] L. de Koning, V. S. Malik, M. D. Kellogg, E. B. Rimm, W. C. Willett, and F. B. Hu, "Sweetened beverage consumption, incident coronary heart disease, and biomarkers of risk in men," *Circulation*, vol. 125, no. 14, pp. 1735–1741, 2012.

[15] L. Vilà, N. Roglans, V. Perna et al., "Liver AMP/ATP ratio and fructokinase expression are related to gender differences in AMPK activity and glucose intolerance in rats ingesting liquid fructose," *Journal of Nutritional Biochemistry*, vol. 22, no. 8, pp. 741–751, 2011.

[16] J. Busserolles, A. Mazur, E. Gueux, E. Rock, and Y. Rayssiguier, "Metabolic syndrome in the rat: females are protected against the pro-oxidant effect of a high sucrose diet," *Experimental Biology and Medicine*, vol. 227, no. 9, pp. 837–842, 2002.

[17] K.-A. Lê, D. Faeh, R. Stettler et al., "A 4-wk high-fructose diet alters lipid metabolism without affecting insulin sensitivity or ectopic lipids in healthy humans," *The American Journal of Clinical Nutrition*, vol. 84, no. 6, pp. 1374–1379, 2006.

[18] S. A. Bayol, B. H. Simbi, J. A. Bertrand, and N. C. Stickland, "Offspring from mothers fed a 'junk food' diet in pregnancy and lactation exhibit exacerbated adiposity that is more pronounced in females," *The Journal of Physiology*, vol. 586, no. 13, pp. 3219–3230, 2008.

[19] A. Soria, A. Chicco, N. Mocchiutti et al., "A sucrose-rich diet affects triglyceride metabolism differently in pregnant and nonpregnant rats and has negative effects on fetal growth," *Journal of Nutrition*, vol. 126, no. 10, pp. 2481–2486, 1996.

[20] M. A. Munilla and E. Herrera, "Maternal hypertriglyceridemia during late pregnancy does not affect the increase in circulating triglycerides caused by the long-term consumption of a sucrose-rich diet by rats," *Journal of Nutrition*, vol. 130, no. 12, pp. 2883–2888, 2000.

[21] K.-L. C. Jen, C. Rochon, S. Zhong, and L. Whitcomb, "Fructose and sucrose feeding during pregnancy and lactation in rats changes maternal and pup fuel metabolism," *Journal of Nutrition*, vol. 121, no. 12, pp. 1999–2005, 1991.

[22] R. H. H. Ching, L. O. Y. Yeung, I. M. Y. Tse, W.-H. Sit, and E. T. S. Li, "Supplementation of bitter melon to rats fed a high-fructose diet during gestation and lactation ameliorates fructose-induced dyslipidemia and hepatic oxidative stress in male offspring," *Journal of Nutrition*, vol. 141, no. 9, pp. 1664–1672, 2011.

[23] S. Rawana, K. Clark, S. Zhong, A. Buison, S. Chackunkal, and K.-L. C. Jen, "Low dose fructose ingestion during gestation and lactation affects carbohydrate metabolism in rat dams and their offspring," *Journal of Nutrition*, vol. 123, no. 12, pp. 2158–2165, 1993.

[24] M. H. Vickers, Z. E. Clayton, C. Yap, and D. M. Sloboda, "Maternal fructose intake during pregnancy and lactation alters placental growth and leads to sex-specific changes in fetal and neonatal endocrine function," *Endocrinology*, vol. 152, no. 4, pp. 1378–1387, 2011.

[25] L. Rodríguez, M. I. Panadero, N. Roglans et al., "Fructose during pregnancy affects maternal and fetal leptin signaling," *The Journal of Nutritional Biochemistry*, vol. 24, no. 10, pp. 1709–1716, 2013.

[26] R. Muniyappa, S. Lee, H. Chen, and M. J. Quon, "Current approaches for assessing insulin sensitivity and resistance in vivo: advantages, limitations, and appropriate usage," *The American Journal of Physiology—Endocrinology and Metabolism*, vol. 294, no. 1, pp. E15–E26, 2008.

[27] S. H. Y. Wong, J. A. Knight, S. M. Hopfer, O. Zaharia, C. N. Leach Jr., and F. W. Sunderman Jr., "Lipoperoxides in plasma as measured by liquid-chromatographic separation of malondialdehyde-thiobarbituric acid adduct," *Clinical Chemistry*, vol. 33, no. 2, part 1, pp. 214–220, 1987.

[28] V. Witko-Sarsat, M. Friedlander, T. N. Khoa et al., "Advanced oxidation protein products as novel mediators of inflammation and monocyte activation in chronic renal failure," *Journal of Immunology*, vol. 161, no. 5, pp. 2524–2532, 1998.

[29] B. Anderstam, B.-H. Ann-Christin, A. Valli, P. Stenvinkel, B. Lindholm, and M. E. Suliman, "Modification of the oxidative stress biomarker AOPP assay: application in uremic samples," *Clinica Chimica Acta*, vol. 393, no. 2, pp. 114–118, 2008.

[30] M. H. Vickers, P. D. Gluckman, A. H. Coveny et al., "Neonatal leptin treatment reverses developmental programming," *Endocrinology*, vol. 146, no. 10, pp. 4211–4216, 2005.

[31] M. Srinivasan, C. Dodds, H. Ghanim et al., "Maternal obesity and fetal programming: effects of a high-carbohydrate nutritional modification in the immediate postnatal life of female rats," *The American Journal of Physiology—Endocrinology and Metabolism*, vol. 295, no. 4, pp. E895–E903, 2008.

[32] M. H. Vickers, S. Reddy, B. A. Ikenasio, and B. H. Breier, "Dysregulation of the adipoinsular axis—a mechanism for the pathogenesis of hyperleptinemia and adipogenic diabetes induced by fetal programming," *Journal of Endocrinology*, vol. 170, no. 2, pp. 323–332, 2001.

[33] A. Shapiro, W. Mu, C. Roncal, K.-Y. Cheng, R. J. Johnson, and P. J. Scarpace, "Fructose-induced leptin resistance exacerbates weight gain in response to subsequent high-fat feeding," *The American Journal of Physiology—Regulatory Integrative and Comparative Physiology*, vol. 295, no. 5, pp. R1370–R1375, 2008.

[34] B. Levi and M. J. Werman, "Long-term fructose consumption accelerates glycation and several age-related variables in male rats," *Journal of Nutrition*, vol. 128, no. 9, pp. 1442–1449, 1998.

[35] M. C. Castro, M. L. Massa, H. del Zotto, J. J. Gagliardino, and F. Francini, "Rat liver uncoupling protein 2: changes induced by a fructose-rich diet," *Life Sciences*, vol. 89, no. 17-18, pp. 609–614, 2011.

[36] Y. Takagi, A. Kashiwagi, Y. Tanaka, T. Asahina, R. Kikkawa, and Y. Shigeta, "Significance of fructose-induced protein oxidation and formation of advanced glycation end product," *Journal of Diabetes and Its Complications*, vol. 9, no. 2, pp. 87–91, 1995.

[37] G. G. Korkmaz, E. Altinoglu, S. Civelek et al., "The association of oxidative stress markers with conventional risk factors in the metabolic syndrome," *Metabolism: Clinical and Experimental*, vol. 62, no. 6, pp. 828–835, 2013.

[38] A. R. Gaby, "Adverse effects of dietary fructose," *Alternative Medicine Review*, vol. 10, no. 4, pp. 294–306, 2005.

[39] T. Nakagawa, H. Hu, S. Zharikov et al., "A causal role for uric acid in fructose-induced metabolic syndrome," *American Journal of Physiology—Renal Physiology*, vol. 290, no. 3, pp. F625–F631, 2006.

[40] R. J. Johnson, T. Nakagawa, L. G. Sanchez-Lozada et al., "Sugar, uric acid, and the etiology of diabetes and obesity," *Diabetes*, vol. 62, no. 10, pp. 3307–3315, 2013.

5

Performance Enhancing Diets and the PRISE Protocol to Optimize Athletic Performance

Paul J. Arciero,[1] Vincent J. Miller,[1,2] and Emery Ward[1]

[1]*Human Nutrition and Metabolism Laboratory, Health and Exercise Sciences Department, Skidmore College, Saratoga Springs, NY 12866, USA*
[2]*College of Graduate Health Studies, A. T. Still University, Mesa, AZ 85206, USA*

Correspondence should be addressed to Paul J. Arciero; parciero@skidmore.edu

Academic Editor: Pedro Moreira

The training regimens of modern-day athletes have evolved from the sole emphasis on a single fitness component (e.g., endurance athlete or resistance/strength athlete) to an integrative, multimode approach encompassing all four of the major fitness components: resistance (R), interval sprints (I), stretching (S), and endurance (E) training. Athletes rarely, if ever, focus their training on only one mode of exercise but instead routinely engage in a multimode training program. In addition, timed-daily protein (P) intake has become a hallmark for all athletes. Recent studies, including from our laboratory, have validated the effectiveness of this multimode paradigm (RISE) and protein-feeding regimen, which we have collectively termed PRISE. Unfortunately, sports nutrition recommendations and guidelines have lagged behind the PRISE integrative nutrition and training model and therefore limit an athletes' ability to succeed. Thus, it is the purpose of this review to provide a clearly defined roadmap linking specific performance enhancing diets (PEDs) with each PRISE component to facilitate optimal nourishment and ultimately optimal athletic performance.

1. Introduction

At every level of athletic competition, the drive to succeed is a natural competitive instinct that requires an appropriate amount, type, and timing of exercise training and nutrient intake. This balance is important because the difference between winning and losing largely depends on the training and nutritional status of the athlete. Thus, in order for any athlete to be successful, proper training and nourishment must be a daily priority.

Specific training regimens for elite athletes are often based on the same science used to formulate exercise and nutrition recommendations for the general public. For example, governing organizations in sports medicine (American College of Sports Medicine, ACSM) and healthcare (American Heart Association, AHA; Centers for Disease Control, CDC; World Health Organization, WHO) generally promote an exercise regimen that includes a combination of (i) cardiorespiratory (aerobic) (150 minutes/week of 30–60 minutes moderate-intensity 5 days/week or 20–60 minutes vigorous-intensity

exercise 3 days/week); (ii) resistance (major muscle groups 2-3 days/week of 2–4 sets and 8–20 repetitions); (iii) flexibility (stretches held for 10–30 seconds, repeated 2–4 times 2-3 days/week); and (iv) neuromotor/functional exercise (balance, agility, coordination 20–30 minutes/day 2-3 days/week).

While the intent of these exercise recommendations is noble, the majority of the US population (>60%) falls short in achieving them [1–3], especially among youth. It may very well be the case, exercise compliance and adherence suffers because the current recommendations are not realistic (up to 7 days of exercise per week) or compatible with many lifestyles. An additional concern with current exercise guidelines is they often lack a clear and specific connection to appropriate dietary intake recommendations.

Interestingly, the contemporary athlete (competitive and noncompetitive) no longer adheres to the traditional, narrowly defined training regimen focused on only one mode of exercise (e.g., only endurance or only resistance) but instead adheres to a multimode, integrative training model. Indeed, the challenge for most athletes today is finding the

TABLE 1: PRISE protocol.

	Exercise	Type	Work	RPE	Monday	Tuesday	Wednesday	Thursday	Friday
PRISE	Protein-pacing (P)	P, A	—	—	20 grams × 5 servings	20 grams × 5 servings	20 grams × 5 servings	20 grams × 5 servings	20 grams × 5 servings
	Resistance (R)	WB	2 sets/exercise 10–15 reps	7–9	WB	—	—	—	—
	Intervals (I)	C	5–7 sets 30 s/4 min rest	10/3	—	X	REST	—	—
	Stretching (S)	S	≤60 min	7–9	—	—		WB	—
	Endurance (E)	C	≥60 min	6	—	—		—	X

Note: P: plant-based; A: animal-based; RPE: rating of perceived effort; RT: resistance training; Sprint: sprint interval training; C: choice of exercise modality; WB: whole body exercise; S: stretching exercise; X: exercise day. Exercise modalities available for C include walking, jogging, running, cycling, swimming, elliptical, rowing, rollerblading, and cross-country skiing.

balance (time and energy) to incorporate all of the fitness components (resistance, anaerobic, aerobic, and flexibility training) into their regular training regimen, recognizing the vital importance each one contributes to their overall success. Thus, herein we propose a scientifically validated model that embraces a holistic and integrative model of exercise training that all athletes are encouraged to follow, termed "PRISE" (Table 1) [4]. The "P" is timed-daily protein-pacing intake; the "R" is resistance training; "I" is interval anaerobic sprint training; "S" is stretching (flexibility, restorative) training; and "E" is endurance aerobic training and is based on 4 days of structured exercise per week (Tables 2 and 3; Figures 1 and 2). This novel paradigm of exercise training integrates the four major fitness components into the training regimen of all athletes, regardless of sport, while still allowing for an athlete to emphasize sport-specific training.

Perhaps equally, if not more, important for athletic performance is proper nourishment, including the type, timing, and amount of specific food and dietary supplement sources. Currently, there is disconnect between sports nutrition guidelines and the progressive multicomponent exercise training regimen (PRISE) that many athletes follow. As an example, most endurance athletes (marathoners, triathletes, etc.) are encouraged to follow a consistent diet of relatively high carbohydrate intake (60–70% of total kcals). However, most endurance athletes adhere to a PRISE training schedule, including resistance (R), interval (I), and stretching (S) training, and therefore need to adapt their nourishment to match this integrative training paradigm in order to achieve success and the same applies to the sprint-type athlete.

It is clear that our current exercise training and nutrition practices need to be readjusted to meet the needs of the evolving athlete. Thus, the major objective of the current sports nutrition review is to establish a clear rationale and link between a scientifically proven integrative model of exercise training (PRISE) performed four days per week and a matching sports performance enhancing diet (PED), to maximize athletic performance. We advocate following the PRISE protocol and linking the prescribed PED to each component for that day to maximize the physiological, biochemical, and hormonal responses. The advantage of incorporating these nutritional strategies on a temporal basis allows the body to avoid repeated long-term exposure and

thus potential for adverse side effects, downregulation (i.e., decreased cellular sensitivity), and tolerance to occur. In addition, athletes should follow a balanced, protein-rich diet that incorporates 20–30 grams of high-quality protein evenly spaced throughout the day (~every 3 hours), including nonexercising days.

2. Timed-Daily Protein-Pacing (P) Intake

Protein is arguably the most crucial nutrient for general health and athletic performance because of its role in protein synthesis, energy metabolism, body composition (optimal lean muscle mass and fat mass), immune support, and satiation. Further, research supports timed-daily protein feedings throughout the day to maximize protein synthesis and thus lean muscle mass accretion [5–7]. Dietary guidelines have consistently encouraged a higher carbohydrate (CHO) intake (up to 65% of total kcals), moderate fat (20–35% of total kcals), and 10–35% of intake as protein (PRO) for proper weight control [8]. However, recent data suggests that consuming protein at the higher acceptable range (~25–35%) enhances energy expenditure [9–11] and body composition [4, 7, 12–14] and may do so independent of inducing weight loss [15]. This is important because it will have important implications for athletes attempting to improve health and performance outcomes without undergoing caloric restriction and weight reduction. Recent data also shows that the combined effects of increased dietary PRO and reduced glycemic index (GI) diets enhances weight loss maintenance [16] and improves body composition [17, 18].

Meal frequency (number of meals eaten) is another important factor for optimization of body composition and athletic performance. Several studies have suggested meal frequency is inversely related to body weight [19, 20].

Mechanisms. It is well established that energy expenditure and metabolism differ greatly in response to macronutrient intake of isoenergetic meals. For example, protein intake elicits the greatest thermogenic response compared to carbohydrate and fat [21–23] and this may be related to increased satiation [21]. In addition, compelling evidence favors dietary proteins containing a full complement of essential amino acids with a high leucine content to maximally stimulate muscle protein

TABLE 2: Resistance exercise (R).

Circle the exercises performed from each category		Reps/time	Resistance
Dynamic warm-up	Perform prior to each workout (5–10 minutes):		
	(1) Pendulum swings (side-to-side) (7) Over-under the fence		
	(2) Pendulum swings (front-to-back) (8) Hip opening/closing		
	(3) High knee (chest) (9) High knees		
	(4) High knee (external rotation) (10) Butt kicks		
	(5) Side shuffle (11) Lunge with twist		
	(6) Carioca (12) Arm windmills		
Footwork and agility	Perform using agility ladder (10 minutes):		
	(1) Forward, double-step (1) Side shuffle		
	(2) Sideways double-step (2) Figure 8's		
	(3) Side-step, double in/out (3) Kangaroo hops 2/1 foot		
	(4) Side shuffle, two-in/out (4) Kangaroo hops, sideways		
	(5) Two leg hops (5) T-drill		
	(6) One leg hops (6) Jump rope		
	(7) Two leg hops, in/out		
	(8) One leg hops, in/out		
	(9) One leg hops, sideways		
Resistance and power exercises	Perform each below (10 minutes): Perform each below (10 minutes):		
	(1) Side-steps toes in/out, ankles/knees (1) Back rows/flys		
	-Side-steps with bands and med ball (2) Pull-ups		
	(2) Forward/backward walk with bands (3) Chest press/fly		
	(3) Squats (4) Pushups (choose one):		
	(4) Lunges with tubing (with med ball) (i) Side walking		
	(5) Lateral lunges (with med ball) (ii) Knees/toes w/physioball		
	Choose 2 below: (iii) Down dog		
	(6) Front step-ups (iv) Side to side (ball)		
	(7) Squat thrusts, med ball throws (v) Heart-to-heart		
	(8) Jump squats (vi) Hi/low		
	(9) Mountain climbers (5) Front/lateral raises		
	(10) Squat-plank-jump squats (6) Biceps curls		
	(11) Lateral step-ups (7) Shoulder press		
	(8) Hyperextensions		
Core Exercises	Perform 4 below (10 minutes): Perform 4 below (5 minutes):		
	(1) Plank knees elbows/hands (1) Knees to chest		
	(2) Plank toes elbows/hands (2) Hyperextension on ball		
	(3) Plank one leg elbows (3) Reverse planks		
	(4) Plank one leg hands on ball (4) Ab hollow		
	(5) Side planks foot-elbow/twist (5) Walking sit-ups		
	(6) Side planks hand stars (6) Crunch bent knee		
	(7) Airplanes (7) Tug-of-war		
	(8) Supermans/womans (8) Side touch/scissors/toe		
	(9) Crunches on ball		
	(10) Plank with ball on knees/toes		

Resistance exercises utilize medicine balls, physioballs, rubber tubes and bands which are incorporated into a dynamic warm-up, footwork and agility drills, resistance and power movements, and core exercises, bodyweight exercises (e.g., lunges, squats, and jumping rope). A 5 minute cool down follows the R routine with gentle stretching. Total R exercise time is 60 minutes.

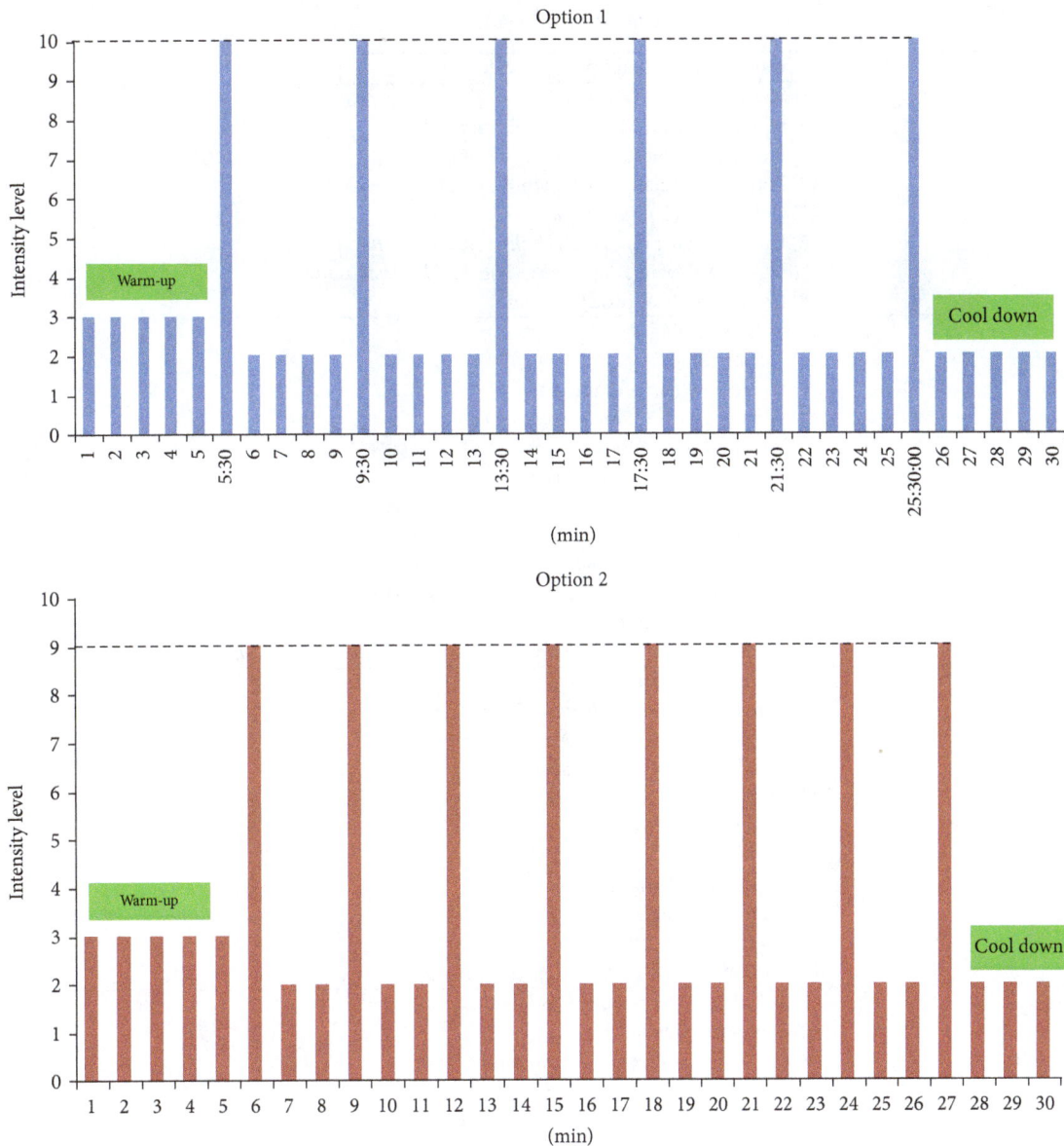

FIGURE 1: Interval exercise (I). Choose an exercise (walking, jogging, running, cycling, swimming, elliptical, snowshoeing, cross-country skiing, jumping rope, rollerblading, rowing, etc.) and one of two options. Option 1: perform 5–7 "all-out" sprint Intervals for 30-seconds at intensity level 10 followed by a 4 minute recovery at intensity Level 2; or Option 2: perform 8–12 sprint "almost all-out" intervals for 60 seconds at intensity level 9 followed by a 2-minute recovery at intensity Level 2. At the beginning and end of each interval session perform a 5-minute dynamic warm-up and gentle stretching cool down, respectively, so that each session is completed within 30–40 minutes.

synthesis [24–26]. In this case, whey protein is considered the ideal protein source. Thus, the precise mechanism responsible for enhanced energy expenditure following macronutrient intake is partly due to an increase in muscle protein synthesis (MPS) that is triggered by protein ingestion. In addition, there is speculation that a frequent macronutrient intake, especially protein-containing meals, favors an anabolic state resulting in an increase in protein synthesis and accretion [5, 26]. Specifically, increased meal frequency (timed-ingestion every 3 hours) of 20 gram servings of whey protein maximizes MPS as well as signaling proteins and transcriptional activity of muscle cells [5]. Indeed, not only does this have beneficial

implications for increased energy expenditure but also for enhanced functional capacity of muscles and an increase in lean body mass, all of which lead to improved body weight control and athletic performance.

Evidence. Our laboratory previously demonstrated that higher PRO (25%, 40%) intakes, including whey protein, more favorably affect body composition compared with a traditional diet (PRO < 20%) consumed over 6 meals per day [12, 13]. In both studies, subjects consuming the higher PRO 6 meals/day lost more body weight, fat mass, and abdominal fat mass and maintained lean body mass. In follow-up to these

TABLE 3: Stretching exercise (S).

Circle the exercises performed from each category		Breaths/time
Sun salutations	(1) Mountain pose (*Tadasana*) (2) Standing forward bend (*Uttanasana*) (3) Plank pose (*Phalakasana*) (4) Four-limbed staff pose (*Chaturanga Dandasana*) (5) Cobra pose (*Bhujangasana*) (6) Upward facing dog pose (*Urdhva Mukha Svanasana*) (7) Downward facing dog pose (*Adho Mukha Svanasana*) (8) Child's pose/rest pose (*Balasana*)	
Standing poses	(1) Neck stretching (2) Side bending (3) Lunge pose (*Anjaneyasana*) (4) Warrior I pose (*Virabhadrasana I*) (5) Warrior II pose (*Virabhadrasana II*) (6) Triangle pose (*Utthita Trikonasana*) (7) Extended side angle pose (*Utthita Parsvakonasana*) (8) Goddess pose (*Utkata Konasana*) (9) Chair pose (*Utkatasana*) (10) Revolved chair pose (*Parivrtta Utkatasana*) (11) Squat pose (*Malasana*) (12) Standing wide-legged forward bend pose (*Prasarita Padottanasana*)	
Balance in motion poses	(1) Tree pose (*Vrksasana*) (2) Warrior III (*Virabhadrasana III*) (3) Lord of the dance pose (*Natarajasana*) (4) Standing one-legged balance (5) Eagle pose (*Garudasana*) (6) Boat pose (*Navasana*) (7) Bicycle pose (8) Bow pose (*Dhanurasana*) (9) Candlestick pose (10) Camel pose (*Ustrasana*) (11) Pigeon pose (*Eka Pada Rajakapotasana*)	
Floor poses	(1) Seated cross-legged pose (*Sukhasana*) (2) Staff pose (*Dandasana*) (3) Seated forward bend (*Paschimottanasana*) (4) Head to knee pose (*Janu Sirsasana*) (5) Wide seated forward bend pose (*Upavistha Konasana*) (6) Table top pose and cat/cow (7) Bridge pose (*Setu Bandhasana*) (9) Butterfly pose (*Baddha Konasana*) (10) Happy baby pose (*Ananda Balasana*) (11) Half twist pose (*Ardha Matsyendrasana*) (12) Head to knee pose (*Janu Sirsasana*) (13) Front split pose (*Hanumanasana*) (14) Frog pose (*Mandukasana*) (15) Spinal twist pose (*Supta Matsyendrasana*) (16) Corpse pose (*Savasana*)	

S is based primarily on traditional yoga "asanas," or poses, with modern elements of Pilates for a total body stretching, flexibility, and strengthening workout. All (S) routines include basic sun salutations, standing poses, balance in motion, a floor core strengthening portion, and a final resting relaxation phase. As participants progress they are instructed to increase the intensity in which they perform the poses so the level of intensity ranges from 7 to 9 on the intensity scale.

investigations, our laboratory recently compared a higher PRO (~35% of kcals) diet (containing ~50% whey protein), moderate in CHO (~40% of kcals) consumed at either 3 or 6 meals/day versus a lower PRO (~15% of kcals) diet, higher in CHO (~60% of kcals) consumed at 3 meals/day, both of which contained complex, low-GI (GI values of <50) CHO's consumed throughout 28 days of energy balance (weight maintenance), and deficit (weight loss), respectively (56 days total) [7]. Our results demonstrated that following the 28-day period of energy balance (weight maintenance)

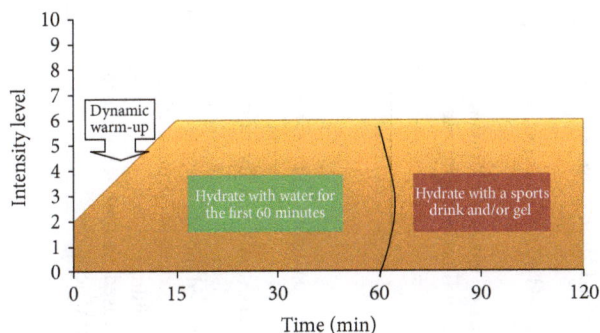

FIGURE 2: Endurance exercise (E). Perform endurance (E) exercise at an intensity level of 6 for 60 minutes or longer using any form of exercise (walking, jogging, running, cycling, swimming, hiking, cross-country skiing, snowshoeing, rollerblading, rowing, etc.). Ideally, perform E outside in nature and in the morning. At the beginning and end of each E session perform a 5-minute dynamic warm-up and a cool-down gentle stretch, respectively.

total and abdominal body fat decreased and lean body mass (LBM) increased in the higher PRO six meals/day (HP6) group versus the 3 meals/day higher PRO and CHO groups. During the 28-day weight loss period, total and abdominal fat continued to decrease and LBM remained elevated only in HP6.

Perhaps most interesting was the finding that postprandial thermogenesis during both weight maintenance and loss was significantly elevated (67–100%) in HP6 compared to the 3 meals per day groups [7]. The increased thermic response in HP6 may partly explain the enhanced total and abdominal fat loss in this group. These findings indicate that macronutrient composition (increased dietary protein), nutrient quality (low glycemic index and unprocessed carbohydrates), and frequency of eating (6x per day) are more important than total energy intake to enhance body composition (reduce abdominal obesity and maintain lean body mass) and enhance postprandial thermogenesis during both weight maintenance and weight loss.

Practical Use. Consuming increased amounts of dietary protein (20–30 grams/serving or 25–35% of total kcal intake), mostly from whey protein sources, more often (4–6 meals meals/day) throughout the day (every 3 hours) decreases abdominal fat and increases postprandial thermogenesis and lean body mass compared to traditional protein and meal frequency intakes. These body composition changes may directly lead to enhanced athletic performance. Importantly, these beneficial improvements are achieved even though total kcals consumed are identical to a traditional feeding pattern. The data from our laboratory indicate, for the first time, that macronutrient composition (increased dietary protein), nutrient quality (low glycemic index and unprocessed carbohydrates), and frequency of eating (4–6x per day) are more important than total energy intake to improve body composition and postprandial thermogenesis and thus athletic performance [7].

3. Resistance (R) and Muscular Performance Training and PEDs

Resistance training (R) is a vital component of every athlete's training regimen given its role in athletic performance. Thus, identifying nutritional strategies that enhance muscle strength, power, and function are essential (Table 4).

3.1. Creatine. Creatine, a component of phosphocreatine, is critical for rapid production of adenosine triphosphate (ATP) [27]. Along with creatine being the most well-researched sports supplement, it has been shown to enhance lean muscle mass, strength, and anaerobic performance and may also improve aerobic endurance [28]. Thus, there is strong evidence it is a potent performance enhancing nutrient.

Mechanisms. Creatine supplementation clearly increases intramuscular creatine and phosphocreatine concentrations [29–33]. Based on the role of phosphocreatine in energy production, this has commonly been proposed as an explanation for creatine's ergogenic effects [31, 34–36]. While one study found creatine to enhance phosphocreatine resynthesis [37], others have not, but have shown the higher phosphocreatine levels to persist throughout contraction and recovery [31, 38, 39]. As such, initial levels of phosphocreatine appear to be more important than its rate of resynthesis.

Protons are consumed when ATP is resynthesized from phosphocreatine [27], which implies that creatine may enhance performance by buffering against intracellular acidosis during exercise [35, 40]. Creatine may also act as a buffer by reducing reliance on glycolysis and the adenylate kinase reaction [35].

Creatine is known to increase intracellular fluid volume [41], which may increase glycogen [42] and protein [43] synthesis, and has been proposed as a mechanism of performance enhancement [41, 44]. However, investigation of creatine's influence on protein synthesis has led to conflicting results in both animals [45, 46] and humans [47–50]. Alternatively, creatine may indirectly increase protein synthesis by facilitating greater training volume [44].

Other possible mechanisms include increased energy efficiency of muscle contraction resulting from a faster relaxation response [51] and enhanced forced production from increased antioxidant capacity [52].

Evidence. In a large meta-analysis, creatine supplementation was found to increase either body weight or lean body mass in 43 of 67 trials [53]. Furthermore, our laboratory has shown creatine supplementation to be effective for increasing lean body mass, particularly when combined with resistance training [34].

Although the influence of creatine supplementation on lean body mass has not received much recent attention, several studies have further supported its benefit. In male professional soccer players, 5 days of creatine loading at $20\,g\cdot d^{-1}$ during typical training and competition led to increases in body mass and jumping power that did not occur with the placebo [54]. Two other recent trials, which did not control for creatine intake, provide some practical

Table 4: Summary of research supporting the PRISE Protocol of performance enhancing diets for athletic performance.

Author group	Nutrient	Number of participants	Duration (days)	Design	Dose	Performance improvements reported
				Resistance		
Antonio and Ciccone, 2013 [55]	Creatine	19	28	Randomized	$5 \cdot g \cdot d^{-1}$	(i) Increased lean body mass (ii) Increased 1RM bench press (iii) Supplementation after workout was more effective than before workout
Gouttebarge et al. 2012 [54]	Creatine	16	5	Double-blind, randomized, placebo-controlled	$20 \cdot g \cdot d^{-1}$	(i) 2.2% increase in body mass (ii) 2.7% increase in vertical jump peak power
Souza-Junior et al. 2011 [56]	Creatine	22	56	Randomized	$20 \cdot g \cdot d^{-1}$ for 7 days $5 \cdot g \cdot d^{-1}$ thereafter (included maltodextrin)	(i) Increased cross sectional area of thigh and arm muscle (ii) Increased 1RM squat and bench press (iii) Comparable results despite reduced training volume due to reduced rest intervals during resistance training
Ispoglou et al. 2011 [87]	Leucine	26	84	Double-blind, placebo-controlled	$4 \cdot g \cdot d^{-1}$	(i) Increased 5RM for 5 of 8 resistance exercises
				Intervals		
de Salles Painelli et al. 2014 [112]	Beta-alanine	40 20 = BA 19 = PL	4 wks	Double-blind	$6.4 \cdot g \cdot d^{-1}$	(i) Increased total work done (ii) Increased mean power output
Ducker et al. 2013 [104]	Beta-alanine	18	28	Randomized, placebo-controlled	$80 \cdot mg \cdot kg^{-1} \cdot BM \cdot d^{-1}$	(i) Improved 800 m track performance
Van Thienen et al. 2009 [105]	Beta-alanine	17	8 wks	Double-blind	$2 \cdot g \cdot d^{-1}$ (days 1–14), then $3 \cdot g \cdot d^{-1}$ (days 15–27), then $4 \cdot g \cdot d^{-1}$ (days 28–56)	(i) Increased sprint performance following a 110 min cycling race
Abian-Vicen et al. 2014 [129]	Caffeine	16	—	Randomized, double-blind, placebo-controlled, crossover	$3 \cdot mg \cdot kg^{-1}$ (as part of energy drink)	(i) Increased single and repeated jump height
Del Coso et al. 2014 [127]	Caffeine	15	—	Randomized, double-blind, placebo-controlled, crossover	$3 \cdot mg \cdot kg^{-1}$ (as part of energy drink)	(i) Increased single and repeated jump height (ii) Increased ball velocity for volleyball spike (iii) Reduced time to complete agility test
Del Coso et al. 2013 [13]	Caffeine	16	—	Randomized, double-blind, placebo-controlled, crossover	$3 \cdot mg \cdot kg^{-1}$ (as part of energy drink)	(i) Increased power output during repeated jumps (ii) Increased running speed during rugby practice games

TABLE 4: Continued.

Author group	Nutrient	Number of participants	Duration (days)	Design	Dose	Performance improvements reported
Del Coso et al. 2013 [132]	Caffeine	26	—	Randomized, double-blind, placebo-controlled, crossover	3 mg·kg⁻¹ (as part of energy drink)	(i) Increased number of sprints and distance covered (total and at running speed above 20 km·h⁻¹) during a simulated rugby match
Duncan et al. 2014 [134]	Caffeine	10	—	Randomized, double-blind, placebo-controlled, crossover	6 mg·kg⁻¹	(i) Increased torque production during isokinetic knee extension at 30, 150, and 300°·s⁻¹.
Lane et al. 2013 [133]	Caffeine	12	—		3 mg·kg⁻¹	(i) 2.8% increase in mean power output during HIIT with normal glycogen levels (ii) 3.5% increase in mean power output during HIIT with low glycogen levels
Lara et al. 2014 [130]	Caffeine	18	—	Randomized, double-blind, placebo-controlled, crossover	3 mg·kg⁻¹ (as part of energy drink)	(i) Increased jump height (ii) Increased sprint speed (iii) Increased number of sprints and distance covered (total and at running speed above 18 km·h⁻¹) during a simulated soccer match
Silva-Cavalcante et al. 2013 [135]	Caffeine	7	—	Randomized, double-blind, placebo-controlled, crossover	5 mg·kg⁻¹	(i) 4.1% reduction in time to complete 4 km cycling time trial with low glycogen levels (ii) 10.8% increase in mean power output during 4 km cycling time trial with low glycogen levels
Camic et al. 2014 [140]	Creatine (polyethylene glycosylated)	77	28	Randomized, double-blind, placebo-controlled	1.25 g·d⁻¹ 2.5 g·d⁻¹	(i) Increase in vertical jump height (ii) Increase in bench press endurance (iii) Reduction in times for shuttle-run and 3-cond drill (iv) Increase in body mass
Oliver et al. 2013 [142]	Creatine	13	6	No control group	20 g·d⁻¹ (included glucose)	(i) Increased power at lactate threshold (P = 0.11), time to fatigue (P = 0.056), and maximal power output (P = 0.082) during incremental cycling to exhaustion
Zuniga et al. 2012 [141]	Creatine	22	7	Randomized, double-blind, placebo-controlled	20 g·d⁻¹	(i) Increased mean power during two Wingate tests separated by 7 minutes
Ducker et al. 2013 [155]	Sodium bicarbonate	24	—	Randomized, blinded, placebo-controlled	0.3 g·kg⁻¹	(i) Reduced total, mean, and best times during repeated maximal running sprints

TABLE 4: Continued.

Author group	Nutrient	Number of participants	Duration (days)	Design	Dose	Performance improvements reported
Mero et al. 2013 [156]	Sodium bicarbonate	13	—	Randomized, double-blind, placebo-controlled, crossover	0.3 g·kg⁻¹	(i) Reduced time to complete second of 2 maximal 100 m freestyle swims separated by 12 minutes
Mueller et al. 2013 [154]	Sodium bicarbonate	8	5	Randomized, double-blind, placebo-controlled, crossover	0.3 g·kg⁻¹	(i) 23.5% increase in time to exhaustion during cycling at critical power (ii) Improved performance maintained throughout 5 consecutive days of supplementation and testing
Stretching						
Black et al. 2010 [177]	Ginger	25	—	Double-blind, crossover study	2 g of raw 2 g of heat-treated	(i) Decreased perception of pain following eccentric exercise
Chuengsamarn et al. 2014 [178]	Curcumin	107 = curcumin 106 = placebo	6 months	Randomized, double-blind, placebo-controlled	250 mg per capsule 6 capsules per day	(i) Decreased pulse wave velocity (ii) Increased adiponectin (iii) Decreased leptin (iv) Decreased HOMA-IR, triglyceride, uric acid, visceral, and total body fat
Takahashi et al. 2014 [191]	Curcumin	10	—	Double-blind, placebo-controlled, counterbalanced crossover	90 mg-single and placebo 180 mg-double	(i) Decreased reactive oxygen metabolites in both groups versus placebo (ii) Increased biological antioxidant potential concentrations in both groups versus placebo
Bloomer et al. 2009 [194]	Omega-3 (EPA:DHA)	14	6 wks	Random order double-blind crossover design study	EPA:DHA 2,224:2,208 mg·d⁻¹,	(i) Decreased resting levels of inflammatory biomarkers (C-reactive protein and TNF-α)
Tartibian et al. 2009 [195]	Omega-3 (EPA:DHA)	27 (n = 9, control) (n = 9, placebo) (n = 9, EPA:DHA)	32	Randomized, double-blinded, repeated measures	324:216 mg·d⁻¹, 30 days and 48 hrs during recovery	(i) Decreased perceived pain and ROM at 48 hours postexercise
Jouris et al. 2011 [196]	Omega-3 (EPA:DHA)	11	7	Repeated measures intervention	2,000:1,000 mg·d⁻¹ for 7 days	(i) Decreased perceived muscle soreness, pain, and swelling.
Smith et al. 2011 [199]	Omega-3 (EPA:DHA)	16	8 wks	Randomized controlled study	EPA:DHA 1.86:1.50 g·d⁻¹	(i) Stimulating protein synthesis through activation of the mTOR-p70s6k signaling pathway in older adults

TABLE 4: Continued.

Author group	Nutrient	Number of participants	Duration (days)	Design	Dose	Performance improvements reported
				Endurance		
Bailey et al. 2009 [218]	Beet root juice	8	6	Double-blind, placebo- (PL-) controlled, crossover study	0.5 liters of BRJ (5.5 mmol/day of NO_3^-)	(i) Single dose BRJ lowered VO_2 during submaximal exercise of 60% maximal work rate (ii) BRJ significantly improved 16.1 km TT performance
Vanhatalo et al. 2010 [219]	Beet root juice	8	15	Balanced crossover	0.5 liters BRJ (5.2 mmol/day NO_3^-)	(i) VO_2 max, peak power output, and work rate associated with anaerobic threshold were higher than placebo and baseline after 15 days of BRJ
Lansley et al. 2011 [217]	Beet root juice	9	6	Randomized, double-blind, crossover design	0.5 liters of BRJ (6.2 mmol/day of NO_3^-)	(i) Reduced the VO_2 for constant-work-rate moderate and severe-intensity running by ~7% (ii) Time to exhaustion was increased during severe-intensity running by ~15% and incremental knee-extension exercise by ~5%
Lansley et al. 2011 [216]	Beet root juice	9	—	Randomized, crossover	0.5 liter BRJ (6.2 mmol of NO_3^-)	(i) Reduced time to completion and significantly increased power output during the 4 km TT (2.8% and 5%, resp.; $P < 0.05$) (ii) Reduced time to completion and significantly increased power output during the 16 km TT (2.7% and 6%, resp.; $P < 0.05$)
Kenjale et al. 2011 [231]	Beet root juice	8	—	Randomized, open-label, crossover study	0.5 liters of BRJ (18.1 mmol/L NO_3^-)	(i) Increased exercise tolerance (walked 18% longer before claudication pain onset and experienced a 17% longer peak walking time) (ii) Decreased fractional O_2 extraction (48% decrease in Hgb peak-curve amplitude)
Murphy et al. 2012 [226]	Beet root juice	11	—	Double-blind placebo-controlled crossover	200 g Beetroot with ≥500 mg NO_3^-	(i) Nonsignificant improvement in running velocity (ii) Running velocity was 5% faster during the last 1.1 miles (1.8 km) of the 5-km run
Hodgson et al. 2013 [137]	Caffeine	8	—	Randomized, single-blind, placebo-controlled, crossover	$5\ mg \cdot kg^{-1}$	(i) 4.9% reduction in cycling time until completion of 70% of maximal work output (ii) Comparable results with coffee as the source of caffeine

TABLE 4: Continued.

Author group	Nutrient	Number of participants	Duration (days)	Design	Dose	Performance improvements reported
Pitchford et al. 2014 [234]	Caffeine	9	—	Randomized, double-blind, placebo-controlled, crossover	3 mg·kg^{-1}	(i) Reduced cycling time to complete work-based time trial in hot conditions ($P = 0.06$)
Spence et al. 2013 [233]	Caffeine	10	—	Randomized, double-blind, placebo-controlled, crossover	200 mg	(i) Reduction of cycling time during second half of 40 km time trial (ii) Insignificant 1.3% reduction in total cycling time during 40 km time trial
Stadheim et al. 2013 [232]	Caffeine	10	—	Randomized, double-blind, placebo-controlled, crossover	6 mg·kg^{-1}	(i) 4% reduction in time to complete 8 km cross-country skiing double-poling time trial (ii) Reduced rating of perceived exertion during 5 minute warm-up intervals at 40, 50, 60, and 70% of aerobic capacity
Stephens et al. 2008 [320]	LMW HMW	8	—		100 g LMS, HMS, or P	(i) Increased performance in LMS and HMS versus placebo (ii) Increased performance in HMS versus LMS
Roberts et al. 2011 [322]	HMS MAT	9	—	Crossover, randomized, double-blind	1 g/kg BM MS 1 g/kg/MD	(i) Decreased glucose and insulin in HMS versus MAT (ii) Increased fat breakdown in HMS versus MAT
Body composition						
Ludy and Mattes 2011 [329]	Capsaicin	25	—	Randomized, crossover	1 g RP after high-FAT diet 1 g RP after high-CHO diet 0 after high-FAT diet 0 after high-CHO diet	(i) Increased EE, core body temperature, and fat oxidation (in oral form) (ii) Decreased energy intake in nonusers, but no Δ in users
Yoneshiro et al. 2012 [337]	Capsaicin	18	—	Single-blind, randomized, placebo-controlled, crossover	9 mg capsinoids (capsules) with 199 mg of rapeseed oil and medium-chain triglycerides 0 (Placebo)	(i) Increased EE through activation of brown adipose tissue in humans
Galgani and Ravussin 2010 [345]	Capsiate	78	4 wks	Parallel-arm double blind, randomized	3 mg·d^{-1} dihydrocapsiate (capsules) 9 mg·d^{-1} dihydrocapsiate (capsules) 0 (Placebo)	(i) Increased RMR when both groups 3 and 9 mg·d^{-1} were combined

TABLE 4: Continued.

Author group	Nutrient	Number of participants	Duration (days)	Design	Dose	Performance improvements reported
Josse et al. 2010 [342]	Capsiate	12	—	Randomized, crossover, double blind	10 mg capsinoids (capsules) 0 (Placebo)	(i) Increased SNSa, energy expenditure, and fat oxidation
Lee et al. 2010 [344]	Capsiate	46	4 wks	Parallel-arm double blind, randomized	3 mg·d^{-1} dihydrocapsiate (capsules) 9 mg·d^{-1} dihydrocapsiate (capsules) 0 (Placebo)	(i) Increased energy expenditure 9 mg·d^{-1} and 3 mg·d^{-1} versus placebo and 9 mg·d^{-1} versus 3 mg·d^{-1}
Snitker et al. 2009 [338]	Capsiate	80	12 wks	Parallel-arm double blind, randomized	6 mg·d^{-1} capsinoids (capsules) 0 (Placebo)	(i) Decreased abdominal adiposity (ii) Tended to increase fat oxidation
Inoue et al. 2007 [343]	Capsiate	44	4 wks	Parallel-arm double blind, randomized	3 mg·d^{-1} capsinoids (capsules) 10 mg·d^{-1} capsinoids (capsules) 0 (Placebo)	(i) Increased VO_2 (10 mg, BMI ≥ 25 kg/m^2)
Stephens et al. 2013 [348]	Carnitine	12	12 wks	Randomized, double-blind	1.36 g L-carnitine + 80 g of CHO 80 g of CHO	(i) Increased muscle carnitine by 20% (ii) Prevented an 18% increase in body fat mass found with the CHO group alone (iii) Increased EE and fat oxidation during low-intensity exercise
Haub et al. 2010 [355]	Resistant starch	11	—	Single-blind randomized, crossover	30 g RS4$_{XL}$ 30 g RS2 30 g DEX 75 g GLU	(i) Lower plasma glucose for RS4$_{XL}$ and RS2 than DEX, and for RS4$_{XL}$ than RS2
Al-Tamimi et al. 2010 [367]	Resistant starch	13	—	Randomized, crossover	65 g of puffed wheat bar (PWB) 80 g of RS4$_X$	(i) Lower glucose 20–60 min and insulin 30–120 min in RS4$_{XL}$ versus PWB and GLU
Shimoyodome et al. 2011 [363]	Resistant starch	10	—	Randomized, crossover	38 g RS4-HDP 38 g RS2-WMS	(i) Lower glucose and insulin, and GIP (ii) Increased fat oxidation and EE

HMW: high molecular weight; LMW: low molecular weight; HMS: hydrothermally modified starch; MAT: maltodextrin.

insights for using creatine to increase lean body mass. In male recreational bodybuilders, 4 weeks of creatine supplementation at 5 g·d^{-1}, combined with resistance training, led to increases in lean body mass and 1 repetition maximum (RM) bench press with indication of greater benefit from postexercise versus preexercise supplementation [55]. The second trial also focused on recreationally trained men but included a creatine loading phase and lasted for 8 weeks [56]. Furthermore, for one of the groups, the rest interval between resistance exercises was progressively decreased by 15 seconds each week, which resulted in a lower training volume [56]. Despite the reduced training volume, increases in muscle cross sectional area for the upper arm and thigh, as well as 1 RM for the squat and bench press, were not different between groups, suggesting that creatine supplementation can be used to increase training efficiency [56]. Other recent trials have shown creatine to reduce postexercise levels of inflammation [57] and muscle damage [58], suggesting it may facilitate recovery.

Practical Use. A common dosage regimen for creatine is 20 g·d^{-1} during the first 4–7 days, followed by 5 g·d^{-1} thereafter [59]. As little as 2 g·d^{-1} has fully [30] or partially [32] maintained the intramuscular creatine levels achieved with loading, and 3 g·d^{-1} for 28 days has produced comparable levels without loading [30]. However, as a part of the PRISE protocol, an acute dosage of 2–5 g 1 hour prior to an R exercise bout may enhance muscular and physical performance. It is unclear if creatine intake from food will provide the same benefits as supplementation. However, herring, salmon, pork, beef, and cod are prominent sources containing 3–10 g·kg^{-1} [60, 61]. Chicken and rabbit are also within this range [62]. Therefore, it is possible to achieve a maintenance dose with whole foods [61], but a loading dose would be much less practical. For example, beef contains approximately 4.5 g·kg^{-1} of creatine [60], which translates to 0.8 g in a single 6 oz. serving.

3.2. Branched-Chain Amino Acids. The branched-chain amino acids (BCAAs), which include leucine, isoleucine, and valine, are essential nutrients involved in muscle protein synthesis and energy metabolism [63]. Leucine is particularly important for stimulating muscle protein synthesis [25], but BCAAs can be used collectively to enhance endurance, reduce muscle breakdown, and stimulate recovery after exercise.

Mechanisms. During exercise, BCAAs are catabolized into succinyl-CoA and acetyl-CoA, both of which can enter the citric acid cycle to support ATP resynthesis [63, 64]. This pathway has a critical role in exercise tolerance [65] and is likely fed by muscle protein breakdown, which can be reduced with BCAA supplementation [66]. Therefore, BCAAs can preserve muscle protein by acting as an energy substrate. Furthermore, BCAAs may enhance exercise performance by reducing central fatigue [67, 68] and enhancing fat oxidation [69–72].

Protein synthesis is the most well-known and arguably the most important mechanism through which BCAAs enhance performance. Although all three of the BCAAs contribute to protein synthesis, leucine is particularly important. This is because leucine activates translation initiation factors and the mammalian target of rapamycin (mTOR), which are influential in the regulation of protein synthesis [73–76].

Evidence. Some trials have shown BCAAs to enhance exercise capacity [72, 77–79] while others have not [80–85]. In a recent trial including 19 untrained males and 8 weeks of resistance training, 9 g·d^{-1} of BCAAs failed to change body composition or improve strength or muscular endurance to a greater extent than the placebo [86]. However, in a similar trial including 26 untrained men and 12 weeks of resistance training, 4 g·d^{-1} of leucine led to greater strength gains [87]. These contrasting results suggest that either leucine alone is more effective, or that 8 weeks is too short of a training period.

Further supporting the importance of leucine, a recent crossover trial including 9 military personnel found that increasing the leucine content of a 10 g essential amino acid (EAA) dose from 1.87 to 3.5 g led to greater muscle protein synthesis and less total-body protein breakdown following 60 minutes of cycle ergometry [88]. Similarly, another recent trial assessed myofibrillar protein synthesis following a bout of resistance exercise and found that increasing the leucine content of 6.25 g of whey protein from 3 to 5 g resulted in the same rate of protein synthesis as 25 g of whey [24]. However, the inclusion of additional BCAAs prevented this outcome, possibly due to increased competition for absorption [24].

Practical Use. As little as 77 mg·kg^{-1} of BCAAs has been shown to reduce muscle protein breakdown during exercise [66]. For EAAs, although 6 g has been shown to enhance protein synthesis [89], 10 g appears to be the optimal dose [26, 90].

While it is generally ideal to consume protein from whole-food sources, EAA supplementation has been suggested as an efficient method of promoting muscle growth while limiting caloric intake [91]. This is particularly relevant to athletes who need to lose or maintain weight. Furthermore, because exercising with a full stomach is generally not desirable [92], supplementation may be more appropriate for preexercise consumption.

A single acute serving of high-quality protein containing the optimal 10 g dose of EAAs contains approximately 1.8 g of leucine [93]. Relative to common protein sources, the leucine content of a 100 g (3.5 oz.) serving of beef, pork, chicken, turkey, salmon, cod, or tuna ranges from approximately 1.3 to 2.3 g [94]. Two eggs or a 100 g serving of haddock, shrimp, or scallops contains slightly less leucine, but still more than 1 g [94].

Finally, liquid sources of protein are known to elevate BCAA, EAA, and leucine concentrations more rapidly [95, 96], which can result in greater protein synthesis [97–99]. Whey [98] and milk [100], if well tolerated, are particularly effective.

4. PEDs for Interval (I) Sprint Training

A growing body of research has documented the benefits of interval sprint training (I) for improved anaerobic and aerobic athletic performance (Table 4). Certain nutritional strategies have proven effective to counter the increased acidic environment induced by I training and thus prolong training time and adaptations, all of which may directly enhance athletic performance.

4.1. Beta-Alanine. Beta-alanine is the rate-limiting precursor in the synthesis of carnosine, a cytoplasmic dipeptide that buffers intracellular H^+ [101]. As such, it may reduce the acidic environment inside the muscle allowing for continued high-intensity anaerobic work performance and therefore may be suitable prior to and during an I exercise session.

Mechanism. Carnosine's role as an H^+ buffer in the muscle is the first line of defense against local changes in pH. The absence of carnosine in isolated muscles leads to acidification and fatigue [101]. Therefore, the use of β-alanine supplementation to buffer H^+ during high intensity exercise that causes muscle acidosis may extend the onset to fatigue by elevating intracellular carnosine concentrations [101] leading to increased work performance.

Evidence. Research has shown improved performance following β-alanine supplementation among different exercise modalities, such as swimming [102], cycling [103], running [104], and sprint performance following long endurance cycling [105]. However, these results are conflicting in nature with others reporting little or no change in performance [106–109] despite elevated carnosine concentrations [107, 110] or resistance to fatigue [107, 108]. Derave et al. [107] reported that four weeks of β-alanine supplementation ($4.8 \, g \cdot d^{-1}$) versus placebo in trained male athletes showed significant improvements in both dynamic knee extension torque (during the fourth and fifth bouts) and carnosine content in the soleus (47%) and gastrocnemius (37%). However, there were no differences in isometric strength or 400 m race time between groups. These findings contradict Ducker et al. [104] who found that male recreational runners improved 800 m race time following 4 weeks of β-alanine supplementation ($6.4 \, g \cdot d^{-1}$) versus placebo. Such contrasting results suggest that differences in training status may limit the effectiveness of β-alanine on improved performance more so than the dosage of β-alanine supplementation.

It is speculated that the effectiveness of β-alanine supplementation may be blunted in trained athletes due to the already elevated muscle buffering capacity from intense exercise training [111]. To compare the effects of β-alanine supplementation and training status, de Salles Painelli et al. [112] tested the effects of β-alanine supplementation in trained and nontrained cyclists. Forty males were separated in two groups based on training status ($N = 20$ endurance trained (T); $N = 20$ nontrained (NT) cyclists). Participants performed four 30 s lower-body Wingate bouts separated by 3 min, both before and after 4 weeks of either placebo or β-alanine supplementation ($6.4 \, g \cdot day^{-1}$). The sum of the four bouts represented the total work done (TWD) and the mean power (MPO) and peak power (PPO) output were obtained from each of the four bouts individually. β-Alanine supplementation was shown to significantly increase TWD in both T and NT groups with no significant difference in the T cyclist placebo group. Furthermore, it was found that MPO significantly improved in the T group during bouts 1, 2, and 4 but also improved in bout 4 for the NT group. It was concluded by de Salles Painelli et al. [112] that, despite training status, β-alanine improved both TWD and MPO during high-intensity exercise.

Practical Use. Research has found that β-alanine supplementation of $3–6 \, g \cdot d^{-1}$ ($\sim 40–80 \, mg \cdot kg^{-1} \cdot BW \cdot d^{-1}$) for at least 4 weeks or longer will increase intramuscular concentrations (30–80%) possibly improving muscle buffering capacity [107, 113, 114]. However, a higher intake ($\sim 6 \, g \cdot d^{-1}$) for four weeks has been shown to elicit greater carnosine concentrations and improvements in performance [103, 104, 112]. A single acute dosage prior to an I exercise session may elicit similar favorable buffering capacity.

4.2. Caffeine. Caffeine is the most widely consumed drug in the world and one of the most extensively studied ergogenic aids. It is well known for enhancing endurance [61, 115–118] but has also been shown to improve strength, power, and other aspects of high-intensity exercise [115, 117, 118]. Its effects are acute and peak with 30–60 minutes.

Mechanisms. The performance benefits of caffeine are related to enhanced fat oxidation and glycogen sparing. Caffeine is known to increase energy expenditure and fat oxidation, mostly through sympathetic nervous system activity (SNSa) [119], and other related mechanisms [61, 118, 120]. Inhibition of adenosine receptor activity, resulting from the molecular similarity between caffeine and adenosine, is the primary mechanism [116, 118, 121]. By binding to its receptors, adenosine can promote an increase in perceived pain and a reduction in arousal [116]. Exercise can accentuate this effect through the catabolism of ATP, adenosine diphosphate (ADP), and adenosine monophosphate (AMP) [116]. Therefore, caffeine-induced impairment of adenosine receptor activity may enhance performance by reducing the perception of discomfort and maintaining or enhancing motor unit firing rates [116]. Further supporting the role of pain perception, caffeine has been observed to increase concentrations of β-endorphins during exercise [122].

Peripheral mechanisms are also believed to contribute to caffeine's ergogenic effects [116, 117, 121]. The most notable is enhanced excitation-contraction coupling, resulting from increased potassium transport in muscle by Na^+/K^+ ATPase and increased release of calcium from the sarcoplasmic reticulum [116, 121].

Evidence. In a meta-analysis including 40 trials, caffeine had mild benefit for high-intensity exercise of short duration and stronger benefit for endurance exercise, but no benefit for graded exercise to exhaustion [123]. In a systematic review of 21 studies involving time trials of at least 5 minutes, caffeine enhanced performance by 2.3–4.3% [124]. For activities lasting 5 minutes or less, another systematic review found caffeine to improve intermittent exercise performance in 11 of 17 trials and strength-related measures in 6 of 11 trials [125]. Finally, in a meta-analysis including 27 trials for strength and 23 trials for muscular endurance, caffeine produced small but significant benefits for each attribute [126].

Trials published since 2013 have shown caffeine to improve agility [127, 128], jump height [127, 129, 130] and power [131], sprint performance [131], and sport-specific performance [127, 130–132] in athletes involved in a variety of sports, including basketball [129], rugby [131, 132], soccer [130], volleyball [127], and several racket sports [128]. Other recent trials have found improvements in cycling power output [133] and isokinetic knee extension torque during resistance exercise [134].

Two recent trials evaluated the influence of caffeine on exercise performed in a glycogen depleted state. In the first, which was a crossover with 12 competitive cyclists, 3 mg·kg^{-1}·BW of caffeine resulted in similar power output during high-intensity interval training (HIIT) compared to the placebo with normal glycogen levels, indicating that caffeine attenuates the performance decline caused by glycogen depletion [133]. In the second crossover trial, which included 7 amateur cyclists, 5 mg·kg^{-1}·BW of caffeine led to better 4 km time trial performance compared to the placebo with normal glycogen levels [135]. However, the difference was not significant.

Practical Use. The dosage of caffeine most commonly shown to enhance performance with minimal side effects is 3–6 mg·kg^{-1} [61, 118, 124]. It is possible to consume such a dosage from coffee [61], but the evidence comparing the efficacy of coffee and caffeine is conflicting [136, 137]. This discrepancy may be a result of variability in other coffee constituents [137], suggesting that caffeine is likely to be more reliable.

Athletes who regularly consume caffeine may have a higher tolerance and experience less benefit [138]. Furthermore, cessation of caffeine usage can result in withdrawal symptoms including headaches and impaired performance [138]. Therefore, to maximize benefit, usage should be discontinued at least 7 days prior to an event with a gradual reduction spread over 3-4 days [138]. Finally, because caffeine has been shown to negate the performance benefits of creatine [139], there appears to be little value in using them together.

4.3. Creatine. The mechanisms and practical applications of creatine were previously discussed in relation to resistance (R) training for muscular development. In regard to high-intensity exercise performance, creatine is most commonly recognized for its effect on strength but has also shown potential for enhancing anaerobic endurance.

In a meta-analysis of 7 trials, including a total of 70 subjects, creatine supplementation with concomitant resistance training led to a 6.85 kg greater increase in 1–3 RM bench press [59]. Similarly, among 37 subjects from 4 trials, there was a 9.6 kg greater increase in 1 RM squat [59]. Despite this evidence of enhanced strength, a meta-analysis of 10 trials,

including a total of 92 participants, found no improvement in cycling power output [59]. However, in a larger meta-analysis, performance improvements were reported in 45 of 61 trials for activities lasting 30 seconds or less, in 17 of 25 trials for activities lasting between 30 and 150 seconds, and in 9 of 18 trials for activities lasting longer than 150 seconds [53]. Effect sizes were significant, although modest, for all measures, and were indicative of diminishing performance benefit with increased exercise duration [53].

Recent evidence indicates that creatine supplementation can enhance performance independently of training. In a trial including 77 men, creatine improved vertical jump, 20-yard shuttle run, 3-cone drill, and bench press endurance despite the lack of a training intervention [140]. Similarly, in two other trials lacking a training intervention, 7 days of creatine supplementation improved mean power during two bouts of the Wingate protocol [141], and 6 days of supplementation showed a tendency for increased lactate threshold, power output, and time to fatigue during incremental cycling [142].

4.4. Sodium Bicarbonate. Bicarbonate is a prominent buffer in human physiology. Supplementation with sodium bicarbonate increases blood pH and bicarbonate concentration, is particularly effective for enhancing anaerobic capacity, and may also improve strength and endurance [143, 144].

Mechanisms. Although the mechanisms are not fully understood, intramuscular acidosis has reduced muscle contractile capacity in multiple studies [145]. When exercise creates a demand for ATP that exceeds mitochondrial capacity, accumulation of protons released from glycolysis and ATP hydrolysis promote acidosis [146]. Although intramuscular acidosis has been argued to have a minimal effect on performance [147], sodium bicarbonate is known to increase pH and bicarbonate concentration [148, 149], which has persisted as the most likely mechanism of performance enhancement [148]. During exercise, sodium bicarbonate has been shown to result in higher lactate levels during exercise despite a similar intramuscular pH [150] and promotes greater glycogen and phosphocreatine utilization [151], suggesting increased capacity for anaerobic energy production. Furthermore, exercise-induced acidosis can inhibit oxidative phosphorylation [152], which implies that the buffering effect of sodium bicarbonate may enhance aerobic energy production as well.

Sodium bicarbonate supplementation has led to greater muscle contraction velocity following 50 minutes of high-intensity cycling [153], suggesting it may reduce neuromuscular fatigue in addition to enhancing energy production.

Evidence. In a meta-analysis including 29 trials, sodium bicarbonate was found to increase anaerobic exercise capacity, with the largest improvements observed for time to exhaustion [149]. The greatest benefit was observed in conjunction with larger drops in pH during exercise [149], suggesting that the benefits are most applicable to glycolytic activities. A more recent meta-analysis, including 38 studies, standardized all results as a measure of mean power production during time trial performance and found a clear but modest performance benefit [148]. Improvement increased slightly as exercise duration increased beyond 1 minute, but durations beyond 10 minutes slightly reduced benefit [148].

In a recent trial including 11 well-trained endurance athletes, $0.3\,g\cdot kg^{-1}\cdot BW$ of sodium bicarbonate was consumed prior to exercise for 5 consecutive days [154]. A similar improvement in time to exhaustion was maintained each day [154], suggesting that supplementation is appropriate for multiday events. Two other recent trials evaluated sodium bicarbonate in combination with beta-alanine. A single $0.3\,g\cdot kg^{-1}\cdot BW$ dose of sodium bicarbonate improved repeated sprint performance in team-sport athletes [155] and improved sprint swimming performance in competitive male swimmers [156]. In both trials, however, the addition of beta-alanine failed to further enhance performance.

In another recent trial including well-trained rowers, preexercise consumption of $0.3\,g\cdot kg^{-1}\cdot BW$ of sodium bicarbonate throughout 4 weeks of HIIT failed to improve time trial performance compared to the placebo [157], suggesting that supplementation may not be effective for enhancing training adaptations. However, this is in contrast to a previous trial that observed greater improvements in lactate threshold and time to exhaustion [158].

Practical Use. The dose of sodium bicarbonate most frequently associated with performance enhancement is $0.3\,g\cdot kg^{-1}$ [148, 149]. However, benefits have been observed with as little as $0.15\,g\cdot kg^{-1}\cdot BW$ [159]. Common gastrointestinal symptoms can be avoided during competition by consuming the dose 3 hours prior to initiating I exercise [160]. Consuming the dose with food may also help [161] but increases the importance of distancing intake from the start of competition [92]. Alternatively, smaller doses can be consumed over several days preceding an I event [162].

Although the buffering effect of alkalizing food [163, 164] is unlikely to produce the 0.05 increase in pH or $6\,mmol\cdot L^{-1}$ increase in bicarbonate that appear necessary for performance improvement [161], potential for benefit may still exist. The alkalizing potential of food is primarily attributed to potassium salts, which increase bicarbonate availability when metabolized [165–167]. Vegetables and fruit are most abundant in potassium salts [165] and have the highest alkalizing potential [168]. Therefore, in conjunction with the numerous health benefits of vegetables and fruit [169], as well as potassium [170], high intakes may facilitate performance enhancement. Furthermore, the alkalizing potential of vegetables and fruit can help to offset the acidifying effect of protein [165, 167, 168], which athletes require in greater amounts [92]. In support of this, a high vegetable intake has recently been shown to increase capillary pH in adults during rest and submaximal exercise [171]. Bicarbonate can be more directly incorporated into the diet with mineral water [163, 172] or baking soda.

5. PEDs for Stretching (S) and Restorative Training

It is well known that intense exercise training induces muscle damage, including an imbalanced ratio of protein breakdown to protein synthesis and increased muscle soreness (i.e., perception of pain) and inflammation [173]. A growing number athletes turn to common nonsteroidal anti-inflammatory drugs (NSAIDs) (i.e., ibuprofen) to alleviate or reduce the perception of pain and to attenuate the inflammatory response [174]. Furthermore, many athletes will perform certain modes of S exercise (i.e., yoga, stretching, and massage) as a form of restorative training to actively alleviate pain from previous strenuous exercise [175]. More recently, the combination of both active recovery exercises and nutraceuticals in the form of BCAA [176], ginger [177], turmeric [178], omega-3 (PUFAs) [179], and tart cherry [180, 181] have been suggested as natural alternatives for reducing exercise-induced inflammation (Table 4).

5.1. Ginger. Ginger (*Zingiber officinale*) is one of the ten most commonly used natural complementary and alternative medical treatments in the United States [182] and has been suggested as a possible alternative to pharmaceuticals for reducing pain and/or inflammation [177].

Mechanism. In animal models, ginger and its chemical constituents gingerols, shogaols, paradols, and zingerone are agonists to the transient receptor potential vanilloid subfamily, member 1 (TRPV1) that function in central and peripheral nociceptive signaling by inhibiting the release of prostaglandins and leukotrienes [183–185]. Ginger has been proposed as an effective analgesic based on its evidence as a natural medicinal in reducing pain and inflammation. Moreover, there are inconsistent findings from NSAIDs such as ibuprofen, naproxen, aspirin, and diclofenac as effective analgesics following eccentric exercise [186]. Thus, ginger consumption may be more efficacious for reducing exercise induced pain and inflammation through activation of TRPV1.

Evidence. It has been found that the use of ginger as pain treatment, with smaller dosages (30 to 510 $mg\cdot d^{-1}$) and longer durations (4 to 36 weeks), resulted in reductions in knee or hip pain in individuals with osteoarthritis. Black et al. [177] reported that following eccentric exercise (18 eccentric elbow flexor contractions at 120% of 1 RM) 2 g of both raw and heat-treated ginger for 11 days significantly decreased the perception of pain following exercise. Evidence supports the use of ginger to aid recovery from muscle-damaging exercise and for longer durations of intake (>2 days), as a single-acute dose had no effect on pain perception following low-moderate (60% $VO_{2\,peak}$) intensity cycling [187]. Thus, effectiveness of ginger on pain perception may prove beneficial as treatment for alleviating intense, muscle-damaging (i.e., eccentric) exercise induced pain, more so as an alternative to pharmaceuticals.

Practical Uses. A higher dosage of 6 g of ginger may lead to possible stomach irritation and therefore a lower dose of 2–3 g of ginger is suggested as it has been shown to be effective in reducing both pain following exercise and blood sugar concentrations [177]. This dosing regimen also allows for daily consumption of ginger beyond just on S days and is well-tolerated.

5.2. Curcumin. Curcumin, a polyphenol responsible for the yellow color of turmeric (curry powder), is known to reduce inflammation and influence metabolic function [188]. As such, curcumin has the potential to support recovery and performance on S training days by promoting metabolic health.

Mechanism. Curcumin is known to regulate inflammation and directly interact with adipocytes, pancreatic cells, and muscle cells [188]. Curcumin has been well documented to regulate biochemical and molecular pathways by modulating molecular targets such as transcription factors, cytokines, enzymes, and the genes responsible for both cell proliferation and apoptosis [189].

Evidence. Curcumin has recently been shown to reduce pain associated with delayed onset muscle soreness (DOMS) following downhill running [190]. Chuengsamarn et al. [178] tested the effects of 250 mg of curcumin compared to placebo (corn starch) ingested twice a day for 6 months on atherogenic risks in individuals with type II diabetes mellitus (T2DM). After 6 months of supplementation it was found that curcumin significantly decreased pulse wave velocity, increased adiponectin and decreased leptin, and also decreased homeostasis model assessment-estimated insulin resistance (HOMA-IR), triglycerides, uric acid, and abdominal obesity (visceral fat and total body fat). These findings indicate that daily incorporation of curcumin will significantly alter the proinflammatory cytokine leptin and the anti-inflammatory cytokine adiponectin, as well as reduce abdominal obesity, all of which helps to ameliorate the atherogenic risks of T2DM individuals [178].

Though a plethora of information on the positive effects of curcumin on diseased individuals has been well documented [189], only one study known to date has specifically reviewed the effects of curcumin on oxidative stress following exercise in humans. Takahashi et al. [191] tested the effects of curcumin on oxidative stress and antioxidant capacity following exercise (60 min at 75% of $VO_{2\,max}$) in ten healthy men. The participants completed three trials in a random order of ingesting either placebo, 90 mg of curcumin-single (before exercise only, 2 hr), or 90 mg of curcumin-double (before and immediately after exercise). It was found that immediately following exercise, both the curcumin-single and double groups had significantly lower derivatives of reactive oxygen metabolites and plasma thioredoxin-1 and significantly elevated biological antioxidant potential and reduced glutathione concentrations compared to the placebo group. These results suggest that exercise-induced oxidative stress may be attenuated by increasing blood antioxidant capacity from curcumin supplementation [191].

Practical Use. Though the consumption of curcumin has shown to be safe and has been consumed by ancient people for thousands of years the scientific analysis and understanding of curcumins effects are still being researched. It has been noted that when working with certain diseased populations or those unaccustomed to curcumin lower dosages (<250 mg) have been shown to reduce abdominal fullness or pain. Dosages of 90–250 mg daily, particularly on S training days, may be an effective adjuvant therapy to aid recovery and healing from strenuous exercise. A possible limitation is the relatively low bioavailability of curcumin consumed orally. However, there have been recent modifications in producing a bioavailable and higher orally absorptive curcumin known as Theracurmin [192].

5.3. Omega-3 Poly-Unsaturated Fatty Acids (PUFAs). The main components of omega-3 polyunsaturated fatty acids (PUFAs) found in fish oil are eicosapentaenoic acid (EPA) and docosahexaenoic acid (DHA) and are produced from the omega-3 fatty acid alpha-linolenic acid (ALA).

Mechanism. Because EPA and DHA are not naturally synthesized in the body and the breakdown of ALA to produce EPA and DHA is enzymatically inefficient, the consumption of fish oil through diet or supplementation is important for providing adequate EPA and DHA concentrations. Both EPA and DHA are eicosanoids that have anti-inflammatory, antithrombotic, antiarrhythmic, and vasodilatory properties. The derivative of the longer chain fatty acid linoleic acid (LA) is arachidonic acid, the precursor to the proinflammatory and prothrombotic eicosanoids. Because ALA and LA compete for the same enzymes in the production of the longer chain fatty acids EPA and arachidonic acid, the consumption of fish or fish oil avoids the enzymatic competition to convert ALA to EPA by providing EPA and DHA directly [193].

Evidence. More commonly known for their cardiovascular benefits, EPA and DHA have been documented to reduce inflammation, as well as delayed onset muscle soreness (DOMS) or the perception of pain from exercise [179, 194–196]. When supplementing with EPA and DHA either prior to or during exercise, or the combination of both, research has found decreased resting levels of inflammatory biomarkers (2,224 : 2,208 mg·d^{-1}, 6 wks) [194], decreased acute-phase proteins after exercise (1.75 : 1.05 g·d^{-1}, 3 wks) [179], and improved perceived muscle soreness, pain, and range of motion 48 hrs post exercise (324 : 216 mg·d^{-1}, 30 days and 48 hrs during recovery) [195]. More recently, Jouris et al. [196] reported the attenuation of DOMS when consuming EPA and DHA at a 2 : 1 ratio (2,000 : 1,000 mg·d^{-1}) for 7 days following an eccentric arm-curl exercise protocol. Yet, despite these beneficial findings, there have been reports of little or no change in inflammation or DOMS following exercise [197, 198]. Recently, in addition to ameliorating pain and inflammation, supplementation with omega-3 PUFA for 8 weeks (1.86 : 1.50 g·d^{-1} EPA : DHA) was shown to augment the activation of the mTOR-p70s6k signaling pathway stimulating protein synthesis in older adults [199]. Thus, omega-3 supplementation may also prove beneficial for the prevention or management of sarcopenia or the atrophy of skeletal muscle [199].

Practical Use. It should be noted that fish oil consumption at higher levels (>4 g per day) may increase the risk of bleeding from decreased adherence of blood platelets and lower blood pressure. Hence, individuals with already low blood pressure or increased risk of hemorrhage should consume moderate to lower intakes of omega-3 PUFA. Athletes that wish to mitigate the effects of exercise-induced inflammation and DOMS are suggested to incorporate omega-3 FA in their diet, especially during S days, and are suggested to do so with 1-2 g·d^{-1} of an EPA : DHA ratio of 2 : 1 [200], or 2–4 g·d^{-1} for those with higher blood lipid profiles or rheumatoid arthritis [193]. A designated safe and general consumption dose of omega-3 PUFA (EPA + DHA) for athletes to consume is ≤3,000 mg·d^{-1} (3 g), as recommended by the US Food and Drug Administration [201]. For many individuals omega-3 capsule supplementation is convenient for ensuring adequate consumption of PUFA, and an alternative for vegetarians, but for those who are able to incorporate whole food sources, flax seeds, walnuts, sardines, and salmon are considered excellent sources of rich omega-3 (e.g., EPA, DHA, and ALA). Because of the concern of high levels of mercury the following fishes have been given as examples of 1 g servings of EPA : DHA because of their low mercury content: 4.0 oz. Tuna (Canned, Light), 2.0–3.5 oz. of salmon (Atlantic, wild), 15 oz. of catfish, and 11 oz. of shrimp (mixed species). For more recommendations of grams of EPA and DHA for various types of fishes and servings see the review by Covington [193].

5.4. Tart Cherry. Cherries are known to be a rich source of bioactive compounds with antioxidant and anti-inflammatory effects [202, 203]. Both the antioxidant and anti-inflammatory effects of cherries are believed to contribute to their potential to reduce pain and enhance exercise recovery [202].

Mechanisms. Although the precise mechanisms of how cherry consumption influences exercise recovery are not fully understood, the mechanical muscle damage induced by eccentric contraction is unlikely to be affected [202]. Instead, improvements in recovery are most likely related to the attenuation of secondary oxidative stress and inflammation [202]. The anthocyanins from both sweet and tart cherries are known to inhibit cyclooxygenase-I and cyclooxygenase-II [204], which provides at least a partial explanation for their anti-inflammatory effects. Tart cherries have a more potent effect.

Evidence. Although the influence of tart cherry on exercise recovery has only been investigated to a limited extent, the available evidence is very promising. Connolly et al. [180] assessed the effect of tart cherry juice (TCJ) on recovery from maximal elbow flexion contractions. The trial included 14 men who consumed 12 oz. of tart cherry juice twice per day for 8 days. Eccentric contractions were performed on the 4th day and recovery was assessed during the subsequent 4 days. The TCJ significantly reduced loss of strength and

pain during recovery [180]. However, no differences were observed in tenderness or loss of range of motion [180]. A similar trial was conducted to determine if the response to tart cherry juice differed for well-trained athletes. In this trial, TCJ was administered as TCJ concentrate of 30 mL (1 oz.) twice per day for 7 days before and 2 days after knee extension exercise performed at 80% of maximum voluntary contraction (MVC) [205]. Consistent with the previous trial, TCJ significantly reduced loss of strength during the two days of recovery, but without any differences in muscle tenderness [205]. The TCJ also reduced protein carbonyl levels during recovery, suggesting a reduction in oxidative stress [205]. These data support TCJ as an effective PED aid following intense, muscle damaging R exercise as a result of mitigating the subsequent oxidative damage [205].

Several trials have also focused on recovery from endurance (E) exercise. Howatson et al. [206] evaluated the effect of two 8 oz. servings per day of TCJ supplementation for 5 days prior, the day of, and 2 days following a marathon run. The tart cherry juice resulted in significantly faster recovery of isometric strength, reduced inflammation and oxidative stress, and increased antioxidant capacity during the subsequent 2 day recovery [206]. In another running trial, consumption of tart cherry juice for 7 days prior to and during a 26.3 km relay race significantly reduced perceived pain following the race [207].

Practical Use. Consumption of approximately 45 sweet Bing cherries per day has been shown to reduce markers of inflammation [208, 209]; however, it is not clear whether the antioxidant and anti-inflammatory potentials of sweet cherries are comparable to tart cherries. Furthermore, most of the available evidence indicating a benefit from tart cherries is based on consumption of juice containing the equivalent of 90–120 cherries per day or 12–16 oz [180, 205–207]. As such, practicality and the limited scope of available evidence favor the use of tart cherry juice. The TCJ used in the aforementioned studies was derived directly from fresh cherries in concentrate or juice form making it feasible for most people to consume.

Emerging evidence indicates that oxidative stress is an important signaling mechanism for muscle remodeling [210] and may therefore be necessary for beneficial adaptations to exercise [211]. This concern is supported by evidence of antioxidant supplementation inhibiting adaptation to exercise [211]. Furthermore, anti-inflammatory substances such as nonsteroidal anti-inflammatory drugs present a similar concern. Similar to antioxidant containing foods, such as TCJ, NSAIDs reduce inflammation by inhibiting cyclooxygenase activity. There is indication of this mechanism inhibiting regeneration of muscle [212] and connective tissue [213], which could impair adaptations to exercise [212] and increase injury risk [213]. As such, the long-term use of antioxidants and NSAIDs may be contraindicated for athletes pursuing enhanced muscle mass development. Based on these findings, acute supplementation with TCJ may be most effective endurance sessions or competitions, rather than for continual use.

6. PEDs for Endurance (E) and Aerobic Training

More athletes are choosing nutritional supplements, from both natural and organic sources, to gain a competitive advantage in endurance-based sports. The increased energy demands of endurance activities require fluid, electrolyte, and energy consumption during training and competition (Table 4). Facilitating the delivery of these key nutrients to working muscles is paramount to athletic performance.

6.1. Beet Root Juice (BRJ). Beetroot juice (BRJ) is among the most popular nutritional supplements to improve endurance performance [214]. Much of this is due to an increased consumption of organic and natural foods [215]. Thus, the trend for organic and natural food products is particularly relevant for athletes at all levels of competition. BRJ is particularly popular among endurance (E) athletes, because of its high concentration of nitrate that has been hypothesized to enhance endurance. For example, there is both anecdotal and scientific support for BRJ to improve time-trial endurance [216] and time to exhaustion [217], reduce steady-state oxygen consumption [218], and increase peak power and work rate at the gas exchange threshold [219].

Mechanisms. Several mechanisms have been postulated for the endurance exercise improvement effects of BRJ. A reduction in phosphocreatine (PCr) degradation and the reduction of build-up of ADP and inorganic phosphate (Pi) at the same relative exercise intensity following BRJ consumption [216, 219] are likely mechanisms responsible for the decrease in O_2 cost (oxidative phosphorylation) of exercise and increased time to exercise failure (reduced muscle fatigue). Beetroot has a high nitrate (NO_3^-) content (>250 mg/100 g of fresh weight), among the highest assessed, and contains more than other foods high in NO_3^- including spinach, celery, arugula, and carrot juice [220]. Nitrate is reduced to nitrite via bacteria in the oral cavity and by specific enzymes (e.g., xanthine oxidase) within tissues. There are several pathways to metabolize nitrite to nitric oxide (NO) and other biologically active nitrogen oxides [221]. Nitric oxide is a signaling molecule formed in the endothelium by the enzyme endothelium nitric oxide synthase (eNOS) which triggers the vasculature to relax (vasodilatation) by interacting with vascular smooth muscle leading to increased blood flow [222] at rest [223] and during exercise [224].

Given these properties, NO has gained a lot of attention for possible E exercise improvements including increased O_2, glucose, and other nutrient uptake to better fuel working muscles. Currently there is no means to provide NO supplementation through the diet (as it is a gas), thus BRJ and its high nitrate concentration are used as a means to generate NO endogenously. Indeed, there is an impressive and growing body of scientific data in support of whole food sources of inorganic nitrate, such as that found in BRJ, showing improved athletic performance.

Evidence. While there is very limited scientific data demonstrating BRJ's effect on resistance (R) exercise [225], the vast majority of data strongly supports its beneficial effect on improving E performance. Lansley et al. [217] recruited 9 healthy, physically active men who consumed either 0.5 liters of BRJ (6.2 mmol·d^{-1} of NO_3^-) or 0.5 liters of NO_3^--depleted BRJ placebo (0.0034 mmol·d^{-1} of NO_3^-) for 6 days followed by acute bouts of submaximal and high-intensity (to exhaustion) running and incremental knee-extension exercises. BRJ consumption increased plasma nitrite by 105% and reduced the O_2 cost for constant-work-rate moderate and severe-intensity running by ~7% compared to placebo. In addition, time to exhaustion was increased during severe-intensity running by ~15% and incremental knee-extension exercise by ~5% with BRJ compared to placebo. These findings suggest that performance benefits (oxygen sparing and enhanced exercise tolerance) of consuming BRJ are attributed to its high NO_3^- content. More recently, Murphy et al. [226], using a double-blind placebo-controlled crossover trial, had 11 recreationally fit men and women consume either baked beetroot (200 g with ≥500 mg NO_3^-) or an isocaloric placebo (cranberry relish) 75 minutes prior to performing a 5 km time trial treadmill run to determine whether whole beetroot consumption would improve running performance. They observed a nonsignificant, 41-second faster finishing time (12.3 ± 2.7 versus 11.9 ± 2.6 km·h^{-1}, resp.; $P = 0.06$) following beetroot consumption compared to placebo. Most impressive, during the last 1.1 miles (1.8 km) of the 5 km run, running velocity was 5% faster (12.7 ± 3.0 versus 12.1 ± 2.8 km·h^{-1}, resp.; $P = 0.02$) and rating of perceived exertion was lower (13.0 ± 2.1 versus 13.7 ± 1.9, resp.; $P = 0.04$) during the beetroot trial compared to the placebo. Thus, it appears that the ingestion of whole-foods containing inorganic NO_3^- (such as beetroot or BRJ) increase plasma nitrite and ultimately NO levels which favorably affect the cellular and vasculature pathways which likely result in the observed improvements in endurance athletic performance.

Given the favorable impact of BRJ on E performance, it would seem likely that BRJ would also favorably impact other markers of athletic performance. As such, Lansley et al. [216] examined the effects of BRJ ingestion on power output, oxygen consumption (VO_2), and performance cycling time trials (TT) using nine competitive male cyclists who consumed either 0.5 liters BRJ (6.2 mmol of NO_3^-) or placebo containing nitrate-depleted BRJ (0.0047 mmol of NO_3^-) before each TT of 4 or 16 km. BRJ consumption increased plasma nitrite by 138% and resulted in significantly reduced time to completion and increased power output during both the 4 km (2.8% and 5%, resp.; $P < 0.05$) and 16 km TT (2.7% and 6%, resp.; $P < 0.05$) compared to the placebo treatment.

Similarly, Bailey et al. [218] supplemented eight healthy, recreationally active men with 0.5 liters of BRJ (5.5 mmol·d^{-1} of NO_3^-) or a low-calorie black currant juice cordial (negligible NO_3^- content) for 6 days while performing moderate (80% gas exchange threshold) and intense cycling (70% of the difference between the power output at the gas exchange threshold and $VO_{2\,peak}$) protocols during the last 3 days. BRJ ingestion increased the average plasma nitrite by 96% and increased the time to task failure by ~16% during fixed high intensity exercise. The authors concluded that increased dietary inorganic NO_3^- consumption from BRJ has the potential to improve high-intensity exercise tolerance.

These data confirm that BRJ improves endurance exercise performance; however, the minimal time needed to use BRJ for a performance benefit remains to be elucidated. One attempt to answer this question was reported by Vanhatalo et al. [219] in which they examined the effects of acute (1 and 5 days) and chronic (15 days) BRJ consumption on a moderate-intensity exercise bout (90% gas exchange threshold) and an incremental cycle ergometer ramp test (increasing work rate by 1 W every 2 sec (30 W/min)) to exhaustion.

Eight healthy subjects (5 males, 3 females) consumed either 0.5 liters BRJ ($5.2\,mmol\cdot d^{-1}\,NO_3^-$) or a placebo (blackcurrant juice cordial with negligible NO_3^- content) for 15 days and were exercise tested on days 1, 5, and 15. Plasma nitrite was significantly increased on all test days following BRJ compared to placebo. The O_2 cost of moderate-intensity exercise (increase in VO_2 relative to the increase in external work rate) was lower during BRJ and was maintained throughout the 15 days ($P = 0.002$). $VO_{2\,max}$, peak power output, and the work rate associated with the anaerobic threshold were all higher following 15 days of BRJ consumption compared to placebo and baseline conditions. In addition, BRJ systolic blood pressure was significantly lower at 2.5 hours after ingestion as well as 2, 12, and 15 days after ingestion compared to PL (-3%; $P < 0.05$). Diastolic blood pressure decreased with BRJ compared to PL (-5%; $P < 0.01$).

The authors concluded that acute (1–5 days) dietary NO_3^- supplementation significantly decreased blood pressure and the O_2 cost of submaximal exercise and increased $VO_{2\,max}$ and peak power output and these outcomes were maintained for at least 15 days with continued BRJ supplementation [219]. While most studies agree with these findings [227–229], others note that highly trained athletes (average $VO_{2\,max}$ of $72 \pm 4\,mL\cdot kg^{-1}\cdot min^{-1}$) may not have the same response to BRJ [230], suggesting that the impact of BRJ may be influenced by the training status of the individual.

In nonathletic populations, the impact of BRJ also has a significant positive impact on endurance performance. Kenjale et al. [231] provided 8 patients with peripheral arterial disease either 0.5 liters of BRJ ($18.1\,mmol\cdot L^{-1}\,NO_3^-$) or an isocaloric placebo on two separate occasions while performing an incremental, graded treadmill running test and demonstrated an increased exercise tolerance (walked 18% longer before claudication pain onset and experienced a 17% longer peak walking time), and decreased fractional O_2 extraction. These findings support BRJ to enhance peripheral tissue oxygenation in hypoxic areas and increase exercise tolerance in individuals with peripheral arterial disease. Thus, strong scientific evidence supports BRJ supplementation as an effective ergogenic aid for both athletes and nonathletes alike in order to improve endurance/aerobic exercise performance.

Practical Use. It is important to note that the acute dose of BRJ used in most research studies is approximately 0.5 liters or ~16 fl. oz. There are several ways to incorporate BRJ into an athlete's diet. One strategy is to prepare the BRJ from the whole beets using the following technique: remove the stalks and thoroughly wash the beets, cut into cubes, submerge in water, bring to a boil and then simmer for 45 minutes until beets are tender, allow to cool, pour off the fluid, and place in refrigerator (lasts up to 5 days) or freeze (up to 3 months). Consume 16 fl. oz. alone or mixed with another antioxidant-rich juice (tart cherry, grape, cranberry, and pomegranate) on endurance exercise days. Another technique is to thoroughly blend 2-3 whole beets (stalks removed) in a food processor, blender, or juice compressor.

6.2. Caffeine. The mechanisms through which caffeine enhances performance, as well as the practical considerations for caffeine use, were previously discussed in relation to anaerobic performance. In addition to the potential for caffeine to enhance anaerobic performance, meta-analysis has indicated it is more effective for enhancing aerobic performance [123].

Several recent studies have demonstrated the beneficial influence of caffeine on endurance performance. In 10 well-trained cross-country skiers, $6\,mg\cdot kg^{-1}\cdot BW$ of caffeine consumed 45 minutes prior to exercise led to better performance and reduced rating of perceived exertion (RPE) during an 8 km double-poling time trial [232]. Similarly, in trained cyclists, 200 mg of caffeine consumed 60 minutes before exercise improved 40 km time trial performance [233], and $5\,mg\cdot kg^{-1}\cdot BW$ of caffeine consumed 60 minutes before exercise, either in supplement form or from coffee, improved performance during an approximately 45 minute time trial with a target of 70% of maximal work output [137]. In addition, $3\,mg\cdot kg^{-1}\cdot BW$ of caffeine consumed 90 minutes prior to exercise has been shown to improve performance during an approximately 60 minute work-based time trial during hot conditions [234].

6.3. Carbohydrate and Fat Intake. Although protein contributes to energy production during E exercise, it is a small contribution relative to fat and carbohydrate [235]. As such, optimal macronutrient intake for supporting the energy demands of E athletic performance is primarily related to fat and carbohydrate.

Storage capacity for glycogen is greatly limited compared to fat and is therefore more tightly regulated [236]. Furthermore, reduced glycogen availability is commonly associated with fatigue [237–239], which implies that it may be advantageous to adjust macronutrient intake in a manner that either spares glycogen or reduces dependency on it. Two contrasting strategies for reducing dependency on glycogen are increasing carbohydrate intake to maintain high glucose and glycogen availability or restricting carbohydrate intake to promote adaptations that increase reliance on fat oxidation.

Mechanisms. Despite the well-established associations between glycogen depletion and fatigue, the mechanisms are not well understood [237–239]. However, it is clear that shifts in macronutrient intake alter the balance between fat and carbohydrate oxidation. In contrast to carbohydrate oxidation, which is largely influenced by carbohydrate intake, fat oxidation is influenced more so by carbohydrate intake than fat intake [236, 240]. More specifically, fat oxidation increases as carbohydrate intake decreases [236, 240].

Carbohydrate metabolism inhibits fat oxidation, and one mechanism for this has been eloquently isolated to the carnitine palmitoyltransferase (CPT) system, which transports long-chain fatty acids through the inner mitochondrial membrane. Infusion of glucose and insulin has been shown to inhibit oxidation of long-chain fatty acids, but not medium-chain fatty acids, implicating the CPT system as a location of inhibition [241]. Similar results have been observed in conjunction with elevated levels of malonyl CoA [242], which is known to inhibit the CPT system [243]. Furthermore, insulin activates acetyl CoA carboxylase (ACC), which catalyzes the production of malonyl CoA [244]. As such, carbohydrate intake is likely to inhibit fat oxidation by promoting insulin release, which then increases production of malonyl CoA by ACC and, in turn, inhibits the CPT system from transporting long-chain fatty acids into mitochondria [243, 245].

In addition to acute shifts in substrate selection, consistent changes in macronutrient intake can promote adaptations that may further enhance energy metabolism. For example, 5 days of reduced carbohydrate and increased fat intake in actively training cyclists increased genetic expression in skeletal muscle for fatty acid translocase (FAT), fatty acid binding protein, and β-hydroxyacyl CoA dehydrogenase (β-HAD), all of which are related to fat oxidation [246]. Similarly, 2 days of reduced carbohydrate and increased fat intake following glycogen depleting exercise led to increased expression for FAT and uncoupling protein 3 (UCP3), while a higher carbohydrate diet led to increased expression for glucose transporter type 4 and glycogenin [247]. In contrast, even without a change in regular diet, the consumption of glucose during moderate exercise has been shown to inhibit expression of FAT, UCP3, and CPT1 compared to the same exercise performed in a fasted state [248].

In addition to enzymatic changes, reduced carbohydrate and increased fat intake have been shown to increase intramuscular fat storage in conjunction with a lower respiratory quotient during exercise in both trained [249–251] and untrained [252] subjects. Although reduction of carbohydrate intake typically results in lower glycogen levels compared to a high-carbohydrate intake [251, 252], similar glycogen levels were maintained in one trial in conjunction with increased intramuscular fat and decreased respiratory quotient during exercise [250].

Evidence. It is well established that carbohydrate consumption prior to or during prolonged exercise enhances performance [238, 253–255]. In contrast, a number of trials have shown a high-fat and reduced-carbohydrate diet to increase fat oxidation or reduce reliance on glycogen during exercise, but with mixed effects on performance, ranging from deleterious to advantageous [250, 252, 256–267]. A meta-analysis of 38 trials found high-carbohydrate intake to be more beneficial, but the results were concluded to be unreliable due to heterogeneity [268]. Furthermore, benefit was minimal for trained subjects [268], and the results were skewed by intervention durations of less than 7 days, which

may have not been enough time for adaptation to reduced carbohydrate intake [262, 268].

Given that carbohydrate restriction reduces glycogen [251, 259, 262, 263, 269, 270], limited capacity for high-intensity exercise is expected. However, the evidence is mixed [250, 257–259, 265, 266, 270–272], with indication that it is possible to maintain high-intensity exercise capacity even when carbohydrate is restricted to less than 10% of energy intake [259, 270–272]. Furthermore, normal glycogen levels and thus performance can be maintained with moderate carbohydrate restriction [250, 252], as well as with supplementation of carbohydrate during the exercise bout, if needed.

Practical Use. It is commonly recommended that athletes follow a high-carbohydrate diet to replenish glycogen and maintain blood glucose [92, 273]. However, a high-fat and reduced-carbohydrate diet may be an effective alternative [237, 274, 275]. The equivocal evidence indicates that a wide range in the ratio of fat and carbohydrate intakes can support high-level performance, although the enhancements in fat utilization observed with carbohydrate restriction are unlikely to have ergogenic value beyond that of a high-carbohydrate diet [276–278]. Despite this, fat adaptation can still be compatible with optimal performance and may have beneficial implications for weight management, training adaptation, and metabolic health. A meta-analysis of 87 trials [15] and a number of more recent trials [271, 279, 280] have shown carbohydrate restriction to have a more favorable influence on body composition during weight loss, particularly when combined with resistance training [281–283]. Furthermore, carbohydrate restriction has clearly been shown to have a highly favorable effect on cardiovascular and metabolic risk factors [284–289]. It has been suggested that only a quarter of the population can tolerate the current recommendations for carbohydrate intake without developing signs of metabolic dysfunction [275]. Although athletes are less susceptible [275], the presence of metabolic risk factors in athletes is not rare, especially in sports that favor a heavier body weight [290, 291]. As such, a high-fat and carbohydrate-restricted diet can be a valuable alternative for athletes who need to manage body weight or have signs of metabolic impairment. In addition, the lower glycogen levels that result from carbohydrate restriction may enhance adaptations to endurance training [273, 292, 293].

Overall, a high-carbohydrate intake is not the only way to support optimal endurance performance. However, carbohydrate intake is likely to be more important for high-intensity performance [278, 294]. Furthermore, there may be considerable variation in the optimal macronutrient ratio for each athlete [256, 262]. Given the equivocal evidence, athletes should determine their optimal ratio of fat and carbohydrate intakes based on a combination of factors including the demands of their sport, their individual response to different macronutrient ratios, and any concerns related to health or body composition. Considerations related to health and individuality are especially applicable to nonelite athletes, who have less reason to prioritize performance over wellbeing. One strategy to determine an ideal and individualized macronutrient ratio, which would be best implemented during the off-season, is to restrict carbohydrate for several weeks and then gradually increase carbohydrate intake until the minimal effective dose needed to sustain performance can be identified. This may include carbohydrate supplementation during exercise, if needed. However, it is important to understand that sufficient adaptation to carbohydrate restriction may take 2–4 weeks [262] and that further adaptation may continue beyond the 4th week [295]. Another approach is to restrict carbohydrate intake long enough to promote fat adaptation and then increase carbohydrate intake prior to or during a competition in order to restore glycogen levels. Enhanced fat oxidation has been shown to persist with such a strategy, although to a lesser extent, and with mixed effects on performance [296].

6.4. Fluid Hydration: Glycerol and Electrolytes. Fluid intake and adequate hydration are critical during E training sessions and competition events. Fluid intake helps to maintain hydration, body temperature (thermoregulations), and plasma volume. For events lasting longer than one hour, athletes need fluids containing carbohydrates and electrolytes rather than water alone. Reduction in body water, availability of carbohydrates, and an inadequate electrolyte balance during prolonged exercise events will hamper performance and may lead to serious medical disorders such as heat exhaustion, heat stroke, or hyponatremia. A 1% reduction in body weight due to water loss may evoke undue stress on the cardiovascular system accompanied by increases in heart rate and inadequate heat transfer to the skin and the environment, an increase in plasma osmolality, and a decrease in plasma volume and affect the intracellular and extracellular electrolyte balance [297].

Water loss occurs through respiration, sweat, feces, and urine; however, during prolonged endurance most water is lost in sweat, especially during high environmental temperatures. About 580 kcals are lost for every liter of sweat that is evaporated [298]. Loss of body fluid during endurance exercise can be determined by changes in body weight; each kg of body weight loss accounts for about 1 liter of fluid loss. Sports drinks with adequate concentrations of electrolytes and carbohydrates promotes maintenance of homeostasis, prevents injuries, and maintains optimal performance [299].

Mechanisms. Regulation of fluid balance is a remarkably complex process. Water is lost from the body through the skin, feces, lungs, and kidneys. Water retention by the kidneys is directly controlled by vasopressin produced in the hypothalamus. Production of vasopressin is affected by hypothalamic receptors sensitive to plasma osmolarity and stretch receptors in the atria of the heart, carotid arteries, and aorta.

The kidneys actively reabsorb sodium to regulate extracellular fluid osmolarity and this is largely controlled by aldosterone produced by the adrenal cortex. As serum osmolarity decreases, the adrenal cortex release of aldosterone is triggered resulting in more sodium reabsorbed and an increase in osmolarity. The kidneys also regulate aldosterone production through the rennin-angiotensin mechanism. Receptors in the juxtaglomerular complex of the kidney tubules respond to low volume (pressure) by releasing rennin, which leads to a hormonal cascade effect resulting in production of angiotensin II, a potent vasoconstrictor, which stimulates the release of aldosterone.

Prolonged E exercise significantly taxes the body's ability to regulate hydration status, body temperature, and electrolytes, thus maintaining hydration during exercise is critical to optimal performance. It is recommended that athletes ingest ~500 mL of fluid 1-2 hours prior to performance and continue to consume cool drinks in the amount of 4–6 ounces every 20 minutes during exercise to replace sweat losses [297, 300, 301]. The rate of water ingestion should not exceed the rate of water loss, as it might result in water retention, weight gain, and exercise-associated hyponatremia [301, 302].

Evidence. Consumption of sports beverage drinks during exercise is recommended to meet carbohydrate energy needs and to replace sweat, water, and electrolyte losses [297]. The majority of the literature supports fluid, carbohydrate, and electrolyte replacement during prolonged ($\geq$60 minutes) endurance exercise. Replacement of Na^+ and K^+ are essential to maintain plasma volume and hydration [303]. Different exercise tasks (metabolic requirements, duration, clothing, and equipment), weather conditions, and other factors such as genetic predisposition, heat acclimatization, and training status influence sweating rate and electrolyte concentrations and determine fluid needs [297].

Carbohydrate and electrolyte content, palatability, color, odor, taste, temperature, and texture of a sports drink can increase fluid consumption before, during, and after exercise [297, 304] and therefore improve performance. Athletes should ingest 4 to 8 ounces of a 6%–8% carbohydrate-electrolyte sports drink every 10 to 20 minutes during exercise and avoid carbohydrate concentrations over 8% as this will delay gastric-emptying and should be avoided. Increasing plasma volume can positively affect performance and sodium in sports drinks may help achieve this by improving glucose and water absorption in the small intestine. Sodium is important in rehydration, especially during exercise in the heat [305].

Galloway and Maughan [306] studied six healthy males who cycled to exhaustion while ingesting either no drink, a 15% carbohydrate-electrolyte drink, or a 2% carbohydrate-electrolyte drink. Consumption of the 2% carbohydrate-electrolyte drink leads to a lower serum osmolality and reduced plasma volume deficits. Potassium is important in rehydration after exercise due to the increased retention of fluid in the intracellular space [305]. Numerous recent studies [305, 307–310] have confirmed that during E events, consumption of glucose-electrolyte solutions improved performance greater than water alone.

Several factors including fluid, fuel substrate, and electrolyte depletion have been implicated in the reduction of endurance performance. Recent investigations have suggested that consumption of lactate and fructose in energy-electrolyte hydration beverages improves performance and delays fatigue compared to glucose-electrolyte beverages via increased substrate oxidation and enhanced buffering capacity [311].

Hyperhydration may be induced by the oral consumption of glycerol which induces an osmotic gradient that favors greater renal water absorption. Studies examining the effect of hyperhydration by glycerol consumption on performance are equivocal. Several studies have shown performance enhancements [298, 312, 313] while others have shown no difference when comparing hyperhydration by glycerol consumption to hyperhydration by water or flavored-water consumption [314–316]. Recently, investigations have examined the effect of glycerol ingestion and fluids of varying tonicities (0.9% versus 0.45% NaCl) during the rehydration period following exercise-induced dehydration (~4% of body weight) and prior to exercise in the heat [307, 317]. Rehydration with either a 0.45% or 0.9% NaCl solution resulted in similar fluid restoration, similar cardiovascular, thermoregulatory, and exercise performance responses and were superior to no fluid ingestion [317]. Glycerol ingestion during the rehydration period was found to significantly prolong subsequent exercise time to exhaustion in the heat but was not associated with specific thermoregulatory or cardiovascular advantages compared to rehydration with water alone [307].

Practical Use. The most recent data suggest that multiple-transportable-carbohydrates containing a combination of glucose/maltodextrin + fructose in combination with electrolytes are the most favorable beverages to ingest during endurance exercise to enhance performance. Specifically, a 6–8% carbohydrate mixture of glucose and fructose (GF) plus an electrolyte solution containing NaCl and K will further aid endurance performance [318]. Coconut water is also gaining in popularity due to its high K concentration. Mixing the coconut water with a GF plus NaCL solution may serve as another electrolyte beverage to enhance hydration and performance.

6.5. Modified and Resistant Starches. As previously mentioned (see carbohydrate intake), endurance athletes must maintain blood glucose and replenish glycogen stores during and following longer bouts, respectively [92, 273]. Indeed, the type of CHO (glycemic index and gastric-emptying rate) in relation to the timing of exercise (pre- and during exercise) is critical in the maintenance of blood glucose and insulin, sparing hepatic glycogen stores, and manipulating substrate utilization for endurance exercise. The blood glucose and insulin responses vary depending on CHO digestion and gastric-emptying rate and need to be considered prior to competition. In efforts to minimize and control the spike in blood glucose and insulin from CHO intake prior to exercise, research has turned to the use of modified and resistant starches as CHO alternatives.

Mechanism. Modified starches have gained popularity because of the benefits to digestion and gastric-emptying rate mostly due to the amylose : amylopectin ratio. In general, the higher the ratio of amylose : amylopectin, the greater the resistance to digestion [319], blunting

the initial response of blood glucose and insulin. This spares glycogen stores and enhances fat oxidation. However, it should be noted that despite the amylose : amylopectin ratio the gastric emptying and absorption rates may also be manipulated by modifying the different starches consumed (i.e., hydrothermal modification) [253].

Evidence. Stephens et al. [320] measured the effects of a high molecular weight (HMW) rapidly digested modified starch commercially known as Vitargo, a low molecular weight glucose polymer (LMW) (similar to commercial sports drinks), and sugar-free water (SFW) on blood glucose and insulin for two hours after a glycogen depleting exercise (GDE) (60 min at 75% of VO_{2max}). Following the two-hour postprandial period each individual performed a 15 min "all-out" bout of cycling. Both the HMW and LMW starch elevated blood glucose and insulin during the two-hour recovery versus the SFW, with the initial response (<60 min) of HMW being significantly greater than the LMW. A greater work output (10%) during the 15 min cycle performance was found when consuming HMW compared to LMW and SFW suggesting that the rise in blood glucose and insulin allowed replenishment of glycogen stores from the HMW between exercise sessions [320].

Jozsi et al. [321] tested blood glucose and insulin response to amylose and amylopectin versus glucose and maltodextrin. Amylose in the form of a resistance starch (see Section 7) acts similar to a dietary fiber allowing increased fat oxidation by blunting glucose and insulin prior to exercise, whereas amylopectin (as waxy maize starch) responds similar to a normal CHO (i.e., glucose). In their study [321] male cyclists ($n = 8$) completed a GDE (60 min at 75% of VO_{2max}) where they consumed either resistance starch (100% amylose), waxy maize (100% amylopectin), glucose, or maltodextrin as 65% (~1950 kcals) of a 3,000 kcal diet for twenty-four hours after GDE. Following the GDE and CHO consumption the individuals performed a thirty-minute cycling time trial. Twenty-four hours after the GDE, immediately before exercise the resistance starch (100% amylose) resulted in significantly lower muscle glycogen concentrations compared to waxy maize (100% amylopectin), glucose, and maltodextrin. However, no differences were found between the four types of CHO during the thirty-minute cycling time trial [321]. Therefore, a high amylopectin starch (100%) [321] and a HMW rapidly digesting starch both [320] increase blood glucose and insulin following a glycogen depleting exercise, while high amylose resistant starch (100%) results in a lower blood glucose and insulin response, thereby inhibiting glycogen resynthesis [321].

However, modification of these starches (i.e., hydrothermal modification) may decrease the digestion time altering the response of blood glucose and insulin regardless of the amylose : amylopectin ratio [253]. Roberts et al. [322] measured a hydrothermally modified and slow digesting starch (HMS) consisting primarily of amylopectin (95%), commercially known as UCAN versus maltodextrin ($1\,g \cdot kg^{-1} \cdot BW$) during both steady state (150 min submaximal cycling bout at 70% VO_{2peak}) and exhaustive (100% VO_{2peak}) exercise, as well as during 75 minutes of recovery. There was no significant difference in performance between either HMS or maltodextrin. However, both the initial and recovery periods of glucose and insulin were blunted with HMS compared to maltodextrin, allowing an increase in fat oxidation [322]. Thus, despite the HMS consisting of amylopectin (95%) known to increase postprandial glucose and insulin [321], modification of the HMS resulted in a lower postprandial glucose and insulin, similar to that of a high amylose starch [321]. This finding suggests that chemical modification of amylopectin as HMS may augment fat oxidation and spare muscle glycogen.

Practical Use. Although performance benefits from modified and resistant starches appear to be minimal, the glucose and insulin responses during exercise and recovery are optimal for both fat oxidation and glycogen resynthesis. Therefore, in efforts to minimize glucose and insulin secretions and promote a greater reliance on fat oxidation it may be recommended to consume $(1 \, g \cdot kg^{-1} \cdot BW)$ either a modified-HMW (UCAN) or 100% amylose starch prior to and during exercise. Athletes who will be competing in multiple events or are dependent on the replenishment of glycogen stores between exercises are recommended a high molecular weight rapidly digesting starch, such as Vitargo. Other whole food sources of resistant starches include cooked and cooled potatoes, whole grains (rice, pasta, etc.), and legumes and, thus, should be consumed prior to and between extended bouts of exercise.

7. PEDs for Energy Metabolism and Body Composition

Optimal body composition plays a critical factor in athletic performance and it varies among different types of athletes and sports. It is well known that energy metabolism and body composition are directly related to each other and nutritional factors are the primary determinants of each (Table 4).

7.1. Caffeine. The practical use of caffeine, as well as mechanisms through which it may enhance performance and energy metabolism, was previously discussed in relation to I and E athletic performance. In regard to energy metabolism, a meta-analysis including 6 trials found caffeine consumption to increase daily energy expenditure by approximately 100 kcal [323]. However, fat oxidation was found to only increase when caffeine was combined with catechins [323], indicating that tea may be a favorable source of caffeine for the purpose of weight management. In this meta-analysis, caffeine intake ranged from 150 to 1604 $mg \cdot d^{-1}$ [323]. Consistent with the above results, our laboratory found a 5 $mg \cdot kg^{-1} \cdot FFM$ dose of caffeine to increase energy expenditure in men [324] and women [325]. Although fat oxidation was not measured in the women, it did not change in the men [324]. In addition, the increase in energy expenditure was reduced for older versus younger women [325], but this was not the case with the men [319, 324] suggesting a gender difference in the influence of age on the metabolic response to caffeine.

In contrast to previous results, two recent trials found 5 $mg \cdot kg^{-1} \cdot d^{-1} \cdot BW$ of caffeine, consumed for 4 days, to have no influence on resting, active, or total energy expenditure in young men [326, 327]. However, in both trials, the caffeine was divided into two doses, one of which was consumed with breakfast and the other with lunch. Furthermore, although participation was restricted to individuals who habitually consumed less than 100 $mg \cdot d^{-1}$ of caffeine, actual caffeine consumption prior to the trials was not reported. As such, the previously described evidence, indicating that caffeine does increase energy expenditure, appears to be more reliable.

7.2. Capsinoids. Capsaicin, the known pungent flavor of hot red chili peppers, has become a popularly marketed natural spice for enhancing thermogenesis (i.e., catecholamines, fat oxidation) and improving satiety [328].

Mechanism. Capsaicin will bind to the TRVP1 passively absorbing through the stomach and upper portion of the small intestine. After being released into circulation, capsaicin will be transported to the adrenal gland to release catecholamines, thereby increasing SNSa and energy expenditure [328, 329].

Evidence. Research has shown capsaicin to increase SNSa [330–334], energy expenditure [329, 335–337], and substrate oxidation [329, 330, 335, 336, 338], although these findings are not universal [339, 340]. The effectiveness of capsaicin to increase thermogenesis and satiety may differ due to varying dosages and with individuals who frequently consume capsaicin compared to nonusers. More recently, Ludy and Mattes [329] tested the effects of capsaicin (1 g red pepper) on energy expenditure, fat oxidation, and satiety in individuals who regularly consume red peppers versus nonusers in both oral and capsule forms. The postprandial energy expenditure increased in both groups following both oral and capsule form. Interestingly, fat oxidation increased only with the oral form with satiety and energy intake decreasing in only the nonusers. Because capsaicin binds to the TRPV1 in the oral cavity activating heat and pain sensitive sensory neurons it is suggested that when consumed orally rather than in capsule form capsaicin's influence on substrate oxidation may be greatest [329].

Because the pungent sensory burn and pain elicited from capsaicin may cause difficulty in palatability, the capsaicin-like compound capsiate, in the form of nonpungent red pepper "CH-19 Sweet," is an alternative for those unaccustomed or opposed to eating spices. Despite the difference of activation sites, both capsaicin and capsiate bind with high affinity to TRPV1 located in the gut, increasing SNSa [338, 341] without the elevated systolic blood pressure and heart rate response reported with capsaicin [334]. As documented with capsaicin, similar supporting research has found capsiate to increase SNSa [334, 342], energy expenditure [342–344] and substrate oxidation [338, 342–344] in higher dosages. Josse et al. [342] found increases in SNSa, energy expenditure, and fat oxidation with 10 mg. Likewise, Galgani and Ravussin [345] found a 54 $kcal \cdot d^{-1}$ increase in RMR when individuals consumed 3 and 9 mg compared to placebo, and Lee et al. [344] reported increases in postprandial energy expenditure and fat oxidation with 3 and 9 mg. Thus, when consumed in higher concentrations (3–10 mg) in capsule form, capsiate appears to elicit the greatest thermogenic response.

Practical Use. Individuals not accustomed or willing to consume spices, capsaicin, and capsiate may serve as effective PED supplements to augment and optimize postprandial thermogenesis and substrate oxidation. Common sources of capsaicin in market stores are chili powder, chili peppers, cayenne pepper, jalapenos, and habaneros. A lower dosage of 2 mg to a higher dosage of 10 mg is recommended for individuals with varying thresholds and tolerance, with capsiate as an alternative to those unwilling or unable to consume the pungent capsaicin. However, those unaccustomed to eating capsaicin should consider timing because consumption of the spice prior to exercise has been reported to cause stomach discomfort, nausea, intestinal cramping, flatulence, and burning bowel movements in male athletes [346].

7.3. Carnitine. Carnitine is naturally synthesized in the body from the essential amino acids lysine and methionine [347]. Based on its role in fatty acid transport, carnitine has the potential to support weight management by facilitating fat oxidation.

Mechanism. Carnitine makes up the substrate to CPT1, a rate-limiting step in fatty acid oxidation within skeletal muscle [348]. It is logical to assume that consuming exogenous carnitine via diet or supplementation would be beneficial by inhibiting carbohydrate utilization and augmenting fatty acid oxidation through enhanced translocation of long-chain acyl groups across the inner mitochondrial membrane [348, 349].

Evidence. Recently, Stephens et al. [348] tested the effects of 12 weeks of L-carnitine (CAR) in combination with a CHO consumed twice daily on muscle expression of genes associated with metabolism, body composition (DEXA), and energy expenditure during low intensity exercise (50% of $VO_{2\,max}$ for 30 min) in twelve males. Participants ingested either the CAR + CHO, n = 6 (1.36 g + 80 g l-CAR + CHO), or CHO, n = 6 (80 g) first thing in the morning and again 4 hours later. Those who consumed CAR + CHO had a 20% and 200% increase in total carnitine and long-chain acyl-CoA, respectively ($P < 0.05$), and elevated expression of genes involved in fatty acid metabolism. There were no changes in body composition in the CAR + CHO group; however over 12 weeks there was a 1.9 and 1.8 kg increase in body mass and whole-body fat mass, respectively, in the CHO group. Furthermore, there were no changes in whole-body energy expenditure in the CHO group, but there was a significant increase of 6% in the CAR + CHO group during the 30 min low-intensity exercise. These findings from Stephens et al. [348] suggest that the consumption of CAR + CHO for twelve weeks prevented increased body fat mass and in turn was associated with a greater energy expenditure and fat oxidation during low-intensity exercise, as supported in previous research [350].

Practical Use. Carnitine is found in abundance throughout the skeletal muscle cells of the body [351] and may be ingested through diet containing red meat, fish, poultry, and dairy products (for amount of carnitine per each nutrient, see Rebouche [352]) or can be biosynthesized within the liver and kidneys primarily from the essential amino acids lysine and methionine [351, 352]. Those who consume a vegetarian diet are estimated to receive ~90% of their total available carnitine from endogenous synthesis due to the lack of available carnitine in plant foods, while omnivores receive one-eighth to one-half of total carnitine through diet [352]. Therefore athletes consuming a predominately plant based diet may consider commercially produced carnitine supplements which have been shown to be safe in humans [353].

7.4. Dietary Fiber and Resistant Starches. Resistance starches (RS) are touted as weight loss wonder foods because they have digestive properties and satiating effects similar to those of dietary fibers [354]. In addition, RS may further facilitate weight management by increasing fat oxidation and total energy expenditure and improving glycemic regulation [354].

Mechanism. In comparison to normal dietary starches (DS), RS lowers the glycemic response by passing digestion in the small intestine and moving directly into the large intestine, where it is fermented into short-chain fatty acids [354, 355]. Of the five types of RS1, RS2, RS3, and RS4 are most commonly measured in humans for their effects on postprandial glycemia/insulinemia responses and gut satiety peptides that influence weight loss or weight maintenance (i.e., energy expenditure). RS1 is physically inaccessible to digestive enzymes from the presence of seed coats (e.g., whole grains), RS2 is a high amylose maize starch comprised primarily of α-1,4 glycosidic links, RS3 is retrograded starch (e.g., pasta or rice that has been cooked then cooled), and RS4 is chemically modified to be resistant to digestion [354, 356].

Evidence. Following consumption of RS2, the postprandial glycemia/insulinemia responses in healthy men and women have been shown to significantly decrease compared to DS [357, 358]. Interestingly though, when RS2 was adjusted as 0%, 2.7%, 5.4%, and 10.7% (percentage of total carbohydrate) at 30% of an individuals' daily energy needs there were no differences in postprandial glycemia/insulinemia, suggesting dosages up to 10% have little impact on glycemia and that a combination of additional ingredients might affect RS2 function [359]. Among the other types of RS, RS3 has been shown to decrease postprandial glycemia [360], and RS4 has resulted in decreases in both postprandial glycemia and insulinemia [361–363].

In addition to improved postprandial glycemic/insulinemic responses, RS2 and 3 have shown to positively alter gut satiating peptides (glucose-dependent insulinotropic peptide (GIP)) [358], suppress energy intake [357], improve satiation/appetite [358, 364], and significantly increase fat oxidation [359], although these findings are not universal [357, 359, 365, 366]. Yet, despite the conflicting results of RS2 and RS3, RS4 has consistently reported favorable postprandial glycemia/insulinemia responses [355, 363, 367] and beneficial increases in energy expenditure in healthy individuals [363].

To determine differences in postprandial glycemia/insulinemia responses between RS (RS2 and RS4 cross-linked, XL) and a normal carbohydrate (dextrose), Haub et al. [355] tested eleven healthy males ($n = 4$) and females ($n = 7$) for two hours after consuming 30 g of $RS4_{XL}$, RS2, or dextrose combined with water. Postprandial glucose and insulin responses were significantly lower in the $RS4_{XL}$ and RS2 compared to dextrose, with $RS4_{XL}$ being significantly lower than that of the RS2 [355]. Because RS is more commonly consumed in combination with foods rather than water alone, Al-Tamimi et al. [367] examined the effects $RS4_{XL}$ in combination with additional ingredients in the form of a nutrition bar and found that $RS4_{XL}$ compared to wheat starch resulted in significantly lower 2 hr postprandial glycemia/insulinemia response.

Shimotoyodome et al. [363] tested RS4 (hydroxypropyl-distarch, HDP) versus waxy maize starch (WMS) using a pancake meal on the 3 hr postprandial glycemic/insulinemic, GIP, and energy expenditure response in healthy lean males. HDP resulted in significantly lower postprandial glycemia/insulinemia and GIP, as well as increased fat oxidation and energy expenditure compared to WMS. Research supports suppressed postprandial glycemia/insulinemia by both RS2 and RS4; interestingly, only RS4 has elicited significant increases in energy expenditure [363].

Practical Use. Current research suggests the effects of RS4 ($RS4_{XL}$ and HDP) on postprandial glycemia/insulinemia, gut satiety peptides GIP, and augmented energy expenditure and fat oxidation [355, 363, 367] are greater than those of RS2. A suggested dosage of 20–40 g of RS4 consumed at breakfast or a late evening snack may facilitate greater appetite suppression, postprandial glycemia and insulinemia, and increase energy expenditure.

7.5. Medium-Chain Triglycerides. Medium-chain triglycerides (MCTs) consist of fatty acids ranging in length from 6 to 12 carbons [368]. Although MCTs appear to have little influence on performance, benefits related to energy balance and weight loss are better supported [369].

Mechanisms. The benefits of MCT consumption may be explained by the unique ways their constituent fatty acids are absorbed and metabolized. Prior to reaching systemic circulation, long-chain fatty acids are reincorporated into triglycerides, assembled into chylomicrons, and released into lymphatic circulation [370]. In contrast, medium-chain fatty acids enter portal circulation directly through the enterocyte [370]. As such, medium-chain fatty acids enter circulation more rapidly and are primarily absorbed by the liver [368]. Once absorbed, medium-chain fatty acids pass through the inner mitochondrial membrane independently of the CPT transport system and can therefore be rapidly oxidized [368]. This, along with the poor binding potential between medium-chain fatty acids and fatty-acid-binding protein, limits the lipogenic potential of MCTs [368]. As such, MCTs are more

likely to be utilized for energy and less likely to be stored as body fat.

MCTs also have the potential to increase energy expenditure [371]. Urinary noradrenaline excretion has been found to increase in conjunction with increased energy expenditure following MCT consumption [372]. Furthermore, in rats fed MCTs, an increase in energy expenditure was prevented with administration of propranolol [373]. Therefore, increased SNSa may be the underlying mechanism.

Evidence. In a recent systematic review [371], 6 of 8 trials found MCT consumption to improve body composition, and 4 of 6 trials identified an increase in energy expenditure [371]. However, an increase in satiety was only observed in 1 of 7 trials [371]. Some trials have shown MCT consumption to increase average daily energy expenditure by more than 100 kcal in overweight [374] and normal-weight [372] men, which could amount to more than 30 lbs of weight loss over a year [375]. However, such an increase is not consistently supported [371] and may be less notable in women [376]. Furthermore, the increase in energy expenditure induced by MCT consumption has been shown to diminish over time [369]. Despite this, superior weight loss has been observed in trials lasting as long as 16 weeks [369]. However, because the increases in energy expenditure and fat oxidation associated with MCT consumption have been inversely correlated with initial body weight, MCT consumption may be more effective for preventing weight gain than promoting weight loss [377].

In a recent trial including 7 normal-weight subjects, a breakfast meal containing 20 g of MCTs was found to increase diet-induced thermogenesis and fat oxidation compared to the same meal with a calorically matched content of sunflower oil [378]. Similarly, in another recent trial, the inclusion of MCTs in a meal replacement shake led to greater diet-induced thermogenesis compared to shakes with lesser amounts of MCTs or no MCTs at all [379]. As such, replacing a portion of dietary fat with MCTs may be an effective strategy for weight maintenance.

Practical Use. Increases in energy expenditure have been observed with MCT intakes ranging from 8 to 35 $g \cdot d^{-1}$ [371]. While MCT oil is readily available, coconut oil and palm kernel oil are two alternatives that more closely resemble whole foods. They contain approximately 63 and 58% MCT, respectively [380]. These oils can easily be incorporated into the diet through cooking or by melting them for use as a sauce or salad dressing. The meat of a coconut, although less practical, is truly a whole food and contains approximately 19% MCT [94].

8. Summary

Advances in athletic performance training and nutrition have prompted a reevaluation of our current practices in order for both (training and nutrition) to work synergistically with each other instead of in isolation to one another. The current review, albeit novel, bridges the gap between athletic performance training and sports nutrition by linking the scientifically validated multicomponent training model (timed-protein feedings; resistance training; interval sprint training; stretching/recovery training; and endurance training; PRISE) employed by most, if not all, athletes with specific performance enhancing diets (PEDs) to foster optimal athletic performance. The goal of this innovative review is to provide a new paradigm of sports nutrition that allows performance training (PRISE) and sports nutrition (PEDs) to complement each other instead of working apart from one another.

Conflict of Interests

Paul J. Arciero is president and founder of Nourishing Science LLC, a company providing nutrition, fitness, and wellness consultative services utilizing certain content contained in this paper. The authors declare no other conflict of interests.

Acknowledgments

The authors are grateful to the following individuals for their assistance with the research conducted on the PRISE protocol: (i) nurses, Patricia Bosen NP and Michelle Lapo, (ii) student researchers, Christopher Darin, Qian Zheng, Kanokwan Bunsawat, JunZhu Zhang, Nicholas Steward, Jake Mendell, Caitlin Ketcham, Steven Brink, Steve Vasquez, Gabriel Zeiff, and Elise Britt, (iii) Scott Connelly MD for financial support, (iv) the hundreds of research study participants for their dedication, cooperation, and strong spirits.

References

[1] L. Dwyer-Lindgren, G. Freedman, R. E. Engell et al., "Prevalence of physical activity and obesity in US counties, 2001–2011: a road map for action," *Population Health Metrics*, vol. 11, no. 1, article 7, 2013.

[2] US Department of Health and Human Services, *2008 Physical Activity Guidelines for Americans*, US Department of Health and Human Services, Washington, DC, USA, 2008.

[3] National Physical Activity Plan Alliance, *2014 United States Report Card on Physical Activity for Children and Youth*, National Physical Activity Plan Alliance, Columbia, SC, USA, 2014.

[4] P. J. Arciero, D. Baur, S. Connelly, and M. J. Ormsbee, "Timed-daily ingestion of whey protein and exercise training reduces visceral adipose tissue mass and improves insulin resistance: the PRISE study," *Journal of Applied Physiology*, vol. 117, no. 1, pp. 1–10, 2014.

[5] J. L. Areta, L. M. Burke, M. L. Ross et al., "Timing and distribution of protein ingestion during prolonged recovery from resistance exercise alters myofibrillar protein synthesis," *Journal of Physiology*, vol. 591, no. 9, pp. 2319–2331, 2013.

[6] S. M. Phillips, D. R. Moore, and J. E. Tang, "A critical examination of dietary protein requirements, benefits, and excesses in athletes," *International Journal of Sport Nutrition and Exercise Metabolism*, vol. 17, supplement, pp. S58–S76, 2007.

[7] P. J. Arciero, M. J. Ormsbee, C. L. Gentile, B. C. Nindl, J. R. Brestoff, and M. Ruby, "Increased protein intake and meal frequency reduces abdominal fat during energy balance and energy deficit," *Obesity*, vol. 21, no. 7, pp. 1357–1366, 2013.

[8] U. S. Department of Agriculture, *Report of the Dietary Guidelines Advisory Committee on the Dietary Guidelines for Americans*, 2010.

[9] G. A. Bray, S. R. Smith, L. de Jonge et al., "Effect of dietary protein content on weight gain, energy expenditure, and body composition during overeating: a randomized controlled trial," *The Journal of the American Medical Association*, vol. 307, no. 1, pp. 47–55, 2012.

[10] C. B. Ebbeling, J. F. Swain, H. A. Feldman et al., "Effects of dietary composition on energy expenditure during weight-loss maintenance," *Journal of the American Medical Association*, vol. 307, no. 24, pp. 2627–2634, 2012.

[11] T. P. Wycherley, L. J. Moran, P. M. Clifton, M. Noakes, and G. D. Brinkworth, "Effects of energy-restricted high-protein, low-fat compared with standard-protein, low-fat diets: a meta-analysis of randomized controlled trials," *American Journal of Clinical Nutrition*, vol. 96, no. 6, pp. 1281–1298, 2012.

[12] P. J. Arciero, C. L. Gentile, R. Martin-Pressman et al., "Increased dietary protein and combined high intensity aerobic and resistance exercise improves body fat distribution and cardiovascular risk factors," *International Journal of Sport Nutrition and Exercise Metabolism*, vol. 16, no. 4, pp. 373–392, 2006.

[13] P. J. Arciero, C. L. Gentile, R. Pressman et al., "Moderate protein intake improves total and regional body composition and insulin sensitivity in overweight adults," *Metabolism: Clinical and Experimental*, vol. 57, no. 6, pp. 757–765, 2008.

[14] M. S. Westerterp-Plantenga, S. G. Lemmens, and K. R. Westerterp, "Dietary protein—its role in satiety, energetics, weight loss and health," *British Journal of Nutrition*, vol. 108, supplement 2, pp. S105–S112, 2012.

[15] J. W. Krieger, H. S. Sitren, M. J. Daniels, and B. Langkamp-Henken, "Effects of variation in protein and carbohydrate intake on body mass and composition during energy restriction: a meta-regression," *The American Journal of Clinical Nutrition*, vol. 83, no. 2, pp. 260–274, 2006.

[16] T. M. Larsen, S.-M. Dalskov, M. Van Baak et al., "Diets with high or low protein content and glycemic index for weight-loss maintenance," *The New England Journal of Medicine*, vol. 363, no. 22, pp. 2102–2113, 2010.

[17] F. Isken, S. Klaus, K. J. Petzke, C. Loddenkemper, A. F. H. Pfeiffer, and M. O. Weickert, "Impairment of fat oxidation under high- vs. low-glycemic index diet occurs before the development of an obese phenotype," *The American Journal of Physiology—Endocrinology and Metabolism*, vol. 298, no. 2, pp. E287–E295, 2010.

[18] S. Soenen, E. A. P. Martens, A. Hochstenbach-waelen, S. G. T. Lemmens, and M. S. Westerterp-plantenga, "Normal protein intake is required for body weight loss and weight maintenance, and elevated protein intake for additional preservation of resting energy expenditure and fat free mass," *The Journal of Nutrition*, vol. 143, no. 5, pp. 591–596, 2013.

[19] S. E. Drummond, N. E. Crombie, M. C. Cursiter, and T. R. Kirk, "Evidence that eating frequency is inversely related to body weight status in male, but not female, non-obese adults reporting valid dietary intakes," *International Journal of Obesity*, vol. 22, no. 2, pp. 105–112, 1998.

[20] P. Fábry, Z. Hejl, J. Fodor, T. Braun, and K. Zvolánková, "The frequency of meals. Its relation to overweight, hypercholesterolaemia, and decreased glucose tolerance," *The Lancet*, vol. 284, no. 7360, pp. 614–615, 1964.

[21] R. Crovetti, M. Porrini, A. Santangelo, and G. Testolin, "The influence of thermic effect of food on satiety," *European Journal of Clinical Nutrition*, vol. 52, no. 7, pp. 482–488, 1998.

[22] J. O. Hill, S. B. Heymsfield, C. McMannus III, and M. DiGirolamo, "Meal size and thermic response to food in male subjects as a function of maximum aerobic capacity," *Metabolism*, vol. 33, no. 8, pp. 743–749, 1984.

[23] K. S. Nair, D. Halliday, and J. S. Garrow, "Thermic response to isoenergetic protein, carbohydrate or fat meals in lean and obese subjects," *Clinical Science*, vol. 65, no. 3, pp. 307–312, 1983.

[24] T. A. Churchward-Venne, L. Breen, D. M. di Donato et al., "Leucine supplementation of a low-protein mixed macronutrient beverage enhances myofibrillar protein synthesis in young men: a double-blind, randomized trial," *The American Journal of Clinical Nutrition*, vol. 99, no. 2, pp. 276–286, 2014.

[25] S. M. Phillips, "A brief review of critical processes in exercise-induced muscular hypertrophy," *Sports Medicine*, vol. 44, supplement 1, pp. S71–S77, 2014.

[26] D. R. Moore, M. J. Robinson, J. L. Fry et al., "Ingested protein dose response of muscle and albumin protein synthesis after resistance exercise in young men," *The American Journal of Clinical Nutrition*, vol. 89, no. 1, pp. 161–168, 2009.

[27] L. Guimarães-Ferreira, "Role of the phosphocreatine system on energetic homeostasis in skeletal and cardiac muscles," *Einstein (São Paulo)*, vol. 12, no. 1, pp. 126–131, 2014.

[28] R. Cooper, F. Naclerio, J. Allgrove, and A. Jimenez, "Creatine supplementation with specific view to exercise/sports performance: an update," *Journal of the International Society of Sports Nutrition*, vol. 9, no. 1, article 33, 2012.

[29] R. C. Harris, K. Soderlund, and E. Hultman, "Elevation of creatine in resting and exercised muscle of normal subjects by creatine supplementation," *Clinical Science*, vol. 83, no. 3, pp. 367–374, 1992.

[30] E. Hultman, K. Söderlund, J. A. Timmons, G. Cederblad, and P. L. Greenhaff, "Muscle creatine loading in men," *Journal of Applied Physiology*, vol. 81, no. 1, pp. 232–237, 1996.

[31] R. Kreis, M. Kamber, M. Koster et al., "Creatine supplementation—part II: in vivo magnetic resonance spectroscopy," *Medicine and Science in Sports and Exercise*, vol. 31, no. 12, pp. 1770–1777, 1999.

[32] L. J. C. van Loon, A. M. Oosterlaar, F. Hartgens, M. K. C. Hesselink, R. J. Snow, and A. J. M. Wagenmakers, "Effects of creatine loading and prolonged creatine supplementation on body composition, fuel selection, sprint and endurance performance in humans," *Clinical Science (Lond)*, vol. 104, no. 2, pp. 153–162, 2003.

[33] J. S. Volek, N. D. Duncan, S. A. Mazzetti et al., "Performance and muscle fiber adaptations to creatine supplementation and heavy resistance training," *Medicine & Science in Sports & Exercise*, vol. 31, no. 8, pp. 1147–1156, 1999.

[34] P. J. Arciero, N. S. Hannibal III, B. C. Nindl, C. L. Gentile, J. Hamed, and M. D. Vukovich, "Comparison of creatine ingestion and resistance training on energy expenditure and limb blood flow," *Metabolism: Clinical and Experimental*, vol. 50, no. 12, pp. 1429–1434, 2001.

[35] J. L. M. Mesa, J. R. Ruiz, M. M. González-Gross, Á. Gutiérrez Sáinz, and M. J. Castillo Garzón, "Oral creatine supplementation and skeletal muscle metabolism in physical exercise," *Sports Medicine*, vol. 32, no. 14, pp. 903–944, 2002.

[36] R. L. Terjung, P. Clarkson, E. R. Eichner et al., "The American College of Sports Medicine Roundtable on the physiological and health effects of oral creatine supplementation," *Medicine and Science in Sports and Exercise*, vol. 32, no. 3, pp. 706–717, 2000.

[37] P. L. Greenhaff, K. Bodin, K. Soderlund, and E. Hultman, "Effect of oral creatine supplementation on skeletal muscle phosphocreatine resynthesis," *The American Journal of Physiology*, vol. 266, no. 5, pp. E725–E730, 1994.

[38] A. M. Jones, D. P. Wilkerson, and J. Fulford, "Influence of dietary creatine supplementation on muscle phosphocreatine kinetics during knee-extensor exercise in humans," *American Journal of Physiology: Regulatory Integrative and Comparative Physiology*, vol. 296, no. 4, pp. R1078–R1087, 2009.

[39] K. Vandenberghe, P. van Hecke, M. van Leemputte, F. Vanstapel, and P. Hespel, "Phosphocreatine resynthesis is not affected by creatine loading," *Medicine & Science in Sports & Exercise*, vol. 31, no. 2, pp. 236–242, 1999.

[40] T. W. Demant and E. C. Rhodes, "Effects of creatine supplementation on exercise performance," *Sports Medicine*, vol. 28, no. 1, pp. 49–60, 1999.

[41] T. N. Ziegenfuss, L. M. Lowery, and P. W. R. Lemon, "Acute fluid volume changes in men during three days of creatine supplementation," *Journal of Exercise Physiology*, vol. 1, no. 3, 1998.

[42] S. Y. Low, M. J. Rennie, and P. M. Taylor, "Modulation of glycogen synthesis in rat skeletal muscle by changes in cell volume," *Journal of Physiology*, vol. 495, part 2, pp. 299–303, 1996.

[43] B. Stoll, W. Gerok, F. Lang, and D. Haussinger, "Liver cell volume and protein synthesis," *Biochemical Journal*, vol. 287, no. 1, pp. 217–222, 1992.

[44] J. S. Volek and E. S. Rawson, "Scientific basis and practical aspects of creatine supplementation for athletes," *Nutrition*, vol. 20, no. 7-8, pp. 609–614, 2004.

[45] D. M. Fry and M. F. Morales, "A reexamination of the effects of creatine on muscle protein synthesis in tissue culture," *Journal of Cell Biology*, vol. 84, no. 2, pp. 294–297, 1980.

[46] J. S. Ingwall, C. D. Weiner, M. F. Morales, E. Davis, and F. E. Stockdale, "Specificity of creatine in the control of muscle protein synthesis," *Journal of Cell Biology*, vol. 62, no. 1, pp. 145–151, 1974.

[47] M. Louis, J. R. Poortmans, M. Francaux et al., "No effect of creatine supplementation on human myofibrillar and sarcoplasmic protein synthesis after resistance exercise," *The American Journal of Physiology—Endocrinology and Metabolism*, vol. 285, no. 5, pp. E1089–E1094, 2003.

[48] M. Louis, J. R. Poortmans, M. Francaux et al., "Creatine supplementation has no effect on human muscle protein turnover at rest in the postabsorptive or fed states," *The American Journal of Physiology—Endocrinology and Metabolism*, vol. 284, no. 4, pp. E764–E770, 2003.

[49] G. Pakise, S. Mihic, D. MacLennan, K. E. Yakasheski, and M. A. Tarnopolsky, "Effects of acute creatine monohydrate supplementation on leucine kinetics and mixed-muscle protein synthesis," *Journal of Applied Physiology*, vol. 91, no. 3, pp. 1041–1047, 2001.

[50] D. S. Willoughby and J. Rosene, "Effects of oral creatine and resistance training on myosin heavy chain expression," *Medicine and Science in Sports and Exercise*, vol. 33, no. 10, pp. 1674–1681, 2001.

[51] M. van Leemputte, K. Vandenberghe, and P. Hespel, "Shortening of muscle relaxation time after creatine loading," *Journal of Applied Physiology*, vol. 86, no. 3, pp. 840–844, 1999.

[52] J. M. Lawler, W. S. Barnes, G. Wu, W. Song, and S. Demaree, "Direct antioxidant properties of creatine," *Biochemical and Biophysical Research Communications*, vol. 290, no. 1, pp. 47–52, 2002.

[53] J. D. Branch, "Effect of creatine supplementation on body composition and performance: a meta-analysis," *International Journal of Sport Nutrition and Exercise Metabolism*, vol. 13, no. 2, pp. 198–226, 2003.

[54] V. Gouttebarge, H. Inklaar, and C. A. Hautier, "Short-term oral creatine supplementation in professional football players: a randomized placebo-controlled trial," *European Journal of Sports & Exercise Science*, vol. 1, no. 2, p. 7, 2012.

[55] J. Antonio and V. Ciccone, "The effects of pre versus post workout supplementation of creatine monohydrate on body composition and strength," *Journal of the International Society of Sports Nutrition*, vol. 10, article 36, 2013.

[56] T. P. Souza-Junior, J. M. Willardson, R. Bloomer et al., "Strength and hypertrophy responses to constant and decreasing rest intervals in trained men using creatine supplementation," *Journal of the International Society of Sports Nutrition*, vol. 8, no. 1, article 17, 2011.

[57] R. Deminice, F. T. Rosa, G. S. Franco, A. A. Jordao, and E. C. de Freitas, "Effects of creatine supplementation on oxidative stress and inflammatory markers after repeated-sprint exercise in humans," *Nutrition*, vol. 29, no. 9, pp. 1127–1132, 2013.

[58] K. F. T. Veggi, M. Machado, A. J. Koch, S. C. Santana, S. S. Oliveira, and M. J. Stec, "Oral creatine supplementation augments the repeated bout effect," *International Journal of Sport Nutrition and Exercise Metabolism*, vol. 23, no. 4, pp. 378–387, 2013.

[59] R. L. Dempsey, M. F. Mazzone, and L. N. Meurer, "Does oral creatine supplementation improve strength? A meta-analysis," *Journal of Family Practice*, vol. 51, no. 11, pp. 945–951, 2002.

[60] P. D. Balsom, K. Soderlund, and B. Ekblom, "Creatine in humans with special reference to creatine supplementation," *Sports Medicine*, vol. 18, no. 4, pp. 268–280, 1994.

[61] M. A. Tarnopolsky, "Caffeine and creatine use in sport," *Annals of Nutrition and Metabolism*, vol. 57, supplement 2, pp. 1–8, 2011.

[62] R. C. Harris, J. A. Lowe, K. Warnes, and C. E. Orme, "The concentration of creatine in meat, offal and commercial dog food," *Research in Veterinary Science*, vol. 62, no. 1, pp. 58–62, 1997.

[63] Y. Shimomura, Y. Yamamoto, G. Bajotto et al., "Nutraceutical effects of branched-chain amino acids on skeletal muscle," *The Journal of Nutrition*, vol. 136, no. 2, pp. 529S–532S, 2006.

[64] L. L. Tatpati, B. A. Irving, A. Tom et al., "The effect of branched chain amino acids on skeletal muscle mitochondrial function in young and elderly adults," *Journal of Clinical Endocrinology and Metabolism*, vol. 95, no. 2, pp. 894–902, 2010.

[65] P. She, Y. Zhou, Z. Zhang, K. Griffin, K. Gowda, and C. J. Lynch, "Disruption of BCAA metabolism in mice impairs exercise metabolism and endurance," *Journal of Applied Physiology*, vol. 108, no. 4, pp. 941–949, 2010.

[66] D. A. MacLean, T. E. Graham, and B. Saltin, "Branched-chain amino acids augment ammonia metabolism while attenuating protein breakdown during exercise," *The American Journal of Physiology*, vol. 267, no. 6, pp. E1010–E1022, 1994.

[67] E. Blomstrand, "A role for branched-chain amino acids in reducing central fatigue," *Journal of Nutrition*, vol. 136, no. 2, pp. 544S–547S, 2006.

[68] E. A. Newsholme and E. Blomstrand, "Branched-chain amino acids and central fatigue," *Journal of Nutrition*, vol. 136, no. 1, supplement, pp. 274S–276S, 2006.

[69] B. K. Greer, J. P. White, E. M. Arguello, and E. M. Haymes, "Branched-chain amino acid supplementation lowers perceived exertion but does not affect performance in untrained males," *Journal of Strength and Conditioning Research*, vol. 25, no. 2, pp. 539–544, 2011.

[70] A. B. Gualano, T. Bozza, P. de Lopes Campos et al., "Branched-chain amino acids supplementation enhances exercise capacity and lipid oxidation during endurance exercise after muscle glycogen depletion," *The Journal of Sports Medicine and Physical Fitness*, vol. 51, no. 1, pp. 82–88, 2011.

[71] H. Kainulainen, J. J. Hulmi, and U. M. Kujala, "Potential role of branched-chain amino acid catabolism in regulating fat oxidation," *Exercise and Sport Sciences Reviews*, vol. 41, no. 4, pp. 194–200, 2013.

[72] K. Matsumoto, T. Koba, K. Hamada, H. Tsujimoto, and R. Mitsuzono, "Branched-chain amino acid supplementation increases the lactate threshold during an incremental exercise test in trained individuals," *Journal of Nutritional Science and Vitaminology*, vol. 55, no. 1, pp. 52–58, 2009.

[73] E. Blomstrand, J. Eliasson, H. K. R. Karlssonr, and R. Köhnke, "Branched-chain amino acids activate key enzymes in protein synthesis after physical exercise," *Journal of Nutrition*, vol. 136, no. 1, supplement, pp. 269S–273S, 2006.

[74] S. R. Kimball and L. S. Jefferson, "Signaling pathways and molecular mechanisms through which branched-chain amino acids mediate translational control of protein synthesis," *Journal of Nutrition*, vol. 136, no. 1, supplement, pp. 227S–231S, 2006.

[75] L. E. Norton and D. K. Layman, "Leucine regulates translation initiation of protein synthesis in skeletal muscle after exercise," *Journal of Nutrition*, vol. 136, no. 2, pp. 533S–537S, 2006.

[76] S. M. Pasiakos and J. P. Mcclung, "Supplemental dietary leucine and the skeletal muscle anabolic response to essential amino acids," *Nutrition Reviews*, vol. 69, no. 9, pp. 550–557, 2011.

[77] E. Blomstrand, P. Hassmen, B. Ekblom, and E. A. Newsholme, "Administration of branched-chain amino acids during sustained exercise—effects on performance and on plasma concentration of some amino acids," *European Journal of Applied Physiology and Occupational Physiology*, vol. 63, no. 2, pp. 83–88, 1991.

[78] M. J. Crowe, J. N. Weatherson, and B. F. Bowden, "Effects of dietary leucine supplementation on exercise performance," *European Journal of Applied Physiology*, vol. 97, no. 6, pp. 664–672, 2006.

[79] K. D. Mittleman, M. R. Ricci, and S. P. Bailey, "Branched-chain amino acids prolong exercise during heat stress in men and women," *Medicine and Science in Sports and Exercise*, vol. 30, no. 1, pp. 83–91, 1998.

[80] J. M. Davis, R. S. Welsh, K. L. de Volve, and N. A. Alderson, "Effects of branched-chain amino acids and carbohydrate on fatigue during intermittent, high-intensity running," *International Journal of Sports Medicine*, vol. 20, no. 5, pp. 309–314, 1999.

[81] K. Madsen, D. A. Maclean, B. Kiens, and D. Christensen, "Effects of glucose, glucose plus branched-chain amino acids, or placebo on bike performance over 100 km," *Journal of Applied Physiology*, vol. 81, no. 6, pp. 2644–2650, 1996.

[82] H. T. Pitkänen, S. S. Oja, H. Rusko et al., "Leucine supplementation does not enhance acute strength or running performance

but affects serum amino acid concentration," *Amino Acids*, vol. 25, no. 1, pp. 85–94, 2003.

[83] H. K. Strüder, W. Hollmann, P. Platen, M. Donike, A. Gotzmann, and K. Weber, "Influence of paroxetine, branched-chain amino acids and tyrosine on neuroendocrine system responses and fatigue in humans," *Hormone and Metabolic Research*, vol. 30, no. 4, pp. 188–194, 1998.

[84] G. Van Hall, J. S. H. Raaymakers, W. H. M. Saris, and A. J. M. Wagenmakers, "Ingestion of branched-chain amino acids and tryptophan during sustained exercise in man: failure to affect performance," *Journal of Physiology*, vol. 486, no. 3, pp. 789–794, 1995.

[85] P. Watson, S. M. Shirreffs, and R. J. Maughan, "The effect of acute branched-chain amino acid supplementation on prolonged exercise capacity in a warm environment," *European Journal of Applied Physiology*, vol. 93, no. 3, pp. 306–314, 2004.

[86] M. Spillane, C. Emerson, and D. S. Willoughby, "The effects of 8 weeks of heavy resistance training and branched-chain amino acid supplementation on body composition and muscle performance," *Nutrition and Health*, vol. 21, no. 4, pp. 263–273, 2012.

[87] T. Ispoglou, R. F. G. J. King, R. C. J. Polman, and C. Zanker, "Daily L-leucine supplementation in novice trainees during a 12-week weight training program," *International Journal of Sports Physiology and Performance*, vol. 6, no. 1, pp. 38–50, 2011.

[88] S. M. Pasiakos, H. L. McClung, J. P. McClung et al., "Leucine-enriched essential amino acid supplementation during moderate steady state exercise enhances postexercise muscle protein synthesis," *The American Journal of Clinical Nutrition*, vol. 94, no. 3, pp. 809–818, 2011.

[89] E. Børsheim, K. D. Tipton, S. E. Wolf, and R. R. Wolfe, "Essential amino acids and muscle protein recovery from resistance exercise," *The American Journal of Physiology—Endocrinology and Metabolism*, vol. 283, no. 4, pp. E648–E657, 2002.

[90] D. J. Cuthbertson, J. Babraj, K. Smith et al., "Anabolic signaling and protein synthesis in human skeletal muscle after dynamic shortening or lengthening exercise," *The American Journal of Physiology—Endocrinology and Metabolism*, vol. 290, no. 4, pp. E731–E738, 2006.

[91] E. Volpi, H. Kobayashi, M. Sheffield-Moore, B. Mittendorfer, and R. R. Wolfe, "Essential amino acids are primarily responsible for the amino acid stimulation of muscle protein anabolism in healthy elderly adults," *The American Journal of Clinical Nutrition*, vol. 78, no. 2, pp. 250–258, 2003.

[92] N. R. Rodriguez, N. M. DiMarco, and S. Langley, "Position of the American Dietetic Association, Dietitians of Canada, and the American College of Sports Medicine: nutrition and athletic performance," *Journal of the American Dietetic Association*, vol. 109, no. 3, pp. 509–527, 2009.

[93] E. L. Glynn, C. S. Fry, M. J. Drummond et al., "Excess leucine intake enhances muscle anabolic signaling but not net protein anabolism in young men and women," *Journal of Nutrition*, vol. 140, no. 11, pp. 1970–1976, 2010.

[94] U. S. Department of Agriculture, *USDA National Nutrient Database for Standard Reference*, Release 26, U. S. Department of Agriculture, 2013.

[95] L. M. Burke, J. A. Winter, D. Cameron-Smith, M. Enslen, M. Farnfield, and J. Decombaz, "Effect of intake of different dietary protein sources on plasma amino acid profiles at rest and after exercise," *International Journal of Sport Nutrition and Exercise Metabolism*, vol. 22, no. 6, pp. 452–462, 2012.

[96] T. B. Conley, J. W. Apolzan, H. J. Leidy, K. A. Greaves, E. Lim, and W. W. Campbell, "Effect of food form on postprandial plasma amino acid concentrations in older adults," *British Journal of Nutrition*, vol. 106, no. 2, pp. 203–207, 2011.

[97] R. Koopman, N. Crombach, A. P. Gijsen et al., "Ingestion of a protein hydrolysate is accompanied by an accelerated in vivo digestion and absorption rate when compared with its intact protein," *The American Journal of Clinical Nutrition*, vol. 90, no. 1, pp. 106–115, 2009.

[98] J. E. Tang, D. R. Moore, G. W. Kujbida, M. A. Tarnopolsky, and S. M. Phillips, "Ingestion of whey hydrolysate, casein, or soy protein isolate: effects on mixed muscle protein synthesis at rest and following resistance exercise in young men," *Journal of Applied Physiology*, vol. 107, no. 3, pp. 987–992, 2009.

[99] D. W. D. West, N. A. Burd, V. G. Coffey et al., "Rapid aminoacidemia enhances myofibrillar protein synthesis and anabolic intramuscular signaling responses after resistance exercise," *The American Journal of Clinical Nutrition*, vol. 94, no. 3, pp. 795–803, 2011.

[100] S. B. Wilkinson, M. A. Tarnopolsky, M. J. MacDonald, J. R. Mac-Donald, D. Armstrong, and S. M. Phillips, "Consumption of fluid skim milk promotes greater muscle protein accretion after resistance exercise than does consumption of an isonitrogenous and isoenergetic soy-protein beverage," *American Journal of Clinical Nutrition*, vol. 85, no. 4, pp. 1031–1040, 2007.

[101] R. M. Hobson, B. Saunders, G. Ball, R. C. Harris, and C. Sale, "Effects of beta-alanine supplementation on exercise performance: a meta-analysis," *Amino Acids*, vol. 43, no. 1, pp. 25–37, 2012.

[102] V. de Salles Painelli, H. Roschel, F. de Jesus et al., "The ergogenic effect of beta-alanine combined with sodium bicarbonate on high-intensity swimming performance," *Applied Physiology, Nutrition and Metabolism*, vol. 38, no. 5, pp. 525–532, 2013.

[103] C. Sale, B. Saunders, S. Hudson, J. A. Wise, R. C. Harris, and C. D. Sunderland, "Effect of β-alanine plus sodium bicarbonate on high-intensity cycling capacity," *Medicine and Science in Sports and Exercise*, vol. 43, no. 10, pp. 1972–1978, 2011.

[104] K. J. Ducker, B. Dawson, and K. E. Wallman, "Effect of beta-alanine supplementation on 800-m running performance," *International Journal of Sport Nutrition and Exercise Metabolism*, vol. 23, no. 6, pp. 554–561, 2013.

[105] R. van Thienen, K. van Proeyen, B. V. Eynde, J. Puype, T. Lefere, and P. Hespel, "β-alanine improves sprint performance in endurance cycling," *Medicine & Science in Sports & Exercise*, vol. 41, no. 4, pp. 898–903, 2009.

[106] B. Saunders, C. Sale, R. C. Harris, and C. Sunderland, "Effect of beta-alanine supplementation on repeated sprint performance during the Loughborough Intermittent Shuttle Test," *Amino Acids*, vol. 43, no. 1, pp. 39–47, 2012.

[107] W. Derave, M. S. Özdemir, R. C. Harris et al., "beta-Alanine supplementation augments muscle carnosine content and attenuates fatigue during repeated isokinetic contraction bouts in trained sprinters," *Journal of Applied Physiology*, vol. 103, no. 5, pp. 1736–1743, 2007.

[108] S. T. Howe, P. M. Bellinger, M. W. Driller, C. M. Shing, and J. W. Fell, "The effect of beta-alanine supplementation on isokinetic force and cycling performance in highly trained cyclists," *International Journal of Sport Nutrition and Exercise Metabolism*, vol. 23, no. 6, pp. 562–570, 2013.

[109] K. M. Sweeney, G. A. Wright, A. Glenn Brice, and S. T. Doberstein, "The effect of beta-alanine supplementation on power performance during repeated sprint activity," *Journal of*

Strength and Conditioning Research, vol. 24, no. 1, pp. 79–87, 2010.

[110] W. Chung, A. Baguet, T. Bex, D. J. Bishop, and W. Derave, "Doubling of muscle carnosine concentration does not improve laboratory 1-hr cycling time-trial performance," *International Journal of Sport Nutrition and Exercise Metabolism*, vol. 24, no. 3, pp. 315–324, 2014.

[111] P. M. Bellinger, "β-Alanine supplementation for athletic performance: an update," *The Journal of Strength & Conditioning Research*, vol. 28, no. 6, pp. 1751–1770, 2014.

[112] V. de Salles Painelli, B. Saunders, C. Sale et al., "Influence of training status on high-intensity intermittent performance in response to β-Alanine supplementation," *Amino Acids*, vol. 46, no. 5, pp. 1207–1215, 2014.

[113] A. Baguet, H. Reyngoudt, A. Pottier et al., "Carnosine loading and washout in human skeletal muscles," *Journal of Applied Physiology*, vol. 106, no. 3, pp. 837–842, 2009.

[114] R. C. Harris, M. J. Tallon, M. Dunnett et al., "The absorption of orally supplied beta-alanine and its effect on muscle carnosine synthesis in human vastus lateralis," *Amino Acids*, vol. 30, no. 3, pp. 279–289, 2006.

[115] L. M. Burke, "Caffeine and sports performance," *Applied Physiology, Nutrition, and Metabolism*, vol. 33, no. 6, pp. 1319–1334, 2008.

[116] J. K. Davis and J. M. Green, "Caffeine and anaerobic performance: ergogenic value and mechanisms of action," *Sports Medicine*, vol. 39, no. 10, pp. 813–832, 2009.

[117] E. R. Goldstein, T. Ziegenfuss, D. Kalman et al., "International society of sports nutrition position stand: caffeine and performance," *Journal of the International Society of Sports Nutrition*, vol. 7, article 5, 2010.

[118] T. E. Graham, "Caffeine and exercise metabolism, endurance and performance," *Sports Medicine*, vol. 31, no. 11, pp. 785–807, 2001.

[119] K. J. Acheson, G. Gremaud, I. Meirim et al., "Metabolic effects of caffeine in humans: lipid oxidation or futile cycling?" *The American Journal of Clinical Nutrition*, vol. 79, no. 1, pp. 40–46, 2004.

[120] T. E. Graham, "Caffeine, coffee and ephedrine: Impact on exercise performance and metabolism," *Canadian Journal of Applied Physiology*, vol. 26, no. 6, supplement, pp. S103–S119, 2001.

[121] M. A. Tarnopolsky, "Effect of caffeine on the neuromuscular system—potential as an ergogenic aid," *Applied Physiology, Nutrition and Metabolism*, vol. 33, no. 6, pp. 1284–1289, 2008.

[122] D. Laurent, K. E. Schneider, W. K. Prusaczyk et al., "Effects of caffeine on muscle glycogen utilization and the neuroendocrine axis during exercise," *Journal of Clinical Endocrinology and Metabolism*, vol. 85, no. 6, pp. 2170–2175, 2000.

[123] M. Doherty and P. M. Smith, "Effects of caffeine ingestion on exercise testing: a meta-analysis," *International Journal of Sport Nutrition and Exercise Metabolism*, vol. 14, no. 6, pp. 626–646, 2004.

[124] M. S. Ganio, J. F. Klau, D. J. Casa, L. E. Armstrong, and C. M. Maresh, "Effect of caffeine on sport-specific endurance performance: a systematic review," *Journal of Strength and Conditioning Research*, vol. 23, no. 1, pp. 315–324, 2009.

[125] T. A. Astorino and D. W. Roberson, "Efficacy of acute caffeine ingestion for short-term high-intensity exercise performance: a systematic review," *Journal of Strength and Conditioning Research*, vol. 24, no. 1, pp. 257–265, 2010.

[126] G. L. Warren, N. D. Park, R. D. Maresca, K. I. McKibans, and M. L. Millard-Stafford, "Effect of caffeine ingestion on muscular strength and endurance: a meta-analysis," *Medicine and Science in Sports and Exercise*, vol. 42, no. 7, pp. 1375–1387, 2010.

[127] A. Pérez-López, J. J. Salinero, J. Abian-Vicen et al., "Caffeinated energy drinks improve volleyball performance in elite female players," *Medicine & Science in Sports & Exercise*, vol. 47, no. 4, pp. 850–856, 2015.

[128] T. Nicholson, G. Middleton, and T. I. Gee, "Does caffeine have an ergogenic effect on sports-specific agility in competitive male racquet sport players?" in *Proceedings of the UKSCA 9th Annual Conference*, Nottingham, UK, 2013.

[129] J. Abian-Vicen, C. Puente, J. J. Salinero et al., "A caffeinated energy drink improves jump performance in adolescent basketball players," *Amino Acids*, vol. 46, no. 5, pp. 1333–1341, 2014.

[130] B. Lara, C. Gonzalez-Millán, J. J. Salinero et al., "Caffeine-containing energy drink improves physical performance in female soccer players," *Amino Acids*, vol. 46, no. 5, pp. 1385–1392, 2014.

[131] J. del Coso, J. Portillo, G. Muñoz, J. Abián-Vicén, C. Gonzalez-Millán, and J. Muñoz-Guerra, "Caffeine-containing energy drink improves sprint performance during an international rugby sevens competition," *Amino Acids*, vol. 44, no. 6, pp. 1511–1519, 2013.

[132] J. del Coso, J. A. Ramírez, G. Muñoz et al., "Caffeine-containing energy drink improves physical performance of elite rugby players during a simulated match," *Applied Physiology, Nutrition and Metabolism*, vol. 38, no. 4, pp. 368–374, 2013.

[133] S. C. Lane, J. L. Areta, S. R. Bird et al., "Caffeine ingestion and cycling power output in a low or normal muscle glycogen state," *Medicine and Science in Sports and Exercise*, vol. 45, no. 8, pp. 1577–1584, 2013.

[134] M. J. Duncan, C. D. Thake, and P. J. Downs, "Effect of caffeine ingestion on torque and muscle activity during resistance exercise in men," *Muscle & Nerve*, vol. 50, no. 4, pp. 523–527, 2014.

[135] M. D. Silva-Cavalcante, C. R. Correia-Oliveira, R. A. Santos et al., "Caffeine increases anaerobic work and restores cycling performance following a protocol designed to lower endogenous carbohydrate availability," *PLoS ONE*, vol. 8, no. 8, Article ID e72025, 2013.

[136] T. E. Graham, E. Hibbert, and P. Sathasivam, "Metabolic and exercise endurance effects of coffee and caffeine ingestion," *Journal of Applied Physiology*, vol. 85, no. 3, pp. 883–889, 1998.

[137] A. B. Hodgson, R. K. Randell, and A. E. Jeukendrup, "The metabolic and performance effects of caffeine compared to coffee during endurance exercise," *PLoS ONE*, vol. 8, no. 4, Article ID e59561, 2013.

[138] B. Sökmen, L. E. Armstrong, W. J. Kraemer et al., "Caffeine use in sports: considerations for the athlete," *Journal of Strength and Conditioning Research*, vol. 22, no. 3, pp. 978–986, 2008.

[139] K. Vandenberghe, N. Gillis, M. van Leemputte, P. van Hecke, F. Vanstapel, and P. Hespel, "Caffeine counteracts the ergogenic action of muscle creatine loading," *Journal of Applied Physiology*, vol. 80, no. 2, pp. 452–457, 1996.

[140] C. L. Camic, T. J. Housh, J. M. Zuniga et al., "The effects of polyethylene glycosylated creatine supplementation on anaerobic performance measures and body composition," *The Journal of Strength & Conditioning Research*, vol. 28, no. 3, pp. 825–833, 2014.

[141] J. M. Zuniga, T. J. Housh, C. L. Camic et al., "The effects of creatine monohydrate loading on anaerobic performance and

one-repetition maximum strength," *Journal of Strength and Conditioning Research*, vol. 26, no. 6, pp. 1651–1656, 2012.

[142] J. M. Oliver, D. P. Joubert, S. E. Martin, and S. F. Crouse, "Oral creatine supplementation's decrease of blood lactate during exhaustive, incremental cycling," *International Journal of Sport Nutrition and Exercise Metabolism*, vol. 23, no. 3, pp. 252–258, 2013.

[143] L. R. McNaughton, J. Siegler, and A. Midgley, "Ergogenic effects of sodium bicarbonate," *Current Sports Medicine Reports*, vol. 7, no. 4, pp. 230–236, 2008.

[144] B. Requena, M. Zabala, P. Padial, and B. Feriche, "Sodium bicarbonate and sodium citrate: ergogenic aids?" *Journal of Strength and Conditioning Research*, vol. 19, no. 1, pp. 213–224, 2005.

[145] S. P. Cairns, "Lactic acid and exercise performance: culprit or friend?" *Sports Medicine*, vol. 36, no. 4, pp. 279–291, 2006.

[146] R. A. Robergs, F. Ghiasvand, and D. Parker, "Biochemistry of exercise-induced metabolic acidosis," *American Journal of Physiology: Regulatory Integrative and Comparative Physiology*, vol. 287, no. 3, pp. R502–R516, 2004.

[147] D. G. Allen, G. D. Lamb, and H. Westerblad, "Skeletal muscle fatigue: cellular mechanisms," *Physiological Reviews*, vol. 88, no. 1, pp. 287–332, 2008.

[148] A. J. Carr, W. G. Hopkins, and C. J. Gore, "Effects of acute alkalosis and acidosis on performance: a meta-analysis," *Sports Medicine*, vol. 41, no. 10, pp. 801–814, 2011.

[149] L. G. Matson and Z. V. Tran, "Effects of sodium bicarbonate ingestion on anaerobic performance: a meta-analytic review," *International Journal of Sport Nutrition*, vol. 3, no. 1, pp. 2–28, 1993.

[150] D. Bishop, J. Edge, C. Davis, and C. Goodman, "Induced metabolic alkalosis affects muscle metabolism and repeated-sprint ability," *Medicine and Science in Sports and Exercise*, vol. 36, no. 5, pp. 807–813, 2004.

[151] M. G. Hollidge-Horvat, M. L. Parolin, D. Wong, N. L. Jones, and G. J. F. Heigenhauser, "Effect of induced metabolic alkalosis on human skeletal muscle metabolism during exercise," *American Journal of Physiology: Endocrinology and Metabolism*, vol. 278, no. 2, pp. E316–E329, 2000.

[152] S. A. Jubrias, G. J. Crowther, E. G. Shankland, R. K. Gronka, and K. E. Conley, "Acidosis inhibits oxidative phosphorylation in contracting human skeletal muscle in vivo," *Journal of Physiology*, vol. 553, no. 2, pp. 589–599, 2003.

[153] A. M. Hunter, G. de Vito, C. Bolger, H. Mullany, and S. D. R. Galloway, "The effect of induced alkalosis and submaximal cycling on neuromuscular response during sustained isometric contraction," *Journal of Sports Sciences*, vol. 27, no. 12, pp. 1261–1269, 2009.

[154] S. M. Mueller, S. M. Gehrig, S. Frese, C. A. Wagner, U. Boutellier, and M. Toigo, "Multiday acute sodium bicarbonate intake improves endurance capacity and reduces acidosis in men," *Journal of the International Society of Sports Nutrition*, vol. 10, no. 1, article 16, 2013.

[155] K. J. Ducker, B. Dawson, and K. E. Wallman, "Effect of beta alanine and sodium bicarbonate supplementation on repeated-sprint performance," *Journal of Strength and Conditioning Research*, vol. 27, no. 12, pp. 3450–3460, 2013.

[156] A. A. Mero, P. Hirvonen, J. Saarela, J. J. Hulmi, J. R. Hoffman, and J. R. Stout, "Effect of sodium bicarbonate and beta-alanine supplementation on maximal sprint swimming," *Journal of the International Society of Sports Nutrition*, vol. 10, no. 1, article 52, 2013.

[157] M. W. Driller, J. R. Gregory, A. D. Williams, and J. W. Fell, "The effects of chronic sodium bicarbonate ingestion and interval training in highly trained rowers," *International Journal of Sport Nutrition and Exercise Metabolism*, vol. 23, no. 1, pp. 40–47, 2013.

[158] J. Edge, D. Bishop, and C. Goodman, "Effects of chronic NaHCO$_3$ ingestion during interval training on changes to muscle buffer capacity, metabolism, and short-term endurance performance," *Journal of Applied Physiology*, vol. 101, no. 3, pp. 918–925, 2006.

[159] D. C. McKenzie, K. D. Coutts, D. R. Stirling, H. H. Hoeben, and G. Kuzara, "Maximal work production following two levels of artificially induced metabolic alkalosis," *Journal of sports sciences*, vol. 4, no. 1, pp. 35–38, 1986.

[160] J. C. Siegler, P. W. M. Marshall, J. Bray, and C. Towlson, "Sodium bicarbonate supplementation and ingestion timing: does it matter?" *The Journal of Strength & Conditioning Research*, vol. 26, no. 7, pp. 1953–1958, 2012.

[161] A. J. Carr, G. J. Slater, C. J. Gore, B. Dawson, and L. M. Burke, "Effect of sodium bicarbonate on [HCO3-], pH, and gastrointestinal symptoms," *International Journal of Sport Nutrition and Exercise Metabolism*, vol. 21, no. 3, pp. 189–194, 2011.

[162] L. M. Burke and D. B. Pyne, "Bicarbonate loading to enhance training and competitive performance," *International Journal of Sports Physiology and Performance*, vol. 2, no. 1, pp. 93–97, 2007.

[163] T. Buclin, M. Cosma, M. Appenzeller et al., "Diet acids and alkalis influence calcium retention in bone," *Osteoporosis International*, vol. 12, no. 6, pp. 493–499, 2001.

[164] I. Kurtz, T. Maher, H. N. Hulter, M. Schambelan, and A. Sebastian, "Effect of diet on plasma acid-base composition in normal humans," *Kidney International*, vol. 24, no. 5, pp. 670–680, 1983.

[165] C. Demigné, H. Sabboh, C. Puel, C. Rémésy, and V. Coxam, "Organic anions and potassium salts in nutrition and metabolism," *Nutrition Research Reviews*, vol. 17, no. 2, pp. 249–258, 2004.

[166] L. A. Frassetto, K. M. Todd, R. C. Morris Jr., and A. Sebastian, "Estimation of net endogenous noncarbonic acid production in humans from diet potassium and protein contents," *The American Journal of Clinical Nutrition*, vol. 68, no. 3, pp. 576–583, 1998.

[167] M. L. Halperin, "Metabolism and acid-base physiology," *Artificial Organs*, vol. 6, no. 4, pp. 357–362, 1982.

[168] T. Remer and F. Manz, "Potential renal acid load of foods and its influence on urine pH," *Journal of the American Dietetic Association*, vol. 95, no. 7, pp. 791–797, 1995.

[169] H. Boeing, A. Bechthold, A. Bub et al., "Critical review: vegetables and fruit in the prevention of chronic diseases," *European Journal of Nutrition*, vol. 51, no. 6, pp. 637–663, 2012.

[170] F. J. He and G. A. MacGregor, "Beneficial effects of potassium on human health," *Physiologia Plantarum*, vol. 133, no. 4, pp. 725–735, 2008.

[171] E.-M. Hietavala, J. R. Stout, J. J. Hulmi et al., "Effect of diet composition on acid-base balance in adolescents, young adults and elderly at rest and during exercise," *European Journal of Clinical Nutrition*, vol. 69, no. 3, pp. 399–404, 2014.

[172] M. van der Aa, "Classification of mineral water types and comparison with drinking water standards," *Environmental Geology*, vol. 44, no. 5, pp. 554–563, 2003.

[173] G. Paulsen, U. R. Mikkelsen, T. Raastad, and J. M. Peake, "Leucocytes, cytokines and satellite cells: what role do they play in muscle damage and regeneration following eccentric exercise?" *Exercise Immunology Review*, vol. 18, pp. 42–97, 2012.

[174] J. L. Ziltener, S. Leal, and P. E. Fournier, "Non-steroidal anti-inflammatory drugs for athletes: an update," *Annals of Physical and Rehabilitation Medicine*, vol. 53, no. 4, pp. 278–288, 2010.

[175] F. Bieuzen, C. M. Bleakley, and J. T. Costello, "Contrast water therapy and exercise induced muscle damage: a systematic review and meta-analysis," *PLoS ONE*, vol. 8, no. 4, Article ID e62356, 2013.

[176] G. Howatson, M. Hoad, S. Goodall, J. Tallent, P. G. Bell, and D. N. French, "Exercise-induced muscle damage is reduced in resistance-trained males by branched chain amino acids: a randomized, double-blind, placebo controlled study," *Journal of the International Society of Sports Nutrition*, vol. 9, article 20, 2012.

[177] C. D. Black, M. P. Herring, D. J. Hurley, and P. J. O'Connor, "Ginger (*Zingiber officinale*) reduces muscle pain caused by eccentric exercise," *The Journal of Pain*, vol. 11, no. 9, pp. 894–903, 2010.

[178] S. Chuengsamarn, S. Rattanamongkolgul, B. Phonrat, R. Tungtrongchitr, and S. Jirawatnotai, "Reduction of atherogenic risk in patients with type 2 diabetes by curcuminoid extract: a randomized controlled trial," *Journal of Nutritional Biochemistry*, vol. 25, no. 2, pp. 144–150, 2014.

[179] E. Ernst, T. Saradeth, and G. Achhammer, "n-3 fatty acids and acute-phase proteins," *European Journal of Clinical Investigation*, vol. 21, no. 1, pp. 77–82, 1991.

[180] D. A. J. Connolly, M. P. McHugh, and O. I. Padilla-Zakour, "Efficacy of a tart cherry juice blend in preventing the symptoms of muscle damage," *British Journal of Sports Medicine*, vol. 40, no. 8, pp. 679–683, 2006.

[181] P. G. Bell, I. H. Walshe, G. W. Davison, E. Stevenson, and G. Howatson, "Montmorency cherries reduce the oxidative stress and inflammatory responses to repeated days high-intensity stochastic cycling," *Nutrients*, vol. 6, no. 2, pp. 829–843, 2014.

[182] S. Bent, "Herbal medicine in the United States: review of efficacy, safety, and regulation—grand Rounds at University of California, San Francisco Medical Center," *Journal of General Internal Medicine*, vol. 23, no. 6, pp. 854–859, 2008.

[183] J. A. O. Ojewole, "Analgesic, antiinflammatory and hypoglycaemic effects of ethanol extract of *Zingiber officinale* (Roscoe) rhizomes (Zingiberaceae) in mice and rats," *Phytotherapy Research*, vol. 20, no. 9, pp. 764–772, 2006.

[184] H.-Y. Young, Y.-L. Luo, H.-Y. Cheng, W.-C. Hsieh, J.-C. Liao, and W.-H. Peng, "Analgesic and anti-inflammatory activities of [6]-gingerol," *Journal of Ethnopharmacology*, vol. 96, no. 1-2, pp. 207–210, 2005.

[185] W. G. Cho and J. G. Valtschanoff, "Vanilloid receptor TRPV1-positive sensory afferents in the mouse ankle and knee joints," *Brain Research*, vol. 1219, pp. 59–65, 2008.

[186] D. A. Connolly, S. P. Sayers, and M. P. McHugh, "Treatment and prevention of delayed onset muscle soreness," *The Journal of Strength & Conditioning Research*, vol. 17, no. 1, pp. 197–208, 2003.

[187] C. D. Black and P. J. O'Connor, "Acute effects of dietary ginger on quadriceps muscle pain during moderate-intensity cycling exercise," *International Journal of Sport Nutrition and Exercise Metabolism*, vol. 18, no. 6, pp. 653–664, 2008.

[188] B. B. Aggarwal, "Targeting lammation-induced obesity and metabolic diseases by curcumin and other nutraceuticals," *Annual Review of Nutrition*, vol. 30, pp. 173–199, 2010.

[189] S. C. Gupta, S. Patchva, W. Koh, and B. B. Aggarwal, "Discovery of curcumin, a component of golden spice, and its miraculous biological activities," *Clinical and Experimental Pharmacology and Physiology*, vol. 39, no. 3, pp. 283–299, 2012.

[190] F. Drobnic, J. Riera, G. Appendino et al., "Reduction of delayed onset muscle soreness by a novel curcumin delivery system (Meriva(R)): a randomised, placebo-controlled trial," *Journal of the International Society of Sports Nutrition*, vol. 11, no. 1, article 31, 2014.

[191] M. Takahashi, K. Suzuki, H. K. Kim et al., "Effects of curcumin supplementation on exercise-induced oxidative stress in humans," *International Journal of Sports Medicine*, vol. 35, no. 6, pp. 469–475, 2014.

[192] H. Sasaki, Y. Sunagawa, K. Takahashi et al., "Innovative preparation of curcumin for improved oral bioavailability," *Biological and Pharmaceutical Bulletin*, vol. 34, no. 5, pp. 660–665, 2011.

[193] M. B. Covington, "Omega-3 fatty acids," *American Family Physician*, vol. 70, no. 1, pp. 133–140, 2004.

[194] R. J. Bloomer, D. E. Larson, K. H. Fisher-Wellman, A. J. Galpin, and B. K. Schilling, "Effect of eicosapentaenoic and docosahexaenoic acid on resting and exercise-induced inflammatory and oxidative stress biomarkers: a randomized, placebo controlled, cross-over study," *Lipids in Health and Disease*, vol. 8, article 36, 2009.

[195] B. Tartibian, B. H. Maleki, and A. Abbasi, "The effects of ingestion of omega-3 fatty acids on perceived pain and external symptoms of delayed onset muscle soreness in untrained men," *Clinical Journal of Sport Medicine*, vol. 19, no. 2, pp. 115–119, 2009.

[196] K. B. Jouris, J. L. McDaniel, and E. P. Weiss, "The effect of omega-3 fatty acid supplementation on the inflammatory response to eccentric strength exercise," *Journal of Sports Science and Medicine*, vol. 10, no. 3, pp. 432–438, 2011.

[197] J. Lenn, T. Uhl, C. Mattacola et al., "The effects of fish oil and isoflavones on delayed onset muscle soreness," *Medicine & Science in Sports & Exercise*, vol. 34, no. 10, pp. 1605–1613, 2002.

[198] A. D. Toft, M. Thorn, K. Ostrowski et al., "N-3 polyunsaturated fatty acids do not affect cytokine response to strenuous exercise," *Journal of Applied Physiology*, vol. 89, no. 6, pp. 2401–2406, 2000.

[199] G. I. Smith, P. Atherton, D. N. Reeds et al., "Dietary omega-3 fatty acid supplementation increases the rate of muscle protein synthesis in older adults: a randomized controlled trial," *American Journal of Clinical Nutrition*, vol. 93, no. 2, pp. 402–412, 2011.

[200] A. P. Simopoulos, "Omega-3 fatty acids and athletics," *Current Sports Medicine Reports*, vol. 6, no. 4, pp. 230–236, 2007.

[201] U. S. Food and Drug Administration Center for Food Safety and Applied Nutrition, *Agency Response Letter. GRAS Notice No. GRN 000105*, 2002.

[202] P. G. Bell, M. P. Mchugh, E. Stevenson, and G. Howatson, "The role of cherries in exercise and health," *Scandinavian Journal of Medicine and Science in Sports*, vol. 24, no. 3, pp. 477–490, 2014.

[203] L. M. McCune, C. Kubota, N. R. Stendell-Hollis, and C. A. Thomson, "Cherries and health: a review," *Critical Reviews in Food Science and Nutrition*, vol. 51, no. 1, pp. 1–12, 2011.

[204] N. P. Seeram, R. A. Momin, M. G. Nair, and L. D. Bourquin, "Cyclooxygenase inhibitory and antioxidant cyanidin glycosides in cherries and berries," *Phytomedicine*, vol. 8, no. 5, pp. 362–369, 2001.

[205] J. L. Bowtell, D. P. Sumners, A. Dyer, P. Fox, and K. N. Mileva, "Montmorency cherry juice reduces muscle damage caused by intensive strength exercise," *Medicine and Science in Sports and Exercise*, vol. 43, no. 8, pp. 1544–1551, 2011.

[206] G. Howatson, M. P. McHugh, J. A. Hill et al., "Influence of tart cherry juice on indices of recovery following marathon running," *Scandinavian Journal of Medicine and Science in Sports*, vol. 20, no. 6, pp. 843–852, 2010.

[207] K. S. Kuehl, E. T. Perrier, D. L. Elliot, and J. C. Chesnutt, "Efficacy of tart cherry juice in reducing muscle pain during running: a randomized controlled trial," *Journal of the International Society of Sports Nutrition*, vol. 7, article 17, 2010.

[208] D. S. Kelley, Y. Adkins, A. Reddy, L. R. Woodhouse, B. E. Mackey, and K. L. Erickson, "Sweet bing cherries lower circulating concentrations of markers for chronic inflammatory diseases in healthy humans," *The Journal of Nutrition*, vol. 143, no. 3, pp. 340–344, 2013.

[209] D. S. Kelley, R. Rasooly, R. A. Jacob, A. A. Kader, and B. E. Mackey, "Consumption of bing sweet cherries lowers circulating concentrations of inflammation markers in healthy men and women," *Journal of Nutrition*, vol. 136, no. 4, pp. 981–986, 2006.

[210] S. K. Powers, J. Duarte, A. N. Kavazis, and E. E. Talbert, "Reactive oxygen species are signalling molecules for skeletal muscle adaptation," *Experimental Physiology*, vol. 95, no. 1, pp. 1–9, 2010.

[211] K. Fisher-Wellman and R. J. Bloomer, "Acute exercise and oxidative stress: a 30 year history," *Dynamic Medicine*, vol. 8, article 1, 2009.

[212] B. J. Schoenfeld, "The use of nonsteroidal anti-inflammatory drugs for exercise-induced muscle damage: implications for skeletal muscle development," *Sports Medicine*, vol. 42, no. 12, pp. 1017–1028, 2012.

[213] S. J. Warden, "Prophylactic use of NSAIDs by athletes: a risk/benefit assessment," *Physician and Sportsmedicine*, vol. 38, no. 1, pp. 132–138, 2010.

[214] M. J. Ormsbee, J. Lox, and P. J. Arciero, "Beetroot juice and exercise performance," *Nutrition and Dietary Supplements*, vol. 5, pp. 27–35, 2013.

[215] "Nutrition business journal's supplement business report," *Nutrition Business Journal*, vol. 3, p. 28, 2010.

[216] K. E. Lansley, P. G. Winyard, S. J. Bailey et al., "Acute dietary nitrate supplementation improves cycling time trial performance," *Medicine and Science in Sports and Exercise*, vol. 43, no. 6, pp. 1125–1131, 2011.

[217] K. E. Lansley, P. G. Winyard, J. Fulford et al., "Dietary nitrate supplementation reduces the O_2 cost of walking and running: a placebo-controlled study," *Journal of Applied Physiology*, vol. 110, no. 3, pp. 591–600, 2011.

[218] S. J. Bailey, P. Winyard, A. Vanhatalo et al., "Dietary nitrate supplementation reduces the O_2 cost of low-intensity exercise and enhances tolerance to high-intensity exercise in humans," *Journal of Applied Physiology*, vol. 107, no. 4, pp. 1144–1155, 2009.

[219] A. Vanhatalo, S. J. Bailey, J. R. Blackwell et al., "Acute and chronic effects of dietary nitrate supplementation on blood pressure and the physiological responses to moderate-intensity and incremental exercise," *American Journal of Physiology—Regulatory Integrative and Comparative Physiology*, vol. 299, no. 4, pp. R1121–R1131, 2010.

[220] N. G. Hord, Y. Tang, and N. S. Bryan, "Food sources of nitrates and nitrites: the physiologic context for potential health benefits," *The American Journal of Clinical Nutrition*, vol. 90, no. 1, pp. 1–10, 2009.

[221] F. J. Larsen, T. A. Schiffer, S. Borniquel et al., "Dietary inorganic nitrate improves mitochondrial efficiency in humans," *Cell Metabolism*, vol. 13, no. 2, pp. 149–159, 2011.

[222] A. J. Webb, N. Patel, S. Loukogeorgakis et al., "Acute blood pressure lowering, vasoprotective, and antiplatelet properties of dietary nitrate via bioconversion to nitrite," *Hypertension*, vol. 51, no. 3, pp. 784–790, 2008.

[223] R. C. Hickner, J. S. Fisher, A. A. Ehsani, and W. M. Kohrt, "Role of nitric oxide in skeletal muscle blood flow at rest and during dynamic exercise in humans," *American Journal of Physiology—Heart and Circulatory Physiology*, vol. 273, no. 1, part 2, pp. H405–H410, 1997.

[224] D. M. Gilligan, J. A. Panza, C. M. Kilcoyne, M. A. Waclawiw, P. R. Casino, and A. A. Quyyumi, "Contribution of endothelium-derived nitric oxide to exercise-induced vasodilation," *Circulation*, vol. 90, no. 6, pp. 2853–2858, 1994.

[225] S. J. Bailey, J. Fulford, A. Vanhatalo et al., "Dietary nitrate supplementation enhances muscle contractile efficiency during knee-extensor exercise in humans," *Journal of Applied Physiology*, vol. 109, no. 1, pp. 135–148, 2010.

[226] M. Murphy, K. Eliot, R. M. Heuertz, and E. Weiss, "Whole beetroot consumption acutely improves running performance," *Journal of the Academy of Nutrition and Dietetics*, vol. 112, no. 4, pp. 548–552, 2012.

[227] J. Kelly, A. Vanhatalo, D. P. Wilkerson, L. J. Wylie, and A. M. Jones, "Effects of nitrate on the power-duration relationship for severe-intensity exercise," *Medicine and Science in Sports and Exercise*, vol. 45, no. 9, pp. 1798–1806, 2013.

[228] D. J. Muggeridge, C. C. F. Howe, O. Spendiff, C. Pedlar, P. E. James, and C. Easton, "A single dose of beetroot juice enhances cycling performance in simulated altitude," *Medicine and Science in Sports and Exercise*, vol. 46, no. 1, pp. 143–150, 2014.

[229] L. J. Wylie, J. Kelly, S. J. Bailey et al., "Beetroot juice and exercise: pharmacodynamic and dose-response relationships," *Journal of Applied Physiology*, vol. 115, no. 3, pp. 325–336, 2013.

[230] P. M. Christensen, M. Nyberg, and J. Bangsbo, "Influence of nitrate supplementation on VO_2 kinetics and endurance of elite cyclists," *Scandinavian Journal of Medicine & Science in Sports*, vol. 23, no. 1, pp. e21–e31, 2013.

[231] A. A. Kenjale, K. L. Ham, T. Stabler et al., "Dietary nitrate supplementation enhances exercise performance in peripheral arterial disease," *Journal of Applied Physiology*, vol. 110, no. 6, pp. 1582–1591, 2011.

[232] H. K. Stadheim, B. Kvamme, R. Olsen, C. A. Drevon, J. L. Ivy, and J. Jensen, "Caffeine increases performance in cross-country double-poling time trial exercise," *Medicine and Science in Sports and Exercise*, vol. 45, no. 11, pp. 2175–2183, 2013.

[233] A. L. Spence, M. Sim, G. Landers, and P. Peeling, "A comparison of caffeine versus pseudoephedrine on cycling time-trial performance," *International Journal of Sport Nutrition and Exercise Metabolism*, vol. 23, no. 5, pp. 507–512, 2013.

[234] N. W. Pitchford, J. W. Fell, M. D. Leveritt, B. Desbrow, and C. M. Shing, "Effect of caffeine on cycling time-trial performance in the heat," *Journal of Science and Medicine in Sport*, vol. 17, no. 4, pp. 445–449, 2014.

[235] G. L. Dohm, "Protein as a fuel for endurance exercise," *Exercise and Sport Sciences Reviews*, vol. 14, pp. 143–173, 1986.

[236] J. P. Flatt, "Use and storage of carbohydrate and fat," *The American Journal of Clinical Nutrition*, vol. 61, no. 4, supplement, pp. 952S–959S, 1995.

[237] R. K. Conlee, "Muscle glycogen and exercise endurance: a twenty-year perspective," *Exercise and Sport Sciences Reviews*, vol. 15, pp. 1–28, 1987.

[238] A. D. Karelis, J. E. W. Smith, D. H. Passe, and F. Pronnet, "Carbohydrate administration and exercise performance: what are the potential mechanisms involved?" *Sports Medicine*, vol. 40, no. 9, pp. 747–763, 2010.

[239] N. Ørtenblad, H. Westerblad, and J. Nielsen, "Muscle glycogen stores and fatigue," *The Journal of Physiology*, vol. 591, no. 18, pp. 4405–4413, 2013.

[240] A. M. Prentice, "Manipulation of dietary fat and energy density and subsequent effects on substrate flux and food intake," *The American Journal of Clinical Nutrition*, vol. 67, no. 3, supplement, pp. 535S–541S, 1998.

[241] L. S. Sidossis, C. A. Stuart, G. I. Shulman, G. D. Lopaschuk, and R. R. Wolfe, "Glucose plus insulin regulate fat oxidation by controlling the rate of fatty acid entry into the mitochondria," *Journal of Clinical Investigation*, vol. 98, no. 10, pp. 2244–2250, 1996.

[242] B. B. Rasmussen, U. C. Holmbäck, E. Volpi, B. Morio-Liondore, D. Paddon-Jones, and R. R. Wolfe, "Malonyl coenzyme A and the regulation of functional carnitine palmitoyltransferase-1 activity and fat oxidation in human skeletal muscle," *The Journal of Clinical Investigation*, vol. 110, no. 11, pp. 1687–1693, 2002.

[243] V. A. Zammit, "The malonyl-CoA-long-chain acyl-CoA axis in the maintenance of mammalian cell function," *Biochemical Journal*, vol. 343, no. 3, pp. 505–515, 1999.

[244] L. A. Witters and B. E. Kemp, "Insulin activation of acetyl-CoA carboxylase accompanied by inhibition of the 5'-AMP-activated protein kinase," *The Journal of Biological Chemistry*, vol. 267, no. 5, pp. 2864–2867, 1992.

[245] J. Denis McGarry, "Dysregulation of fatty acid metabolism in the etiology of type 2 diabetes," *Diabetes*, vol. 51, no. 1, pp. 7–18, 2002.

[246] D. Cameron-Smith, L. M. Burke, D. J. Angus et al., "A short-term, high-fat diet up-regulates lipid metabolism and gene expression in human skeletal muscle," *The American Journal of Clinical Nutrition*, vol. 77, no. 2, pp. 313–318, 2003.

[247] M. J. Arkinstall, R. J. Tunstall, D. Cameron-Smith, and J. A. Hawley, "Regulation of metabolic genes in human skeletal muscle by short-term exercise and diet manipulation," *The American Journal of Physiology—Endocrinology and Metabolism*, vol. 287, no. 1, pp. E25–E31, 2004.

[248] A. E. Civitarese, M. K. C. Hesselink, A. P. Russell, E. Ravussin, and P. Schrauwen, "Glucose ingestion during exercise blunts exercise-induced gene expression of skeletal muscle fat oxidative genes," *The American Journal of Physiology—Endocrinology and Metabolism*, vol. 289, no. 6, pp. E1023–E1029, 2005.

[249] N. A. Johnson, S. R. Stannard, K. Mehalski et al., "Intramyocellular triacylglycerol in prolonged cycling with high- and low-carbohydrate availability," *Journal of Applied Physiology*, vol. 94, no. 4, pp. 1365–1372, 2003.

[250] M. Vogt, A. Puntschart, H. Howald et al., "Effects of dietary fat on muscle substrates, metabolism, and performance in athletes," *Medicine and Science in Sports and Exercise*, vol. 35, no. 6, pp. 952–960, 2003.

[251] T. W. Zderic, C. J. Davidson, S. Schenk, L. O. Byerley, and E. F. Coyle, "High-fat diet elevates resting intramuscular triglyceride concentration and whole body lipolysis during exercise," *American Journal of Physiology—Endocrinology and Metabolism*, vol. 286, no. 2, pp. E217–E225, 2004.

[252] J. W. Helge, B. Wulff, and B. Kiens, "Impact of a fat-rich diet on endurance in man: role of the dietary period," *Medicine and Science in Sports and Exercise*, vol. 30, no. 3, pp. 456–461, 1998.

[253] M. J. Ormsbee, C. W. Bach, and D. A. Baur, "Pre-exercise nutrition: the role of macronutrients, modified starches and supplements on metabolism and endurance performance," *Nutrients*, vol. 6, no. 5, pp. 1782–1808, 2014.

[254] J. Temesi, N. A. Johnson, J. Raymond, C. A. Burdon, and H. T. O'Connor, "Carbohydrate ingestion during endurance exercise improves performance in adults," *Journal of Nutrition*, vol. 141, no. 5, pp. 890–897, 2011.

[255] T. J. Vandenbogaerde and W. G. Hopkins, "Effects of acute carbohydrate supplementation on endurance performance: A meta-analysis," *Sports Medicine*, vol. 41, no. 9, pp. 773–792, 2011.

[256] L. M. Burke, D. J. Angus, G. R. Cox et al., "Effect of fat adaptation and carbohydrate restoration on metabolism and performance during prolonged cycling," *Journal of Applied Physiology*, vol. 89, no. 6, pp. 2413–2421, 2000.

[257] J. Fleming, M. J. Sharman, N. G. Avery et al., "Endurance capacity and high-intensity exercise performance responses to a high fat diet," *International Journal of Sport Nutrition and Exercise Metabolism*, vol. 13, no. 4, pp. 466–478, 2003.

[258] J. H. Goedecke, C. Christie, G. Wilson et al., "Metabolic adaptations to a high-fat diet in endurance cyclists," *Metabolism: Clinical and Experimental*, vol. 48, no. 12, pp. 1509–1517, 1999.

[259] E. V. Lambert, D. P. Speechly, S. C. Dennis, and T. D. Noakes, "Enhanced endurance in trained cyclists during moderate intensity exercise following 2 weeks adaptation to a high fat diet," *European Journal of Applied Physiology and Occupational Physiology*, vol. 69, no. 4, pp. 287–293, 1994.

[260] D. M. Muoio, J. J. Leddy, P. J. Horvath, A. B. Awad, and D. R. Pendergast, "Effect of dietary fat on metabolic adjustments to maximal VO$_2$ and endurance in runners," *Medicine & Science in Sports & Exercise*, vol. 26, no. 1, pp. 81–88, 1994.

[261] K. A. O'Keeffe, R. E. Keith, G. D. Wilson, and D. L. Blessing, "Dietary carbohydrate intake and endurance exercise performance of trained female cyclists," *Nutrition Research*, vol. 9, no. 8, pp. 819–830, 1989.

[262] S. D. Phinney, B. R. Bistrian, W. J. Evans, E. Gervino, and G. L. Blackburn, "The human metabolic response to chronic ketosis without caloric restriction: preservation of submaximal exercise capability with reduced carbohydrate oxidation," *Metabolism*, vol. 32, no. 8, pp. 769–776, 1983.

[263] S. D. Phinney, E. S. Horton, E. A. H. Sims, J. S. Hanson, E. Danforth Jr., and B. M. LaGrange, "Capacity for moderate exercise in obese subjects after adaptation to a hypocaloric, ketogenic diet," *The Journal of Clinical Investigation*, vol. 66, no. 5, pp. 1152–1161, 1980.

[264] A. L. Robins, D. M. Davies, and G. E. Jones, "The effect of nutritional manipulation on ultra-endurance performance: a case study," *Research in Sports Medicine*, vol. 13, no. 3, pp. 199–215, 2005.

[265] D. S. Rowlands and W. G. Hopkins, "Effects of high-fat and high-carbohydrate diets on metabolism and performance in cycling," *Metabolism: Clinical and Experimental*, vol. 51, no. 6, pp. 678–690, 2002.

[266] N. K. Stepto, A. L. Carey, H. M. Staudacher, N. K. Cummings, L. M. Burke, and J. A. Hawley, "Effect of short-term fat adaptation on high-intensity training," *Medicine & Science in Sports & Exercise*, vol. 34, no. 3, pp. 449–455, 2002.

[267] J. L. Walberg, V. K. Ruiz, S. L. Tarlton, D. E. Hinkle, and R. W. Thye, "Exercise capacity and nitrogen loss during a high or low carbohydrate diet," *Medicine and Science in Sports and Exercise*, vol. 20, no. 1, pp. 34–43, 1988.

[268] M. Erlenbusch, M. Haub, K. Munoz, S. MacConnie, and B. Stillwell, "Effect of high-fat or high-carbohydrate diets on endurance exercise: a meta-analysis," *International Journal of Sport Nutrition and Exercise Metabolism*, vol. 15, no. 1, pp. 1–14, 2005.

[269] E. Hultman and J. Bergström, "Muscle glycogen synthesis in relation to diet studied in normal subjects.," *Acta Medica Scandinavica*, vol. 182, no. 1, pp. 109–117, 1967.

[270] J. D. Symons and I. Jacobs, "High-intensity exercise performance is not impaired by low intramuscular glycogen," *Medicine and Science in Sports and Exercise*, vol. 21, no. 5, pp. 550–557, 1989.

[271] A. Paoli, K. Grimaldi, D. D'Agostino et al., "Ketogenic diet does not affect strength performance in elite artistic gymnasts," *Journal of the International Society of Sports Nutrition*, vol. 9, no. 1, article 34, 2012.

[272] J. C. Sawyer, R. J. Wood, P. W. Davidson et al., "Effects of a short-term carbohydrate restricted diet on strength and power performance," *The Journal of Strength & Conditioning Research*, vol. 27, no. 8, pp. 2255–2262, 2013.

[273] L. M. Burke, "Fueling strategies to optimize performance: training high or training low?" *Scandinavian Journal of Medicine and Science in Sports*, vol. 20, supplement 2, pp. 48–58, 2010.

[274] L. M. Burke and J. A. Hawley, "Effects of short-term fat adaptation on metabolism and performance of prolonged exercise," *Medicine and Science in Sports and Exercise*, vol. 34, no. 9, pp. 1492–1498, 2002.

[275] J. S. Volek and S. D. Phinney, *The Art and Science of Low Carbohydrate Performance. A Revolutionary Program to Extend Your Physical and Mental Performance Envelope*, Beyond Obesity LLC, Miami, Fla, USA, 2012.

[276] L. M. Burke and B. Kiens, "'Fat adaptation' for athletic performance: the nail in the coffin?" *Journal of Applied Physiology*, vol. 100, no. 1, pp. 7–8, 2006.

[277] J. W. Helge, "Adaptation to a fat-rich diet: effects on endurance performance in humans," *Sports Medicine*, vol. 30, no. 5, pp. 347–357, 2000.

[278] J. W. Helge, "Long-term fat diet adaptation effects on performance, training capacity, and fat utilization," *Medicine & Science in Sports & Exercise*, vol. 34, no. 9, pp. 1499–1504, 2002.

[279] C. D. Gardner, A. Kiazand, S. Alhassan et al., "Comparison of the Atkins, Zone, Ornish, and LEARN diets for change in weight and related risk factors among overweight premenopausal women: the A to Z weight loss study: a randomized trial," *Journal of the American Medical Association*, vol. 297, no. 9, pp. 969–977, 2007.

[280] S. S. Summer, B. J. Brehm, S. C. Benoit, and D. A. D'Alessio, "Adiponectin changes in relation to the macronutrient composition of a weight-loss diet," *Obesity*, vol. 19, no. 11, pp. 2198–2204, 2011.

[281] P. T. Jabekk, I. A. Moe, S. E. Tomten, and A. T. Høstmark, "Resistance training in overweight women on a ketogenic diet conserved lean body mass while reducing body fat," *Nutrition and Metabolism*, vol. 7, article 17, 2010.

[282] D. K. Layman, E. Evans, J. I. Baum, J. Seyler, D. J. Erickson, and R. A. Boileau, "Dietary protein and exercise have additive effects on body composition during weight loss in adult women," *Journal of Nutrition*, vol. 135, no. 8, pp. 1903–1910, 2005.

[283] J. S. Volek, E. E. Quann, and C. E. Forsythe, "Low-carbohydrate diets promote a more favorable body composition than low-fat diets," *Strength and Conditioning Journal*, vol. 32, no. 1, pp. 42–47, 2010.

[284] O. Ajala, P. English, and J. Pinkney, "Systematic review and meta-analysis of different dietary approaches to the management of type 2 diabetes," *American Journal of Clinical Nutrition*, vol. 97, no. 3, pp. 505–516, 2013.

[285] R. D. Feinman and J. S. Volek, "Carbohydrate restriction as the default treatment for type 2 diabetes and metabolic syndrome," *Scandinavian Cardiovascular Journal*, vol. 42, no. 4, pp. 256–263, 2008.

[286] T. Hu, K. T. Mills, L. Yao et al., "Effects of low-carbohydrate diets versus low-fat diets on metabolic risk factors: a meta-analysis of randomized controlled clinical trials," *American Journal of Epidemiology*, vol. 176, supplement 7, pp. S44–S54, 2012.

[287] A. J. Nordmann, A. Nordmann, M. Briel et al., "Effects of low-carbohydrate vs low-fat diets on weight loss and cardiovascular risk factors: a meta-analysis of randomized controlled trials," *Archives of Internal Medicine*, vol. 166, no. 3, pp. 285–293, 2006.

[288] F. L. Santos, S. S. Esteves, A. D. C. Pereira, W. S. Yancy Jr., and J. P. L. Nunes, "Systematic review and meta-analysis of clinical trials of the effects of low carbohydrate diets on cardiovascular risk factors," *Obesity Reviews*, vol. 13, no. 11, pp. 1048–1066, 2012.

[289] J. S. Volek and R. D. Feinman, "Carbohydrate restriction improves the features of Metabolic Syndrome. Metabolic Syndrome may be defined by the response to carbohydrate restriction," *Nutrition and Metabolism*, vol. 2, article 31, pp. 1–17, 2005.

[290] J. Guo, X. Zhang, L. Wang, Y. Guo, and M. Xie, "Prevalence of metabolic syndrome and its components among Chinese professional athletes of strength sports with different body weight categories," *PLoS ONE*, vol. 8, no. 11, Article ID e79758, 2013.

[291] G. D. Steffes, A. E. Megura, J. Adams et al., "Prevalence of metabolic syndrome risk factors in high school and NCAA division I football players," *Journal of Strength & Conditioning Research*, vol. 27, no. 7, pp. 1749–1757, 2013.

[292] K. Baar and S. McGee, "Optimizing training adaptations by manipulating glycogen," *European Journal of Sport Science*, vol. 8, no. 2, pp. 97–106, 2008.

[293] J. A. Hawley, K. D. Tipton, and M. L. Millard-Stafford, "Promoting training adaptations through nutritional interventions," *Journal of Sports Sciences*, vol. 24, no. 7, pp. 709–721, 2006.

[294] S. D. Phinney, "Ketogenic diets and physical performance," *Nutrition and Metabolism*, vol. 1, article 2, 2004.

[295] S. D. Phinney, B. R. Bistrian, R. R. Wolfe, and G. L. Blackburn, "The human metabolic response to chronic ketosis without caloric restriction: physical and biochemical adaptation," *Metabolism*, vol. 32, no. 8, pp. 757–768, 1983.

[296] W. K. Yeo, A. L. Carey, L. Burke, L. L. Spriet, and J. A. Hawley, "Fat adaptation in well-trained athletes: effects on cell metabolism," *Applied Physiology, Nutrition & Metabolism*, vol. 36, no. 1, pp. 12–22, 2011.

[297] American College of Sports Medicine, M. N. Sawka, L. M. Burke et al., "American College of Sports Medicine position stand. Exercise and fluid replacement," *Medicine & Science in Sport & Exercise*, vol. 39, no. 2, pp. 377–390, 2007.

[298] M. R. Naghii, "The significance of water in sport and weight control," *Nutrition and Health*, vol. 14, no. 2, pp. 127–132, 2000.

[299] S. K. Powers, J. Lawler, S. Dodd, R. Tulley, G. Landry, and K. Wheeler, "Fluid replacement drinks during high intensity exercise effects on minimizing exercise-induced disturbances in homeostasis," *European Journal of Applied Physiology and Occupational Physiology*, vol. 60, no. 1, pp. 54–60, 1990.

[300] E. F. Coyle, "Fluid and fuel intake during exercise," *Journal of Sports Sciences*, vol. 22, no. 1, pp. 39–55, 2004.

[301] S. P. von Duvillard, W. A. Braun, M. Markofski, R. Beneke, and R. Leithäuser, "Fluids and hydration in prolonged endurance performance," *Nutrition*, vol. 20, no. 7-8, pp. 651–656, 2004.

[302] T. D. Noakes, "Drinking guidelines for exercise: what evidence is there that athletes should drink 'as much as tolerable', 'to replace the weight lost during exercise' or 'ad libitum'?" *Journal of Sports Sciences*, vol. 25, no. 7, pp. 781–796, 2007.

[303] B. Sanders, T. D. Noakes, and S. C. Dennis, "Sodium replacement and fluid shifts during prolonged exercise in humans," *European Journal of Applied Physiology*, vol. 84, no. 5, pp. 419–425, 2001.

[304] D. Bernardot, *Advanced Sports Nutrition*, Human Kinetics, Champaign, Ill, USA, 2006.

[305] S. M. Shirreffs, L. E. Armstrong, and S. N. Cheuvront, "Fluid and electrolyte needs for preparation and recovery from training and competition," *Journal of Sports Sciences*, vol. 22, no. 1, pp. 57–63, 2004.

[306] S. D. R. Galloway and R. J. Maughan, "The effects of substrate and fluid provision on thermoregulatory, cardiorespiratory and metabolic responses to prolonged exercise in a cold environment in man," *Experimental Physiology*, vol. 83, no. 3, pp. 419–430, 1998.

[307] S. A. Kavouras, L. E. Armstrong, C. M. Maresh et al., "Rehydration with glycerol: endocrine, cardiovascular, and thermoregulatory responses during exercise in the heat," *Journal of Applied Physiology*, vol. 100, no. 2, pp. 442–450, 2006.

[308] J. L. J. Bilzon, J. L. Murphy, A. J. Allsopp, S. A. Wootton, and C. Williams, "Influence of glucose ingestion by humans during recovery from exercise on substrate utilisation during subsequent exercise in a warm environment," *European Journal of Applied Physiology*, vol. 87, no. 4-5, pp. 318–326, 2002.

[309] D. M. J. Vrijens and N. J. Rehrer, "Sodium-free fluid ingestion decreases plasma sodium during exercise in the heat," *Journal of Applied Physiology*, vol. 86, no. 6, pp. 1847–1851, 1999.

[310] R. J. Maughan, L. R. Bethell, and J. B. Leiper, "Effects of ingested fluids on exercise capacity and on cardiovascular and metabolic responses to prolonged exercise in man," *Experimental Physiology*, vol. 81, no. 5, pp. 847–859, 1996.

[311] J. L. Azevedo Jr., E. Tietz, T. Two-Feathers, J. Paull, and K. Chapman, "Lactate, fructose and glucose oxidation profiles in sports drinks and the effect on exercise performance," *PLoS ONE*, vol. 2, no. 9, article e927, 2007.

[312] M. J. Anderson, J. D. Cotter, A. P. Garnham, D. J. Casley, and M. A. Febbraio, "Effect of glycerol-induced hyperhydration on thermoregulation and metabolism during exercise in the heat," *International Journal of Sport Nutrition*, vol. 11, no. 3, pp. 315–333, 2001.

[313] S. Hitchins, D. T. Martin, L. Burke et al., "Glycerol hyperhydration improves cycle time trial performance in hot humid conditions," *European Journal of Applied Physiology and Occupational Physiology*, vol. 80, no. 5, pp. 494–501, 1999.

[314] W. A. Latzka, M. N. Sawka, S. J. Montain et al., "Hyperhydration: thermoregulatory effects during compensable exercise-heat stress," *Journal of Applied Physiology*, vol. 83, no. 3, pp. 860–866, 1997.

[315] W. A. Latzka, M. N. Sawka, S. J. Montain et al., "Hyperhydration: tolerance and cardiovascular effects during uncompensable exercise-heat stress," *Journal of Applied Physiology*, vol. 84, no. 6, pp. 1858–1864, 1998.

[316] F. E. Marino, D. Kay, and J. Cannon, "Glycerol hyperhydration fails to improve endurance performance and thermoregulation in humans in a warm humid environment," *Pflugers Archiv*, vol. 446, no. 4, pp. 455–462, 2003.

[317] R. W. Kenefick, C. M. Maresh, L. E. Armstrong, D. Riebe, M. E. Echegaray, and J. W. Castellani, "Rehydration with fluid of varying tonicities: effects on fluid regulatory hormones and exercise performance in the heat," *Journal of Applied Physiology*, vol. 102, no. 5, pp. 1899–1905, 2007.

[318] N. M. Cermak and L. J. C. van Loon, "The use of carbohydrates during exercise as an ergogenic aid," *Sports Medicine*, vol. 43, no. 11, pp. 1139–1155, 2013.

[319] R. F. Tester, J. Karkalas, and X. Qi, "Starch—composition, fine structure and architecture," *Journal of Cereal Science*, vol. 39, no. 2, pp. 151–165, 2004.

[320] F. Stephens, M. Roig, G. Armstrong, and P. L. Greenhaff, "Post-exercise ingestion of a unique, high molecular weight glucose polymer solution improves performance during a subsequent bout of cycling exercise," *Journal of Sports Sciences*, vol. 26, no. 2, pp. 149–154, 2008.

[321] A. C. Jozsi, T. A. Trappe, R. D. Starling et al., "The influence of starch structure on glycogen resynthesis and subsequent cycling performance," *International Journal of Sports Medicine*, vol. 17, no. 5, pp. 373–378, 1996.

[322] M. D. Roberts, C. Lockwood, V. J. Dalbo, J. Volek, and C. M. Kerksick, "Ingestion of a high-molecular-weight hydrothermally modified waxy maize starch alters metabolic responses to prolonged exercise in trained cyclists," *Nutrition*, vol. 27, no. 6, pp. 659–665, 2011.

[323] R. Hursel, W. Viechtbauer, A. G. Dulloo et al., "The effects of catechin rich teas and caffeine on energy expenditure and fat oxidation: a meta-analysis," *Obesity Reviews*, vol. 12, no. 7, pp. e573–e581, 2011.

[324] P. J. Arciero, A. W. Gardner, J. Calles-Escandon, N. L. Benowitz, and E. T. Poehlman, "Effects of caffeine ingestion on NE kinetics, fat oxidation, and energy expenditure in younger and older men," *The American Journal of Physiology—Endocrinology and Metabolism*, vol. 268, no. 6, pp. E1192–E1198, 1995.

[325] P. J. Arciero, C. L. Bougopoulos, B. C. Nindl, and N. L. Benowitz, "Influence of age on the thermic response to caffeine in women," *Metabolism: Clinical and Experimental*, vol. 49, no. 1, pp. 101–107, 2000.

[326] P. B. Júdice, J. P. Magalhães, D. A. Santos et al., "A moderate dose of caffeine ingestion does not change energy expenditure but decreases sleep time in physically active males: a double-blind randomized controlled trial," *Applied Physiology, Nutrition and Metabolism*, vol. 38, no. 1, pp. 49–56, 2013.

[327] P. B. Júdice, C. N. Matias, D. A. Santos et al., "Caffeine intake, short bouts of physical activity, and energy expenditure: a double-blind randomized crossover trial," *PLoS ONE*, vol. 8, no. 7, Article ID e68936, 2013.

[328] R. Hursel and M. S. Westerterp-Plantenga, "Thermogenic ingredients and body weight regulation," *International Journal of Obesity*, vol. 34, no. 4, pp. 659–669, 2010.

[329] M. J. Ludy and R. D. Mattes, "The effects of hedonically acceptable red pepper doses on thermogenesis and appetite," *Physiology and Behavior*, vol. 102, no. 3-4, pp. 251–258, 2011.

[330] K. Lim, M. Yoshioka, S. Kikuzato et al., "Dietary red pepper ingestion increases carbohydrate oxidation at rest and during exercise in runners," *Medicine and Science in Sports and Exercise*, vol. 29, no. 3, pp. 355–361, 1997.

[331] M. Yoshioka, S. St-Pierre, V. Drapeau et al., "Effects of red pepper on appetite and energy intake," *British Journal of Nutrition*, vol. 82, no. 2, pp. 115–123, 1999.

[332] M. Yoshioka, M. Imanaga, H. Ueyama et al., "Maximum tolerable dose of red pepper decreases fat intake independently of spicy sensation in the mouth," *British Journal of Nutrition*, vol. 91, no. 6, pp. 991–995, 2004.

[333] T. Matsumoto, C. Miyawaki, H. Ue, T. Yuasa, A. Miyatsuji, and T. Moritani, "Effects of capsaicin-containing yellow curry sauce on sympathetic nervous system activity and diet-induced thermogenesis in lean and obese young women," *Journal of Nutritional Science and Vitaminology*, vol. 46, no. 6, pp. 309–315, 2000.

[334] S. Hachiya, F. Kawabata, K. Ohnuki et al., "Effects of CH-19 Sweet, a non-pungent cultivar of red pepper, on sympathetic nervous activity, body temperature, heart rate, and blood pressure in humans," *Bioscience, Biotechnology and Biochemistry*, vol. 71, no. 3, pp. 671–676, 2007.

[335] M. Yoshioka, S. St-Pierre, M. Suzuki, and A. Tremblay, "Effects of red pepper added to high-fat and high-carbohydrate meals on energy metabolism and substrate utilization in Japanese women," *British Journal of Nutrition*, vol. 80, no. 6, pp. 503–510, 1998.

[336] M. P. G. M. Lejeune, E. M. R. Kovacs, and M. S. Westerterp-Plantenga, "Effect of capsaicin on substrate oxidation and weight maintenance after modest body-weight loss in human subjects," *British Journal of Nutrition*, vol. 90, no. 3, pp. 651–659, 2003.

[337] T. Yoneshiro, S. Aita, Y. Kawai, T. Iwanaga, and M. Saito, "Nonpungent capsaicin analogs (capsinoids) increase energy expenditure through the activation of brown adipose tissue in humans," *The American Journal of Clinical Nutrition*, vol. 95, no. 4, pp. 845–850, 2012.

[338] S. Snitker, Y. Fujishima, H. Shen et al., "Effects of novel capsinoid treatment on fatness and energy metabolism in humans: possible pharmacogenetic implications," *The American Journal of Clinical Nutrition*, vol. 89, no. 1, pp. 45–50, 2009.

[339] K. D. K. Ahuja, I. K. Robertson, D. P. Geraghty, and M. J. Ball, "Effects of chili consumption on postprandial glucose, insulin, and energy metabolism," *The American Journal of Clinical Nutrition*, vol. 84, no. 1, pp. 63–69, 2006.

[340] A. J. Smeets and M. S. Westerterp-Plantenga, "The acute effects of a lunch containing capsaicin on energy and substrate utilisation, hormones, and satiety," *European Journal of Nutrition*, vol. 48, no. 4, pp. 229–234, 2009.

[341] K. Iwai, A. Yazawa, and T. Watanabe, "Roles as metabolic regulators of the non-nutrients, capsaicin and capsiate, supplemented to diets," *Proceedings of the Japan Academy Series B: Physical and Biological Sciences*, vol. 79, no. 7, pp. 207–212, 2003.

[342] A. R. Josse, S. S. Sherriffs, A. M. Holwerda, R. Andrews, A. W. Staples, and S. M. Phillips, "Effects of capsinoid ingestion on energy expenditure and lipid oxidation at rest and during exercise," *Nutrition and Metabolism*, vol. 7, article 65, 2010.

[343] N. Inoue, Y. Matsunaga, H. Satoh, and M. Takahashi, "Enhanced energy expenditure and fat oxidation in humans with high BMI scores by the ingestion of novel and non-pungent capsaicin analogues (capsinoids)," *Bioscience, Biotechnology and Biochemistry*, vol. 71, no. 2, pp. 380–389, 2007.

[344] T. A. Lee, Z. Li, A. Zerlin, and D. Heber, "Effects of dihydrocapsiate on adaptive and diet-induced thermogenesis with a high protein very low calorie diet: a randomized control trial," *Nutrition and Metabolism*, vol. 7, article 78, 2010.

[345] J. E. Galgani and E. Ravussin, "Effect of dihydrocapsiate on resting metabolic rate in humans," *The American Journal of Clinical Nutrition*, vol. 92, no. 5, pp. 1089–1093, 2010.

[346] M. N. Opheim and J. W. Rankin, "Effect of capsaicin supplementation on repeated sprinting performance," *Journal of Strength and Conditioning Research*, vol. 26, no. 2, pp. 319–326, 2012.

[347] J. Bremer, "Carnitine—metabolism and functions," *Physiological Reviews*, vol. 63, no. 4, pp. 1420–1480, 1983.

[348] F. B. Stephens, B. T. Wall, K. Marimuthu et al., "Skeletal muscle carnitine loading increases energy expenditure, modulates fuel metabolism gene networks and prevents body fat accumulation in humans," *Journal of Physiology*, vol. 591, no. 18, pp. 4655–4666, 2013.

[349] I. B. Fritz and K. T. Yue, "Acyltransferase and the role of acylcarnitine derivatives in the catalytic increase of fatty acid oxidation induced by carnitine," *Journal of Lipid Research*, vol. 4, pp. 279–288, 1963.

[350] B. T. Wall, F. B. Stephens, D. Constantin-Teodosiu, K. Marimuthu, I. A. Macdonald, and P. L. Greenhaff, "Chronic oral ingestion of L-carnitine and carbohydrate increases muscle carnitine content and alters muscle fuel metabolism during exercise in humans," *Journal of Physiology*, vol. 589, part 4, pp. 963–973, 2011.

[351] G. E. Orer and N. A. Guzel, "The effects of acute l-carnitine supplementation on endurance performance of athletes," *Journal of Strength and Conditioning Research*, vol. 28, no. 2, pp. 514–519, 2014.

[352] C. J. Rebouche, "Carnitine function and requirements during the life cycle," *The FASEB Journal*, vol. 6, no. 15, pp. 3379–3386, 1992.

[353] M. R. Rubin, J. S. Volek, A. L. Gomez et al., "Safety measures of L-carnitine L-tartrate supplementation in healthy men," *The Journal of Strength and Conditioning Research*, vol. 15, no. 4, pp. 486–490, 2001.

[354] J. A. Higgins, "Resistant starch and energy balance: impact on weight loss and maintenance," *Critical Reviews in Food Science and Nutrition*, vol. 54, no. 9, pp. 1158–1166, 2014.

[355] M. D. Haub, K. L. Hubach, E. K. Al-Tamimi, S. Ornelas, and P. A. Seib, "Different types of resistant starch elicit different glucose reponses in humans," *Journal of Nutrition and Metabolism*, vol. 2010, Article ID 230501, 4 pages, 2010.

[356] K. N. Englyst, S. Liu, and H. N. Englyst, "Nutritional characterization and measurement of dietary carbohydrates," *European Journal of Clinical Nutrition*, vol. 61, supplement 1, pp. S19–S39, 2007.

[357] C. L. Bodinham, G. S. Frost, and M. D. Robertson, "Acute ingestion of resistant starch reduces food intake in healthy adults," *British Journal of Nutrition*, vol. 103, no. 6, pp. 917–922, 2010.

[358] A. Raben, A. Tagliabue, N. J. Christensen, J. Madsen, J. J. Holst, and A. Astrup, "Resistant starch: the effect on postprandial glycemia, hormonal response, and satiety," *American Journal of Clinical Nutrition*, vol. 60, no. 4, pp. 544–551, 1994.

[359] J. A. Higgins, D. R. Higbee, W. T. Donahoo, I. L. Brown, M. L. Bell, and D. H. Bessesen, "Resistant starch consumption promotes lipid oxidation," *Nutrition & Metabolism*, vol. 1, article 8, 2004.

[360] A. S. Klosterbuer, W. Thomas, and J. L. Slavin, "Resistant starch and pullulan reduce postprandial glucose, insulin, and GLP-1, but have no effect on satiety in healthy humans," *Journal of Agricultural and Food Chemistry*, vol. 60, no. 48, pp. 11928–11934, 2012.

[361] M. D. Haub, J. A. Louk, and T. C. Lopez, "Novel resistant potato starches on glycemia and satiety in humans," *Journal of*

Nutrition and Metabolism, vol. 2012, Article ID 478043, 4 pages, 2012.

[362] P. M. Heacock, S. R. Hertzler, and B. Wolf, "The glycemic, insulinemic, and breath hydrogen responses in humans to a food starch esterified by 1-octenyl succinic anhydride," *Nutrition Research*, vol. 24, no. 8, pp. 581–592, 2004.

[363] A. Shimotoyodome, J. Suzuki, Y. Kameo, and T. Hase, "Dietary supplementation with hydroxypropyl-distarch phosphate from waxy maize starch increases resting energy expenditure by lowering the postprandial glucose-dependent insulinotropic polypeptide response in human subjects," *British Journal of Nutrition*, vol. 106, no. 1, pp. 96–104, 2011.

[364] N. de Roos, M. L. Heijnen, C. de Graaf, G. Woestenenk, and E. Hobbel, "Resistant starch has little effect on appetite, food intake and insulin secretion of healthy young men," *European Journal of Clinical Nutrition*, vol. 49, no. 7, pp. 532–541, 1995.

[365] S. Ranganathan, M. Champ, C. Pechard et al., "Comparative study of the acute effects of resistant starch and dietary fibers on metabolic indexes in men," *The American Journal of Clinical Nutrition*, vol. 59, no. 4, pp. 879–883, 1994.

[366] A. Tagliabue, A. Raben, M. L. Heijnen, P. Deurenberg, E. Pasquali, and A. Astrup, "The effect of raw potato starch on energy expenditure and substrate oxidation," *American Journal of Clinical Nutrition*, vol. 61, no. 5, pp. 1070–1075, 1995.

[367] E. K. Al-Tamimi, P. A. Seib, B. S. Snyder, and M. D. Haub, "Consumption of cross-linked resistant starch ($RS4_{XL}$) on glucose and insulin responses in humans," *Journal of Nutrition and Metabolism*, vol. 2010, Article ID 651063, 6 pages, 2010.

[368] A. C. Bach and V. K. Babayan, "Medium-chain triglycerides: an update," *The American Journal of Clinical Nutrition*, vol. 36, no. 5, pp. 950–962, 1982.

[369] M. E. Clegg, "Medium-chain triglycerides are advantageous in promoting weight loss although not beneficial to exercise performance," *International Journal of Food Sciences and Nutrition*, vol. 61, no. 7, pp. 653–679, 2010.

[370] S. S. Gropper and J. L. Smith, *Advanced Nutrition and Human Metabolism*, Wadsworth Publishing, Belmont, Calif, USA, 6th edition, 2013.

[371] A. C. Rego Costa, E. L. Rosado, and M. Soares-Mota, "Influence of the dietary intake of medium chain triglycerides on body composition, energy expenditure and satiety: a systematic review," *Nutrición Hospitalaria*, vol. 27, no. 1, pp. 103–108, 2012.

[372] A. G. Dulloo, M. Fathi, N. Mensi, and L. Girardier, "Twenty-four-hour energy expenditure and urinary catecholamines of humans consuming low-to-moderate amounts of medium-chain triglycerides: a dose-response study in a human respiratory chamber," *European Journal of Clinical Nutrition*, vol. 50, no. 3, pp. 152–158, 1996.

[373] N. J. Rothwell and M. J. Stock, "Stimulation of thermogenesis and brown fat activity in rats fed medium chain triglyceride," *Metabolism*, vol. 36, no. 2, pp. 128–130, 1987.

[374] M.-P. St-Onge, R. Ross, W. D. Parsons, and P. J. H. Jones, "Medium-chain triglycerides increase energy expenditure and decrease adiposity in overweight men," *Obesity Research*, vol. 11, no. 3, pp. 395–402, 2003.

[375] M.-P. St-Onge and P. J. H. Jones, "Physiological effects of medium-chain triglycerides: potential agents in the prevention of obesity," *Journal of Nutrition*, vol. 132, no. 3, pp. 329–332, 2002.

[376] M. P. St-Onge, C. Bourque, P. J. Jones, R. Ross, and W. E. Parsons, "Medium- versus long-chain triglycerides for 27 days increases fat oxidation and energy expenditure without resulting in changes in body composition in overweight women," *International Journal of Obesity and Related Metabolic Disorders*, vol. 27, no. 1, pp. 95–102, 2003.

[377] M.-P. St-Onge and P. J. H. Jones, "Greater rise in fat oxidation with medium-chain triglyceride consumption relative to long-chain triglyceride is associated with lower initial body weight and greater loss of subcutaneous adipose tissue," *International Journal of Obesity and Related Metabolic Disorders*, vol. 27, no. 12, pp. 1565–1571, 2003.

[378] M. E. Clegg, M. Golsorkhi, and C. J. Henry, "Combined medium-chain triglyceride and chilli feeding increases diet-induced thermogenesis in normal-weight humans," *European Journal of Nutrition*, vol. 52, no. 6, pp. 1579–1585, 2013.

[379] H. Amagase, R. Handel, and D. Nance, "Impact of a combination of medium-chain triglycerides and dietary fiber in a form of meal replacement shake on postprandial energy expenditure," *The FASEB Journal*, vol. 27, no. 854, p. 1, 2013.

[380] V. K. Babayan, "Medium chain triglycerides and structured lipids," *Lipids*, vol. 22, no. 6, pp. 417–420, 1987.

6

Nutrition and Reproductive Health: Sperm versus Erythrocyte Lipidomic Profile and ω-3 Intake

Gabriela Ruth Mendeluk,[1] Mariano Isaac Cohen,[2] Carla Ferreri,[3] and Chryssostomos Chatgilialoglu[4]

[1]*Laboratory of Male Fertility, Hospital de Clínicas "José de San Martín", INFIBIOC, Faculty of Pharmacy and Biochemistry, University of Buenos Aires, 5950-800 Buenos Aires, Argentina*
[2]*Urology Division, Hospital de Clínicas "José de San Martín", University of Buenos Aires, 5950-800 Buenos Aires, Argentina*
[3]*Consiglio Nazionale delle Ricerche (CNR), Istituto per la Sintesi Organica e la Fotoreattività (ISOF), 40129 Bologna, Italy*
[4]*Institute of Nanoscience and Nanotechnology, National Center of Scientific Research "Demokritos", Agia Paraskevi, 15310 Athens, Greece*

Correspondence should be addressed to Gabriela Ruth Mendeluk; gmendeluk@ffyb.uba.ar
and Carla Ferreri; carla.ferreri@isof.cnr.it

Academic Editor: C. S. Johnston

Fatty acid analyses of sperm and erythrocyte cell membrane phospholipids in idiopathic infertile patients evidenced that erythrocyte contents of EPA, DHA, omega-6–omega-3 ratio and arachidonic acid provide a mathematical correspondence for the prediction of EPA level in sperm cells. The erythrocyte lipidomic profile of patients was significantly altered, with signatures of typical Western pattern dietary habits and no fish intake. A supplementation with nutritional levels of EPA and DHA and antioxidants was then performed for 3 months, with the follow-up of both erythrocyte and sperm cell membranes composition as well as conventional sperm parameters. Some significant changes were found in the lipidomic membrane profile of erythrocyte but not in sperm cells, which correspondently did not show significant parameter ameliorations. This is the first report indicating that membrane lipids of different tissues do not equally metabolize the fatty acid elements upon supplementation. Molecular diagnostic tools are necessary to understand the cell metabolic turnover and monitor the success of nutraceuticals for personalized treatments.

1. Introduction

Decreasing the number of men affected by infertility has become a top priority for many health organizations, including Healthy People 2020. Modifiable lifestyles should then be considered before deciding high complexity assisted reproduction techniques clearly bypassing the natural barriers of human development. Age when starting a family, nutrition, weight management, exercise, psychological stress, cigarette smoking, recreational drugs use, medications, alcohol use, caffeine consumption, environmental and occupation exposure, preventive care, clothing choices, hot water, and lubricants are among them [1].

Evidence suggests that male fertility decreases by men being either overweight or underweight (as defined by body mass index [BMI] > 25 kg/m^2 and BMI < 20 kg/m^2, resp.) [2].

Healthy diet and regular exercise are therefore both recommended to maintain BMI between 20 and 25 kg/m^2 [3]. While considering nutrition, correlation among obesity, metabolic syndrome (MS), and reproductive axis was observed [4]. Excess adipose tissue results in increased conversion of testosterone to estradiol, which may lead to secondary hypogonadism through reproductive axis suppression. The benefits of low-fat intake in male fertility have recently been reported [5].

In a preliminary cross-sectional study, high intake of saturated fats was negatively related to sperm concentration whereas higher intake of omega-3 fats was positively related to sperm morphology [6]. On the other hand, the negative impact of nutritional deficiencies on semen quality after bariatric surgery was reported [7–9]. Moreover, an adequate intake of certain fatty acids is necessary for an adequate spermatogenesis [10]. In a recent study, it was found that 20- and 22-carbon fatty acids are present in semen in much larger quantities than thus far reported in any other mammalian tissue [11]. The first mechanism by which ω-3 and ω-6 PUFAs affect spermatogenesis is by the incorporation into spermatozoa cell membrane. ω-3 and ω-6 PUFAs are structural components of cell membranes [12], and the lipid bilayer properties of fluidity and permeability are maintained by the presence of these PUFAs [13]. Indeed, the successful fertilization of spermatozoa depends on the lipids of the spermatozoa membrane [14]. It has been recently been demonstrated that walnuts added to a Western-style diet improved sperm vitality, motility, and morphology [15]. There is no doubt that diet is a modifiable factor that could impact male fertility [16].

The aim of this study was to evaluate erythrocyte and sperm cell membrane fatty acids before and after 3 months of supplementation with omega-3 in parallel with sperm quality parameters in a group of ten "idiopathic infertile" patients, examined by the andrologist and nutritionist staff.

2. Material and Methods

2.1. Study Design. Evaluation of erythrocyte and sperm lipidomic profile from infertile patients was carried out at the beginning of the study while being evaluated by andrologists and nutritionists. In case of fatty acid deficiencies of erythrocyte composition determined in comparison with healthy interval values reported in literature [21], therapeutic supplementation was carried out for three months. Changes after treatment were analysed.

2.2. Patients. A total of 10 "idiopathic infertile" patients were recruited. Varicocele, infection, or genetic diseases were discarded. Demographic, clinical, and nutritional data were recorded in an interview taken by the physician. Height, weight, nationality, smoking addiction, physical activities, and medical anamnesis were registered.

The present study was conducted according to the guidelines laid down in the Declaration of Helsinki and was approved by the Institutional Review Board of The University Clinical Hospital "José de San Martin" (Buenos Aires, Argentina). All the participants received information on the project and gave written informed consent to be included.

2.3. Dietary Measures. Usual diet at baseline was determined using a self-administered nutritional questionnaire guided by trained nutritionists. The frequency per week for each food type and the quantity per portion were registered, underlying dietary patterns.

2.4. Lipidomic Profile. Blood samples, obtained from the patients, were collected in K2-EDTA vacutainers (Becton, Dickinson and Company, Franklin Lakes, NJ, USA). The erythrocyte fatty acid membrane profile analysis was carried out as previously described, using the erythrocyte membrane pellet obtained by standard methods [18].

Briefly, the phospholipid fraction of the membrane pellet was treated with 0.5 M KOH/MeOH for 10 min at room temperature, and the corresponding fatty acid methyl esters (FAMEs) were formed and extracted with *n*-hexane. Fatty acid methyl esters were analyzed by GC (Agilent 6850, Milan) equipped with a 60 m × 0.25 mm × 0.25 μm (50%-cyanopropyl)-methylpolysiloxane column (DB23, Agilent, USA) and a flame ionization detector (FID) with the following oven program: temperature started from 165°C, held for 3 min, followed by an increase of 1°C/min up to 195°C, held for 40 min, followed by a second increase of 10°C/min up to 240°C, and held for 10 min. A constant pressure mode (29 psi) was chosen with helium as carrier gas. Methyl esters were identified by comparison with the retention times of authentic samples. Geometrical TFAs (trans fatty acids) were recognized by comparison with standard references obtained by synthesis, as already described [17]. The amounts of the individual FAME were calculated as a percentage of the total measured FAMEs and their standard deviations calculated in Excel.

For sperm lipidomic profile of the semen, seminal plasma was discarded and the cells were washed in PBS (pH = 7.8). One mL of buffer containing at least 6×10^6 spermatozoa was used to isolate membrane PL according to the described procedure. Two repetitions for the same analysis were performed. Two repetitions for the analysis of the same sample were performed.

2.5. Nutraceutical Supplementation. Commercial soft-gel capsules containing ω-3 fatty acid-based formulas from fish and algal oils were used. Patients visited the physician three times during the study: at baseline and one month and two months later. In each visit the andrologist gave them the capsules for the next month and asked them if they had any problem with the intake. He verified in all the cases that it was correctly performed, receiving the empty blister each month.

2.5.1. Nutritional Values and Analysis for Capsule. A commercial soft-gel capsule containing ω-3 fatty acids from fish oil was used (Lipinutragen srl, Bologna, Italy). The main components and quantities of each capsule are described as follows: 5.31 Kcal; KJ 21.95; protein 0.17 g; 0.04 g carbohydrates; fat 0.510 g of which fatty acids are such as 0.040 g saturated; 0.042 g monounsaturated; and 0.323 g polyunsaturated of which EPA (eicosapentaenoic acid) 0.103 g, DHA (docosahexaenoic acid) 0.070 g, and ALA (alfa-linolenic acid) 0.070 mg; vitamin C 90.00 mg (112.5% RDA); vitamin E 10.00 mg (αTE; 83.3% RDA); astaxanthin 2.00 mg; and L-α-glyceryl phosphoryl choline 90 mg.

2.5.2. Nutritional Values and Analysis for Capsule. A commercial soft-gel capsule containing DHA from algal oil was used. The main components and quantities of each capsule are described as follows: 5.118 Kcal; KJ 21.14; protein 0.18 g; carbohydrates 0.08 g; fat 0.450 g of which fatty acids are such as 0.053 g saturated; 0.037 g monounsaturated; 0.314 g polyunsaturated fatty acids of which DHA is 0.200 g; vitamin C 90.00 mg (112.5% RDA); vitamin E 10.00 mg (αTE; 83.3% RDA); astaxanthin 2.00 mg; and L-α-glyceryl phosphoryl choline 100 mg.

TABLE 1: Demographic data.

Patient	Age	Height (cm)	Weight (kg)	Body mass index (kg/m²)	Smoker	Job type	Physical activity	Sport
1	33	170	71	24.6	Ex	Architect	Light	No
2	31	177	80	25.5	Ex	Builder	Light	No
3	37	178	83	26.2	No	Driver	No	No
4	32	172	75	25.4	Yes	Teacher	Medium	Yes
5	44	180	95	29.3	No	Mechanic	Light	No
6	35	171	84	28.7	Yes	Gardener	Strong	Yes
7	32	170	65	22.5	No	Stonemason	Light	No
8	32	169	92	32.2	Yes	Seller	No	No
9	37	164	89.4	32.2	Yes	Employee	Light	No
10	30	185	120	35.1	Ex	Operator	Light	No

Data recorded after the clinical interview: BMI categories: underweight = 18.5; normal weight = 18.5–24.9; overweight = 25–29.9; obesity = BMI of 30 or greater.

2.6. Sperm Assays. Conventional sperm assay was performed according to WHO criteria (2010) [20]. Samples were obtained by masturbation after a period of 3 to 5 days of sexual abstinence. Subjective motility was evaluated by expert technicians. Sperm motility was classified as progressive motility (PR), nonprogressive motility (NP), and immotility (IM). Sperm vitality was evaluated using eosin Y; the sample was stained with 0.5% eosin Y in 0.9% aqueous sodium chloride solution; bright red (dead) and uncolored (living) cells were scored by light microscope. Finally, semen smears were Papanicolaou stained and morphology was assessed using Kruger's strict criteria. Conventional sperm parameters were also expressed in absolute values as recommended by WHO [20].

A Sperm Class Analyzer CASA system (SCA Microptic SL, Barcelona, Spain) was employed to assess kinetic parameters and sperm count. The basic components of the system are a bright field microscope with phase contrast Ph- (negative phase contrast) microscopy to visualize the sample (Nikon E-200, Japan) lens magnification: 10x, a digital camera to capture images (Basler A312 Inc., Vision-Technology, Germany), and a computer with SCA software installed. Slides with samples were laid on a thermostatic plate at 37°C. A minimum of 400 sperm cell tracks were captured and 25 digitized images per second were analyzed for each sample. The assays were conducted in accordance with instrument's standardization and validation, by using a Leja chamber 10 (10 μm in depth) [22]. A qualified operator validated each analyzed image. The improved Neubauer haemocytometer was used only in cases of very low count. Appropriate dilutions were made in Mac Comber fixative (formaldehyde: 1 mg/mL, NaHCO3: 5 g made up to 100 mL with purified water).

2.7. Statistic. Evaluation results after supplementation were based on Wilcoxon signed ranks test. Multiple linear regression was employed to predict sperm fatty acid profile (Software: Infostat-Universidad Nacional de Córdoba; 2014 version). p values less than 0.05 were considered statistically significant.

3. Results

3.1. Patient's Data. Demographic data are shown in Table 1. Only two patients showed relevant clinical data, one suffering asthma and the other parathyroid cancer. While analyzing the food intake questionnaire, interesting data were recorded; the main data is that neither of the participants consumed fish; their diet was based on red meat and carbohydrates, that is, the typical "Western diet." Their infertility history ranged from secondary infertility to 6 years of primary infertility.

3.2. Red Blood Cell Membrane Lipidomic Profile. In erythrocytes membranes, it was possible to monitor the fatty acid residues of membrane phospholipids, and we were interested in seeing the changes of the EPA and DHA levels in the patients while comparing baseline and postsupplementation profile, with variation of the fatty acid levels after supplementation.

The values were reported as relative percentages of a cohort of 12 fatty acids chosen as the most significant components of the erythrocyte membrane fatty acids [18, 21]. The percentages were obtained from the gas chromatographic analysis (GC) of fatty acid methyl esters derived from the membrane phospholipids, following previously published procedures [17]. GC is the gold standard for the fatty acid analysis and the relative percentages in erythrocyte membranes were used in other widely used lipid biomarker evaluations, such as the ω-3 cardiovascular risk index [19].

The comparison of the found values in the patient cohort could also be made with known interval values registered for the Italian population and integrated with literature data reported for erythrocyte membranes of healthy subjects [21].

At the beginning of the study, all the patients had low levels of EPA and DHA in red blood cells membranes and increased values of the saturated fatty acid/monounsaturated fatty acid and ω6/ω3 ratios (Table 2). Eight of the 10 participants had a high cardiovascular risk index, measured in erythrocyte membranes as % eicosapentaenoic + docosahexaenoic acid (0–4%) [19].

TABLE 2: Fatty acid profile in erythrocyte membranes.

FAME[a]	Subjects before supplementation ($n = 10$; % rel)	Subjects after supplementation ($n = 9$; % rel)	Optimal interval values (% rel)[b]	p value
16:0	26.8 (24.3; 29.8)	27.3 (24.1; 29)	17–27	0.8326
18:0	20.7 (19.1; 21.4)	19.5 (16.9; 21.5)	13–20	0.0294
9cis-16:1	0.4 (0.2; 0.6)	0.6 (0.4; 0.7)	0.2–0.5	0.04
9cis-18:1	15.8 (14.2; 18.3)	16.0 (14.3; 17.7)	9–18	0.2138
11cis-18:1	1.4 (1.1; 1.7)	1.3 (1; 1.6)	0.7–1.3	0.4072
18:2 (omega-6, LA)	10.3 (9.5; 14.3)	12.3 (10.0; 14.7)	9–16	0.0044
20:3 (omega-6, GLA)	2.0 (1.4; 3.4)	2.3 (1.3; 3.3)	1.9–2.4	0.863
20:4 (omega-6, ARA)	18.1 (16.6; 19.1)	15.9 (14.0; 19.2)	13–17	0.0092
20:5 (omega-3, EPA)	0.4 (0.2; 0.5)	0.9 (0.6; 1.2)	0.5–0.9	0.0112
22:6 (omega-3, DHA)	3.3 (2.2; 4.5)	4.3 (2.7; 6.3)	5–7	0.0366
Trans-18:1[c]	0.2 (0.1; 0.3)	0.1 (0; 0.2)[c]	0.1–0.3	0.0306
Total SFA	47.5 (45.0; 50.7)	46.9 (42.6; 48.5)	30–45	0.1636
Total MUFA	17.5 (16.1; 20.6)	18.1 (15.9; 19.7)	13–23	0.2856
Total PUFA	34.2 (32.4; 38.2)	35.5 (31.5; 39.8)	28–39	0.3868
SFA/MUFA	2.7 (2.3; 3.1)	2.5 (2.4; 3.0)	1.7–2	0.0196
Omega-6/omega-3	8.4 (6.2; 13.7)	5.7 ± (4.5; 9.4)	3.5–5.5	0.01
CV risk index[d]	3.6 (2.6; 4.8)	5.2 (3.4; 7.2)	0–4%, high risk 4–8%, intermediate risk >8%, minimal risk	0.0098

The values of the fatty acids are reported as relative percentage (% rel) of the total of 12 fatty acids chosen as representative components of the erythrocyte membrane fatty acids GC analysis as reported in [17]. The values were given as median (min; max). n is the number of subjects. [a]FAME (fatty acid methyl ester) was determined performing membrane phospholipid extraction, derivatization, and GC analysis as described. The identification of the peaks was performed using authentic samples as described [17, 18]. [b]The optimal interval values are reported from [17], as found in a survey of healthy controls reported in the scientific literature and compared with a group of 2500 analyses obtained from the Italian population]. [c]9-trans-18:1 is considered. [d]As described in [19], p values less than 0.05 were considered statistically significant.

From the membrane lipid profile obtained in the patient cohort, it was possible to envisage a decreased functionality due to poor polyunsaturated components such as ω-3 fatty acids. Further evaluation of the erythrocyte functionality will be matter of specific studies. It is worth noting that the fish intake is known to influence the presence of the ω-3 long chain fatty acids and give health benefits [23], and the food questionnaire of these patients showed no fish consumption. Besides being involved as disease risk, the ω-3 fatty acids in membranes are recognized as essential elements for the correct functionality [24]. There is no doubt on the need of recovering the membrane status of these patients; therefore a supplementation was assigned to follow for 3 months.

After supplementation of 1 capsule of omega-3-based nutraceutical and 1 capsule of DHA-based nutraceutical per day to the patients (see Materials and Methods), the erythrocyte membrane profiles were obtained. The effects of supplementation were clearly visible. They were statistically significant for EPA ($p < 0.01$) and DHA ($p < 0.03$). In particular, arachidonic acid levels in the erythrocyte membranes decreased after supplementation ($p < 0.009$). This could be indicated as favourable remodelling of the erythrocyte membranes due to the ω-3 supplementation, with exchange of membrane lipids with the fatty acid pool enriched with the ω-3 supplementation, so that less inflammatory status can be obtained from a better ω-6/ω-3 balance. Considering one patient who exited from the study, the risk of cardiovascular disease after treatment decreased to the intermediate range (4–8%) in 8 out of the 9 studied patients.

3.3. Sperm. Only two out of the studied patients were asthenospermic, while the rest were severely comprised as being oligoasthenoteratospermic. In sperm cells, the fatty acids were also evaluated before and after supplementation. Table 4 reports the sperm cell lipidomic profile with the full recognition of all fatty acids, expressed as relative percentages of the total fatty acid found in the samples. It is important to note that in Table 4 the fatty acids were recognized with the library of cis- and trans fatty acid, as described elsewhere [17, 25]. It is remarkable that while conventional sperm parameters (Table 3) were recorded after treatment of 3 months with the omega-3 formulations and no changes were detected, sperm lipidomic profile after supplementation also did not show significant changes. Due to the low sperm count found in most

TABLE 3: Sperm parameters.

Sperm parameter	Before treatment ($n = 10$)	After treatment ($n = 9$)	p value
Semen volume (mL)	3.0 (1.7; 7.5)	3.5 (1.2; 4.5)	0.52
Sperm concentration ($\times 10^6$/mL)	1.5 (0.05; 53.6)	1.1 (0; 64.3)	0.09
Total sperm count ($\times 10^6$/ejaculate)	3.28 (0.09; 214.4)	4.95 (0; 211.75)	0.85
Progressive motility (%)	4.5 (0; 36.9)	3 (0; 20)	0.32
Total progressive motility ($\times 10^6$/ejaculate)	0.3 (0; 24.45)	0.18 (0; 11.63)	0.60
Normal morphology (%)	2 (0; 14)	0 (0; 11)	0.72
Total normal morphology ($\times 10^6$/ejaculate)	0 (0; 30.02)	0 (0; 23.29)	0.49
Vitality (%)	55 (0; 89)	66 (0; 80)	0.37
Total vitality ($\times 10^6$/ejaculate)	0.63 (0; 190.82)	0.8 (0; 156.7)	0.66
Average path velocity (μm/sec)	33.65 (10.5; 66.8)	20.05 (7.6; 43.3)	0.04
Straight line velocity (μm/sec)	27.2 (5.8; 59.1)	16.7 (6.3; 36.2)	0.16
Curvilinear velocity (μm/sec)	41.6 (19.6; 83.7)	27.5 (9.4; 59.2)	0.04

The values are expressed as median (min; max). Sperm assays were performed according to WHO criteria (2010) [20] ($n = 9$). The kinetic parameters were determined with CASA system (SCA Microptic) ($n = 4$). p values less than 0.05 were considered statistically significant.

TABLE 4: Fatty acid profile in sperm cells isolated from the human.

FAME[a]	Samples before supplementation ($n = 8$; % rel)	Samples after supplementation ($n = 8$; % rel)	p value
14:0	1.4 (0.83; 1.84)	1.41 (0.76; 2.55)	0.61
16:0	33.3 (30.23; 40.78)	33.79 (24.12; 35.41)	0.25
17:0	3.21 (1.28; 3.99)	2.08 (1.25; 2.82)	0.12
18:0	12.67 (7.55; 16.47)	10.33 (8.73; 27.07)	0.75
20:0	2.5 (1.02; 3.05)	1.54 (1.11; 2.63)	0.48
22:0	7.67 (2.05; 10.12)	4.53 (1.78; 12.54)	0.76
TOT. SFA	58.03 (52.06; 64.22)	55.01 (38.72; 66.74)	0.19
6c 16:1	1.38 (1.0; 1.78)	1.9 (0.65; 3.98)	>0.99
9c 16:1	3.99 (2.63; 6.31)	3.68 (1.72; 18.15)	>0.99
9c 18:1	9.85 (6.56; 12.85)	8.52 (7.2; 15.15)	0.44
11c 18:1	6.96 (3.77; 9.72)	4.79 (2.05; 19.27)	0.87
20:1	3.33 (0.57; 6.79)	3.83 (1.15; 11.87)	0.29
22:1	1.25 (0.33; 2.14)	0.88 (0.26; 1.56)	0.12
TOT. MUFA	27.48 (18.96; 30.67)	26.5 (20.29; 52.94)	0.54
SFA/MUFA	2.17 (1.84; 2.81)	2.17 (0.73; 3.01)	0.4
18:2	4.21 (3.32; 5.56)	5.0 (2.5; 6.19)	0.37
20:2	0.66 (0.16; 1.48)	0.36 (0.26; 1.25)	0.48
20:3	1.2 (0.37; 2.44)	1.21 (0.65; 2.54)	0.91
20:4 (arachidonic acid)	1.54 (0.91; 2.47)	1.23 (0.63; 3.21)	0.41
TOT. ω6	7.54 (6.77; 9.71)	8.12 (5.0; 10.27)	0.71
20:5 (EPA)	0.46 (0.16; 1.2)	0.31 (0.17; 0.59)	0.34
22:5	0.95 (0.44; 2.41)	0.62 (0.35; 3.72)	0.15
22:6 (DHA)	3.02 (1.46; 19.47)	3.5 (1.41; 10.4)	0.71
TOT. ω3	4.59 (3.02; 22.06)	4.45 (1.93; 11.79)	0.78
ω6/ω3	1.75 (0.34; 2.57)	1.59 (0.62; 3.3)	0.46
Trans fatty acid (9t-18:1)	0.18 (0; 0.65)	0.12 (0; 0.26)	0.24

The values of the fatty acids are reported as relative percentage (% rel) of the fatty acid peak areas detected in the GC analysis (>98% recognized peaks). The values were given as median (min; max). n is the number of samples. [a]FAME (fatty acid methyl ester) was determined performing membrane lipid extraction, derivatization, and GC analysis as described [17, 18]. p values less than 0.05 were considered statistically significant.

of the patients, sperm kinetics could only be evaluated in four of them. Slight differences could be envisaged in average path and curvilinear velocity after treatment (Table 3).

EPA, DHA, the $\omega6/\omega3$ ratio, and arachidonic acid in erythrocytes were good predictors of EPA in sperm (R^2: 0.975) with $\beta_{coefficient}$: β_{ARA} = 0.23 (p-value = 0.020); β_{EPA} = −1.474 (p value = 0.020); β_{DHA} = 0.871 (p value = 0.004); and $\beta\ \omega6/\omega3$ = 0.189 (p value = 0.014) at the beginning of the study, the prediction equation being

$$EPA_{sperm} = -7.8 + 0.23\left[arachidonic\ acid_{ery}\right]$$
$$- 1.474\left[EPA_{ery}\right] + 0.871\left[DHA_{ery}\right] \tag{1}$$
$$+ 0.189\left[\frac{\omega6}{\omega3}ery\right].$$

After the supplementation, the correspondence between sperm and erythrocyte was completely lost.

4. Discussion

The fatty acid analysis and lipidomic profiles of erythrocytes and sperm cells from infertile patients were performed following published protocols [14, 17]. Nutraceutical treatment was effected with 1 capsule of ω-3 and 1 capsule of DHA per day for 3 months.

After the supplementation, the level of the ω-3 fatty acids in plasma can be obviously found more elevated; however we were interested in following the incorporation in phospholipid structures that can be much less efficient. Phospholipids of red blood cells are of great value for their fatty acid asset as multifaceted expression of *de novo* biosynthesis, nutrient uptake, and exchange for efficient tissue maintenance. The mean erythrocyte lifetime of 120 days renders the membrane lipidome analysis strongly related to the stabilized dietary habits and to the individual metabolic status. The diet evolved with a ratio of ω-6/ω-3 essential fatty acids of approximately 3/1-4/1 along the centuries of human development, whereas in the last century with industrialized diets (denominated "Western diets") the ratio reached 20–30/1 [24]. The change of ω-6/ω-3 ratio has been correlated with the pathogenesis of many diseases including cardiovascular disease, cancer, and inflammatory and autoimmune diseases. Despite the strong correlation between fatty acids and health conditions, a very scarce application of lipidomic tools has been observed for monitoring dietary intakes of fatty acids and their consequent molecular contribution. The new field of nutrilipidomics indicates the strategy of lipidomic molecular diagnostics to envisage the nutritional elements needed to be supplemented to the individual in a personalized manner [18, 21]. This strategy takes into account two main factors: (i) the target of a lipid intervention, addressing membranes as the most important functional compartment of fatty acids in living organisms; (ii) the personalization of the type and dose of fatty acids to be used and the importance to use cocktail of fatty acids with molecules able to control free radical and oxidation processes during biodistribution. The role of multicomponent supplementation containing fatty acids in order

to enhance the bioavailability of the supplementation will be matter of further studies.

Our results were followed up by clinicians who evidenced that the patients regularly took the nutraceuticals and declared to have improved their general status. The lipid composition of the sperm cell membrane has been shown to exert a significant effect upon the functional quality of spermatozoa. Compared with normozoospermic samples, asthenozoospermic samples showed lower levels of PUFA and higher amount of saturated fatty acids [26]. In accordance with data reported by Lenzi et al. [14], we found that palmitic and stearic acids were the most representative saturated fatty acids in whole spermatozoa, oleic acid was the most frequent monounsaturated one, the essential fatty acid, linoleic acid, was the main ω-6 fatty acid, and DHA was the most representative long chain PUFA ω-3 of mature sperm cells. In spite of this general feature, our data on DHA were much lower than the data reported by the authors, being comparable to those found by the same group in immature germ cells. The fact that the patients gave the indication of low fish intake in the diet can be explicative of our findings. A higher DHA content was associated with better sperm morphology and function. On the other hand, it is well known that polyunsaturated fatty acids are particularly susceptible to peroxidation damage by free radicals. Although reactive oxygen species (ROS) play significant role for physiological sperm function, when the production of potentially destructive ROS exceeds the natural antioxidant defences, oxidative stress is connected with cell damage. Moreover, oxidative stress at the level of the testicular microenvironment may result in decreased spermatogenesis and sperm damage. Total antioxidant capacity in seminal plasma was directly related to PUFA, omega-3, and DHA. Detrimental effects of lipid peroxidation should decrease sperm quality and be responsible for fertility problems, so we believe that the higher contents of DHA in good sperm samples are dependent not only on nutrition but also on the overall antioxidant capacity of each person. We did not perform any specific measurement of oxidative stress nor evaluated ROS production in the gamete and neither did we determine the lipid profile in seminal plasma that was not the target of this study, focused on the sperm. The spermatozoa are in tight equilibrium with seminal plasma and are finally the responsible male cell of fertilization outcome.

The supplementation of EPA and DHA was at a dosage that simulated the intake of ω-3 fatty acids from dietary sources. Indeed, several reports regarding ω-3 fatty acid supplementation made use of higher dosage, therefore not simulating dietary intakes and finding correlation between the supplementation and the dosage of ω-3 fatty acids in the body [27]. In our case, the findings that 3 months of supplementation of nutritional dosages of ω-3 fatty acids is successful for erythrocyte membranes are important for further studies on the "effective" dosages to be used in nutraceutical interventions that reach sperm cells.

We could verify that 3 months of supplementation did not change conventional sperm parameters in a significant manner. Data on sperm kinetics, although few, are promising to envisage slight differences in sperm function; they should be further studied. Our results give important and first evidence that there are quite different distribution rates of supplemented fatty acid in the membrane lipids of different tissues, comparing the changes recorded in erythrocytes and sperm cells of the same patient. Therefore, molecular diagnostic tools that are of great help in understanding clinical outcomes can monitor the success of supplementations, as well as of dietary treatments. As it could be expected, the statistical prediction equation could not be applied to the erythrocyte and sperm values after supplementation.

The ω6/ω3 ratio was much higher in our patients than the one found in oligoasthenospermic samples by Aksoy et al. [28]. A balanced ω6/ω3 ratio proved to be necessary in male rat reproduction highlighting the necessity to determine an appropriate ω6/ω3 ratio in man in the future [29].

Spermatogenesis is a very complex, highly organized and regulated process that takes place in the seminiferous epithelium of testis tubules and involves three major fundamental biological processes: the renewal of stem cells and the production and expansion of progenitor cells (mitosis); the reduction, by one-half, of the number of chromosomes in each progenitor cell (meiosis); and the unique differentiation of haploid cells (spermiogenesis). In our experience, three months of nutraceutical supplementation was not sufficient to remodel sperm membrane lipidomics, at least in these patients having also severe alteration in their red blood cell membrane lipidomic profile. Our study encourages longer trials with omega-3 supplementation to allow sperm lipidomic to adjust and then evaluate the treatment in regard to the irreversibility of some teratospermic cases.

It would be interesting to pursue a multicentric survey of lipidomic profiles in order to expand the data from the initial collection of data performed with the Italian population [18, 21] and also compare geographically different patient cohorts with the average of healthy controls. As a matter of fact, the profound deficit of ω-3 EPA and DHA in the patients with increased arachidonic acid presence can create proinflammatory conditions, which are responsible for impaired functionality of cells. The reduced levels of ω-3 express the fish intakes [30], which is in fact absent in the cohort and nowadays is remarked as responsible for health impairment [23, 31]. Overall, the ω-3 contribution to membrane asset and functionality is widely recognized [32]; therefore the clinical outcomes cannot be established without taking into account the personalized fatty acid levels and needs.

We agree with Tavilani et al. [33] who stated that lipid content is regulated locally within the male reproductive tract. To our knowledge this is the first report on a mathematic equation that could predict EPA in sperm by EPA, DHA, the ratio ω-6/ω-3, as compared with optimal values found in literature [18, 21], and arachidonic acid in erythrocyte membranes, thus reflecting the real impact of nutrition on individual male reproductive health in regard to fatty acid intake. We interpret the idea that the equilibrium installed between sperm and red cell lipidomic profile in relation to diet is disrupted by treatment. A first hypothesis is that administration of the capsules for a longer period would have redounded in a new equilibrium favouring spermatogenesis. More generally speaking, the approach of molecular diagnostics together with the parallel analysis of erythrocyte and sperm cell

membranes can be suggested as diagnostic approach helping the clinical observation of diseases such as, in our particular scenario, the "male fertility."

Conflict of Interests

Gabriela Mendeluk and Mariano Cohen declare no conflict of interests. Chryssostomos Chatgilialoglu and Carla Ferreri are cofounders of Lipinutragen srl, spin-off company recognized by the Consiglio Nazionale delle Ricerche (Italy).

Acknowledgments

This research was supported by a Clinical Grant from University of Buenos Aires Science and Technology (UBACYT-CB06). The collaboration between CNR (Bologna, Italy) and the Laboratory of Male Fertility (Buenos Aires, Argentina) was performed in the frame of the European Cooperation in Science and Technology, COST Action 1201: Biomimetic Radical Chemistry. The authors are truthfully grateful to the sponsors of this study, The University of Buenos Aires and the Ministry of Science, Technology and Productive Innovation of Argentina. They wish to thank Daniel Ruffini for his technical support, Dr. Norma Pugliese and Dr. Julia Ariagno for helping in the logistics of the biological samples shipping, Dr. Maria Lujan Calcagno for statistical advice, and Dr. Patricia Chenlo for statistical analysis. Also they would like to express their appreciation to the nutritionists who guided the nutritional questionnaire and in particular to all the staff of the Laboratory of Male Fertility and Dr. Sunda and Mr. Deplano (Lipinutragen srl, Bologna) for technical assistance in lipidomic analysis.

References

[1] Y. Barazani, B. F. Katz, H. M. Nagler, and D. S. Stember, "Lifestyle, environment, and male reproductive health," *Urologic Clinics of North America*, vol. 41, no. 1, pp. 55–66, 2014.

[2] L. A. Wise, K. J. Rothman, E. M. Mikkelsen, H. T. Sørensen, A. Riis, and E. E. Hatch, "An internet-based prospective study of body size and time-to-pregnancy," *Human Reproduction*, vol. 25, no. 1, pp. 253–264, 2010.

[3] S. C. I. Moore, A. V. Patel, C. E. Matthews et al., "Leisure time physical activity of moderate to vigorous intensity and mortality: a large pooled cohort analysis," *PLoS Medicine*, vol. 9, no. 11, Article ID e1001335, 2012.

[4] K. Michalakis, G. Mintziori, A. Kaprara, B. C. Tarlatzis, and D. G. Goulis, "The complex interaction between obesity, metabolic syndrome and reproductive axis: a narrative review," *Metabolism: Clinical and Experimental*, vol. 62, no. 4, pp. 457–478, 2013.

[5] M. C. Afeiche, N. D. Bridges, P. L. Williams et al., "Dairy intake and semen quality among men attending a fertility clinic," *Fertility and Sterility*, vol. 101, no. 5, pp. 1280–1287, 2014.

[6] J. A. Attaman, T. L. Toth, J. Furtado, H. Campos, R. Hauser, and J. E. Chavarro, "Dietary fat and semen quality among men attending a fertility clinic," *Human Reproduction*, vol. 27, no. 5, pp. 1466–1474, 2012.

[7] A. S. di Frega, B. Dale, L. Di Matteo, and M. Wilding, "Secondary male factor infertility after Roux-en-Y gastric bypass for

morbid obesity: case report," *Human Reproduction*, vol. 20, no. 4, pp. 997–998, 2005.

[8] N. Sermondade, N. Massin, F. Boitrelle et al., "Sperm parameters and male fertility after bariatric surgery: three case series," *Reproductive BioMedicine Online*, vol. 24, no. 2, pp. 206–210, 2012.

[9] L. Lazaros, E. Hatzi, S. Markoula et al., "Dramatic reduction in sperm parameters following bariatric surgery: report of two cases," *Andrologia*, vol. 44, no. 6, pp. 428–432, 2012.

[10] M. R. Safarinejad and S. Safarinejad, "The roles of omega-3 and omega-6 fatty acids in idiopathic male infertility," *Asian Journal of Andrology*, vol. 14, no. 4, pp. 514–515, 2012.

[11] B. Ahluwalia and R. T. Holman, "Fatty acid composition of lipids of bull, boar, rabbit and human semen," *Journal of Reproduction and Fertility*, vol. 18, no. 3, pp. 431–437, 1969.

[12] M. Mazza, M. Pomponi, L. Janiri, P. Bria, and S. Mazza, "Omega-3 fatty acids and antioxidants in neurological and psychiatric diseases: an overview," *Progress in Neuro-Psychopharmacology and Biological Psychiatry*, vol. 31, no. 1, pp. 12–26, 2007.

[13] A. A. Farooqui, L. A. Horrocks, and T. Farooqui, "Glycerophospholipids in brain: their metabolism, incorporation into membranes, functions, and involvement in neurological disorders," *Chemistry and Physics of Lipids*, vol. 106, no. 1, pp. 1–29, 2000.

[14] A. Lenzi, L. Gandini, V. Maresca et al., "Fatty acid composition of spermatozoa and immature germ cells," *Molecular Human Reproduction*, vol. 6, no. 3, pp. 226–231, 2000.

[15] W. A. Robbins, L. Xun, L. Z. FitzGerald, S. Esguerra, S. M. Henning, and C. L. Carpenter, "Walnuts improve semen quality in men consuming a western-style diet: randomized control dietary intervention trial," *Biology of Reproduction*, vol. 87, no. 4, Article ID Article 101, pp. 101–108, 2012.

[16] J. E. Chavarro, L. Mínguez-Alarcón, J. Mendiola, A. Cutillas-Tolín, J. J. López-Espín, and A. M. Torres-Cantero, "*Trans* fatty acid intake is inversely related to total sperm count in young healthy men," *Human Reproduction*, vol. 29, no. 3, pp. 429–440, 2014.

[17] A. Ghezzo, P. Visconti, P. M. Abruzzo et al., "Oxidative stress and erythrocyte membrane alterations in children with autism: correlation with clinical features," *PLoS ONE*, vol. 8, no. 6, Article ID e66418, 2013.

[18] C. Ferreri and C. Chatgilialoglu, *Membrane Lipidomics for Personalized Health*, John Wiley & Sons, Chichester, UK, 2015.

[19] W. S. Harris and C. Von Schacky, "The Omega-3 index: a new risk factor for death from coronary heart disease?" *Preventive Medicine*, vol. 39, no. 1, pp. 212–220, 2004.

[20] World Health Organization, *WHO Laboratory Manual for the Examination of Human Semen*, 5th edition, 2010.

[21] C. Ferreri and C. Chatgilialoglu, "Role of fatty acid-based functional lipidomics in the development of molecular diagnostic tools," *Expert Review of Molecular Diagnostics*, vol. 12, no. 7, pp. 767–780, 2012.

[22] P. H. Chenlo, J. I. Ariagno, M. N. Pugliese et al., "Study of human semen: implementation of an objective method," *Acta Bioquímica Clínica Latinoamericana*, vol. 47, pp. 61–69, 2013.

[23] D. Mozaffarian and E. B. Rimm, "Fish intake, contaminants, and human health: evaluating the risks and the benefits," *The Journal of the American Medical Association*, vol. 296, no. 15, pp. 1885–1899, 2006.

[24] A. P. Simopoulos, "Essential fatty acids in health and chronic disease," *The American Journal of Clinical Nutrition*, vol. 70, supplement 3, pp. 560S–569S, 1999.

[25] C. Chatgilialoglu, C. Ferreri, M. Melchiorre, A. Sansone, and A. Torreggiani, "Lipid geometrical isomerism: from chemistry to biology and diagnostics," *Chemical Reviews*, vol. 114, no. 1, pp. 255–284, 2014.

[26] H. Tavilani, M. Doosti, K. Abdi, A. Vaisiraygani, and H. R. Joshaghani, "Decreased polyunsaturated and increased saturated fatty acid concentration in spermatozoa from asthenozoospermic males as compared with normozoospermic males," *Andrologia*, vol. 38, no. 5, pp. 173–178, 2006.

[27] M. L. Nording, J. Yang, K. Georgi et al., "Individual variation in lipidomic profiles of healthy subjects in response to omega-3 fatty acids," *PLoS ONE*, vol. 8, no. 10, Article ID e76575, 2013.

[28] Y. Y. Aksoy, H. Aksoy, K. Altinkaynak, H. R. Aydin, and A. Özkan, "Sperm fatty acid composition in subfertile men," *Prostaglandins Leukotrienes and Essential Fatty Acids*, vol. 75, no. 2, pp. 75–79, 2006.

[29] L. Yan, X.-L. Bai, Z.-F. Fang, L.-Q. Che, S.-Y. Xu, and D. Wu, "Effect of different dietary omega-3/omega-6 fatty acid ratios on reproduction in male rats," *Lipids in Health and Disease*, vol. 12, no. 1, article 33, 2013.

[30] S. A. Sands, K. J. Reid, S. L. Windsor, and W. S. Harris, "The impact of age, body mass index, and fish intake on the EPA and DHA content of human erythrocytes," *Lipids*, vol. 40, no. 4, pp. 343–347, 2005.

[31] Z. S. Tan, W. S. Harris, A. S. Beiser et al., "Red blood cell omega-3 fatty acid levels and markers of accelerated brain aging," *Neurology*, vol. 78, no. 9, pp. 658–664, 2012.

[32] R. C. Valentine and D. L. Valentine, "Omega-3 fatty acids in cellular membranes: a unified concept," *Progress in Lipid Research*, vol. 43, no. 5, pp. 383–402, 2004.

[33] H. Tavilani, A. Vatannejad, M. Akbarzadeh, M. Atabakhash, S. Khosropour, and A. Mohaghgeghi, "Correlation between lipid profile of sperm cells and seminal plasma with lipid profile of serum in infertile men," *Avicenna Journal of Medical Biochemistry*, vol. 2, no. 1, Article ID e19607, 2014.

Plasma Fatty Acids in Zambian Adults with HIV/AIDS: Relation to Dietary Intake and Cardiovascular Risk Factors

Christopher K. Nyirenda,[1,2,3] **Edmond K. Kabagambe,**[3,4] **John R. Koethe,**[3,5]
James N. Kiage,[4] **Benjamin H. Chi,**[6,7] **Patrick Musonda,**[6] **Meridith Blevins,**[3,8]
Claire N. Bosire,[9] **Michael Y. Tsai,**[10] **and Douglas C. Heimburger**[3,4]

[1] *Ndola Central Hospital, School of Medicine, 10101 Ndola, Zambia*

[2] *School of Medicine, Copperbelt University, 10101 Ndola, Zambia*

[3] *Vanderbilt Institute for Global Health, Vanderbilt University, Nashville, TN 37203, USA*

[4] *Division of Epidemiology, Department of Medicine, Vanderbilt University Medical Center, Nashville, TN 37203, USA*

[5] *Division of Infectious Diseases, Department of Medicine, Vanderbilt University Medical Center, Nashville, TN 37232, USA*

[6] *Centre for Infectious Disease Research in Zambia, 10101 Lusaka, Zambia*

[7] *Department of Obstetrics and Gynecology, University of North Carolina at Chapel Hill, Chapel Hill, NC 27599, USA*

[8] *Department of Biostatistics, Vanderbilt University Medical Center, Nashville, TN 37203, USA*

[9] *Division of Cancer Epidemiology and Genetics, National Cancer Institute, Nutritional Epidemiology Branch,*
Bethesda, MD 20850, USA

[10] *Department of Laboratory Medicine and Pathology, University of Minnesota Medical School, Minneapolis, MN 55455, USA*

Correspondence should be addressed to Edmond K. Kabagambe; edmond.kabagambe@vanderbilt.edu

Academic Editor: Duo Li

Objective. To determine whether 24 hr dietary recalls (DR) are a good measure of polyunsaturated fatty acid (PUFA) intake when compared to plasma levels, and whether plasma PUFA is associated with markers of HIV/AIDS progression and cardiovascular disease (CVD) risk. *Methods.* In a cross-sectional study among 210 antiretroviral therapy-naïve HIV-infected adults from Lusaka, Zambia, we collected data on medical history and dietary intake using 24 hr DR. We measured fatty acids and markers of AIDS progression and CVD risk in fasting plasma collected at baseline. *Results.* PUFA intakes showed modest correlations with corresponding plasma levels; Spearman correlations were 0.36 ($p < 0.01$) for eicosapentaenoic acid and 0.21 ($p = 0.005$) for docosahexaenoic acid. While there were no significant associations ($p > 0.05$) between total plasma PUFA and C-reactive protein (CRP) or lipid levels, plasma arachidonic acid was inversely associated with CRP and triglycerides and positively associated with HDL-C, CD4+ T-cell count, and plasma albumin ($p < 0.05$). Plasma saturated fatty acids (SFA) were positively associated with CRP ($\beta = 0.24$; 95% CI: 0.08 to 0.40, $p = 0.003$) and triglycerides ($\beta = 0.08$; 95% CI: 0.03 to 0.12, $p < 0.01$). *Conclusions.* Our data suggest that a single DR is inadequate for assessing PUFA intake and that plasma arachidonic acid levels may modulate HIV/AIDS progression and CVD risk.

1. Introduction

HIV/AIDS is a major cause of morbidity and mortality in Zambia, where in 2009 the national HIV prevalence in the 15–49 year age group was estimated at 13.5% [1], but effective treatment is complicated by a high degree of malnutrition in the HIV-infected population. A review of HIV-infected adults starting antiretroviral therapy (ART) in Lusaka, Zambia, found 33% of patients were undernourished by World Health Organization criteria (e.g., a body mass index (BMI) < 18.5 kg/m²), and 9% were severely malnourished (BMI < 16 kg/m²) [2]. Mortality in the early ART treatment period is higher among patients with low BMI, low CD4+ T-cell counts, low hemoglobin, low serum albumin and phosphate levels, advanced HIV/AIDS stage, and immune reconstitution inflammatory syndrome [3–5].

In Zambia, our group recently reported a higher risk of early ART mortality among individuals with moderately elevated triglyceride concentrations, a finding which may be related to the interaction of fatty acids and systemic inflammation [6]. The essential fatty acids (n-3 and n-6 polyunsaturated fatty acids (PUFA)) are nutrients of primary importance for health, and many research works in the last decades have shown the role of an adequate intake of n-3 and n-6 PUFA in the prevention of several diseases, in particular of cardiovascular diseases [7–9]. These studies were done mainly in non-HIV populations. Omega 3 fatty acids have been known to modulate biomarkers such as C-reactive protein (CRP) and CD4 count which are important determinants in the progression of HIV disease and other inflammatory conditions [10, 11]. To our knowledge, apart from one small clinical trial that assessed the benefit of combining fenofibrate with n-3 fatty acids in improving HIV-related clinical outcomes [12], studies on effects of fatty acids on metabolic parameters associated with morbidity and mortality in HIV are lacking, especially in resource-limited settings.

Reliable identification of individuals with key nutrient deficiencies that could be improved with nutritional support would inform the design of nutritional rehabilitation programs to reduce morbidity and mortality in HIV/AIDS patients. The Zambian diet is mainly composed of cereals, predominantly maize, starchy roots, and, to a lesser extent, fruits and vegetables. Cereals provide almost two-thirds of the dietary energy supply.

In urban areas of Zambia food consumption patterns are changing: rice and sweet potatoes are gaining importance [13]. Urbanization and globalization are responsible for changes in dietary patterns, as consumption is shifting from fresh and minimally processed traditional foods to imported processed foods acquired from supermarkets [14]. Because fish intake in much of Zambia is relatively low [15], we hypothesized that dietary intake of long-chain PUFA may be inadequate in HIV-infected adults and could influence markers of CVD risk and HIV disease progression. In this study, we sought to determine whether 24-hr dietary recalls (DR) conducted in an urban ART clinic setting in Zambia could be used to effectively estimate PUFA intake and whether PUFA and saturated fatty acids (SFA) measured in plasma are associated with markers of cardiovascular disease risk.

2. Methods

2.1. Study Population.
We enrolled HIV-infected adults (age 16–60 years) eligible for ART initiation into the diet, genetic polymorphisms in lipid-metabolizing enzyme genes, and antiretroviral therapy-related dyslipidemia (DGPLEAD) study at Chawama Clinic, a government health centre in Lusaka, Zambia, between January and December 2007. All participants had a BMI $\geq 16 \, \text{kg/m}^2$ and CD4+ lymphocyte count $\geq 50 \, \text{cells}/\mu\text{L}$, and were ART eligible according to the Zambia national HIV guidelines. The DGPLEAD study has been described in detail previously [6]. Dietary and metabolic

profiles described in this study were assessed before initiating ART. Written informed consent was obtained from all participants during the primary study. The research protocol was approved by the Vanderbilt University Institutional Review Board in Nashville Tennessee, USA, and the University of Zambia Biomedical Research Ethics Committee in Lusaka, Zambia.

2.2. Data Collection.
Data collection was conducted by a study nurse, a clinical officer, and a supervising physician (CKN). At the first encounter, medical history and physical examination with anthropometric measurements were performed. A fasting blood sample was drawn on the same day as the 24 hr dietary recalls (DR). Intake of fatty acids as well as that of total energy and other nutrients was assessed using a 24 hr DR that was administered by a study nurse or a clinical officer, each with training by a registered dietitian and aided by commercial food models. Foods and amounts consumed over the preceding 24 hours were recorded by the interviewer. Food composition and nutrient quantity were computed from a modified food composition database using NDS-R software (Nutrition Data System for research software version 2006, developed by the Nutrition Coordinating Center, University of Minnesota, Minneapolis, MN, USA (http://www.ncc.umn.edu/)) [16]. The NDS-R nutrient database was supplemented with nutrient composition data for local staple foods already published by the Zambian National Food and Nutrition Commission (available from http://www.nfnc.org.zm/).

2.3. Laboratory Assays.
Participants were asked to come to the clinic after an 8-hour fast before their blood draw. Participants who reported being in a nonfasting state were rescheduled to return after fasting. Plasma and serum specimens were collected from each participant and used to determine total cholesterol (TC), high density lipoprotein-cholesterol (HDL-C), low-density lipoprotein-cholesterol (LDL-C), triglycerides, insulin, glucose, creatinine, CRP, and albumin. Methods for metabolic assays in DGPLEAD have been described previously [6, 17].

Lipid profiles and glucose were measured using a Roche Cobas Integra 400+ autoanalyzer (Roche Diagnostics, Indianapolis, IN, USA). Triglycerides, LDL-C, and TC concentrations were measured using an enzymatic colorimetric assay while HDL-C was measured using a homogeneous enzymatic colorimetric assay. Serum creatinine, CRP, and albumin concentrations were determined on a Roche Modular P analyzer using bromocresol purple assay for albumin and immunoturbidimetric assay for CRP (Roche Diagnostics, Indianapolis, IN, USA).

Fatty acid concentrations were measured in the phospholipid fraction of plasma using gas chromatography. The extraction and quantification of fatty acids were performed at the University of Minnesota using a standard validated assay that has been described in detail [18–20]. This assay identifies 29 individual fatty acids ranging from 12:0 through 24:1n9.

2.4. Exposure and Dependent Variable Definitions. For the first objective, we focused on pairwise Spearman rank correlations between diet and plasma PUFA measurements. In the second objective, total plasma PUFA and SFA were the main exposure variables. The dependent variables were markers of HIV/AIDS disease progression (i.e., BMI, CD4+ cell count, and serum albumin) and cardiovascular disease risk (i.e., CRP, triglycerides, HDL-C, and LDL-C). In secondary analyses with plasma arachidonic acid as the main independent variable we investigated BMI, CD4+ cell count, plasma albumin, CRP, triglycerides, HDL-C, LDL-C, and plasma albumin as dependent variables.

2.5. Statistical Analysis. Of the 210 participants, 90% had complete data on fatty acids in plasma. Fatty acids in plasma were expressed as a percentage of the total fatty acids analyzed while dietary fatty acids were expressed as a percentage of total energy per day [21]. To validate fatty acid intakes, we computed pairwise Spearman correlation coefficients for each PUFA estimated from 24 hr DR against a corresponding PUFA measured in plasma. The exception was plasma α-linolenic acid in plasma which was tested for correlation with total dietary linolenic acid since α- and γ-linolenic acid could not be separated in our 24 hour DR.

Next we determined whether plasma fatty acids are associated with CRP and lipid profiles using multivariable linear regression models. For each of the dependent variables, that is, CRP, triglycerides, HDL-C, and LDL-C we estimated associations with the exposure variables, namely, total plasma PUFA and SFA adjusted for age, sex, BMI, plasma monounsaturated fatty acids (MUFA), *trans* fatty acid, alcohol use, and smoking.

Additional analyses were conducted to understand whether individual PUFA was associated with markers of CVD risk and HIV disease progression (e.g., CD4+ counts and plasma albumin). In this analysis, Spearman rank correlations between plasma AA and each of the markers of CVD risk and HIV disease progression (e.g., CRP, lipids, CD4+ count, albumin, and BMI) were determined. We then distributed plasma AA concentrations into quartiles and used ANOVA with robust variance estimator to determine whether markers of CVD risk and HIV/AIDS disease progression significantly varied by quartiles of AA before and after adjustment for age, sex, smoking, and alcohol consumption.

Data were analyzed using SAS version 9.4 (Cary, NC) and STATA version 12.1 (College Station, TX).

3. Results

3.1. Description of the Population. The characteristics of the study population are shown in Table 1. Most participants were women (54%) and relatively young, with a median age of 32 years for women and 35 years for men. The proportion of participants with BMI <18.5 kg/m^2 was 32%. Although the median CD4+ cell count was somewhat higher in women (143 cells/μL) compared to men (129 cells/μL), this difference did not reach statistical significance ($p > 0.05$). Concentrations

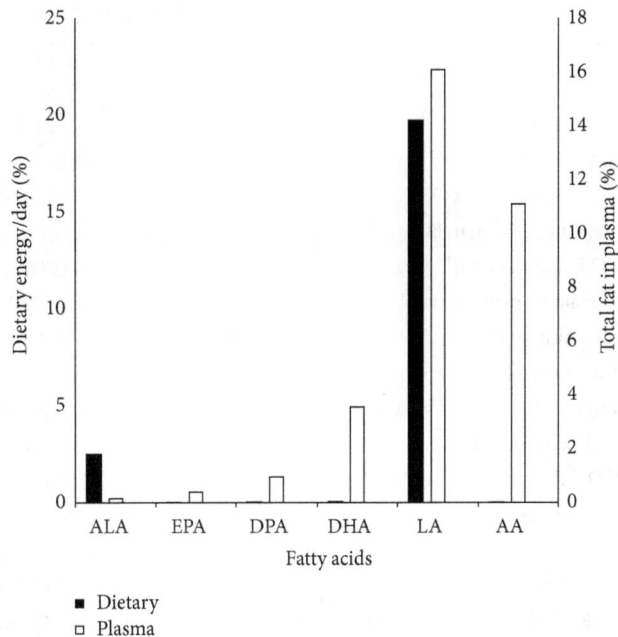

FIGURE 1: Proportions of major n-3 and n-6 fatty acids in diet and plasma of HIV/AIDS patients enrolled from January to December, 2007, in Lusaka, Zambia. Values are median intakes for 24-hour dietary recalls and median proportions for plasma fatty acids. ALA = α-Linolenic acid; EPA = eicosapentaenoic acid; DPA = docosapentaenoic acid; DHA = docosahexaenoic acid; LA = linoleic acid; AA = arachidonic acid.

of CRP also tended to be higher in women than in men but the differences were not statistically significant ($p > 0.05$). The frequency of smoking was quite low in our study population with only one (0.9%) current smoker among women and 10 (11%) among men ($p < 0.05$).

3.2. Validation of the Fatty Acid Estimates from 24 hr DR. The proportional distributions of fatty acids in the diet expressed as percent energy intake per day and in the plasma as percent of total fat are shown in Figure 1. The highest median percent energy was from linoleic acid (LA) (11.5%) and the lowest was from EPA (0.007%). The highest median percent of total fat in the plasma was from LA (16.1%) and the lowest was from α-linolenic acid (ALA) (0.15%).

Table 2 shows the Spearman correlation coefficients between plasma and corresponding dietary PUFA. The correlations are for the major n-3 and n-6 fatty acids. Two of the six fatty acids examined showed statistically significant correlations. The correlations ranged from very low to moderately low. The highest significant correlation was for EPA ($r = 0.36$, $p < 0.01$) and the second highest was for DHA ($r = 0.21$, $p = 0.005$). The correlations for ALA, DPA, LA, and AA were not statistically significant.

In multivariable linear regression analyses, total plasma PUFA concentrations were not significantly associated with CRP on a log scale ($\beta = -0.10$; 95% CI: -0.22 to 0.02, $p = 0.09$) in analyses adjusting for age, sex, BMI, MUFA, *trans* fatty acids, current smoking, and alcohol status (data not

TABLE 1: Baseline characteristics of the study population.

Variable	Women $n = 113$	Men $n = 97$	p
Age, years	32 (27, 37)	35 (31, 40)	0.003
BMI, kg/m^2	19.6 (18.0, 21.7)	19.6 (18.0, 21.3)	0.62
Current smoker, n (%)	1 (0.93)	10 (10.5)	0.003
Current drinker, n (%)	10 (9.26)	12 (12.6)	0.41
CD4, count, cell/μL	143 (108, 175)	129 (88, 168)	0.07
C-reactive protein, mg/L	12.6 (2.01, 32.2)	5.71 (1.51, 32.1)	0.24
Albumin, g/dL	3.00 (2.60, 3.40)	3.20 (2.60, 3.70)	0.09
Total cholesterol, mmol/L	3.56 (2.94, 4.11)	3.21 (2.68, 3.75)	0.01
Triglycerides, mmol/L	1.10 (0.89, 1.51)	1.05 (0.81, 1.43)	0.39
LDL-cholesterol, mmol/L	2.13 (1.60, 2.50)	1.77 (1.34, 2.38)	0.02
HDL-cholesterol, mmol/L	0.81 (0.47, 1.09)	0.71 (0.52, 1.04)	0.42
TC : HDL-C, ratio	4.49 (3.44, 7.00)	4.73 (3.32, 6.12)	0.47
Total energy intake, kcal/day	1588 (1179, 1956)	1777 (1356, 2305)	0.03
Total fat, % energy/day	34.1 (27.8, 42.2)	28.9 (20.5, 35.5)	<0.0001
Total *trans* fatty acids, % energy/day	0.27 (0.09, 0.51)	0.30 (0.12, 0.63)	0.37
Total saturated fatty acids, % energy/day	6.52 (5.12, 8.00)	5.52 (3.74, 7.64)	0.02
Total MUFA, % energy/day	9.47 (7.84, 11.8)	8.05 (5.72, 10.9)	0.01
Total dPUFA, % energy/day	15.1 (11.4, 19.2)	11.1 (8.0, 14.8)	<0.0001
n-3 dPUFAs, % energy/day	1.89 (1.30, 2.60)	1.40 (0.92, 1.99)	<0.0001
n-6 dPUFAs, % energy/day	14.6 (10.9, 18.6)	10.9 (7.8, 14.0)	<0.0001
Total pPUFA, % total fat	37.1 (35.2, 39.1)	36.2 (34.4, 38.4)	0.07
n-3 pPUFA, % total fat	5.51 (4.57, 5.98)	4.85 (4.28, 5.49)	0.01
n-6 pPUFA, % total fat	31.9 (29.9, 33.1)	31.7 (29.6, 33.3)	0.38

Values are median (25th, 75th percentile) unless otherwise stated.
BMI, body mass index; HDL, high density lipoprotein; LDL, low density lipoprotein; MUFA, monounsaturated fatty acids; PUFA, polyunsaturated fatty acids; dPUFA, PUFA from dietary sources; pPUFA, PUFA levels determined from plasma; TC, total cholesterol.

TABLE 2: Spearman correlation coefficients between plasma and dietary polyunsaturated fatty acids.

PUFA type	Plasma composition (% of total fat)	Dietary composition (% energy/day)	r	p
ALA (18:3n-3)	0.15 (0.12, 0.20)	1.50 (0.96, 1.92)	−0.05	0.50
EPA (20:5n-3)	0.39 (0.28, 0.51)	0.01 (0, 0.12)	0.36	<0.01
DPA (22:5n-3)	0.95 (0.77, 1.13)	0.01 (0, 0.03)	−0.01	0.90
DHA (22:6n-3)	3.54 (2.98, 4.27)	0.03 (0, 0.17)	0.21	0.01
LA (18:2n-6)	16.1 (14.6, 18.0)	11.5 (8.10, 14.9)	0.06	0.09
AA (20:4n-6)	11.1 (9.58, 12.3)	0.04 (0.01, 0.08)	0.09	0.21

Plasma and dietary composition values are median (25th, 75th percentile).
AA, arachidonic acid; ALA, α-linolenic acid; DHA, docosahexaenoic acid; DPA, docosapentaenoic acid; EPA, eicosapentaenoic acid; LA, linoleic acid; PUFA, polyunsaturated fatty acid.

shown). A positive, though weak, association was observed between total plasma PUFA and triglycerides on a log scale ($\beta = 0.03$; 95% CI: 0.002 to 0.06, $p = 0.04$). No significant associations were observed between total plasma PUFA and HDL-C or LDL-C.

Plasma AA was inversely correlated with CRP (spearman correlation coefficient, $r = -0.19$, $p = 0.02$) and triglycerides ($r = -0.14$, $p = 0.05$) and positively associated with CD4+ count ($r = 0.16$, $p = 0.02$), albumin ($r = 0.44$, $p < 0.001$), HDL-C ($r = 0.26$, $p = 0.01$), and LDL-C ($r = 0.29$, $p = 0.001$). As shown in Table 3, the associations between AA and CRP, serum albumin, triglycerides, HDL-C, and LDL-C remained significant after adjustment for age, sex, smoking, and alcohol consumption in ANOVA models.

In multivariable regression, a positive association was also observed between total plasma SFA and CRP ($\beta = 0.24$; 95% CI: 0.08 to 0.40, $p = 0.003$) and between SFA and triglyceride concentrations ($\beta = 0.08$; 95% CI: 0.03 to 0.12, $p < 0.01$). We did not detect an association with HDL-C and LDL-C.

4. Discussion

We observed modest correlations between EPA and DHA estimated from 24 hr DR and corresponding measures in plasma but weak correlations for other PUFAs. We also observed null associations between total plasma PUFA and markers of HIV/AIDS disease progression and CVD-risk, but significant beneficial associations were observed between

TABLE 3: Quartiles of plasma arachidonic acid (AA) levels and markers of cardiovascular disease and HIV/AIDS progression.

Variable	Model	1st Quartile ($n = 49$)	2nd ($n = 49$)	3rd ($n = 49$)	4th ($n = 49$)	p
BMI, kg/m^2	1	19.6 ± 0.34	19.9 ± 0.37	20.3 ± 0.41	20.5 ± 0.46	0.38
	2	20.1 ± 0.5	20.5 ± 0.5	20.8 ± 0.6	20.9 ± 0.6	0.39
CD4, cells/μL	1	130 ± 9	122 ± 8	145 ± 7	147 ± 7	0.02
	2	133 ± 12	126 ± 13	145 ± 12	150 ± 13	0.04
Albumin, g/dL	1	2.6 ± 0.1	3.1 ± 0.1	3.2 ± 0.1	3.4 ± 0.1	<0.0001
	2	2.7 ± 0.1	3.2 ± 0.2	3.3 ± 0.1	3.5 ± 0.2	<0.0001
C-reactive protein, mg/L	1	32.2 ± 8.1	25.6 ± 6.7	30.2 ± 7.2	11.4 ± 2.4	0.004
	2	26.9 ± 8.2	19.9 ± 6.5	24.8 ± 7.5	4.3 ± 5.6	0.01
Triglycerides, mmol/L	1	1.30 ± 0.09	1.27 ± 0.10	1.37 ± 0.09	1.04 ± 0.06	0.01
	2	1.26 ± 0.15	1.24 ± 0.16	1.37 ± 0.19	0.98 ± 0.15	0.003
HDL-C, mmol/L	1	0.61 ± 0.06	0.81 ± 0.07	0.79 ± 0.06	0.99 ± 0.09	0.004
	2	0.68 ± 0.09	0.86 ± 0.11	0.84 ± 0.09	1.05 ± 0.12	0.01
LDL-C, mmol/L	1	1.66 ± 0.13	1.99 ± 0.10	2.17 ± 0.09	2.12 ± 0.08	0.01
	2	1.78 ± 0.18	2.12 ± 0.15	2.24 ± 0.14	2.22 ± 0.15	0.02

Model 1: unadjusted; Model 2: means and standard errors are adjusted for age, sex, and smoking and alcohol consumption. p-values are from ANOVA models that use robust standard errors.

plasma AA and these markers. Higher plasma AA concentrations were significantly associated with higher HDL-C, LDL-C, CD4+ cell counts, and serum albumin, a profile consistent with improved patient survival.

The finding that ALA provided the highest median percent energy intake for the n-3 dietary fatty acids (1.5% or 2.5 g/day) is consistent with the notion that ALA is the major form of n-3 fatty acids obtained from dietary sources [22, 23]. The reported intake for ALA in our study is slightly higher than the range reported among western adults (0.5 to 2 g/d) [24], probably due to higher content of foods from plant sources in Zambia compared to western countries. The highest median percent proportion of total fat in the plasma among the n-3 fatty acids was from DHA (3.5%). The observation that LA contributed the highest median dietary percent energy (11.5% or 19.7 g/day) is consistent with LA's position as the major n-6 fatty acid obtained from the diet [22, 23]. As in the diet, LA was the major n-6 PUFA in plasma. Dietary LA undergoes a metabolic process which converts it to various n-6 fatty acids including AA. This may explain why even if our study reported very low median dietary intake of AA (0.04%) compared to LA (11.5%), the plasma proportions were closer (AA 11.1%, LA 16.1%).

The sources of ALA and n-6 PUFA in the Zambian diet are likely to include vegetable oils such as soybean and sunflower oil. Other sources are nuts, meats, and eggs. These food items are readily available and generally affordable in urban areas, which may explain the relatively high consumption levels reported in our study. Fish and other forms of seafood are much better sources of very long chain n-3 fatty acids such as EPA and DHA [24] but are not readily available. In Zambia, despite the availability of large bodies of fresh waters from rivers and lakes, the fish industry is not well developed; very little fish farming is practiced [15]. Thus, consumption of long-chain fatty acids in Zambia (0.08 g/day for EPA, DPA, and DHA) is lower than what is reported among adults in Northern and Eastern Europe, North America, and Australasia (mean intakes ~0.15–0.25 g/d) [25].

The correlation coefficients for EPA and DHA measured in 24 hr DR and plasma in our study were 0.36 and 0.21, respectively. These correlations were somewhat higher than those observed for EPA ($r = 0.21$) and lower than that for DHA (r ranges from 0.42 to 0.48), LA ($r = 0.25$), and ALA ($r = 0.23$) as measured from food frequency questionnaire and plasma studies [26, 27]. The correlations were also higher than those observed for EPA and DHA measured from a food frequency questionnaire and adipose tissue ($r = 0.15$ and 0.18, resp.) [21]. Comparable correlations of 0.15 and 0.61 for EPA and higher correlations of 0.47 and 0.57 for DHA in men and women, respectively, have been reported in a previous study [28]. In another diet-adipose tissue study, the correlation for PUFA was 0.4 and that for linolenic was 0.34 [29].

The finding from our study that total plasma PUFA was not associated with lower CRP and lipid profiles may be a result of opposing actions among fatty acids when combined as total PUFA, in contrast with the roles they play individually. For example, supplementation with fish oil rich in EPA and DHA (n-3 PUFAs) has been reported to reduce inflammation in conditions such as colitis in both animal and clinical studies [30, 31]. In contrast, AA (an n-6 PUFA) is known to exert both proinflammatory and anti-inflammatory effects through its metabolites [32, 33].

Our finding that AA was inversely associated with CRP and triglycerides and positively associated with HDL-C, CD4+ cell count and plasma albumin suggests that individual fatty acids may influence clinical outcomes in HIV/AIDS patients. However, this observation needs further investigation.

The observation that SFA was positively associated with CRP ($\beta = 0.24$, $p = 0.003$) is consistent with findings from a previous study in which, after adjusting for other covariates,

SFA emerged as the single most important nutrient contributing to an increase in serum CRP levels [34]. The finding that SFA was positively associated with serum triglyceride concentration (β = 0.08, p < 0.01) supports evidence from previous studies that found SFA to be positively associated with coronary heart disease risk [35, 36].

In a clinical trial, daily supplementation with 1 g of n-3 fatty acids did not reduce the rate of cardiovascular events in patients at high risk for cardiovascular events [37]. However, the study reported a significant reduction in triglyceride levels (0.16 mmol/L), more among patients receiving n-3 fatty acids than among those receiving placebo (p < 0.001), without a significant effect on other lipids. Similarly, in our secondary analyses total n-3 fatty acids were not associated with reductions in CRP or improved lipid profiles and this may justify the need to explore the role of individual fatty acids in improving the CVD risk profiles.

The cross-sectional nature of our study limits causal inference between the exposure variables and the dependent variables of interest. The study could have been prone to interviewer bias arising from inconsistency in the way the DR was administered by different interviewers. The study could also have been prone to reporting bias which may have arisen from the participants being inclined to report healthier foods more than the less healthy foods. To mitigate these potential biases, the interviewers were specifically trained to elicit complete dietary histories from all participants. Lastly, the study population of adult HIV patients was recruited at a single health facility, which could limit generalizability to HIV-infected individuals in other settings. We also acknowledge that because the study was done among patients not yet on ART, we do not know the associations between fatty acids and markers of HIV/AIDS disease progression or CVD risk among those already on ART. Studies that assess effects of ART before and during ART are warranted so as to determine whether supplementation with fatty acids will modulate outcomes from ART in resource-limited settings.

5. Conclusions

The significant but generally low diet-plasma long-chain PUFA correlations could suggest that a single self-reported 24 hr DR may be inadequate for assessing PUFA intake in HIV/AIDS patients in Zambia. The study also suggests that SFA, which were positively related to markers of CVD risk, could play a role in HIV-related cardiovascular disease. Our study has found no evidence that total PUFA are inversely associated with CVD risk markers in HIV patients. However, there was evidence from secondary analyses that individual fatty acids, particularly AA, may play a role in improving CVD risk profiles and markers of HIV/AIDS disease progression.

Disclaimer

The content is solely the responsibility of the authors and does not necessarily represent the official views of the National Institutes of Health or other funding agencies.

Conflict of Interests

The authors declare that there is no conflict of interests regarding the publication of this paper.

Authors' Contribution

Study conception and design were performed by Edmond K. Kabagambe, Christopher K. Nyirenda, and Douglas C. Heimburger. Data acquisition was conducted by Edmond K. Kabagambe, Douglas C. Heimburger, Claire N. Bosire, Christopher K. Nyirenda, Benjamin H. Chi, and Michael Y. Tsai. Data analysis, interpretation, and paper drafting or critical review were accomplished by Christopher K. Nyirenda, Edmond K. Kabagambe, Claire N. Bosire, Patrick Musonda, Meridith Blevins, James N. Kiage, John R. Koethe, Benjamin H. Chi, and Douglas C. Heimburger.

Acknowledgments

This research was conducted to fulfill the requirements for CN's Master of Public Health degree, and the authors acknowledge the support and guidance of Sten H. Vermund, M.D., Ph.D. (Director, Vanderbilt Institute for Global Health), Holly Cassell, MPH (Director of Operations and Management, Vanderbilt Institute for Global Health), William O. Cooper, M.D., MPH (former Director of the Vanderbilt MPH program), and Kasonde Bowa, MBChB, M.S. MMed (Inaugural Dean of the Copperbelt University, School of Medicine, Ndola, Zambia). This study was supported by the Nutrition Obesity Research Center (through Grant DK056336), the Department of Epidemiology, and Center for AIDS Research at the University of Alabama at Birmingham. Additional support was obtained from the Fulbright Scholars Program, United States Department of State, the Fogarty International Center, and the National Institute of Allergy and Infectious Diseases of the National Institutes of Health under Award nos. D43 TW001035, R24 TW 007988, and R21 AI 076430.

References

[1] UNAIDS, *Report on the Global AIDS Epidemic 2010, Joint United Nations Programme on HIV/AIDS*, UNAIDS, 2010, http://www.unaids.org/globalreport/documents/20101123_GlobalReport_full_en.pdf.

[2] J. R. Koethe, A. Lukusa, M. J. Giganti et al., "Association between weight gain and clinical outcomes among malnourished adults initiating antiretroviral therapy in Lusaka, Zambia," *Journal of Acquired Immune Deficiency Syndromes*, vol. 53, no. 4, pp. 507–513, 2010.

[3] J. S. A. Stringer, I. Zulu, J. Levy et al., "Rapid scale-up of antiretroviral therapy at primary care sites in Zambia: feasibility and early outcomes," *Journal of the American Medical Association*, vol. 296, no. 7, pp. 782–793, 2006.

[4] D. C. Heimburger, J. R. Koethe, C. Nyirenda et al., "Serum phosphate predicts early mortality in adults starting antiretroviral therapy in Lusaka, Zambia: a prospective cohort study," *PLoS ONE*, vol. 5, no. 5, Article ID e10687, 2010.

[5] M. Müller, S. Wandel, R. Colebunders, S. Attia, H. Furrer, and M. Egger, "Immune reconstitution inflammatory syndrome in patients starting antiretroviral therapy for HIV infection: a systematic review and meta-analysis," *The Lancet Infectious Diseases*, vol. 10, no. 4, pp. 251–261, 2010.

[6] J. N. Ngu, D. C. Heimburger, D. K. Arnett et al., "Fasting triglyceride concentrations are associated with early mortality following antiretroviral therapy in Zambia," *North American Journal of Medical Sciences*, vol. 3, pp. 79–88, 2010.

[7] D. N. Carrol and M. T. Roth, "Evidence for the cardioprotective effects of omega-3 fatty acids," *Annals of Pharmacotherapy*, vol. 36, no. 12, pp. 1950–1956, 2002.

[8] L. Djoussé, S. C. Hunt, D. K. Arnett, M. A. Province, J. H. Eckfeldt, and R. C. Ellison, "Dietary linolenic acid is inversely associated with plasma triacylglycerol: the National Heart, Lung, and Blood Institute Family Heart Study," *American Journal of Clinical Nutrition*, vol. 78, no. 6, pp. 1098–1102, 2003.

[9] L. Djoussé, A. R. Folsom, M. A. Province, S. C. Hunt, and R. C. Ellison, "Dietary linolenic acid and carotid atherosclerosis: the National Heart, Lung, and Blood Institute Family Heart Study," *The American Journal of Clinical Nutrition*, vol. 77, no. 4, pp. 819–825, 2003.

[10] Y. Okamoto, K. Okano, K. Izuishi, H. Usuki, H. Wakabayashi, and Y. Suzuki, "Attenuation of the systemic inflammatory response and infectious complications after gastrectomy with preoperative oral arginine and ω-3 fatty acids supplemented immunonutrition," *World Journal of Surgery*, vol. 33, no. 9, pp. 1815–1821, 2009.

[11] I. Reinders, J. K. Virtanen, I. A. Brouwer, and T.-P. Tuomainen, "Association of serum n-3 polyunsaturated fatty acids with C-reactive protein in men," *European Journal of Clinical Nutrition*, vol. 66, no. 6, pp. 736–741, 2012.

[12] J. G. Gerber, D. W. Kitch, C. J. Fichtenbaum et al., "Fish oil and fenofibrate for the treatment of hypertriglyceridemia in HIV-infected subjects on antiretroviral therapy: results of ACTG A5186," *Journal of Acquired Immune Deficiency Syndromes*, vol. 47, no. 4, pp. 459–466, 2008.

[13] W. S. Siamusantu, *Zambia Nutrition Profile—Nutrition and Consumer Protection Division*, FAO, 2009.

[14] K. S. Reddy, *Chronic Disease Epidemics in Developing Countries: Can We Telescope Transition?* University of Zambia Medical Library, 2001.

[15] D. B. Nyirenda, M. Musukwa, R. H. Mugode, and J. Shindano, *Revised Food Composition Tables of Zambia*, Zambia Ministry of Health, 2005.

[16] J. R. Koethe, M. Blevins, C. Bosire et al., "Self-reported dietary intake and appetite predict early treatment outcome among low-BMI adults initiating HIV treatment in sub-Saharan Africa," *Public Health Nutrition*, vol. 16, no. 3, pp. 549–558, 2013.

[17] J. N. Kiage, D. C. Heimburger, C. K. Nyirenda et al., "Cardiometabolic risk factors among HIV patients on antiretroviral therapy," *Lipids in Health and Disease*, vol. 12, no. 1, article 50, 2013.

[18] E. K. Kabagambe, M. Y. Tsai, P. N. Hopkins et al., "Erythrocyte fatty acid composition and the metabolic syndrome: a National Heart, Lung, and Blood Institute GOLDN study," *Clinical Chemistry*, vol. 54, no. 1, pp. 154–162, 2008.

[19] E. K. Kabagambe, J. M. Ordovas, P. N. Hopkins, M. Y. Tsai, and D. K. Arnett, "The relation between erythrocyte trans fat and triglyceride, VLDL- and HDL-cholesterol concentrations

depends on polyunsaturated fat," *PLoS ONE*, vol. 7, no. 10, Article ID e47430, 2012.

[20] J. Cao, K. A. Schwichtenberg, N. Q. Hanson, and M. Y. Tsai, "Incorporation and clearance of omega-3 fatty acids in erythrocyte membranes and plasma phospholipids," *Clinical Chemistry*, vol. 52, no. 12, pp. 2265–2272, 2006.

[21] A. Baylin, E. K. Kabagambe, X. Siles, and H. Campos, "Adipose tissue biomarkers of fatty acid intake," *American Journal of Clinical Nutrition*, vol. 76, no. 4, pp. 750–757, 2002.

[22] P. M. Kris-Etherton, D. S. Taylor, S. Yu-Poth et al., "Polyunsaturated fatty acids in the food chain in the United States," *American Journal of Clinical Nutrition*, vol. 71, no. 1, pp. 179S–188S, 2000.

[23] DHA/EPA Omega 3-Institute, *Metabolism of Omega-6 and Omega-3 Fatty Acids and the Omega 6: Omega 3 Ratio*, DHA/EPA Omega 3-Institute, 2013.

[24] British Nutrition Foundation, *Briefing Paper: n-3 Fatty Acids and Health*, British Nutrition Foundation, 1999.

[25] Scientific Advisory Committee on Nutrition (SACN), *Advice on Fish Consumption: Benefits and Risks*, Scientific Advisory Committee on Nutrition (SACN), London, UK, 2004.

[26] Q. Sun, J. Ma, H. Campos, S. E. Hankinson, and F. B. Hu, "Comparison between plasma and erythrocyte fatty acid content as biomarkers of fatty acid intake in US women," *The American Journal of Clinical Nutrition*, vol. 86, no. 1, pp. 74–81, 2007.

[27] P. Berstad, E. Thiis-Evensen, M. H. Vatn, and K. Almendingen, "Fatty acids in habitual diet, plasma phospholipids, and tumour and normal colonic biopsies in young colorectal cancer patients," *Journal of Oncology*, vol. 2012, Article ID 254801, 9 pages, 2012.

[28] A. Tjonneland, K. Overvad, E. Thorling, and M. Ewertz, "Adipose tissue fatty acids as biomarkers of dietary exposure in Danish men and women," *American Journal of Clinical Nutrition*, vol. 57, no. 5, pp. 629–633, 1993.

[29] M. Garland, F. M. Sacks, G. A. Colditz et al., "The relation between dietary intake and adipose tissue composition of selected fatty acids in US women," *American Journal of Clinical Nutrition*, vol. 67, no. 1, pp. 25–30, 1998.

[30] D. Camuesco, J. Gálvez, A. Nieto et al., "Dietary olive oil supplemented with fish oil, rich in EPA and DHA (n-3) polyunsaturated fatty acids, attenuates colonic inflammation in rats with DSS-induced colitis," *The Journal of Nutrition*, vol. 135, no. 4, pp. 687–694, 2005.

[31] R. Lorenz, P. C. Weber, P. Szimnau, W. Heldwein, T. Strasser, and K. Loeschke, "Supplementation with n-3 fatty acids from fish oil in chronic inflammatory bowel disease—a randomized, placebo-controlled, double-blind cross-over trial," *Journal of Internal Medicine*, vol. 225, no. 731, pp. 225–232, 1989.

[32] E. Ricciotti and G. A. FitzGerald, "Prostaglandins and inflammation," *Arteriosclerosis, Thrombosis, and Vascular Biology*, vol. 31, pp. 986–1000, 2011.

[33] C. N. Serhan, S. Krishnamoorthy, A. Recchiuti, and N. Chiang, "Novel anti-inflammatory—pro-resolving mediators and their receptors," *Current Topics in Medicinal Chemistry*, vol. 11, no. 6, pp. 629–647, 2011.

[34] S. Arya, S. Isharwal, A. Misra et al., "C-reactive protein and dietary nutrients in urban Asian Indian adolescents and young adults," *Nutrition*, vol. 22, no. 9, pp. 865–871, 2006.

[35] D. C. Heimburger and J. D. Ard, *Handbook of Clinical Nutrition*, Mosby, 4th edition, 2006.

[36] K. T. Khaw, M. D. Friesen, E. Riboli, R. Luben, and N. Wareham, "Plasma phospholipid fatty acid concentration and incident coronary heart disease in men and women: the EPIC-Norfolk prospective study," *PLoS Medicine*, vol. 9, no. 7, Article ID e1001255, 2012.

[37] J. Bosch, H. C. Gerstein, G. R. Dagenais et al., "n-3 fatty acids and cardiovascular outcomes in patients with dysglycemia," *The New England Journal of Medicine*, vol. 367, no. 4, pp. 309–318, 2012.

8

Prevalence of Overweight and Obesity among Students in the Kumasi Metropolis

D. B. Kumah, K. O. Akuffo, J. E. Abaka-Cann, D. E. Affram, and E. A. Osae

Department of Optometry and Visual Science, College of Science,
Kwame Nkrumah University of Science and Technology, Kumasi, Ghana

Correspondence should be addressed to D. B. Kumah; ben56kay@gmail.com

Academic Editor: Michael B. Zemel

The aim was to determine the prevalence of obesity and overweight among students in the Kumasi metropolis. In a descriptive cross-sectional study, 500 students aged 10 to 20 years were examined from two junior high schools selected by multistage sampling technique and three randomly selected senior high schools. Height and weight were measured in all participants and the body mass index (BMI) of each individual was calculated. Body mass index classes were calculated according to the International Obesity Task Force standards. Out of the 500 students, 290 (58.00%) were males and 210 (42.00%) were females. The prevalence of underweight, normal weight, overweight, and obesity was 7.40%, 79.60%, 12.20%, and 0.80%, respectively. Overweight was more prevalent among students than obesity. There is therefore the need to establish effective public health promotion campaigns among students in order to curtail future implications on health.

1. Introduction

Obesity and overweight have both been described as anomalous accumulation of excessive body fat which may be harmful to health [1]. There is no single cause to explain all cases of obesity and overweight but most studies implicate imbalance in the amounts of calories consumed and those expended [1]. Energy breakdown is said to be less than energy buildup. The disruption of the normal satiety feedback mechanisms, hyperinsulinism, insulin resistance, and genetics are some of the biophysiological causes of obesity and overweight [2].

Some researchers also attribute obesity and overweight to obesogenic environments where people are frequently exposed to and consume savory foods with hidden fats and sugars that can impair metabolism and lead to obesity [3]. Some public health experts also associate the development of obesity and overweight with socioeconomic status, urban lifestyle, family size, physical inactivity, educational status, cultural factors, and poor eating habits [4–6]. Persons who spend their leisure inactively such as in prolonged watching of television and playing of video games have been said to be at risk of obesity [7].

In addition to being pathologically chronic, obesity and overweight have been associated with other comorbidities in both younger and older populations. Heart disease, vision problems, cancer, hepatic impairment, diabetes, and other disease conditions as well as the economic burden involved in managing these conditions have been associated with obesity and overweight [8, 9]. Obese and overweight individuals are also stigmatized; in some societies, people portray a disease stigma towards those suffering from obesity and overweight and see them as immoral, lazy, unclean, and voracious [10].

While the majority of the researches done highlight obesity and overweight as problems of the developed countries, recent studies also show that the third world countries are no exception [11, 12]. Reports show that the issue of childhood and adolescence obesity in third world countries requires public health attention [12, 13]. During the time of this research, there were limited published findings on childhood and adolescence obesity and overweight in Ghana.

This study aimed to determine the prevalence of obesity and overweight among students aged 10–20 years in the Kumasi metropolis of Ghana, to sensitize the public on the emerging trend of childhood and adolescence obesity and overweight in Ghana and to provide data for public health professionals and policy planners.

2. Methods

2.1. Sampling. A descriptive cross-sectional study of 500 students aged 10 to 20 years was carried out from May 2010 to June 2010 in the Kumasi metropolis, Ghana. Simple random sampling was used to select three senior high schools (SHS). Multistage sampling technique was used to select 2 junior high schools (JHS). Simple random sampling was used to select one submetropolis from a total of 10 submetropolises. Within the selected submetropolis, simple random sampling was used to select two junior high schools. In each selected school, one class out of each educational level was randomly selected. Students included in the study were apparently healthy.

2.2. Sample Demographics. Kumasi is an urban and cosmopolitan settlement. Therefore, students recruited for the study came from diverse ethnic backgrounds; most of them were Ashantis and others belonged to represented minor groups from Northern Ghana. The study respondents were largely from middle class homes. Parents of the recruited students had little or no formal education and worked as tradesmen and tradeswomen and skilled professionals.

2.3. Anthropometric Measurement. Height and weight were measured using standardized protocols [14]. Weight was measured without shoes to the nearest 0.1 kg using a single previously standardized portable weighing scale. Height was measured without shoes and recorded to the nearest 0.1 cm with a height rod fixed on a wall. The body mass index (BMI) of each individual was calculated as weight in kilograms divided by height in metres squared. The BMI classes we adopted are the same as those used by the International Obesity Task Force for the international overweight and obesity cutoffs for children [15, 16].

2.4. Data Analysis. The statistical package for social scientists (SPSS) software, version 16.0, Chicago, USA, was used to analyze the collected data. Descriptive statistics were used to examine anthropometric characteristics. Independent *t*-test was used to test gender differences in anthropometric characteristics.

2.5. Ethical Consideration. This research was conducted with approval from the heads of the selected junior and senior high schools and from the teachers. Informed consent forms were signed by parents to approve the participation of their wards.

3. Results and Discussion

A total of 500 students aged 10 to 20 years took part in the study. Two hundred and ninety (58.00%) males and 210 (42.00%) females were included in the study. The overall mean age was 15.91 ± 1.78 years (16.00 ± 2.00 years for the males and 15.00 ± 2.00 years for the females).

TABLE 1: Summary of anthropometric measurements per sex of study subjects.

Parameters	Males Mean ± SD [95% CI]	Females Mean ± SD [95% CI]
Weight (kg)	54.71 ± 9.27 [53.64, 55.78]	51.12 ± 10.36 [49.71, 52.52]
Height (cm)	160.85 ± 9.03 [159.80, 161.89]	157.24 ± 8.97 [156.02, 158.46]
BMI (kg/m^2)	21.04 ± 2.64 [20.74, 21.35]	20.57 ± 3.27 [20.13, 21.02]

CI: confidence interval; BMI: body mass index.

Table 1 shows a summary of the measured anthropometric parameters per sex of study subjects. Males were significantly heavier and taller than females ($P < 0.05$) but there were no differences in BMI ($P > 0.05$).

The overall prevalence of underweight, normal weight, overweight, and obesity was 7.40%, 79.60%, 12.20%, and 0.80%, respectively. Among males and females, prevalence of overweight was 6.80% and 5.40%, respectively, while the prevalence of obesity was 0.20% and 0.60%, respectively. These results are slightly close to those reported in a study done among students in Nigeria [17] where there was a high prevalence of overweight and a low prevalence of obesity.

Comparable to other studies done in Ghana and other second world countries [18, 19], this study showed that the prevalence rate of overweight was higher among female students than male students. There are however divergent findings in studies done in developed nations [20, 21] where prevalence rates of overweight among male study subjects were higher than among females. For this study, the difference in the prevalence of obesity and overweight between the genders was insignificant ($P > 0.05$); the apparently higher values for males could be confounded by a greater number of male respondents (290) as compared to the female respondents (210).

While challenges to our study design did not allow for understanding reasons for the distribution of obesity and overweight among the genders, other studies report cultural attributions as to why females may be more overweight or obese than their male counterparts [17]. A study showed that obese females considered putting on weight as a sign of affluence and happiness. Others also believed that being obese afforded the needed strength in their sport and accorded them the necessary respect [22].

This is supported by undocumented reports from some areas in Ghana, where excessive weight gain is considered to be the result of eating good and healthy food. Table 2 summarizes the prevalence of underweight, normal weight, overweight, and obesity according to the gender of the students.

Findings from this study must be considered in the light of certain challenges to the study design. The study depended on anthropometric measurements to establish whether or not the students were of the following categories: underweight, normal weight, overweight, or obesity. A study conducted in

TABLE 2: Prevalence of underweight, normal weight, overweight, and obesity according to gender.

Gender	Prevalence rate (%)				
	Underweight	Normal weight	Overweight	Obesity	Total
Male	2.60	48.20	6.80	0.20	**57.80**
Female	4.80	31.40	5.40	0.60	**42.20**
Total	**7.40**	**79.60**	**12.20**	**0.80**	**100.00**

Ghana extensively looked at how the various weight statuses were linked to socioeconomic status of subjects and level of education of students' parents [23]. Another study also found significant associations between smoking and overweight and obesity [17] while other reports reveal positive associations between some psychosocial factors (e.g., loneliness and social isolation) [23, 24].

Even though this study did not look at socioeconomic status, smoking, dietary factors, loneliness, behavioural patterns, age of menarche, and order of births of the students and how these may influence their different weight statuses, it is reasonable from other studies [17, 23–26] to say that a complex interplay of these factors could contribute to our findings in this study.

To the best of our knowledge, the study has provided an insight into the prevalence of obesity and overweight among children and adolescents aged 10–20 years in the Kumasi metropolis of Ghana.

4. Conclusion

The study revealed a high prevalence of overweight (12.20%) among the students. Therefore, there is a need to establish effective prevention and health promotion programmes among the students. This would enable maintaining healthy weights and avoiding the possible immediate and long-term health complications associated with overweight and obesity.

Conflict of Interests

The authors declare that there is no conflict of interests regarding the publication of this paper.

References

[1] World Health Organization, *Obesity and Overweight*, 2009, http://www.who.int/mediacentre/factsheets/fs311/en/index .html.

[2] P. Codogno and A. J. Meijer, "Autophagy: a potential link between obesity and insulin resistance," *Cell Metabolism*, vol. 11, no. 6, pp. 449–451, 2010.

[3] H. David, "An integrative view of obesity," *The American Journal of Clinical Nutrition*, vol. 91, supplement, pp. 280S–283S, 2010.

[4] H. M. Al-Hazzaa and A. A. Al-Rasheedi, "Adiposity and physical activity levels among preschool children in Jeddah, Saudi Arabia," *Saudi Medical Journal*, vol. 28, no. 5, pp. 766–773, 2007.

[5] A. Fazah, C. Jacob, E. Moussa, R. El-Hage, H. Youssef, and P. Delamarche, "Activity, inactivity and quality of life among Lebanese adolescents," *Pediatrics International*, vol. 52, no. 4, pp. 573–578, 2010.

[6] A. A. Al-Nuaim, Y. Al-Nakeeb, M. Lyons et al., "The prevalence of physical activity and sedentary behaviours relative to obesity among adolescents from Al-ahsa, Saudi Arabia: rural versus urban variations," *Journal of Nutrition and Metabolism*, vol. 2012, Article ID 417589, 9 pages, 2012.

[7] M. K. Sodhi, "TV viewing versus play—trends and impact on obesity," *Online Journal of Health and Allied Sciences*, vol. 9, no. 2, 2010.

[8] P. W. Franks, R. L. Hanson, W. C. Knowler, M. L. Sievers, P. H. Bennett, and H. C. Looker, "Childhood obesity, other cardiovascular risk factors, and premature death," *The New England Journal of Medicine*, vol. 362, no. 6, pp. 485–493, 2010.

[9] Y. C. Wang, K. McPherson, T. Marsh, S. L. Gortmaker, and M. Brown, "Health and economic burden of the projected obesity trends in the USA and the UK," *The Lancet*, vol. 378, no. 9793, pp. 815–825, 2011.

[10] R. M. Puhl and C. A. Heuer, "Obesity stigma: important considerations for public health," *The American Journal of Public Health*, vol. 100, no. 6, pp. 1019–1028, 2010.

[11] S. Moore, J. N. Hall, S. Harper, and J. W. Lynch, "Global and national socioeconomic disparities in obesity, overweight, and underweight status," *Journal of Obesity*, vol. 2010, Article ID 514674, 11 pages, 2010.

[12] World Health Organization, *Global Strategy on Diet, Physical Activity and Health: Childhood Overweight and Obesity*, 2010, http://www.who.int/dietphysicalactivity/childhood/en/.

[13] A. K. Ziraba, J. C. Fotso, and R. Ochako, "Overweight and obesity in urban Africa: a problem of the rich or the poor?" *BMC Public Health*, vol. 9, article 465, 2009.

[14] T. Lohman, A. Roche, and R. Martorell, *Anthropometric Standardization Reference Manual*, Human Kinetics Books, Champaign, Ill, USA, 1988.

[15] T. J. Cole, M. C. Bellizzi, K. M. Flegal, and W. H. Dietz, "Establishing a standard definition for child overweight and obesity worldwide: international survey," *British Medical Journal*, vol. 320, no. 7244, pp. 1240–1243, 2000.

[16] T. J. Cole, K. M. Flegal, D. Nicholls, and A. A. Jackson, "Body mass index cut offs to define thinness in children and adolescents: international survey," *British Medical Journal*, vol. 335, no. 7612, pp. 194–197, 2007.

[17] K. Peltzer and S. Pengpid, "Overweight and obesity and associated factors among school-aged adolescents in Ghana and Uganda," *International Journal of Environmental Research and Public Health*, vol. 8, no. 10, pp. 3859–3870, 2011.

[18] M. E. G. Armstrong, M. I. Lambert, K. A. Sharwood, and E. V. Lambert, "Obesity and overweight in South African primary school children—the health of the nation study," *South African Medical Journal*, vol. 96, no. 5, pp. 439–444, 2006.

[19] S. H. Hamaideh, R. Y. Al-Khateeb, and A. B. Al-Rawashdeh, " Overweight and obesity and their correlates among Jordanian adolescents," *Journal of Nursing Scholarship*, vol. 42, no. 4, pp. 387–394, 2010.

[20] C. Curry, S. N. Gabhainn, E. Godeau et al., *Inequalities in Young People's Health: HBSC International Report from the 2005/2006 Survey*, WHO Regional Office for Europe, Copenhagen, Denmark, 2008.

[21] A. Yngve, I. De Bourdeaudhuij, A. Wolf et al., "Differences in prevalence of overweight and stunting in 11-year olds across Europe: the Pro Children study," *European Journal of Public Health*, vol. 18, no. 2, pp. 126–130, 2008.

[22] T. Puoane, L. Tsolekile, and N. Steyn, "Perceptions about body image and sizes among black African girls living in Cape Town," *Ethnicity and Disease*, vol. 20, no. 1, pp. 29–34, 2010.

[23] A.-K. M. Ibrahim, A.-H. M. Ali, and F. E. Sivarajan, "Predictors of obesity in school-aged Jordanian adolescents," *International Journal of Nursing Practice*, vol. 16, no. 4, pp. 397–405, 2010.

[24] B. K. Potter, L. L. Pederson, S. S. H. Chan, J.-A. L. Aubut, and J. J. Koval, "Does a relationship exist between body weight, concerns about weight, and smoking among adolescents? An integration of the literature with an emphasis on gender," *Nicotine and Tobacco Research*, vol. 6, no. 3, pp. 397–425, 2004.

[25] H. Mohammed and F. Vuvor, "Prevalence of childhood overweight/obesity in basic school in Accra," *Ghana Medical Journal*, vol. 46, no. 3, pp. 124–12, 2012.

[26] R. K. Goyal, V. N. Shah, B. D. Saboo et al., "Prevalence of overweight and obesity in Indian adolescent school going children: its relationship with socioeconomic status and associated lifestyle factors," *Journal of Association of Physicians of India*, vol. 58, no. 3, pp. 151–158, 2010.

9

A Comparative Study of the Metabolic and Skeletal Response of C57BL/6J and C57BL/6N Mice in a Diet-Induced Model of Type 2 Diabetes

Elizabeth Rendina-Ruedy,[1] **Kelsey D. Hembree,**[1] **Angela Sasaki,**[1] **McKale R. Davis,**[1] **Stan A. Lightfoot,**[2] **Stephen L. Clarke,**[1] **Edralin A. Lucas,**[1] **and Brenda J. Smith**[1]

[1]Department of Nutritional Sciences, Oklahoma State University, Stillwater, OK 74078, USA
[2]Center for Cancer Prevention and Drug Development, University of Oklahoma Health Sciences Center, Oklahoma City, OK 73104, USA

Correspondence should be addressed to Brenda J. Smith; bjsmith@okstate.edu

Academic Editor: H. K. Biesalski

Type 2 diabetes mellitus (T2DM) represents a complex clinical scenario of altered energy metabolism and increased fracture incidence. The C57BL/6 mouse model of diet-induced obesity has been used to study the mechanisms by which altered glucose homeostasis affects bone mass and quality, but genetic variations in substrains of C57BL/6 may have confounded data interpretation. This study investigated the long-term metabolic and skeletal consequences of two commonly used C57BL/6 substrains to a high fat (HF) diet. Male C57BL/6J, C57BL/6N, and the negative control strain, C3H/HeJ, mice were fed a control or HF diet for 24 wks. C57BL/6N mice on a HF diet demonstrated an increase in plasma insulin and blood glucose as early as 4 wk, whereas these responses were delayed in the C57BL/6J mice. The C57BL/6N mice exhibited more severe hepatic steatosis and inflammation. Only the C57BL/6N mice lost significant trabecular bone in response to the high fat diet. The C3H/HeJ mice were protected from bone loss. The data show that C57BL/6J and C57BL/6N mice differ in their metabolic and skeletal response when fed a HF diet. These substrain differences should be considered when designing experiments and are likely to have implications on data interpretation and reproducibility.

1. Introduction

Increasing prevalence of type 2 diabetes mellitus (T2DM) has stimulated research focused on the pathogenesis and treatment of T2DM and its complications. Initial studies examining fracture as a possible complication of T2DM indicated that type 2 diabetics were not at risk of fracture based on bone mineral density (BMD), the clinical standard for screening [1–3]. However, data analyzed from clinical trials with fracture as an outcome variable instead of BMD revealed that both men and women with T2DM experience an increase in fracture (i.e., 1.5–3-fold) beginning 5–10 years after diagnosis [4–8]. Collectively, the clinical evidence indicates that, independent of BMD, type 2 diabetics are at increased risk of fracture that is exacerbated over time.

Rodent models have enabled investigators to study the molecular mechanisms involved in altering bone quality in T2DM [9]. One of the most commonly utilized models has been the C57BL/6 young growing mouse fed a diet high in total and saturated fat (HF), which exhibits an increase in adiposity, impaired glucose tolerance, and dyslipidemia, similar to prediabetes in humans [10–12]. C57BL/6 mice have been reported to exhibit decreased trabecular bone and either increased, decreased, or no change in cortical bone in response to long-term intake of a HF diet [13–17]. When alterations in bone microarchitecture occur, they are usually accompanied by impaired bone quality as evidenced by compromised biomechanical properties [15, 17–21]. Factors that may contribute to some of the discrepancies in the literature describing the skeletal response to a HF diet could be due to the composition of the diets, the age of the mice, the duration of the study, and, importantly, differences in the C57BL/6 mouse substrain's response (e.g., C57BL/6J or C57BL/6N) [13–16, 21].

A review of published reports revealed that studies utilizing different C57BL/6 substrains (e.g., C57BL/6J and C57BL/6N) are often treated interchangeably without mention of genetic variations that could have important implications on the results and their interpretation. For example, the C57BL/6J mouse has a missense mutation in the gene encoding nicotinamide nucleotide transhydrogenase (*Nnt*) that alters RNA splicing and leads to the deletion of exons 7–11 [22–24]. When C57BL/6J mice are fed a high fat diet, they exhibit impaired glucose tolerance that appears to result from suppressed insulin secretion by pancreatic β-cells [25]. Thus, genetic differences in the C57BL/6J mouse could contribute to some of the discrepancies in the literature in regard to metabolic and skeletal responses reported in the diet-induced obesity model of T2DM.

In contrast to the C57BL/6 mouse, the C3H/HeJ mouse model has been used in mechanistic studies because of its blunted metabolic response to a high fat diet [26, 27]. C3H/HeJ mice have a nonfunctional toll-like receptor (TLR) 4 due to a point mutation in the toll-interleukin 1 receptor domain [26–28]. TLR-4 is expressed on bone cells (i.e., both the osteoblasts and osteoclasts) [29, 30]. Our lab and others [31–34] have shown that TLR-4 ligands (e.g., lipopolysaccharide or LPS and saturated free fatty acids or sFFAs) as well as downstream inflammatory mediators have the potential to uncouple bone turnover. Because of interest in sFFAs and gut-derived LPS in the pathophysiology of T2DM and its complications, the C3H/HeJ strain has become an important research tool to examine the role of TLR-4 in these metabolic responses.

To date, a direct comparison of the long-term metabolic and skeletal responses of the C57BL/6J and C57BL/6N substrains to a HF diet has not been reported in the literature. If, as we hypothesized, the metabolic response to a HF diet in these two substrains differs due to genetic variations, this may alter the inflammatory response, hormones, and adipokines, subsequently affecting the bone. Such differences would be important relative to the interpretation of results and could assist investigators in selecting the most appropriate model. Furthermore, because of our laboratory's interest in TLR-4 and bone, in this study C3H/HeJ mice were used as a negative control for comparative purposes [26, 27].

2. Methods

2.1. Animal Care and Diets. Eight-week-old male mice, C57BL/6N from Charles River (Wilmington, MA) and C57BL/6J and C3H/HeJ mice from Jackson Labs (Bar Harbor, ME), were obtained (n = 30 mice/strain) for these studies. Animals were acclimated for 7 days and then randomly assigned to a control AIN-93M (10% kcals from fat) or a HF (45% kcals from fat; Harlan Teklad, TD.06415) diet for 24 wk. Body weight and food intake were recorded throughout the study. Total feed efficiency was calculated by determining the gain in body weight (mg) per energy unit consumed (kcal)

[35]. Venous tail blood was collected following a 6 hr fast for evaluation of glucose and insulin at 4 wk intervals. After 24 wk, mice were anesthetized (ketamine/xylazine cocktail 70 and 30 mg/kg body weight, resp.) as previously reported and whole body DXA (Lunar PIXI, GE Medical Systems, Madison, WI) scans were performed. Mice were exsanguinated via the carotid artery. An aliquot of blood was collected for total white blood cell (WBC) counts and the remainder processed for plasma in EDTA coated tubes and stored at −80°C. All procedures were approved by the Institutional Animal Care and Use Committee of Oklahoma State University.

2.2. Intraperitoneal Glucose Tolerance Test. One week prior to the end of the study (23rd wk), mice were fasted for 6 hrs and an intraperitoneal (IP) glucose tolerance test (IGTT) was performed. An IP glucose solution (2 g glucose/kg bodyweight) was administered, followed by blood glucose monitoring at 15, 30, 60, 90, and 120 min. Area under the curve (AUC) was determined by calculating the sum of rectangular area between each time point.

2.3. Analysis of Insulin, Adipokines, and Osteocalcin. Plasma insulin was assessed at 4 wk intervals, whereas plasma leptin, adiponectin, and osteocalcin (OCN), both total OCN (Gla-OCN) and undercarboxylated OCN (Glu-OCN), were determined only at the final time point. All assays were performed using commercially available ELISA kits including insulin (Crystal Chem, Downers Grove, IL), leptin and adiponectin (EMD Millipore, Billerica, MA), and Gla-OCN and Glu-OCN (Clontech Takara Bio, Mountain View, CA), following the manufacturer's protocol. Gla-OCN is reported as an indicator of bone turnover and given the importance of the carboxylation status of OCN relative to total OCN on glucose metabolism [20], the ratio of [Glu-OCN]/[Gla-OCN] was calculated to provide insight into the relationship between the skeletal and metabolic response to treatment.

2.4. Body Composition and Bone Densitometry. Whole body DXA scans were performed to determine body composition, bone mineral area (BMA), content (BMC), and BMD. All scans were analyzed using PIXImus Series Software version 1.4x (GE Lunar PIXI, Madison, WI).

2.5. Microcomputerized Tomography (Micro-CT). Micro-CT (micro-CT 40, SCANCO Medical, Switzerland) was used to evaluate bone microarchitecture at the proximal tibia metaphysis, tibia middiaphysis, and 4th lumbar vertebral body. Analysis of trabecular bone was performed at the proximal tibia metaphysis on high resolution scans (2048 × 2048 pixels) and the volume of interest (VOI) included 750 μm of secondary spongiosa. The VOI was analyzed using a threshold of 300, a sigma of 0.7, and support of 1.0. Trabecular bone of the vertebra was assessed on images 80 μm from the dorsal and caudal growth plates at medium resolution (1024 × 1024 pixels) and included only secondary spongiosa. Images generated from the scans of the vertebrae were analyzed at a threshold of 340 and a sigma and support of 1.2 and 2.0, respectively. Trabecular parameters evaluated included trabecular bone volume expressed as a percentage of total volume (BV/TV), trabecular number (Tb.N.), trabecular thickness (Tb.Th.), trabecular separation (Tb.Sp.) connectivity density (ConnDens), and structural model index (SMI).

Cortical bone was evaluated by analyzing a 120 μm section at the mid-diaphysis of the tibia. Assessment of cortical bone parameters included cortical porosity, thickness, area, and medullary area of the tibial middiaphysis. The acquired images were analyzed at a threshold of 300, a sigma of 0.7, and support of 1.0.

2.6. Analysis of Biomechanical Properties of the Tibia. Tibiae were cleaned of soft-adhering tissue and stored in phosphate buffered saline (PBS) at 4°C until analyses were performed. Reference point indentation (RPI) was applied laterally at the tibia-fibula junction using a BioDent (Active Life Scientific, Inc., Santa Barbara, CA), and the first cycle indentation distance and touchdown distance were recorded. Each tibia was subjected to a testing protocol of 2 N force, 2 Hz, and 10 cycles.

2.7. Histology of the Liver. Fixed (10% neutral buffered formalin) liver samples were processed and sectioned (5 μm) for staining with hematoxylin and eosin to assess histological changes associated with nonalcoholic fatty liver disease (NAFLD) that occurs in obesity and/or diabetes. Steatosis and fibrosis were scored on a scale from 0 to 4, with 0 indicating the absence of hepatic lipid droplets or fibrosis, whereas 4 indicated pronounced steatosis or fibrosis. Lobular and portal inflammation was scored using a range of 0–3, with 0 indicating the absence of macrophage infiltration and 3 corresponding to severe inflammation. Balloon degeneration was scored using a 0–2 system, with 0 defined as the lack of degeneration and 2 indicating modest presence of parenchymal cell death. All scoring was performed by the study pathologist who was blinded to treatments.

2.8. RNA Isolation and Gene Expression Analysis. Total RNA was isolated from the liver and bone marrow using TriZol Reagent (Invitrogen, Grand Island, NY) as previously described [36, 37]. cDNA was synthesized following a standardized laboratory protocol and qPCR was performed using SYBR green chemistry (7900HT Fast Real-Time, Applied Biosystems, Foster City, CA). Hepatic genes of interest included fatty acid synthase (*Fasn*), sterol regulatory element-binding protein (*Srebp1c*), glucose transporter 2 or solute carrier family (*Slc2a2*), peroxisome proliferative-activator α (*Ppara*), and glutathione peroxidase (*Gpx1*) and in the bone marrow *Fasn*, *Ppara*, and *Gpx1* (Table S1 available online at http://dx.doi.org/10.1155/2015/758080). All qPCR results were evaluated by the comparative cycle number at threshold (C_Q) method (User Manual #2, Applied Biosystems) using peptidylprolyl isomerase B or cyclophilin B (*Ppib*) as the invariant control.

2.9. Statistical Analysis. Statistical analyses were performed using Statistical Analysis Software version 9.3 (SAS Institute, NC). The primary objective was to determine the difference in response to a HF diet of a given strain and, therefore,

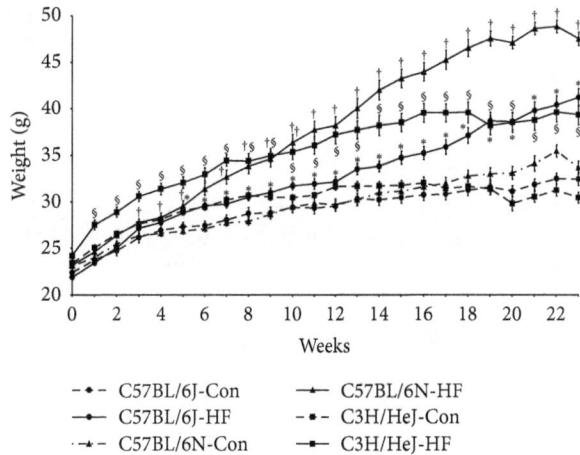

FIGURE 1: Body weights of C57BL/6J, C57BL/6N, and C3H/HeJ mice fed a control (Con; AIN-93M) or a high fat diet (HF; 45% kcal from fat) were recorded weekly. Data is presented as the mean ± SE, $n = 15$ mice in each group. Symbols, * for C57BL/6J, † for C57BL/6N, or § for C3H/HeJ, indicate significant differences ($P < 0.05$) of dietary treatment within a given mouse strain.

Student's paired *t*-test was used unless stated otherwise. However, to further assess differences in the responsiveness between the two C57BL/6 substrains, if a statistical difference ($P < 0.05$) was observed for a given parameter between Con and HF within a given strain, the magnitude of response (i.e., percent change of HF compared to Con) was compared between strains using one-way ANOVA. When the *F* value was <0.05, *post hoc* analyses were performed with Fischer's least square means separation test. Chi-squared tests were used for histological scoring of liver specimens. All data are presented as mean ± standard error (SE) and a $P < 0.05$ was considered statistically significant.

3. Results

3.1. Body and Fat Pad Weight, Body Composition, and Feed Efficiency. At baseline, body weight between strains differed (C3H/HeJ > C57BL/6N > C57BL/6J); however, no differences existed within a given strain between the two dietary treatment groups (i.e., Con versus HF; *data not shown*). After 5 wk on the HF diet, the C57BL/6J exhibited a significant increase in body weight compared to the C57BL/6J Con, whereas the C57BL/6N on the HF diet had a higher ($P < 0.05$) body weight after only 3 wk (Figure 1). The C3H/HeJ mice on the HF diet also exhibited a more rapid increase in body weight after only 1 wk compared to their respective controls (Figure 1). Analysis of body composition revealed the increase in body weight was due to a significant increase in both lean and fat mass for the two C57BL/6 substrains as well as the C3H/HeJ mice (Table 1). The amount of food consumed was less for the mice on the HF diet in each strain (Table 1). However, on a kcal basis the C57BL/6N mice on the HF diet consumed +2.1 kcal/day and the C57BL/6J and C3H/HeJ on the HF diet consumed +1.2 kcal/day compared to their respective controls (*data not shown*). Overall, feed

TABLE 1: Body composition, tissue weights, food intake, total white blood cell counts, and adipokines.

	C57BL/6J			C57BL/6N			C3H/HeJ		
	Con	HF	P values	Con	HF	P values	Con	HF	P values
Final body weight (g)	32.4 ± 0.7	41.2 ± 0.9	**<0.0001**	33.7 ± 0.7	47.5 ± 0.7	**<0.0001**	30.5 ± 0.6	39.3 ± 1.1	**<0.0001**
Body composition									
Lean (g)	21.8 ± 0.4	24.9 ± 0.3	**<0.0001**	22.1 ± 0.3	27.7 ± 0.5	**<0.0001**	19.8 ± 0.3	23.9 ± 0.5	**<0.0001**
Fat (g)	7.1 ± 0.4	12.0 ± 1.2	**0.0006**	9.6 ± 0.6	18.3 ± 0.6	**<0.0001**	6.7 ± 0.3	11.1 ± 0.6	**<0.0001**
Percent fat (%)	24.2 ± 1.0	31.8 ± 1.4	**0.0018**	29.4 ± 1.2	39.7 ± 0.9	**<0.0001**	25.2 ± 0.8	31.5 ± 0.8	**<0.0001**
Fat pad (g)	0.68 ± 0.08	1.67 ± 0.13	**<0.0001**	1.14 ± 0.11	1.69 ± 0.07	**0.0067**	0.52 ± 0.07	0.90 ± 0.06	**0.0051**
Feed efficiency (mg/kcal)	4.63 ± 0.38	8.41 ± 0.43	**0.0011**	5.28 ± 0.34	11.01 ± 0.43	**<0.0001**	3.24 ± 0.26	6.17 ± 0.36	**0.0086**
Food intake (g/day)	3.17 ± 0.04	2.74 ± 0.06	**<0.0001**	3.22 ± 0.04	2.98 ± 0.04	**0.0002**	3.19 ± 0.04	2.78 ± 0.06	**<0.0001**
Thymus (g)	0.042 ± 0.002	0.048 ± 0.003	0.1639	0.055 ± 0.004	0.076 ± 0.006	**0.0096**	0.019 ± 0.002	0.024 ± 0.002	0.0984
Spleen (g)	0.091 ± 0.005	0.098 ± 0.004	0.2754	0.106 ± 0.006	0.132 ± 0.009	**0.0323**	0.101 ± 0.004	0.123 ± 0.006	**0.0066**
WBC (1×10^5)	17.04 ± 1.80	18.55 ± 1.61	0.5350	21.69 ± 1.55	16.14 ± 1.72	**0.0045**	7.60 ± 0.71	11.48 ± 1.05	**0.0252**
Adipokines									
Leptin (ng/mL)	3.64 ± 1.14	16.13 ± 3.67	**0.0045**	8.92 ± 1.87	44.96 ± 5.06	**<0.0001**	0.84 ± 0.15	11.08 ± 2.72	**0.0014**
Adiponectin (µg/mL)	9.94 ± 0.76	7.45 ± 0.72	**0.0068**	6.36 ± 0.34	6.97 ± 0.25	0.1637	5.05 ± 0.40	5.13 ± 0.17	0.8315

Final body weight, body composition, food intake, tissue weight, and total white blood cell counts (WBC) and adipokines after 24 wk on a control (Con = AIN-93M) or a high fat diet (HF = 45% kcal from fat) in C57BL/6J, C57BL/6N, and C3H/HeJ mice. Values are means ± SE, n = 15 mice in each group. P values show the differences between dietary treatments within a given strain.

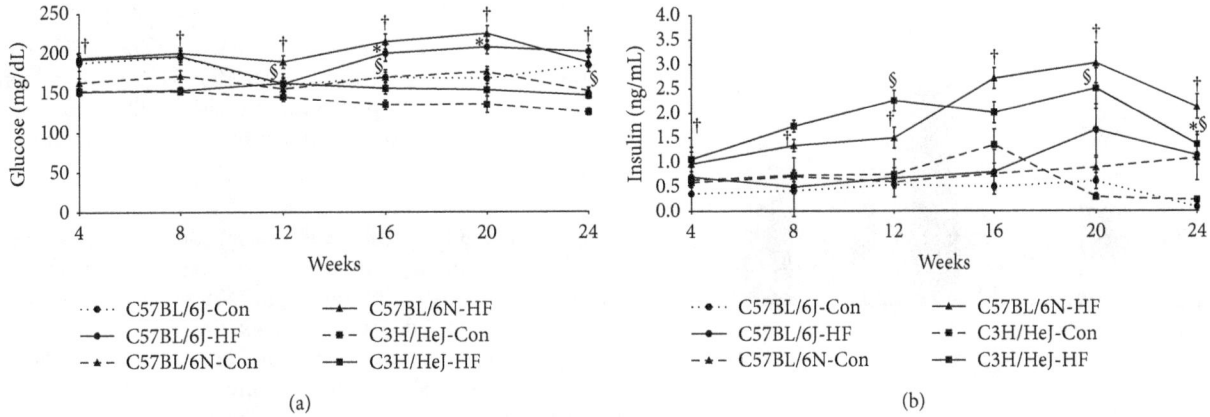

FIGURE 2: Blood glucose (a) and plasma insulin (b) were determined at 4 wk intervals in mice from each of the three strains fed a control (Con; AIN-93M) or a high fat diet (HF; 45% kcal from fat). Symbols, * for C57BL/6J, † for C57BL/6N, or § for C3H/HeJ, indicate significant differences ($P < 0.05$) of dietary treatment for a given mouse strain.

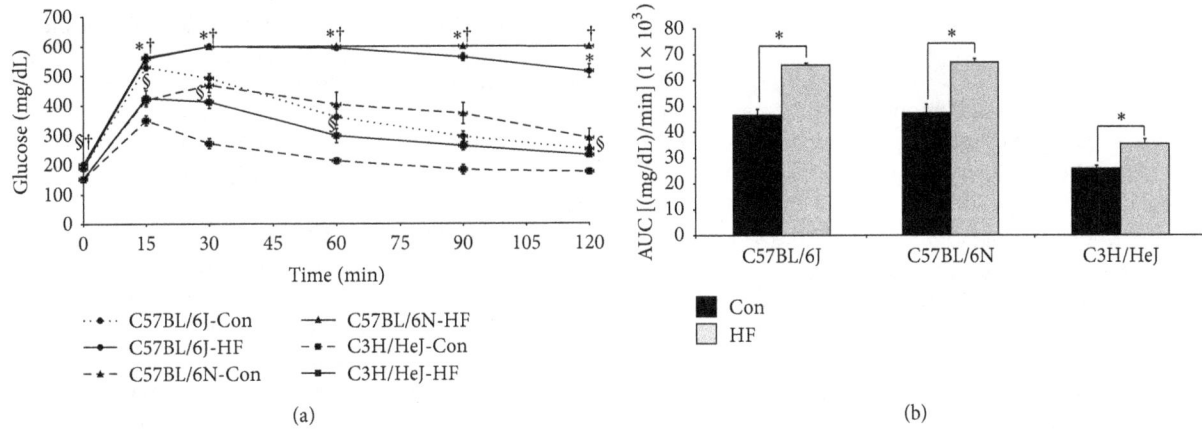

FIGURE 3: Glucose tolerance test results following 24 wk on control or high fat diet. One week prior to the end of the study, an intraperitoneal glucose tolerance test was administered (2 g glucose/kg body weight) in the C57BL/6J, C57BL/6N, and C3H/HeJ mice fed a control (Con; AIN-93M) or a high fat diet (HF; 45% kcal from fat). (a) Tail blood was collected following 15, 30, 60, 90, and 120 min following glucose injection and symbols, * for C57BL/6J, † for C57BL/6N, or § C3H/HeJ, indicate significant differences ($P < 0.05$) of dietary treatment for a given mouse strain. (b) Area under the curve (AUC) was calculated for the IGTT and symbol, *, represents a significant difference ($P < 0.05$) between dietary treatments for a given strain.

efficiency was higher in the C57BL/6N mice than in the C57BL/6J mice on the HF diet (Table 1), further demonstrating the differences in metabolic responsiveness between these two substrains.

3.2. Tissue Weights and White Blood Cells. After 24 wk on a HF diet the C57BL/6N mice exhibited splenomegaly, thymic hypertrophy, and decreased WBC, but the C57BL/6J mice failed to demonstrate these immunological changes (Table 1). C3H/HeJ mice had a similar response to the HF diet in terms of tissue weights (i.e., spleen and thymus) and total WBCs compared to the C57BL/6N mice (Table 1).

3.3. Blood Glucose, Plasma Insulin, and Glucose Tolerance Test. C57BL/6N mice on the HF diet were the only strain that had elevated fasting blood glucose (Figure 2(a)) and plasma insulin (Figure 2(b)) after 4, 8, 12, 16, 20, and 24 wk of

treatment compared to their Con counterparts. The C57BL/6J substrain on the HF diet was hyperglycemic at 16 and 20 wk (Figure 2(a)) and hyperinsulinemic at 24 wk (Figure 2(b)). Importantly, neither substrain achieved a fasting blood glucose consistent with frank diabetes (i.e., >250 mg/dL) that is associated with polyuria and polydipsia [38]. Similar to the C57BL/6J mice, the C3H/HeJ strain on the HF diet exhibited delayed-onset of hyperglycemia (Figure 2(a)), while their plasma insulin was increased at 12, 20, and 24 wk (Figure 2(b)).

At the end of the study, IGTT showed that the C57BL/6J and C57BL/6N as well as the C3H/HeJ mice on the HF diet exhibited impaired glucose intolerance (Figures 3(a) and 3(b)). The percent change in AUC to the HF diet demonstrated that the magnitude of response of the two C57BL/6 substrains was similar (*data not shown*). The C3H/HeJ mice also exhibited impaired glucose intolerance after 24 wk on a

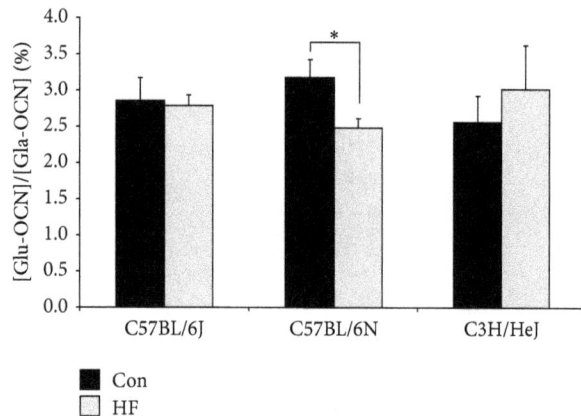

FIGURE 4: Plasma osteocalcin (OCN) expressed as percent of under-carboxylated Glu-OCN per total Gla-OCN in C57BL/6J, C57BL/6N, and C3H/HeJ mice on a control (Con; AIN-93M) or a high fat diet (HF; 45% kcal from fat) after 24 wk. Symbol, *, represents a significant difference ($P < 0.05$) between dietary treatments for a given strain.

HF diet (Figure 3). It should be noted that, despite elevated AUC, the C3H/HeJ mice on the HF diet maintained the ability to restore blood glucose by the final IGTT time point.

3.4. Plasma Adipokines and Osteocalcin. Both the C57BL/6J and C57BL/6N substrains had elevated plasma leptin after 24 wk on a HF diet (Table 1), but there was no significant difference in the magnitude of the response between the two substrains. Similarly, the C3H/HeJ mice on the HF diet also had higher plasma leptin (Table 1). Interestingly, at 24 wk the C57BL/6J mice, but not the C57BL/6N substrain, exhibited a decrease in plasma adiponectin in response to a HF diet (Table 1).

After 24 wk on a HF diet, there were no differences in Gla-OCN as an indicator of bone turnover due to diet in the C57BL/6 substrains or C3H/HeJ mice (data not show). The carboxylation status of OCN (i.e., Glu/Gla-OCN ratio), which has been shown to influence insulin sensitivity and systemic energy metabolism, was reduced only in the C57BL/6N mice after 24 wk on a HF diet (Figure 4).

3.5. Histological Evaluation of Hepatic Tissue. Representative micrographs of liver sections from each group show that the C57BL/6J and C57BL/6N strains as well as the C3H/HeJ strain experienced some degree of hepatic steatosis in response to the HF diet (Figure 5). The C57BL/6N mice on the HF diet had a significantly higher lobular and portal inflammation mean score compared to the Con (Table 2). Although the C57BL/6J mice on the HF diet had more lobular inflammation than their respective controls ($P = 0.0038$), the frequency of the inflammatory response was markedly lower in this substrain compared to the C57BL/6N (i.e., lobular inflammation in 92% C57BL/6N versus 54% C57BL/6J and portal inflammation in 77% C57BL/6N versus 23% C57BL/6J) (Table 2). While none of the C57BL/6J mice on the HF diet exhibited liver fibrosis, 23% of the treated

C57BL/6N mice had fibrotic changes (Table 2). Balloon degeneration was also more severe in the C57BL/6N mice on the HF diet compared to the C57BL/6J (Table 2). Despite a lack of lobular and portal inflammation and fibrosis in the C3H/HeJ mice, balloon degeneration was severe in this strain (Table 2).

3.6. Whole Body Bone Densitometry. Both the C57BL/6J and C57BL/6N mice demonstrated a decrease in whole body BMC and BMA, but no change in whole body BMD in response to the HF diet after 24 wk (Table 3). When BMD was expressed relative to body weight, differences due to diet were observed suggesting that the bone density did not increase relative to the increase in body weight (Table 3).

3.7. Microarchitectural Changes in Trabecular and Cortical Bone. Micro-CT analyses of the lumbar vertebra revealed significant loss of trabecular bone or BV/TV with the HF diet in C57BL/6N, while the skeletal response of the C57BL/6J mice did not reach the level of statistical significance ($P < 0.0579$) (Figure 6(a)). As expected, the C3H/HeJ mice were protected from vertebral bone loss (Figure 6) or nonmorpho-metric parameters with HF diet (Table 3). Both the C57BL/6J and C57BL/6N mice on the HF diet had a higher SMI indicative of a weaker, more rod-like trabecular bone in the vertebra (Table 3).

In contrast to the vertebra, no changes were observed in trabecular or cortical parameters analyzed at the proximal tibial metaphysis or the tibial middiaphysis in the C57BL/6J or the C57BL/6N mice. The C3H/HeJ mice failed to demonstrate alterations in trabecular bone of the proximal tibia but did exhibit an increase in the medullary area at the middiaphysis (Table 3).

3.8. Changes in Biomechanical Properties of the Tibia. Based on reference point indentation testing on cortical bone at the tibia-fibula junction, no changes were observed in first cycle indentation distance or touchdown distance in any strain following 24 wk on a HF diet when compared to their respective Con (Table 3).

3.9. Characterization of Genes Involved in Energy Metabolism and Inflammation from the Liver and Bone Marrow. Determination of genes involved in hepatic metabolism and inflammation revealed that the C57BL/6N mice on the HF diet had altered metabolic processes, including the upregulation of glucose uptake (Slc2a2), triglyceride storage (Fasn and Srebp1c) and adipogenesis (Ppara), as well as antioxidant capacity (Gpx1) (Table 4). Interestingly, none of these alterations in gene expression were observed in the C57BL/6J mice after 24 wk on the HF diet.

To determine the degree to which oxidative stress and adipogenesis contributed to bone loss with the HF diet model, Gpx1 and Pparg mRNA abundance was determined in the bone marrow. Similar to the hepatic tissue, the abundance of Gpx1 mRNA was increased in the C57BL/6N mice on the HF diet, suggesting an increase in antioxidant capacity (Table 4). In contrast, the C57BL/6J mice on the HF diet

TABLE 2: Pathological scoring of hepatic tissue after 24 wk on a control or high fat diet.

	C57BL/6J							C57BL/6N							C3H/HeJ						
	0	1	2	3	4	Mean	P value	0	1	2	3	4	Mean	P value	0	1	2	3	4	Mean	P value
Steatosis																					
Con	10	1	0	0	0	0.09	*0.0028*	6	2	3	0	0	0.73	*0.0020*	10	0	0	0	0	0.00	*0.0001*
HF	2	4	6	1	0	1.46		0	2	1	2	8	3.23		1	4	5	4	0	1.86	
Fibrosis																					
Con	0	0	0	0	0	0.00	—	11	0	0	0	0	0.00	0.0885	10	0	0	0	0	0.00	—
HF	0	0	0	0	0	0.00		10	3	0	0	0	0.23		14	0	0	0	0	0.00	
Lobular inflammation																					
Con	11	0	0	0		0.00	*0.0038*	11	0	0	0		0.00	*<0.0001*	10	0	0	0		0.00	0.0641
HF	6	7	0	0		0.54		1	12	0	0		0.92		10	4	0	0		0.29	
Portal inflammation																					
Con	11	0	0	0		0.00	0.0885	11	0	0	0		0.00	*0.0001*	10	0	0	0		0.00	—
HF	10	3	0	0		0.23		3	10	0	0		0.77		14	0	0	0		0.00	
Balloon degeneration																					
Con	0	0	0			0.00	*0.0062*	11	0	0			0.00	*<0.0001*	10	0	0			0.00	*<0.0001*
HF	5	7	1			0.69		0	0	13			2.00		1	3	10			1.64	

Frequency of steatosis (0–4), lobular and portal inflammation (0–3), fibrosis (0–4), and balloon degeneration (0–2), along with mean scores and P values for control versus high fat diet within a given strain based on Chi-squared statistical analyses.

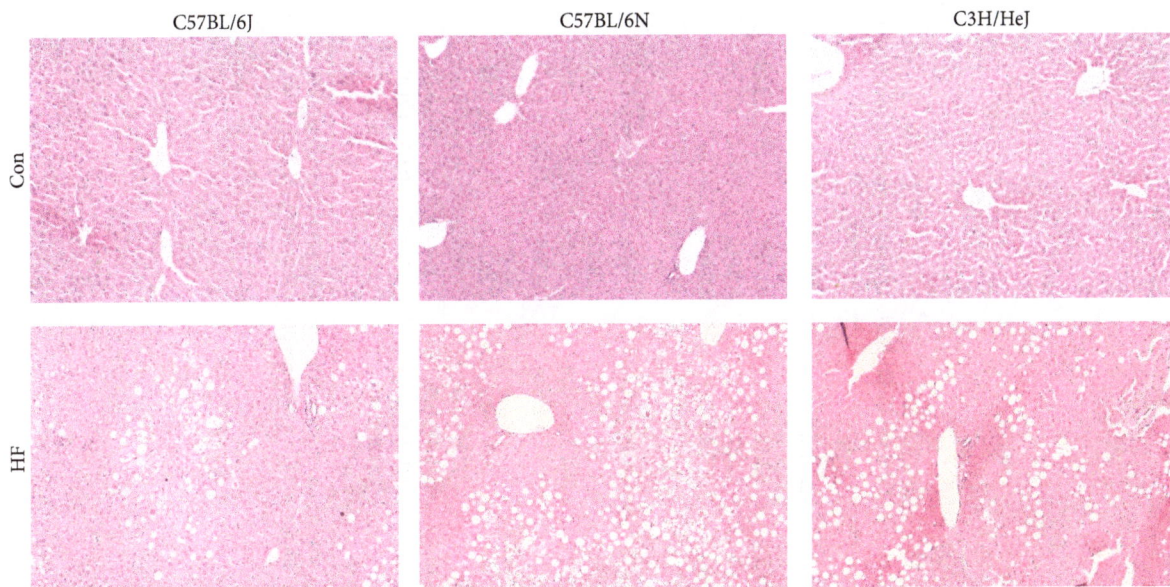

FIGURE 5: Representative micrographs of liver histology sections stained with hematoxylin and eosin from C57BL/6J, C57BL/6N, and C3H/HeJ mice following 24 wk on a control (Con; AIN-93M) or a high fat diet (HF; 45% kcal from fat). Representative images were photographed and are presented at a 10x magnification.

demonstrated a decrease in the relative abundance of *Gpx1* (Table 4). Additionally, no alterations were observed in the transcriptional regulator of adipogenesis, *Pparg*, in any strain after 24 wk (Table 4).

4. Conclusions

The findings of this study show that the C57BL/6J and the C57BL/6N mouse differ in their metabolic response to a HF diet over a 24 wk study period. Discrepancies in the metabolic response between the two strains may be attributed in part to the missense mutation (M35T) in exon 1 and a multiexon deletion of *Nnt* in the C57BL/6J mice [39, 40]. This mutation in *Nnt* has been reported to uncouple β-cell mitochondrial metabolism leading to less ATP production in pancreatic islets, enhanced K_{ATP} channel activity, and, consequently, impaired glucose-stimulated insulin secretion [23, 39, 41]. Only fasting insulin was assessed in the current

TABLE 3: Bone densitometry of the whole body along with bone microarchitectural and biomechanical parameters of the spine and tibia.

	C57BL/6J			C57BL/6N			C3H/HeJ		
	Con	HF	P values	Con	HF	P values	Con	HF	P values
Whole body bone densitometry									
BMC (mg)	731.2 ± 30.3	622.8 ± 23.6	*0.0101*	614.5 ± 24.1	492.9 ± 11.8	*<0.0001*	878.5 ± 25.1	810.1 ± 14.3	*<0.0001*
BMA (cm^2)	12.31 ± 0.40	10.87 ± 0.42	*0.0149*	11.03 ± 0.33	8.91 ± 0.12	*<0.0001*	12.90 ± 0.15	11.76 ± 0.22	*<0.0001*
BMD (mg/cm^2)	59.0 ± 0.8	57.2 ± 0.6	0.0866	55.9 ± 0.8	55.4 ± 0.6	0.6817	69.5 ± 0.7	68.6 ± 0.7	0.3777
BMD/BW [(mg/cm^2)/g]	1.84 ± 0.05	1.39 ± 0.05	*<0.0001*	1.67 ± 0.04	1.17 ± 0.02	*<0.0001*	2.29 ± 0.05	1.76 ± 0.05	*<0.0001*
Lumbar vertebra trabecular									
Conn density (1/mm^3)	232.45 ± 13.37	249.77 ± 10.96	0.3398	221.21 ± 14.25	214.62 ± 16.73	0.7766	145.78 ± 18.26	116.81 ± 10.83	0.2023
SMI	0.30 ± 0.09	0.67 ± 0.02	*0.0028*	0.59 ± 0.10	1.00 ± 0.09	*0.0136*	1.10 ± 0.11	1.07 ± 0.09	0.8508
App density (mg HA/ccm)	395.86 ± 20.97	364.38 ± 7.04	0.1852	358.84 ± 7.84	318.54 ± 8.85	*0.0087*	279.49 ± 10.48	287.93 ± 11.99	0.6070
Mat density (mg HA/ccm)	1093.1 ± 6.7	1084.0 ± 8.9	0.4366	1078.5 ± 8.3	1076.9 ± 8.2	0.9970	1123.5 ± 10.2	1135.5 ± 11.1	0.4464
Tibial midshaft									
Cortical porosity (%)	4.30 ± 0.26	4.35 ± 0.13	0.2026	4.50 ± 0.14	4.71 ± 0.10	0.8831	3.07 ± 0.08	3.69 ± 0.45	0.0505
Cortical thickness (mm)	0.17 ± 0.04	0.19 ± 0.04	0.0796	0.14 ± 0.03	0.16 ± 0.03	0.8602	0.18 ± 0.05	0.20 ± 0.04	0.1381
Cortical area (mm^2)	0.011 ± 0.001	0.010 ± 0.001	0.0535	0.010 ± 0.001	0.011 ± 0.001	0.8756	0.014 ± 0.001	0.013 ± 0.001	0.3270
Medullary area (mm^2)	0.035 ± 0.001	0.032 ± 0.001	0.8879	0.034 ± 0.004	0.037 ± 0.002	0.7738	0.030 ± 0.001	0.034 ± 0.001	*0.0218*
Indentation distance (μm)	34.7 ± 1.5	32.9 ± 0.9	0.3231	29.9 ± 0.7	34.9 ± 4.5	0.2603	35.3 ± 1.1	38.4 ± 3.1	0.3587
Touchdown distance (μm)	162.1 ± 12.9	150.2 ± 8.2	0.4465	148.2 ± 6.5	179.6 ± 14.7	0.0615	158.8 ± 10.8	157.6 ± 6.7	0.4269

Whole body bone densitometry and trabecular and cortical bone microarchitecture, along with reference point indentation (RPI) after 24 wk on a control (AIN-93M) or a HF (45% kcal from fat). Values are means ± SE, n = 10–15 mice per group for bone densitometry and n = 6 mice per group for micro-CT data. P values show the differences between dietary treatments within a given strain. Values are means ± SE.

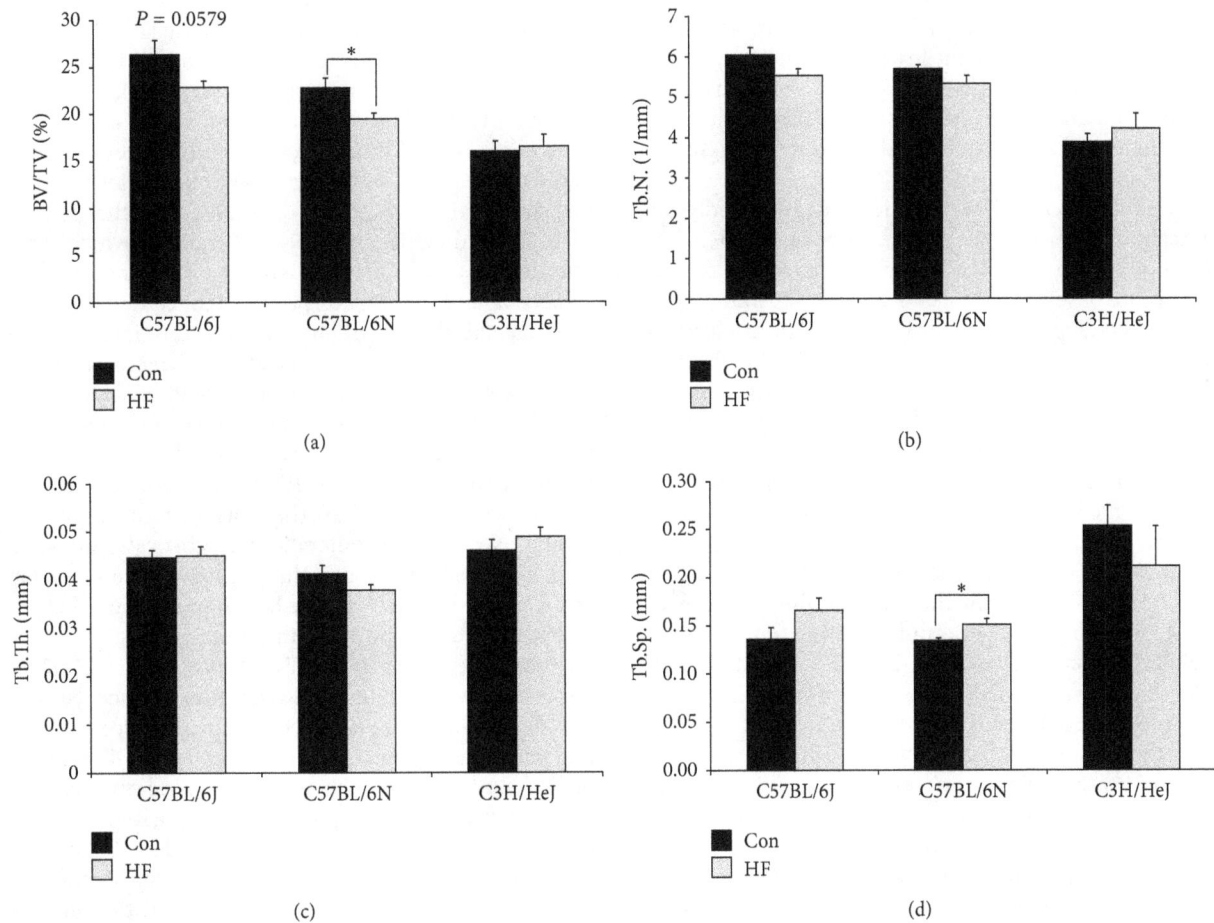

FIGURE 6: Micro-CT analyses of trabecular bone in the lumbar vertebra (L4) in C57BL/6J, C57BL/6N, and C3H/HeJ mice on a control (Con; AIN-93M) or a high fat diet (HF; 45% kcal from fat) for 24 wk. Parameters include (a) bone volume/total volume (BV/TV), (b) trabecular number (Tb.N.), (c) trabecular thickness (Tb.Th.), and (d) trabecular separation (Tb.Sp.). Symbol, *, represents a significant difference ($P <$ 0.05) between dietary treatments for a given strain.

TABLE 4: Relative fold change of gene expression in the liver and bone marrow in mice fed a high fat diet compared to the control diet.

	C57BL/6J		C57BL/6N		C3H/HeJ	
	HF	P value	HF	P value	HF	P value
Liver						
Fasn	1.55 ± 0.31	*0.1748*	$2.20 \pm 0.36^*$	***0.0248***	1.23 ± 0.35	*0.5934*
Gpx1	0.89 ± 0.12	*0.6197*	$1.54 \pm 0.15^*$	***0.0127***	1.31 ± 0.14	*0.1653*
Ppara	1.10 ± 0.16	*0.5427*	$1.41 \pm 0.11^*$	***0.0478***	0.98 ± 0.10	*0.8582*
Slc2a2	0.99 ± 0.10	*0.9792*	$1.56 \pm 0.14^*$	***0.0096***	1.08 ± 0.21	*0.7756*
Srebp1c	1.20 ± 0.26	*0.4985*	2.07 ± 0.51	*0.0997*	0.93 ± 0.21	*0.8329*
Tnf	1.76 ± 0.40	*0.2237*	1.12 ± 0.28	*0.7238*	2.09 ± 0.51	*0.0703*
Bone marrow						
Fasn	0.94 ± 0.05	*0.6811*	0.94 ± 0.07	*0.5619*	1.20 ± 0.15	*0.2388*
Gpx1	$0.56 \pm 0.08^*$	***0.0053***	$1.45 \pm 0.13^*$	***0.0228***	1.05 ± 0.18	*0.7983*
Pparg	1.35 ± 0.39	*0.4349*	0.78 ± 0.10	*0.1829*	1.21 ± 0.18	*0.3279*

Mean fold regulation of genes involved in systemic metabolism and inflammation is presented for the animals on the high fat diet (45% kcal from fat) relative to their control (AIN-93M). All target genes were normalized to invariant control (*Ppib*). Symbol, *, represents a significant difference ($P = 0.05$) between dietary treatments within a given strain.

study; however, early onset of hyperinsulinemia with HF diet was only observed in the C57BL/6N mice with an intact, functional *Nnt*. The coincident lower feed efficiency in the C57BL/6J compared to the C57BL/6N mice resulted in a delay in the development of hyperglycemia and shorter duration of exposure to conditions associated with impaired glucose tolerance.

The C57BL/6J and C57BL/6N mice had a markedly different hepatic response to the HF diet after 24 wk. Increased mRNA abundance of *Fasn* and a modest increase in *Srebpc1* in the presence of severe liver steatosis in the C57BL/6N mice on the HF diet suggest an increase in hepatic triglyceride synthesis and storage. Conversely, the C57BL/6J substrain, which has lower glucokinase activity and thus impaired glucose sensing, may explain the lack of transcriptional regulation of *Fasn* and *Srebp1c* [39]. Furthermore, C57BL/6N mice on the HF diet demonstrated an increase in *Slc2a2* gene expression, which encodes the non-insulin-sensitive glucose transporter 2 and has been reported to be upregulated in response to a HF diet [42]. Histological evaluation suggests that the C57BL/6N mice on the HF diet also experienced the most pronounced hepatic inflammation, compared to the C57BL/6J mice. Interestingly, the HF-induced NAFLD that develops in the C57BL/6J was previously attributed, in part, to the spontaneous mutation in the *Nnt* [39, 43]; however, the data demonstrate that the C57BL/6N mice, with a functional *Nnt*, develop more severe NAFLD in response to HF diet feeding. Given the more pronounced metabolic phenotype in the C57BL/6N mice on the HF diet, it was counterintuitive that only the C57BL/6J mice demonstrated the anticipated decrease in plasma adiponectin following a HF diet. Although the mechanism for reduced adiponectin during obesity and T2DM has been attributed, in part, to an increase in local TNF-α in adipose tissue [44], strain-related variability of adiponectin gene expression and plasma has been previously documented [45]. As mice continue to be used for models of impaired glucose homeostasis, further investigation is warranted to fully understand these strain differences relative to metabolic handling. The findings of this study also demonstrate that the C3H/HeJ mice may not be completely resistant to diet-induced obesity and the subsequent metabolic changes. Differences in the C3H/HeJ strain's response to a high fat diet compared to previous reports may be attributed to the difference in the control strain used [26, 28]. Specifically, previous studies have compared the C3H/H3J response to HF diet to C3H/HeOuJ or C3H/HeN, both of which have a functional TLR-4 [26, 28]. However, in these studies, no comparisons were made with C3H/HeJ mice on a control diet. Therefore, it is not possible to determine if the differences in metabolic response to a HF diet are a result of TLR-4 or genetic variability in the control substrain background. Furthermore, the C3H/HeJ strain has recently been shown to have a genetic variation in the leptin receptor gene (*Lepr*) [46]. This mutation could account for the impaired metabolic response observed in the C3H/HeJ strain on a HF diet due to the central role leptin has on regulating energy intake and expenditure. However, in the present study, food intake in the C3H/HeJ strain was not significantly altered and the absence of a leptin-mediated effect on food or energy intake does not rule out implications of the *Lepr* defect on the skeletal response [18, 19, 47].

In conjunction with the metabolic comparisons, the other primary objective of this study was to compare the skeletal response to a HF diet in two commonly used C57BL/6 substrains. The C57BL/6J mice on the HF diet experienced a 13.6% reduction in trabecular bone of the vertebra, although not statistically significant. C57BL/6N was the only strain that exhibited significant trabecular bone loss which occurred only in the vertebra. In the absence of alterations in tibia trabecular and cortical bone microarchitecture, it is conceivable that the absence of alterations in the tibia could result from site-specific changes associated with increased adiposity and greater weight-bearing, which could offset some of the negative effects of glucose intolerance on bone [48, 49]. Based on reports in the literature that bone biomechanical properties are changed in T2DM independent of alterations in bone mass [8, 50, 51], the tibia was subjected to RPI testing. No detectable alterations in cortical bone strength were observed after 6 months in the absence of structural changes. The C57BL/6N mice had more prolonged exposure to hyperglycemia and hyperinsulinemia in response to the HF diet compared to the C57BL/6J substrain, and C57BL/6N were the only substrain to lose significant trabecular bone in the spine. While there have been conflicting reports on how a HF diet impacts bone in C57BL/6 mice [12, 14–17], the results of this study indicate the skeletal response may be linked to the duration of disrupted insulin signaling and glucose intolerance. This idea is further supported by the response of the C3H/HeJ mice in which case an attenuated glucose, leptin, and insulin response to the HF diet failed to induce bone loss. Several reports have shown that a high fat diet uncouples bone turnover by increasing bone resorption and decreasing bone formation in various rodent models [16, 17, 21]. In this study, osteocalcin which is considered a marker of bone turnover was the only bone marker assessed and it was not altered at the end of the study. Although this does not rule out alterations in bone metabolism occurring earlier, future studies are needed to investigate the mechanism involved in the site-specific loss of bone observed in this animal model.

Based on recent literature describing the hormone OCN as a regulator of systemic energy metabolism [52–54], the role of OCN on both metabolic and skeletal changes induced by HF was investigated. After 24 wk, the C57BL/6N mice on the HF diet had a lower ratio of plasma Glu-OCN/Gla-OCN. Because circulating undercarboxylated (Glu-OCN) can act directly on pancreatic β-cells to stimulate insulin secretion [20, 55–57], it would be expected that a reduction in Glu-OCN would lead to a decrease in insulin secretion. Instead, fasting plasma insulin was elevated in the C57BL/6N mice on the HF diet compared to their respective controls. Alternatively, these results could indicate the direct effects impaired glucose tolerance has on osteoclastogenesis and bone resorption [58]. In this regard, the acidic milieu of the osteoclast-resorption lacunae is capable of decarboxylating OCN, resulting in its release from the bone matrix and its subsequent circulation in the blood [55]. The complexity of OCN's role on bone and energy metabolism during glucose intolerance, as well as these implications in various mouse strains, warrants further investigation. Additional studies are also needed to determine how genes involved in the gamma carboxylation of OCN (i.e., *Ggcx* and *Esp1*) by osteoblasts as well as the role of decarboxylation of OCN by osteoclast are regulated in response to changes energy homeostasis [55, 59].

To date, this is the first study to directly compare the C57BL/6J and C57BL/6N substrains' response to a HF diet from a metabolic and skeletal perspective. Although neither substrain developed frank T2DM, the data presented here show that C57BL/6N mice exhibit an earlier metabolic response consistent with impaired glucose tolerance or pre-diabetes to the HF diet compared to the C57BL/6J mice. Moreover, the skeletal response followed that of the metabolic changes; this was demonstrated by the fact that significant trabecular bone loss occurred in the C57BL/6N mice, which demonstrated the robust metabolic alterations associated with clinical T2DM. Given the observed differences in feed efficiency between the C57BL/6 substrains, further research is also warranted to identify the mechanisms underlying altered energy utilization. In contrast to the C57BL/6 mice, the C3H/HeJ strain was protected from the metabolic and skeletal changes induced by a HF diet. While a number of questions remain including how bone metabolism is being altered in response to a high fat diet on a molecular level, this study highlights the need to consider not only the most appropriate strain but also the most appropriate substrain of mouse when designing experiments. Other important factors to consider when studying the relationship between glucose intolerance and bone include the site-specific skeletal response and the study duration. These decisions could significantly impact data interpretation and the translational implications as they relate to understanding how bone metabolism is altered in the context of T2DM.

Conflict of Interests

The authors declare that there is no conflict of interests regarding the publication of this paper.

Acknowledgments

The authors would like to express their gratitude to Kristen Hester and Sandra Peterson (Department of Nutritional Sciences, Oklahoma State University, Stillwater, OK 74078) for their technical assistance. This work was generously supported by Oklahoma Center for the Advancement of Science and Technology (HR10-068); Oklahoma Agriculture Experiment Station (OKL02867); and United States Department of Agriculture (2012-67011-19906).

References

[1] U. Rishaug, K. I. Birkeland, J. A. Falch, and S. Vaaler, "Bone mass in non-insulin-dependent diabetes mellitus," *Scandinavian Journal of Clinical and Laboratory Investigation*, vol. 55, no. 3, pp. 257–262, 1995.

[2] P. L. A. van Daele, R. P. Stolk, H. Burger et al., "Bone density in non-insulin-dependent diabetes mellitus: the Rotterdam study," *Annals of Internal Medicine*, vol. 122, no. 6, pp. 409–414, 1995.

[3] R. P. Stolk, P. L. A. van Daele, H. A. P. Pols et al., "Hyperinsulinemia and bone mineral density in an elderly population: the Rotterdam study," *Bone*, vol. 18, no. 6, pp. 545–549, 1996.

[4] K. K. Nicodemus and A. R. Folsom, "Type 1 and type 2 diabetes and incident hip fractures in postmenopausal women," *Diabetes Care*, vol. 24, no. 7, pp. 1192–1197, 2001.

[5] L. J. Melton III, C. L. Leibson, S. J. Achenbach, T. M. Therneau, and S. Khosla, "Fracture risk in type 2 diabetes: update of a population-based study," *Journal of Bone and Mineral Research*, vol. 23, no. 8, pp. 1334–1342, 2008.

[6] A. V. Schwartz, D. E. Sellmeyer, K. E. Ensrud et al., "Older women with diabetes have an increased risk of fracture: a prospective study," *Journal of Clinical Endocrinology and Metabolism*, vol. 86, no. 1, pp. 32–38, 2001.

[7] M. Janghorbani, D. Feskanich, W. C. Willett, and F. Hu, "Prospective study of diabetes and risk of hip fracture: the nurses' health study," *Diabetes Care*, vol. 29, no. 7, pp. 1573–1578, 2006.

[8] A. V. Schwartz, E. Vittinghoff, D. C. Bauer et al., "Association of BMD and FRAX score with risk of fracture in older adults with type 2 diabetes," *Journal of the American Medical Association*, vol. 305, no. 21, pp. 2184–2192, 2011.

[9] R. J. Fajardo, L. Karim, V. I. Calley, and M. L. Bouxsein, "A review of rodent models of type 2 diabetic skeletal fragility," *Journal of Bone and Mineral Research*, vol. 29, no. 5, pp. 1025–1040, 2014.

[10] R. S. Surwit, C. M. Kuhn, C. Cochrane, J. A. McCubbin, and M. N. Feinglos, "Diet-induced type II diabetes in C57BL/6J mice," *Diabetes*, vol. 37, no. 9, pp. 1163–1167, 1988.

[11] S. Collins, T. L. Martin, R. S. Surwit, and J. Robidoux, "Genetic vulnerability to diet-induced obesity in the C57BL/6J mouse: physiological and molecular characteristics," *Physiology & Behavior*, vol. 81, no. 2, pp. 243–248, 2004.

[12] C. Gallou-Kabani, A. Vigé, M.-S. Gross et al., "C57BL/6J and A/J mice fed a high-fat diet delineate components of metabolic syndrome," *Obesity*, vol. 15, no. 8, pp. 1996–2005, 2007.

[13] J. J. Cao, B. R. Gregoire, and H. Gao, "High-fat diet decreases cancellous bone mass but has no effect on cortical bone mass in the tibia in mice," *Bone*, vol. 44, no. 6, pp. 1097–1104, 2009.

[14] J. M. Patsch, F. W. Kiefer, P. Varga et al., "Increased bone resorption and impaired bone microarchitecture in short-term and extended high-fat diet-induced obesity," *Metabolism: Clinical and Experimental*, vol. 60, no. 2, pp. 243–249, 2011.

[15] S. S. Ionova-Martin, J. M. Wade, S. Tang et al., "Changes in cortical bone response to high-fat diet from adolescence to adulthood in mice," *Osteoporosis International*, vol. 22, no. 8, pp. 2283–2293, 2011.

[16] F. Parhami, Y. Tintut, W. G. Beamer, N. Gharavi, W. Goodman, and L. L. Demer, "Atherogenic high-fat diet reduces bone mineralization in mice," *Journal of Bone and Mineral Research*, vol. 16, no. 1, pp. 182–188, 2001.

[17] G. V. Halade, M. M. Rahman, P. J. Williams, and G. Fernandes, "High fat diet-induced animal model of age-associated obesity and osteoporosis," *The Journal of Nutritional Biochemistry*, vol. 21, no. 12, pp. 1162–1169, 2010.

[18] C. M. Steppan, D. T. Crawford, K. L. Chidsey-Frink, H. Ke, and A. G. Swick, "Leptin is a potent stimulator of bone growth in ob/ob mice," *Regulatory Peptides*, vol. 92, no. 1–3, pp. 73–78, 2000.

[19] P. Ducy, M. Amling, S. Takeda et al., "Leptin inhibits bone formation through a hypothalamic relay: a central control of bone mass," *Cell*, vol. 100, no. 2, pp. 197–207, 2000.

[20] N. K. Lee, H. Sowa, E. Hinoi et al., "Endocrine regulation of energy metabolism by the skeleton," *Cell*, vol. 130, no. 3, pp. 456–469, 2007.

[21] X. M. Lu, H. Zhao, and E. H. Wang, "A high-fat diet induces obesity and impairs bone acquisition in young male mice," *Molecular Medicine Reports*, vol. 7, no. 4, pp. 1203–1208, 2013.

[22] K. Mekada, K. Abe, A. Murakami et al., "Genetic differences among C57BL/6 substrains," *Experimental Animals*, vol. 58, no. 2, pp. 141–149, 2009.

[23] H. C. Freeman, A. Hugill, N. T. Dear, F. M. Ashcroft, and R. D. Cox, "Deletion of nicotinamide nucleotide transhydrogenase: a new quantitive trait locus accounting for glucose intolerance in C57BL/6J mice," *Diabetes*, vol. 55, no. 7, pp. 2153–2156, 2006.

[24] T.-T. Huang, M. Naeemuddin, S. Elchuri et al., "Genetic modifiers of the phenotype of mice deficient in mitochondrial superoxide dismutase," *Human Molecular Genetics*, vol. 15, no. 7, pp. 1187–1194, 2006.

[25] B. Ahrén and G. Pacini, "Insufficient islet compensation to insulin resistance vs. reduced glucose effectiveness in glucose-intolerant mice," *American Journal of Physiology: Endocrinology and Metabolism*, vol. 283, no. 4, pp. E738–E744, 2002.

[26] M. Poggi, D. Bastelica, P. Gual et al., "C3H/HeJ mice carrying a toll-like receptor 4 mutation are protected against the development of insulin resistance in white adipose tissue in response to a high-fat diet," *Diabetologia*, vol. 50, no. 6, pp. 1267–1276, 2007.

[27] D. M. L. Tsukumo, M. A. Carvalho-Filho, J. B. C. Carvalheira et al., "Loss-of-function mutation in toll-like receptor 4 prevents diet-induced obesity and insulin resistance," *Diabetes*, vol. 56, no. 8, pp. 1986–1998, 2007.

[28] H. Nakamura, Y. Fukusaki, A. Yoshimura et al., "Lack of Toll-like receptor 4 decreases lipopolysaccharide-induced bone resorption in C3H/HeJ mice *in vivo*," *Oral Microbiology and Immunology*, vol. 23, no. 3, pp. 190–195, 2008.

[29] M. Rauner, W. Sipos, and P. Pietschmann, "Osteoimmunology," *International Archives of Allergy and Immunology*, vol. 143, no. 1, pp. 31–48, 2007.

[30] K. Bandow, A. Maeda, K. Kakimoto et al., "Molecular mechanisms of the inhibitory effect of lipopolysaccharide (LPS) on osteoblast differentiation," *Biochemical and Biophysical Research Communications*, vol. 402, no. 4, pp. 755–761, 2010.

[31] K. Gunaratnam, C. Vidal, R. Boadle, C. Thekkedam, and G. Duque, "Mechanisms of palmitate-induced cell death in human osteoblasts," *Biology Open*, vol. 2, no. 12, pp. 1382–1389, 2013.

[32] E. A. Droke, K. A. Hager, M. R. Lerner et al., "Soy isoflavones avert chronic inflammation-induced bone loss and vascular disease," *Journal of Inflammation*, vol. 4, article 17, 2007.

[33] M. R. Hill, K. D. Denson, S. Y. Bu et al., "LPS-mediated chronic inflammation increases circulating RANKL coincident with bone loss in C57BL/6 mice," *Journal of Bone and Mineral Research*, abstract S410, 2006.

[34] S.-R. Oh, O.-J. Sul, Y.-Y. Kim et al., "Saturated fatty acids enhance osteoclast survival," *Journal of Lipid Research*, vol. 51, no. 5, pp. 892–899, 2010.

[35] R. S. Surwit, M. N. Feinglos, J. Rodin et al., "Differential effects of fat and sucrose on the development of obesity and diabetes in C57BL/6J and A/J mice," *Metabolism*, vol. 44, no. 5, pp. 645–651, 1995.

[36] E. Rendina, Y. F. Lim, D. Marlow et al., "Dietary supplementation with dried plum prevents ovariectomy-induced bone loss while modulating the immune response in C57BL/6J mice," *Journal of Nutritional Biochemistry*, vol. 23, no. 1, pp. 60–68, 2012.

[37] E. Rendina, K. D. Hembree, M. R. Davis et al., "Dried plum's unique capacity to reverse bone loss and alter bone metabolism in postmenopausal osteoporosis model," *PLoS ONE*, vol. 8, no. 3, Article ID e60569, 2013.

[38] K. L. Svenson, R. von Smith, P. A. Magnani et al., "Multiple trait measurements in 43 inbred mouse strains capture the phenotypic diversity characteristic of human populations," *Journal of Applied Physiology*, vol. 102, no. 6, pp. 2369–2378, 2007.

[39] A. A. Toye, J. D. Lippiat, P. Proks et al., "A genetic and physiological study of impaired glucose homeostasis control in C57BL/6J mice," *Diabetologia*, vol. 48, no. 4, pp. 675–686, 2005.

[40] A. Nicholson, P. C. Reifsnyder, R. D. Malcolm et al., "Diet-induced obesity in two C57BL/6 substrains with intact or mutant nicotinamide nucleotide transhydrogenase (Nnt) gene," *Obesity*, vol. 18, no. 10, pp. 1902–1905, 2010.

[41] H. Freeman, K. Shimomura, E. Horner, R. D. Cox, and F. M. Ashcroft, "Nicotinamide nucleotide transhydrogenase: a key role in insulin secretion," *Cell Metabolism*, vol. 3, no. 1, pp. 35–45, 2006.

[42] E. Riu, T. Ferre, A. Hidalgo et al., "Overexpression of c-myc in the liver prevents obesity and insulin resistance," *The FASEB Journal*, vol. 17, no. 12, pp. 1715–1717, 2003.

[43] Q. M. Anstee and R. D. Goldin, "Mouse models in non-alcoholic fatty liver disease and steatohepatitis research," *International Journal of Experimental Pathology*, vol. 87, no. 1, pp. 1–16, 2006.

[44] G. R. Hajer, T. W. van Haeften, and F. L. J. Visseren, "Adipose tissue dysfunction in obesity, diabetes, and vascular diseases," *European Heart Journal*, vol. 29, no. 24, pp. 2959–2971, 2008.

[45] M. Haluzik, C. Colombo, O. Gavrilova et al., "Genetic background (C57BL/6J Versus FVB/N) strongly influences the severity of diabetes and insulin resistance in ob/ob mice," *Endocrinology*, vol. 145, no. 7, pp. 3258–3264, 2004.

[46] S. Kapur, M. Amoui, C. Kesavan et al., "Leptin receptor (Lepr) is a negative modulator of bone mechanosensitivity and genetic variations in Lepr may contribute to the differential osteogenic response to mechanical stimulation in the C57BL/6J and C3H/HeJ pair of mouse strains," *The Journal of Biological Chemistry*, vol. 285, no. 48, pp. 37607–37618, 2010.

[47] A. Hamann and S. Matthaei, "Regulation of energy balance by leptin," *Experimental and Clinical Endocrinology & Diabetes*, vol. 104, no. 4, pp. 293–300, 1996.

[48] A. C. Looker, K. M. Flegal, and L. J. Melton III, "Impact of increased overweight on the projected prevalence of osteoporosis in older women," *Osteoporosis International*, vol. 18, no. 3, pp. 307–313, 2007.

[49] S. J. Kuruvilla, S. D. Fox, D. M. Cullen, and M. P. Akhter, "Site specific bone adaptation response to mechanical loading," *Journal of Musculoskeletal Neuronal Interactions*, vol. 8, no. 1, pp. 71–78, 2008.

[50] T. Yamaguchi, I. Kanazawa, M. Yamamoto et al., "Associations between components of the metabolic syndrome versus bone mineral density and vertebral fractures in patients with type 2 diabetes," *Bone*, vol. 45, no. 2, pp. 174–179, 2009.

[51] J. N. Farr, M. T. Drake, S. Amin, L. J. Melton III, L. K. McCready, and S. Khosla, "In vivo assessment of bone quality in postmenopausal women with type 2 diabetes," *Journal of Bone and Mineral Research*, vol. 29, no. 4, pp. 787–795, 2014.

[52] T. L. Clemens and G. Karsenty, "The osteoblast: an insulin target cell controlling glucose homeostasis," *Journal of Bone and Mineral Research*, vol. 26, no. 4, pp. 677–680, 2011.

[53] M. Ferron, E. Hinoi, G. Karsenty, and P. Ducy, "Osteocalcin differentially regulates β cell and adipocyte gene expression and affects the development of metabolic diseases in wild-type mice," *Proceedings of the National Academy of Sciences of the United States of America*, vol. 105, no. 13, pp. 5266–5270, 2008.

[54] E. Hinoi, N. Gao, D. Y. Jung et al., "An osteoblast-dependent mechanism contributes to the leptin regulation of insulin secretion," *Annals of the New York Academy of Sciences*, vol. 1173, supplement 1, pp. E20–E30, 2009.

[55] M. Ferron, J. Wei, T. Yoshizawa et al., "Insulin signaling in osteoblasts integrates bone remodeling and energy metabolism," *Cell*, vol. 142, no. 2, pp. 296–308, 2010.

[56] M. Ferron, M. D. McKee, R. L. Levine, P. Ducy, and G. Karsenty, "Intermittent injections of osteocalcin improve glucose metabolism and prevent type 2 diabetes in mice," *Bone*, vol. 50, no. 2, pp. 568–575, 2012.

[57] M.-T. Rached, A. Kode, B. C. Silva et al., "FoxO1 expression in osteoblasts regulates glucose homeostasis through regulation of osteocalcin in mice," *Journal of Clinical Investigation*, vol. 120, no. 1, pp. 357–368, 2010.

[58] K. I. Larsen, M. Falany, W. Wang, and J. P. Williams, "Glucose is a key metabolic regulator of osteoclasts; glucose stimulated increases in ATP/ADP ratio and calmodulin kinase II activity," *Biochemistry and Cell Biology*, vol. 83, no. 5, pp. 667–673, 2005.

[59] M. Haraikawa, N. Tsugawa, N. Sogabe et al., "Effects of gamma-glutamyl carboxylase gene polymorphism (R325Q) on the association between dietary vitamin K intake and gamma-carboxylation of osteocalcin in young adults," *Asia Pacific Journal of Clinical Nutrition*, vol. 22, no. 4, pp. 646–654, 2013.

Efficacy of Multiple Micronutrients Fortified Milk Consumption on Iron Nutritional Status in Moroccan Schoolchildren

Imane El Menchawy, Asmaa El Hamdouchi, Khalid El Kari, Naima Saeid, Fatima Ezzahra Zahrou, Nada Benajiba, Imane El Harchaoui, Mohamed El Mzibri, Noureddine El Haloui, and Hassan Aguenaou

Joint Unit of Nutrition and Food Research (URAC39), Ibn Tofaïl University-CNESTEN, Regional Designated Center for Nutrition (AFRA/IAEA), Kenitra, 14000 Rabat, Morocco

Correspondence should be addressed to Imane El Menchawy; imenchawy@gmail.com

Academic Editor: Christel Lamberg-Allardt

Iron deficiency constitutes a major public health problem in Morocco, mainly among women and children. The aim of our paper is to assess the efficacy of consumption of multiple micronutrients (MMN) fortified milk on iron status of Moroccan schoolchildren living in rural region. Children ($N = 195$), aged 7 to 9 y, were recruited from schools and divided into two groups: the nonfortified group (NFG) received daily a nonfortified Ultra-High-Temperature (UHT) milk and the fortified group received (FG) daily UHT milk fortified with multiple micronutrients including iron sulfate. Blood samples were collected at baseline (T0) and after 9 months (T9). Hemoglobin (Hb) was measured *in situ* by Hemocue device; ferritin and C Reactive Protein were assessed in serum using ELISA and nephelometry techniques, respectively. Results were considered significant when the p value was <0.05. At T9 FG showed a reduction of iron deficiency from 50.9% to 37.2% ($p = 0.037$). Despite the low prevalence of iron deficiency anemia (1.9%); more than 50% of children in our sample suffered from iron deficiency at baseline. The consumption of fortified milk reduced the prevalence of iron deficiency by 27% in schoolchildren living in high altitude rural region of Morocco. *Clinical Trial Registration.* Our study is registered in the Pan African Clinical Trial Registry with the identification number PACTR201410000896410.

1. Introduction

Anemia is recognized as the most common nutritional deficiency worldwide. There are 2 billion people (>30% of the world's population) suffering from anemia [1]. Infants and preschoolers are at major risk, especially in the developing countries. Iron deficiency (ID) is the major cause of anemia and both anemia and iron deficiency in infants and young children are associated with adverse effects on neural development [2]. Inadequate diet due to low iron intake and/or bioavailability is its main etiology [3]. In Morocco, according to a Sentinel Survey for Monitoring and Evaluation of the Fortification Process conducted in 2006–2008, 31.5% of children under 5 y of age suffered from anemia [4]. This prevalence did not improve since the last national survey conducted in 2000 where 31.6% of children aged 6 m–5 y were anemic [5].

In 2001 the Moroccan Ministry of Health had developed and implemented the National Program of Fight against Micronutrients Deficiencies (NPFMD), including iron deficiency, which consisted of iron supplementation of children suffering from anemia, nutritional education, and fortification of staple foods commonly consumed by the entire population.

Supplementation programs and health education to change dietary practices in preschool children have achieved limited success [6]; hence, wheat flour was fortified with elemental electrolytic iron and B group vitamins to improve iron status among Moroccan population [7]. The impact study conducted in 2006–2008 revealed that the fortification of flour with elemental iron did not have a significant effect on the reduction of the prevalence of iron deficiency and iron deficiency anemia in children aged 2–5 y. This was mainly due to several factors, for example, the weak bioavailability of iron used in fortification, Moroccan culinary habits, and the widespread use of flour produced by artisanal mills that are not complying with fortification strategy [8]. It has been therefore recommended to replace the form of iron used for wheat flour fortification by one that is more bioavailable [9].

In 2011, the Ministry of Health launched the National Nutrition Strategy (NNS) 2011–2019 with the aim to improve the nutritional status of the Moroccan population. The objective set for iron deficiency anemia was to reduce its prevalence by 1/3 by 2019 compared to its level in 2011 [10].

However, the accomplishment of the objectives of NNS and NPFMD would not be possible without the participation of other actors in order to target specific vulnerable populations like schoolchildren. Therefore, the Foundation for Child Nutrition (FCN), in partnership with the Ministry of National Education (MNE) and the Ministry of Health (MH), started distributing milk fortified with multiple micronutrients to schoolchildren living in rural regions in Morocco most affected by malnutrition. More than 23.500 children benefited from this intervention.

Hence, in partnership with the FCN, the MNE, and the MH, our team undertook a study to assess the efficacy of the consumption of MMN fortified milk (including iron sulfate) on iron nutritional status of schoolchildren living in rural mountainous regions of Morocco.

2. Materials and Methods

2.1. Study Design. The study is a longitudinal interventional, placebo-controlled double blind one conducted among schoolchildren (N = 195), aged from 7 to 9 years. It lasted for 9 months from February to October 2012. Children were eligible for the study if they were 7 to 9 years old and were not taking supplements during the period of study. Children presenting signs of severe malnutrition or anemia (Hb < 9 mg/dL) were excluded from the study (and transferred to a local health center for follow-up).

The study got the approval of the Ministry of National Education. The purpose and the protocol of the study were presented and explained to the local authorities, regional medical representatives, school Head Masters, teaching staff, and parent's union representatives in schools who in turn explained clearly the benefits of the study to children's parents.

Subsequently, oral and written consents were obtained from children and their parents, respectively, before the beginning of the survey.

Trained medical technicians were recruited from the regional local facilities to help with the samples collection.

2.2. Site of the Study. The region where the study took place is situated in the center of Morocco, it is 400 to 700 m above sea level, the climate is continental, and the rainfall varies between 300 and 750 mm per year. Farming is the dominant activity of the population (78.2% of labor force is rural in 2011) [11].

The region is also known to have low-income communities and high prevalence of micronutrient deficiencies [12].

More than one-third of children aged under 5 y suffer from stunting whereas the national prevalence is 23.7% [13].

2.3. Selection of Schools. The schoolchildren were recruited from primary schools that were selected on the basis of the following criteria: accessibility to our field team and to the milk distributors, large attendance of schoolchildren enough to cover the required number for the study age range, climatic conditions mainly to avoid interruption of milk distribution by unforeseeable weather, and similarity of socioeconomic and living conditions. Schoolchildren were divided into two groups to receive either the fortified or the nonfortified milk. A distance of 52 km separated the two sites to avoid errors of distribution and/or exchange of milk batches between schoolchildren.

2.4. Sample Size. This paper represents one arm of a multiple armed survey aiming to evaluate the nutritional status of schoolchildren with regard to several micronutrients, namely, vitamins A and D, iron, and iodine. Accordingly, the calculation of the sample size was based on the standard deviation (0.6 μmol/L) of the serum level of vitamin A previously determined in a reference regional study on the impact of the consumption of oil fortified with vitamin A on the nutritional status of childbearing women done in 2006–2008 in Morocco. To observe a difference of 0.4 μmol/L with 5% level of significance and 80% power between the intervention and control groups, and after accounting for 20% dropouts, a sample size of 43 children per group was required.

2.5. Milk Composition. Two batches of milk were developed and produced for the purpose of the survey. Both fortified and nonfortified milk were identical in appearance, taste, and smell and had the same packages. The only difference was in the nutritional composition as presented in Table 1.

The amount of fortificant added to the fortified milk to obtain the 30% coverage of RDI was determined based on the guidelines of the European Council 2008/100/CE relative to nutritional food labeling [14].

The macro- and micronutrient contents of each batch of milk were doubly checked by Aquanal (Laboratoire Aquitaine Analyses) in France and LOARC (Laboratoire Officiel d'Analyses et de Recherches Chimiques, Casablanca) in Morocco before the beginning and in midsurvey.

2.6. Allocation of Groups. 195 children were assigned to one of the two groups to receive either fortified milk or nonfortified milk by random drawing of schools.

Children in both Fortified Milk Group (FG) and Nonfortified Milk Group (NFG) received daily 200 mL of whole UHT fortified or nonfortified milk, respectively, during the 9 months of the survey (including weekends and vacation days).

Each child was attributed a code and received daily the corresponding type of milk. The distribution of milk was supervised either by the school principal or the teacher in charge. A separate list was prepared for absent children and their milk was delivered to them at the end of the day. Before

TABLE 1: Composition of nonfortified and fortified milk.

Nutritional composition	Nonfortified milk		Fortified milk	
	Amount/200 mL serving	% RDI[a] children 7–9 y	Amount/200 mL serving	% RDI[a] children 7–9 y
Energy (Kcal)	154.8	—	154.8	—
Fat (%)	5.8	—	5.8	—
Protein (g)	5.8	—	5.8	—
Lipids (g)	6	—	6	—
Carbohydrates (g)	19.44	—	19.44	—
Calcium (mg)	240	30	240	30
Iron (mg)	<0.4	<3	4.2	30
Iodine (μg)	20.8	<14	45	30
Vitamin D3 (μg)	<1	<10	3	30
Vitamin A (μg)	54	<7	240	30

RDI: recommended dietary intake.
[a]The values were based on the guidelines of the European Council 2008/100/CE relative to nutritional food labeling.

weekends and vacation days, a quantity of milk sufficient to cover the period was delivered to the parents of the children.

2.7. Data Collection

2.7.1. Socioeconomic Questionnaire (SES). The data on socioeconomic standards and living conditions of the children and their families were collected at baseline by interviewing the parents. We used an adequate questionnaire that was adapted from other questionnaires used nationally to serve the purpose of our survey. Information collected included the level of education of parents, household size, household monthly global expenses, and alimentary expenses.

2.7.2. Anthropometric Measurements. Anthropometric measurements were taken following standard procedures [15] at baseline. Height was recorded to the nearest 0.1 cm using a stadiometer (Fazzini-2 meters) and weight was recorded to the nearest 0.1 kg using a portable scale (Seca 750-Germany). Stunting and thinness were defined as Height-for-Age (HAZ) and Body Mass Index-for-Age (BAZ) Z-scores < −2, respectively, according to the World Health Organization (WHO) [16].

2.8. Biochemical Analyses

2.8.1. Hb Measurements. Hb analysis was done at baseline and end line. It was performed in situ using the Hemocue portable spectrophotometer (HemoCue AB, Angelholm, Sweden) on a drop of venous blood withdrawn while doing the blood sampling. Anemia was defined as Hb levels <11.5 mg/dL [16]. Hb values measured were adjusted for altitude [17].

2.8.2. Blood Sampling. Whole blood (8 mL) was collected in dry tubes from nonfasting children at baseline and endpoint by venipuncture. Directly after collection, the samples were centrifuged at 5000 rpm for 5 mn and serum was aliquoted

in Cryovial tubes and transferred in isothermic box under 4–8°C to the laboratory and then stored at −80°C until analysis of serum Ferritin (SF) and C Reactive Protein (CRP). All the analyses were performed in laboratories of UMRNA (Unité Mixte de Recherche en Nutrition et Alimentation URAC39, Université Ibn Tofail-CNESTEN, Kénitra-Rabat, Morocco).

(1) Assessment of Serum Ferritin. Quantitative determination of ferritin level in the serum was performed in the laboratory using a colorimetric immunoenzymatic method type ELISA sandwich (NovaTec Immundiagnostica GMBH, Germany). Iron deficiency was defined as serum ferritin <15 μg/L and iron deficiency anemia was defined as iron deficiency along with anemia [16].

(2) Serum Concentrations of CRP. The level of CRP in the serum was determined by nephelometry using the Minineph kit (MININEPH, Références, ZK044.L.R, The Binding Site, Birmingham, UK). In our survey CRP serves mainly as a biomarker of inflammation or subclinical infection on days of blood sampling. A cutoff of >10 mg/L was used for abnormal serum CRP concentrations [18].

2.9. Statistical Analysis. Data analysis was done by the software IBM SPSS Statistics version 20 (Statistical Package for the Social Sciences). Anthropometric measurements were analyzed by Anthro+ (WHO standards) [16]. The distribution normality of the quantitative variables was tested by Kolmogorov-Smirnov test. The variables normally distributed were presented as mean ± standard deviation and those nonnormally distributed as median (interquartile range). ANOVA was used to compare variances between independent samples. The homogeneity of variances was tested using Leven's test, and the correction of Welch was used in the case of nonhomogeneous variances. Mann-Whitney test was used to compare independent samples for variables nonnormally distributed. Wilcoxon test was used to compare the relation between T0 and T9 within the same group. The nominal variables were presented as proportion and 95%

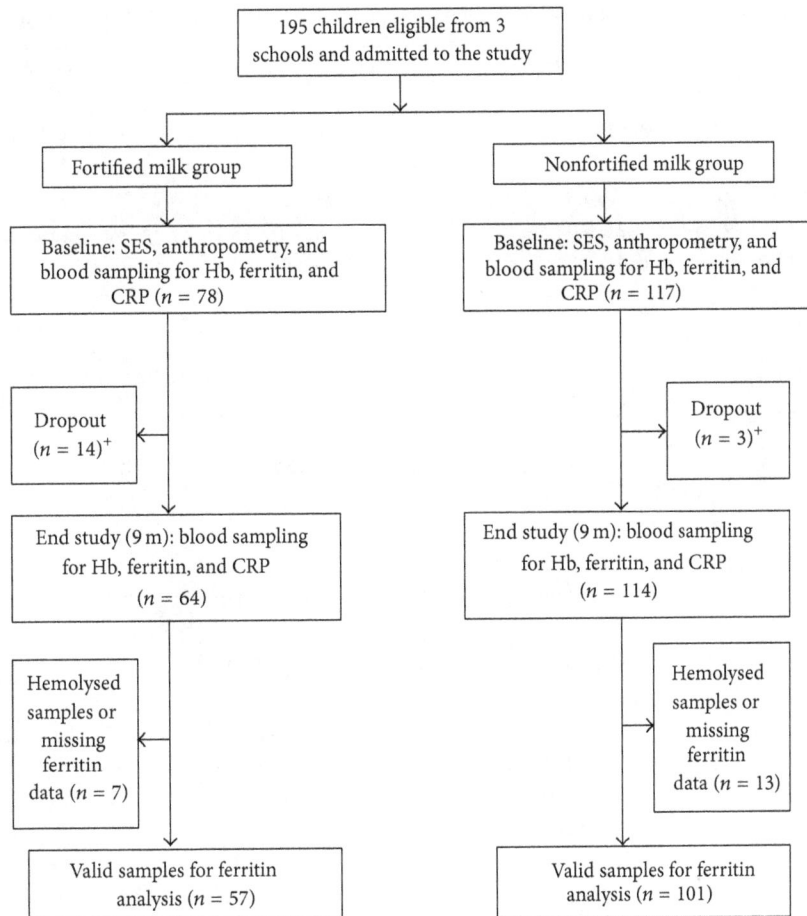

FIGURE 1: Participant Flowchart: [+]dropouts: participants who either refused to give blood or were absent on the day of samples withdrawal, changed school or were relocated during study, or were excluded due to severe anemia (Hb < 9 mg/dL). SES: Socioeconomic Status, Hb: Hemoglobin, and CRP: C Reactive Protein.

Confidence Interval (Lower-Upper). Chi-square test was used to test independence between nominal variables. Chi-square value was corrected for cells with a theoretical frequency less than 5; if a theoretical $n < 5$ we take the p value of Fisher, and 95% Confidence Intervals were determined using the Bootstrap technique based on 1000 bootstrap samples. The correlation between high CRP values (>10 mg/L) and high ferritin level was tested using Bivariate Correlations test. A difference was considered as statistically significant if $p < 0.05$.

3. Results

Figure 1 represents the participant flowchart. The rate of compliance was not the same in both groups; we observed a larger number of dropouts in the fortified group that was due to participants' refusal to continue the survey (children refused to give blood samples or were absent on the day of blood withdrawal), change of school, or relocation out of the study area. Nevertheless, the size of the sample in FG was still statistically valid. No adverse events because of the intervention were reported during the course of the study.

The growth parameters and socioeconomic characteristics are presented in Table 2.

There were no significant differences in baseline anthropometric measurements or socioeconomic characteristics of children between the NFG and the FG.

In general, the prevalence of illiteracy in mothers for both groups was high compared to fathers with 95.2% and 60.7%, respectively.

91.1% of households spend less than 195 US$/m for food compared to 54.4% for general expenditure. 195 US$ is the equivalent of the guaranteed minimum wage for governmental employees.

To assess the dietary habits of our population, we used a food frequency questionnaire that was filled by the children's mothers at baseline. The preliminary analysis of the data collected showed that the majority of children consumed foods rich in iron or that stimulate its absorption (e.g., meat, legumes, and fruits). Dairy products consumption was moderate for yogurt and low for cheese. On the other hand, more than 90% of children consumed tea at least once per week, which could be the reason behind the high prevalence of ID among children. Both groups had similar food trends and

TABLE 2: Baseline demographic, anthropometric, and socioeconomic characteristics of schoolchildren enrolled in the study.

	Total ($N = 195$)		NFG ($N = 117$)		FG ($N = 78$)		p value
General characteristics							
Age (y) (mean ± SD)	8.0 ± 0.7		8.0 ± 0.7		7.9 ± 0.8		0.371
Baseline anthropometry							
Height (cm) (mean ± SD)	122.3 ± 6.1		121.9 ± 6.3		122.8 ± 5.6		0.352
Weight (kg) (mean ± SD)	23.2 ± 3.0		23.1 ± 3.0		23.2 ± 2.9		0.483
BMI (kg/m^2) (mean ± SD)	15.4 ± 1.1		15.5 ± 1.0		15.4 ± 1.2		0.358
Nutritional status							
Stunting[a] HAZ <−2 SD (%)	8.4		6.8		10.3		0.219
Thinness[a] BAZ <−2 SD (%)	2.1		0		5.1		
	%	95% CI	%	95% CI	%	95% CI	p value
Sex							
Female	50.6	(42.9–58.3)	52.4	(42.7–61.2)	47.7	(35.4–60.0)	0.815
Male	49.4	(41.7–57.1)	47.6	(38.8–57.3)	52.3	(40.0–64.6)	
Level of education							
Mother							
Illiterate	95.2	(91.7–98.2)	98.1	(95.1–100.0)	90.8	(83.1–96.9)	
Primary	3.6	(1.2–6.5)	1.0	(0.0–2.9)	7.7	(1.5–15.4)	0.069
Secondary	1.2	(0.0–3.6)	1.0	(0.0–2.9)	1.5	(0.0–4.6)	
Father							
Illiterate	60.7	(53.6–68.5)	60.2	(51.5–68.9)	61.5	(49.2–73.8)	
Primary	31.5	(24.4–38.7)	32.0	(23.3–40.8)	30.8	(18.5–43.1)	0.562
Secondary	7.1	(3.6–11.3)	7.8	(2.9–13.6)	6.2	(1.5–12.3)	
College	0.6	(0.0–1.8)	0.0		1.5	(0.0–4.6)	
Household size							
<6 persons	48.8	(41.1–56.5)	49.5	(39.8–59.2)	47.7	(35.4–60.0)	0.944
6 to 10 persons	51.2	(43.5–58.9)	50.5	(40.8–60.2)	52.3	(40.0–64.6)	
Total monthly expense							
<122US$	25.6	(19.6–32.7)	28.2	(19.4–37.8)	21.5	(12.3–32.3)	
122–195US$	28.6	(21.4–35.7)	21.4	(14.6–29.1)	40.0	(27.7–52.3)	
196–244US$	26.2	(20.2–32.7)	31.1	(22.3–39.8)	18.5	(9.2–29.2)	0.117
245–366US$	11.9	(7.1–16.7)	11.7	(5.8–18.4)	12.3	(4.6–20.0)	
>367US$	7.7	(4.2–11.9)	7.8	(2.9–13.6)	7.7	(1.5–13.8)	
Monthly expense for food							
<110US$	63.1	(55.4–70.2)	60.2	(50.5–69.9)	67.7	(55.4–78.5)	
110–147US$	18.5	(12.5–24.4)	23.3	(15.5–31.1)	10.8	(3.1–18.5)	
148–195US$	9.5	(5.4–14.3)	6.8	(2.9–12.6)	13.8	(6.2–23.0)	0.076
196–305US$	8.3	(4.8–12.5)	9.7	(3.9–16.5)	6.2	(1.5–12.3)	
>306US$	0.6	(0.0–1.8)	0.0		1.5	(0.0–4.6)	

BMI: Body Mass Index; [a]HAZ and BAZ were calculated by *Anthropo+*.

the difference in dietary behaviors between FG and NFG was not statistically significant ($p > 0.05$).

3.1. Iron Deficiency (Tables 3 and 4). At T9 FG showed a reduction of the prevalence of iron deficiency (serum ferritin < 15 μg/L) in comparison with T0 from 50.9% (95% CI: 38.6–63.2) to 37.2% (95% CI: 23.3–51.2), while for the NFG it remained stable between T0 and T9 at 56% and 56.4%, respectively. The difference between FG and NFG was statistically significant at T9 ($p = 0.035$). The p value calculated within the same groups between T0 and T9 was significant for

the FG ($p = 0.037$) and nonsignificant for the NFG ($p = 0.927$).

The median (interquartile) of serum ferritin increased in the FG from 14.0 (9.0; 20.0) at T0 to 17.0 (11.0; 26.0) at T9, while it remained stable in NFG. The difference between the two groups was statistically significant at T9 ($p = 0.019$).

3.2. Anemia and Iron Deficiency Anemia (Tables 3 and 4). The prevalence of iron deficiency anemia (IDA) (Hb < 11.5 mg/dL and ferritin < 15 μg/L) for the FG dropped from 1.8% (95% CI: 0.0–5.3) at T0 to 0.0% at T9. For the NFG,

TABLE 3: Iron status at baseline and endpoint in the study groups.

Biochemical parameters	Total N	Total Mean ± SD	NFG N	NFG Mean ± SD	FG N	FG Mean ± SD	p value*
Hemoglobin (mg/dL)							
Baseline	178	14.45 ± 1.46	114	14.58 ± 1.58	64	14.22 ± 1.21	0.090
End line	178	14.88 ± 1.35	114	15.05 ± 1.43	64	14.59 ± 1.16	0.213
	N	Median; interquartile	N	Median; interquartile	N	Median; interquartile	p value*
Serum ferritin (µg/L)							
Baseline	158	13.0 (9.0; 21.0)	101	13.0 (8.0; 21.0)	57	14.0 (9.0; 20.0)	0.610
End line	144	14.0 (9.0; 22.7)	101	13.0 (8.0; 20.0)	43	17.0 (11.0; 26.0)	0.019

*p value by one way ANOVA for means and Mann-Whitney test for medians.

TABLE 4: Prevalence of anemia, iron deficiency anemia, and iron deficiency at baseline and endpoint in both groups.

	Total %	Total 95% CI N = 195	NFG %	NFG 95% CI N = 117	FG %	FG 95% CI N = 78	p value*
Anemia[a] (Hb < 11.5 mg/dL)							
Baseline	2.2	(0.6–4.5)	2.6	(0.0–6.1)	1.6	(0.0–4.7)	0.999
End line	2.2	(0.6–4.5)	2.6	(0.0–6.1)	1.6	(0.0–4.7)	0.999
Iron deficiency anemia[b] (Hb < 11.5 mg/dL and fe < 15 µg/L)							
Baseline	1.9	(0.0–4.5)	2.0	(0.0–5.0)	1.8	(0.0–5.3)	0.760
End line	1.4	(0.0–3.5)	2.0	(0.0–5.0)	0.0	(0.0–0.0)	0.064
Iron deficiency[c] (Fe < 15 µg/L)							
Baseline		N = 158		N = 101		N = 57	
	54.1	(45.9–62.4)	56.0	(47.0–65.0)	50.9	(38.6–63.2)	0.536
End line		N = 144		N = 101		N = 43	
	50.7	(43.1–59.0)	56.4	(45.6–66.3)	37.2	(23.3–51.2)	0.035
p value** for iron deficiency within same group				0.927		0.037	

*p value for comparing deficiency prevalence among study groups using χ^2-test.
**p value for comparing deficiency prevalence within same study group using Wilcoxon test.
[a]Anemia was defined as Hb levels <11.5 mg/dL. [b]Iron deficiency anemia was defined as iron deficiency along with anemia by the above-mentioned criteria.
[c]Iron deficiency was defined as serum ferritin <15 µg/L.

it remained unchanged at 2.0% both at baseline and end line. The difference between both groups was not statistically significant at T9 ($p = 0.064$).

The mean Hb increased slightly in both groups. At T0 it was 14.58 ± 1.58 and 14.22 ± 1.21 for NFG and FG, respectively, whereas at T9 it became 15.05 ± 1.43 and 14.59 ± 1.16.

The difference between the two groups at T9 was not statistically significant ($p = 0.213$).

4. Discussion

Our results showed that the consumption of milk fortified with ferrous sulfate and other micronutrients is efficacious in reducing the prevalence of iron deficiency and improving iron status indicators in a sample of children 7–9 y of age. Authors from different countries previously published results of efficacy interventions using fortified milk and reported varying degrees of success in reducing the iron deficiency depending on the dose and duration of intervention.

In Chile, two studies conducted in infants confirmed the efficacy of iron-fortified milk with ferrous sulfate combined with ascorbic acid [19, 20]. While in India, a trial conducted among children aged 1–4 years for a period of one year demonstrated the efficacy of a multiple micronutrients (including iron and zinc) fortified milk on growth, body iron stores, and anemia [21].

In 2003, a study done in Morocco to assess the effect of a dual-fortified salt (DFS) containing iodine and microencapsulated iron on nutritional status of schoolchildren showed that the prevalence of IDA in the fortified group decreased from 35% at baseline to 8% after 40 weeks of intervention ($p < 0.001$) [22].

While two other surveys conducted in Brazil and Sweden revealed a lesser efficacy of fortified milk. In the first one 185 Brazilian children with mild or severe anemia received milk fortified with 3 mg/L of iron amino acid chelate. After 222 days of intervention, 43% remained anemic. The reduced efficacy in this study was attributed to the low level of iron fortification (3 mg/L) [23], while in Sweden, a controlled trial in 36 children treated for 6 months with fortified milk with 7.0 or 14.9 mg/L of iron reported no significant effects on hematological and iron status indicators and this has been explained by the fact that these children had a good baseline iron status; thus, noticeable changes in Hemoglobin or iron status should not be expected [24].

In our trial, we observed an increase of median ferritin levels and a marked reduction in the prevalence of iron deficiency (27%) in FG, compared to NFG, and this has been reported by other fortification trials conducted in low-income countries [25, 26] which highlights a specific effect attributable to the intervention. Also availability of vitamin A, essential for erythropoiesis, could have resulted in a better overall improvement of iron status [27].

However, it is worthy to emphasize that our milk contained naturally a high calcium level (240 mg of calcium per 200 mL) and it is well known that calcium in milk interferes significantly with the absorption of iron. The mechanism of action for absorption inhibition is unknown. Recent analyses of the dose-effect relationship show that the first 40 mg of calcium in a meal does not inhibit absorption of haem and nonhaem iron. Above this level of calcium intake, a sigmoid relationship develops, and at levels of 300–600 mg calcium, it reaches a 60% maximal inhibition of iron absorption [28]. Thus, the effects on the prevalence of anemia and iron status herein described were most probably modulated by the interaction of both iron and calcium at the mucosal cell (at the intestinal level), resulting in a less pronounced efficacy in improving iron than fortification without a high level of calcium. The efficacy could have also been more evident if enhancers of iron absorption had been added to this milk. Ferrous sulfate along with vitamin C, to potentiate bioavailability of iron, added to milk proved to be more effective in reducing the prevalence of anemia in other studies [21, 29, 30].

5. Conclusions

This study provides evidence that delivery of iron via a food-based vehicle, milk in this instance, is a feasible option and produces a positive effect on iron status among schoolchildren. It provides a potential strategy for achieving Millennium Development Goals targeting reduction in mortality, morbidity, and malnutrition among children, constituting an example of how the use of research can directly benefit the design of successful public nutrition programs such as the National Nutrition Strategy, the National Program of Fight against Micronutrient Deficiencies, and the application of the recommendations of the second International Conference on Nutrition (ICN2). Indeed, our work may represent a solution at the national level, encouraging the generalized distribution of fortified breakfasts sponsored by the MNE in rural schools. There are indications (according to FCN) that such distribution may result in a reduction of school dropout rates too.

Limitations of the Study

Because of *a priori* criteria of selection of the schools, we were unable to recruit an equal number of children in both groups (as shown in Figure 1). In spite of this and the small size of the study population our findings were statistically valid. Nevertheless, future studies should try to overcome these limitations.

Abbreviations

BAZ: Body Mass Index-for-Age Z-scores
CRP: C Reactive Protein
DFS: Dual-fortified salt
FCN: Foundation for Child Nutrition
FG: Fortified Milk Group
HAZ: Height for age Z-scores
ID: Iron deficiency
IDA: Iron deficiency anemia
MH: Ministry of Health
MMN: Multiple micronutrients
MNE: Ministry of National Education
NFG: Nonfortified Milk Group
NNS: National Nutrition Strategy
NPFMD: National Program of Fight against Micronutrients Deficiencies
RDI: Recommended dietary intake
SD: Standard deviation
SES: Socioeconomic Status
SF: Serum ferritin
SPSS: Statistical Package for the Social Sciences
UHT: Ultra-High-Temperature
WAZ: Weight for age Z-scores.

Conflict of Interests

The authors declare having no conflict of interests. None of the authors was affiliated in any way with an entity involved with the manufacture or marketing of milk.

Acknowledgments

The authors would like to gratefully acknowledge the contributions of schoolchildren who participated in this study, their parents, teachers, health workers, local authorities, and other support staff. We are also grateful to acknowledge the contribution of Foundation for Child Nutrition for providing UHT milk used in the survey.

References

[1] World Health Organization, "Global targets 2025. To improve maternal, Infant and young child nutrition," October 2014, http://www.who.int/nutrition/topics/nutrition_globaltargets2025/en/.

[2] World Health Organization, *The World Health Report 2001: Mental Health: New Understanding, New Hope*, World Health Organization, Geneva, Switzerland, 2001.

[3] L. Allen, B. de Benoist, O. Dary, and R. Hurrell, Eds., *Guidelines on Food Fortification with Micronutrients*, World Health Organization/Food and Agriculture Organization of the United Nations, Geneva, Switzerland, 2006.

[4] Ministère de la Santé, *Système Sentinelle de Suivi et Évaluation du Processus de la Fortification et son impact sur l'état nutritionnel de la Population. Projet GAIN/Composante Suivi et Évaluation*, Rabat Rapport, Ministère de la Santé, Direction de la Population, 2008.

[5] Ministère de la Santé, "Enquête nationale sur l'anémie par carence en fer, la supplémentation et la couverture des ménages par le sel iodé," Rapport Ministère de la Santé, Ministère de la Santé, Rabat, Morocco, 2000.

[6] L. H. Allen, "Iron supplements: scientific issues concerning efficacy and implications for research and programs," Journal of Nutrition, vol. 132, no. 4, pp. 813S–819S, 2002.

[7] H. Aguenaou, "La malnutrition invisible ou la ńfaim cachéeż au Maroc et les stratégies de lute," Biomatec Echo, vol. 5, no. 2, pp. 158–164, 2007.

[8] A. El Hamdouchi, K. El Kari, E. A. Rjimati, M. El Mzibri, N. Mokhtar, and H. Aguenaou, "Does flour fortification with electrolytic elemental iron improve the prevalence of iron deficiency anaemia among women in childbearing age and preschool children in Morocco?" Mediterranean Journal of Nutrition and Metabolism, vol. 6, no. 1, pp. 73–78, 2013.

[9] N. Mokhtar, MS/UNICEF, Evaluation de la situation nutritionnelle au Maroc, Mars 2010 (communication personnelle).

[10] MS/UNICEF, "La Stratégie Nationale de la Nutrition 2011–2019," http://www.unicef.org/morocco/french/Strategie_Nationale_de_Nutrition_.pdf.

[11] HCP-Direction régionale Tadla Azilal-Monographie régionale 2012, http://www.hcp.ma/downloads/Monographies-regionales_tl1957.html.

[12] MS—Enquête Nationale sur la Population et la Santé ENPS-II, 1992, http://www.sante.gov.ma/Publications/Etudes_enquete.

[13] Enquête sur la Population et la Santé Familiale EPSF, 2003/2004, http://www.sante.gov.ma/Publications/Etudes_enquete.

[14] Directive 2008/100/CE de la Commission modifiant la directive 90/496/CEE du Conseil relative à l'étiquetage nutritionnel des denrées alimentaires en ce qui concerne les apports journaliers recommandés, les coefficients de conversion pour le calcul de la valeur énergétique et les définitions, http://www.legifrance.gouv.fr/.

[15] T. G. Lohman, A. F. Roche, and R. Martorell, Anthropometric Standardization Reference Manual, Human Kinetics, Champaign, Ill, USA, 1988.

[16] WHO Multicentre Growth Reference Study Group, WHO Child Growth Standards: 406 Length/Height-for-Age, Weight-for-Age, Weight-for-Length, Weight-for-Height and Body 407 Mass Index-for-Age: Methods and Development, WHO, Geneva, Switzerland, 2006, http://www.who.int/growthref/en/.

[17] H. Dirren, M. H. G. M. Logman, D. V. Barclay, and W. B. Freire, "Altitude correction for hemoglobin," European Journal of Clinical Nutrition, vol. 48, no. 9, pp. 625–632, 1994.

[18] V. Q. Bui, A. D. Stein, A. M. DiGirolamo et al., "Associations between serum C-reactive protein and serum zinc, ferritin, and copper in Guatemalan school children," Biological Trace Element Research, vol. 148, no. 2, pp. 154–160, 2012.

[19] A. Stekel, M. Olivares, M. Cayazzo, P. Chadud, S. Llaguno, and F. Pizarro, "Prevention of iron deficiency by milk fortification. II A field trial with a full-fat acidified milk," American Journal of Clinical Nutrition, vol. 47, no. 2, pp. 265–269, 1988.

[20] E. Hertrampf, M. Olivares, T. Walter et al., "Iron-deficiency anemia in the nursing infant: its elimination with iron-fortified milk," Revista Médica de Chile, vol. 118, no. 12, pp. 1330–1337, 1990.

[21] S. Sazawal, U. Dhingra, P. Dhingra et al., "Micronutrient fortified milk improves iron status, anemia and growth among children 1–4 years: a double masked, randomized, controlled trial," PLoS ONE, vol. 5, no. 8, Article ID e12167, 2010.

[22] M. B. Zimmermann, C. Zeder, N. Chaouki, A. Saad, T. Torresani, and R. F. Hurrell, "Dual fortification of salt with iodine and microencapsulated iron: a randomized, double-blind, controlled trial in Moroccan schoolchildren," American Journal of Clinical Nutrition, vol. 77, no. 2, pp. 425–432, 2003.

[23] C. Iost, J. J. Name, R. B. Jeppsen, and H. D. Ashmead, "Repleting hemoglobin in iron deficiency anemia in young children through liquid milk fortification with bioavailable iron amino acid chelate," Journal of the American College of Nutrition, vol. 17, no. 2, pp. 187–194, 1998.

[24] M. A. Virtanen, C. J. E. Svahn, L. U. Viinikka, N. C. R. Räihä, M. A. Siimes, and I. E. M. Axelsson, "Iron-fortified and unfortified cow's milk: effects on iron intakes and iron status in young children," Acta Paediatrica, vol. 90, no. 7, pp. 724–731, 2001.

[25] D. M. Ash, S. R. Tatala, E. A. Frongillo Jr., G. D. Ndossi, and M. C. Latham, "Randomized efficacy trial of a micronutrient-fortified beverage in primary school children in Tanzania," The American Journal of Clinical Nutrition, vol. 77, no. 4, pp. 891–898, 2003.

[26] M. Faber, J. D. Kvalsvig, C. J. Lombard, and A. J. S. Benadé, "Effect of a fortified maize-meal porridge on anemia, micronutrient status, and motor development of infants," American Journal of Clinical Nutrition, vol. 82, no. 5, pp. 1032–1039, 2005.

[27] C. M. Smuts, M. A. Dhansay, M. Faber et al., "Efficacy of multiple micronutrient supplementation for improving anemia, micronutrient status, and growth in South African infants," Journal of Nutrition, vol. 135, no. 3, pp. 653S–659S, 2005.

[28] World Health Organization, Vitamin and Mineral Requirements in Human Nutrition: Report of a Joint FAO/WHO Expert Consultation, World Health Organization, Bangkok, Thailand, 1998, http://whqlibdoc.who.int/publications/2004/9241546123.pdf.

[29] A. Stekel, M. Olivares, F. Pizarro, P. Chadud, I. Lopez, and M. Amar, "Absorption of fortification iron from milk formulas in infants," The American Journal of Clinical Nutrition, vol. 43, no. 6, pp. 917–922, 1986.

[30] A. Stekel, M. Olivares, F. Pizarro et al., "Prevention of iron deficiency in infants by fortified milk. Field study of a low-fat milk," Archivos Latinoamericanos de Nutrición, vol. 36, pp. 654–661, 1986.

11

Fasting Blood Glucose Profile among Secondary School Adolescents in Ado-Ekiti, Nigeria

I. O. Oluwayemi,[1] S. J. Brink,[2] E. E. Oyenusi,[3] O. A. Oduwole,[3] and M. A. Oluwayemi[4]

[1]Department of Paediatrics, College of Medicine, Ekiti State University, Ado-Ekiti, Ekiti State, Nigeria
[2]New England Diabetes and Endocrine Center, USA
[3]Paediatric Endocrinology Training Centre for West Africa, Lagos University Teaching Hospital, Idi-Araba, Lagos, Nigeria
[4]Clinical Nursing Services, Ekiti State University Teaching Hospital, Ado-Ekiti, Ekiti State, Nigeria

Correspondence should be addressed to I. O. Oluwayemi; dareoluwayemi@gmail.com

Academic Editor: Pedro Moreira

Background. Over the past two decades there has been an increase in type 2 diabetes mellitus (T2DM) in children. Baseline data is needed to assess the impact of changing lifestyles on Ado-Ekiti, a previously semiurban community in Southwest Nigeria. This study was therefore conducted to assess the fasting blood glucose (FBG) of adolescents in Ado-Ekiti, Nigeria. *Methodology.* This was a cross-sectional study involving 628 adolescents from three different secondary schools in Ado-Ekiti, Nigeria. With parental consent, volunteers completed a structured questionnaire, and an overnight FBG was measured. *Results.* There were 346 males and 282 females (male : female ratio = 1.2 : 1). Their ages ranged from 10 to 19 years (mean age: 14.2 ± 1.7 years). Four hundred and forty-four (70.7%) had normal FBG, while 180 (28.7%) and 4 (0.6%) had FBG in the prediabetic and diabetic range, respectively. Female gender, age group 10–14 years, and family history of obesity were significantly associated with impaired FBG (P value <0.001, <0.001, and 0.045, resp.). *Conclusion.* Impaired FBG is common among secondary school adolescents and it is more prevalent among younger female adolescents (10–14 years) with positive family history of obesity.

1. Introduction

There is rising incidence of noncommunicable diseases like diabetes, hypertension, and coronary heart diseases globally [1–3]. Increased risk of impaired glucose tolerance, insulin resistance, and type 2 diabetes (T2DM) has been found to be associated with obesity in adolescents [4]. The American Academy of Paediatrics and the American Diabetes Association have recommended that children aged 10 years or at the onset of puberty who are overweight and have at least two other risk factors should be tested every two years for T2DM [5, 6]. The risk factors for developing T2DM include family history of T2DM in first- or second-degree relative, belonging to certain ethnic groups (i.e., Native American, African American, Hispanic, Japanese, or other Asian/Pacific Islanders), or having signs associated with insulin resistance (hypertension, dyslipidemia, acanthosis nigricans, or polycystic ovarian syndrome) [5, 7]. Similarly other endogenous

populations, for instance, in Canada, Australia, Russia, and Latin America, may share similar genetic predispositions [8].

Beck-Nielsen and Groop [9] proposed a three-stage model for the development of T2DM. Stage 1 includes fasting hyperinsulinemia with normal or slightly increased blood glucose, especially mild fasting hyperglycemia. Stage 2 is characterized by prediabetic glucose intolerance with insulin resistance, and Stage 3 is development of classical symptomatic or nonsymptomatic T2DM with more persistent hyperglycemia present. Many of the macrovascular changes associated with diabetes and related to cardiovascular disease (CVD) begin in Stages 1 and 2, well before overt diagnosis [10].

Adolescents are a dependent population and development of diabetes mellitus or other noncommunicable diseases will pose a burden to parents and society at large,, hence the need to continually assess the FBG of adolescents in rapidly changing communities. This will help in early

TABLE 1: Fasting blood glucose grading according to age group.

| Age group | Fasting blood glucose (FBG) grading | | | Total |
	Euglycemia 2.8–5.5 mmol/L (50–99 mg/dL)	Impaired FBG (prediabetic) 5.6–6.9 mmol/L (100–125 mg/dL)	Diabetes FBG ≥7 mmol/L (126 mg/dL)	
10–14 years	232	121	4	357
15–19 years	212	59	0	271
Total (%)	444 (70.7)	180 (28.7)	4 (0.6)	628 (100)

detection and hopefully the control of the prediabetic phase through education and lifestyle modification. It will also help in prompt commencement of treatment in those with established diabetes to improve quality of life and prevent complications. Data on FBG among Nigerian adolescents are scarce, hence the need for the present study especially in Ado-Ekiti, a relatively new and rapidly developing state capital in Nigeria with changes in lifestyle associated with urbanization such as inappropriate dietary practices (fast food consumption, low fruit consumption) and low physical activity [11, 12].

2. Material and Methods

Secondary school adolescents of both sexes who satisfied the inclusion criteria were recruited for the study from three different schools in Ado-Ekiti, the capital of Ekiti State, Nigeria. Ethical clearance and permission to enter the schools were obtained from the Research and Ethics Committee of the Ekiti State University Teaching Hospital, Ado-Ekiti, and the state's Ministry of Education, respectively. Inclusion criteria for the study were apparently healthy secondary school adolescents aged between 10 and 19 years, with no history of diabetes. With parental consent, volunteers completed a structured questionnaire, and FBG was measured after an overnight fast. Capillary blood sample was collected for FBG measurement using Accu-Chek Active glucometer, after the thumb or index finger had been cleaned with wet (water) cotton wool. Data was entered and analyzed using SPSS 16.0 for Windows (SPSS Inc., Chicago, USA). Subjects were grouped based on their age, gender, family history of obesity and diabetes, and fasting blood glucose (FBG). Cross tabulation and tests for association with chi square (χ^2) were done and P values less than 0.05 were regarded as significant.

3. Results

The subjects in the present study comprised 346 male and 282 female adolescents (male to female ratio of 1.2 : 1) with mean age of 14.2 ± 1.7 years and age range of 10 to 19 years. Four (0.6%) adolescents had FBG in diabetic range; 180 (28.7%) had impaired FBG and 121 (67%) of these were in the 10–14-year age group. Also, all the four adolescents who had diabetic FBG range were in the same 10–14-year age group (Table 1). There were 77 (41.8%) males and 107 (58.2%) females in the 184 (29.3%) adolescents with high FBG (180 in prediabetic

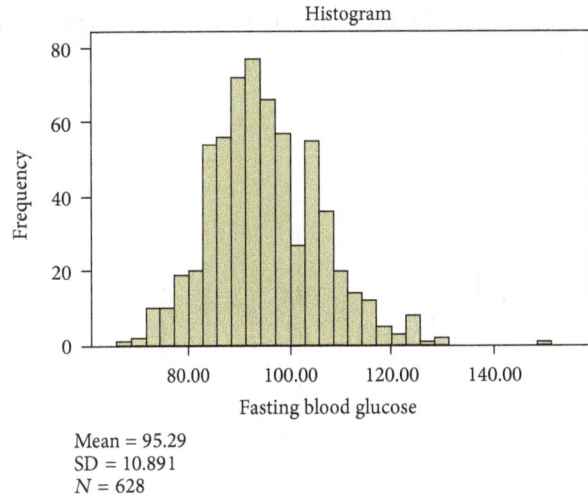

FIGURE 1: Fasting blood glucose (mg/dL) pattern in 628 secondary school adolescents.

and 4 in diabetic range) among the studied 628 adolescents giving a male to female ratio of 0.7 : 1. The proportion of males with high FBG of the overall studied male population of 346 was 77 (22.3%) while the proportion of females with high FBG of the overall female population of 282 was 107 (37.9%). This difference was statistically significant (χ^2 = 18.462, df = 1, P = 0.000). Figure 1 shows the FBG profile in the studied sample of secondary school adolescents in Ado-Ekiti. The mean (SD) for their FBG was 95.3 (10.9), median was 94.0, and it ranged from 68 to 149 mg/dL. The variability of FBG in both genders was comparable except that the median FBG for females was slightly higher (Figure 2). The mean (SD) FBG for males was 93.4 (10.8) and for females was 97.6 (10.5). The FBG chart for age (Figure 3) showed two distinct peaks which occurred at 12 and 17 years though there appears to be a little rise at the age of 14 years. Female gender, early adolescence (age group 10–14 years), and family history of obesity have statistically significant influence on FBG levels (P value <0.001, <0.001, and 0.045, resp.).

4. Discussion

The review of the fasting blood glucose of the adolescents in Ado-Ekiti showed that the majority (70.7%) had normal FBG, and a rather large number, about a third (28.7%),

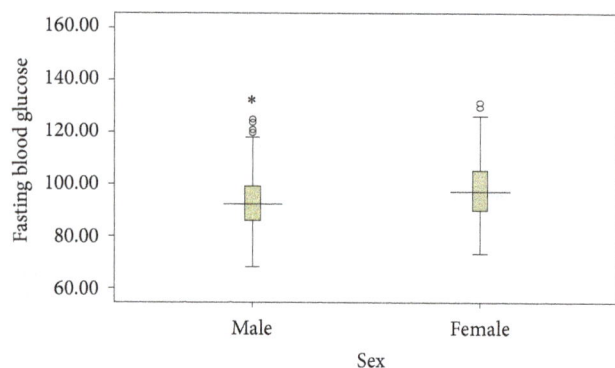

FIGURE 2: Comparison of FBG (mg/dL) pattern in male and female population.

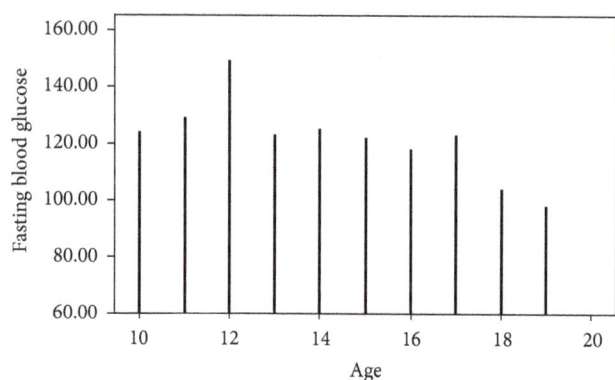

FIGURE 3: Profile of fasting blood glucose (mg/dL) according to age (years) of adolescents in Ado-Ekiti.

had impaired FBG in the prediabetic range and 0.6% of them had presumed diabetic FBG range. The possibility of secondary school adolescents falsely denying taking some juice or snack before the test cannot be completely ruled out and this may contribute to the unusually high proportion of those with impaired FBG in this study. There may also be some bias using plasma meter BG readings rather than more centralized sampling systems as reported in other studies [2, 13]. Some nervousness of participants not used to finger pricks may also have contributed to some false elevations and repeat testing will be proposed to double-check these results. The FBG profile of the adolescents plotted against their ages showed two distinct peaks which occurred at 12 and 17 years (Figure 3); this is similar to the peaks in their BMI which occurred at the same ages of 12 and 17 years. This finding is in concordance with findings among adolescents in Beijing area of China [14] whose FBG peaked at 12-13 years and at 17 years of age. Increased secretion of growth hormone and adrenocortical and gonadal hormones during puberty usually causes increase in insulin resistance and this could explain the peaks in FBG of adolescents at 12 and 17 years which roughly correspond to early and late phase of puberty in both genders combined [13]. It was also demonstrated in this index study that early adolescence (age group 10–14 years), female sex, and family history of obesity

have statistically significant association with impaired fasting blood glucose and this finding agrees with findings from previous studies [13, 14]. Moreover, the mean FBG in the male and female adolescents in index study (93.4 ± 10.8; 97.6 ± 10.5 mg/dL, resp.) is slightly higher than those of Moroccan [2] male and female adolescents of a similar age (92 ± 14.0; 89 ± 10.0 mg/dL, resp.). This difference may be due to different methods used for analysis of FBG. In the Moroccan study, Mehdad et al. [2] used hexokinase method to determine FBG while the current study assessed capillary FBG with Accu-Chek Active glucometer. Also, the Moroccan study [2] had a smaller and disproportionate study population (44 males; 123 females) compared to the index study of 346 male and 282 female adolescents.

Strength and Limitation of the Study

(1) The sample size for this study is adequate and both genders were well represented.

(2) The population of adolescents available for this study from private secondary school was small compared to those from public secondary schools, hence preventing meaningful comparison.

5. Conclusion

The present study among secondary school adolescents in Ado-Ekiti, Nigeria, showed that FBG was normal in two-thirds of them while the remaining one-third had impaired FBG. Also, impaired FBG is significantly more common among younger female adolescents (10–14 years) with positive family history of obesity.

Conflict of Interests

The authors declare that there is no conflict of interests regarding the publication of this paper.

Acknowledgments

The authors wish to sincerely appreciate all who participated and assisted in making this study a reality. Their special thanks also go to ISPAD (International Society of Paediatric and Adolescent Diabetes) for providing a grant for this study.

References

[1] World Health Organization, "Obesity and overweight," Fact sheet No 311, 2012.

[2] S. Mehdad, A. Hamrani, K. El Kari et al., "Body mass index, waist circumference, body fat, fasting blood glucose in a sample of Moroccan adolescents aged 11–17 years," *Journal of Nutrition and Metabolism*, vol. 2012, Article ID 510458, 7 pages, 2012.

[3] A. O. Akinpelu, O. O. Oyewole, and K. S. Oritogun, "Overweight and obesity: does it occur in Nigerian adolescents in an urban community?" *International Journal of Biomedical and Healthcare Science*, vol. 4, pp. 11–17, 2008.

[4] E. P. Whitlock, S. B. Williams, R. Gold, P. R. Smith, and S. A. Shipman, "Screening and interventions for childhood

overweight: a summary of evidence for the US Preventive Services Task Force," *Pediatrics*, vol. 116, no. 1, pp. e125–e144, 2005.

[5] P. W. Speiser, M. C. J. Rudolf, H. Anhalt et al., "Consensus statement: childhood obesity," *Journal of Clinical Endocrinology and Metabolism*, vol. 90, no. 3, pp. 1871–1887, 2005.

[6] S. Gahagan, J. Silverstein, and Committee on Native American Child Health and Section on Endocrinology, "Prevention and treatment of type 2 diabetes mellitus in children, with special emphasis on American Indian and Alaska Native children," *Pediatrics*, vol. 112, no. 4, p. e328, 2003.

[7] S. B. Sondike, "Overweight and obesity," in *Pediatric Endocrinology and Inborn Error of Metabolism*, K. Sarafoglou, G. F. Hoffmann, and K. S. Roth, Eds., pp. 275–293, The McGraw-Hill, 2009.

[8] S. J. Brink, "How should children and youth with type 2 diabetes be treated?" in *Theoretical and Practical Update in Paediatric Endocrinology and Diabetes*, I. P. Velea, C. Paul, and S. J. Brink, Eds., pp. 89–126, Mirton, Timişoara, Romania, 2014.

[9] H. Beck-Nielsen and L. C. Groop, "Metabolic and genetic characterization of prediabetic states. Sequence of events leading to non-insulin-dependent diabetes mellitus," *The Journal of Clinical Investigation*, vol. 94, no. 5, pp. 1714–1721, 1994.

[10] A. Jessup and J. S. Harrell, "The metabolic syndrome: look for it in children and adolescents, too!," *Clinical Diabetes*, vol. 23, no. 1, pp. 26–32, 2005.

[11] A. N. Onyiriuka, D. D. Umoru, and A. N. Ibeawuchi, "Weight status and eating habits of adolescent Nigerian urban secondary school girls," *South African Journal of Child Health*, vol. 7, no. 3, pp. 108–112, 2013.

[12] A. K. Singh, A. Maheshwari, N. Sharma, and K. Anand, "Lifestyle associated risk factors in adolescents," *Indian Journal of Pediatrics*, vol. 73, no. 10, pp. 901–906, 2006.

[13] G. Plourde, "Impact of obesity on glucose and lipid profiles in adolescents at different age groups in relation to adulthood," *BMC family Ppractice*, vol. 3, article 18, 2002.

[14] B.-Y. Cao, J. Mi, C.-X. Gong et al., "Blood glucose profile in children and adolescents in Beijing area," *Chinese Journal of Pediatrics*, vol. 46, no. 4, pp. 297–300, 2008.

12

Effects of 8-Prenylnaringenin and Whole-Body Vibration Therapy on a Rat Model of Osteopenia

Daniel B. Hoffmann,[1] Markus H. Griesel,[1] Bastian Brockhusen,[1]
Mohammad Tezval,[1] Marina Komrakova,[1] Bjoern Menger,[1] Marco Wassmann,[2]
Klaus Michael Stuermer,[1] and Stephan Sehmisch[1]

[1]*Department of Trauma and Reconstructive Surgery, University of Goettingen, 37075 Goettingen, Germany*
[2]*Medical Institute of General Hygiene and Environmental Health, University of Goettingen, 37075 Goettingen, Germany*

Correspondence should be addressed to Daniel B. Hoffmann; daniel.hoffmann@med.uni-goettingen.de

Academic Editor: Christel Lamberg-Allardt

Background. 8-Prenylnaringenin (8-PN) is the phytoestrogen with the highest affinity for estrogen receptor-α (ER-α), which is required to maintain BMD. The osteoprotective properties of 8-PN have been demonstrated previously in tibiae. We used a rat osteopenia model to perform the first investigation of 8-PN with whole-body vertical vibration (WBVV). *Study Design.* Ovariectomy was performed on 52 of 64 Sprague-Dawley rats. Five weeks after ovariectomy, one group received daily injections (sc) of 8-PN (1.77 mg/kg) for 10 weeks; a second group was treated with both 8-PN and WBVV (twice a day, 15 min, 35 Hz, amplitude 0.47 mm). Other groups received either only WBVV or no treatment. *Methods.* The rats were sacrificed 15 weeks after ovariectomy. Lumbar vertebrae and femora were removed for biomechanical and morphological assessment. *Results.* 8-PN at a cancer-safe dose did not cause fundamental improvements in osteoporotic bones. Treatment with 8-PN caused a slight increase in uterine wet weight. Combined therapy using WBVV and 8-PN showed no significant improvements in bone structure and biomechanical properties. *Conclusion.* We cannot confirm the osteoprotective effects of 8-PN at a cancer-safe dose in primary affected osteoporotic bones. Higher concentrations of 8-PN are not advisable for safety reasons. Adjunctive therapy with WBVV demonstrates no convincing effects on bones.

1. Introduction

In developed countries, postmenopausal osteoporosis is currently a serious problem that will only escalate in the future. Multiple prognoses and aging populations indicate that there will be a significant increase (more than 100%) in typical osteoporotic fractures, such as proximal femur fractures, over the next several decades [1]. The main cause of postmenopausal osteoporosis is estrogen deficiency, which increases bone resorption and accelerates bone loss [2]. Unfortunately, hormone replacement therapy (HRT), which prevents hip and spinal fractures, is no longer recommended following the 2002 Women's Health Initiative (WHI) study that revealed its life-threatening side effects, including the increased risk of cancer, stroke, and arteriosclerosis [3]. Therefore, safe and effective alternative therapies for osteoporosis are greatly needed.

Interest in phytoestrogens has recently increased. Phytoestrogens are hormonally active plant-derived compounds with estrogen-like effects on estrogen-dependent tissues [4]. More specifically, phytoestrogens interact directly with the α and β estrogen receptors (ERs) [4]. Both of these receptors are expressed in bone cells, including osteoclasts, osteoblasts, osteocytes, and chondrocytes [5]. ER-α is especially important for bone development and maintaining bone mineral density [6]. It has crucial effects on both trabecular and cortical bone [6]. ER-α acts on cells by stimulating target gene transcription through two activation functions (AF1 and AF2) [7]. Unlike ER-α-AF2, the ER-α-AF1 pathway is tissue-specific and essential for trabecular bone growth [7]. Additionally, ER-β only minimally influences cortical bone in mice, as reported previously [8]. Unfortunately, most of the known phytoestrogens primarily interact with ER-β.

Examples of mild osteoprotective phytoestrogens are soy, genistein, daidzein, equol, and 8-prenylnaringenin (8-PN) [4, 9, 10]. Genistein, daidzein, and equol demonstrate higher affinities for ER-β [11], whereas 8-PN has a higher affinity for ER-α [12]. 8-PN is a component of female hop cones, as well as of a crude Thai drug [13], and is therefore a component of beer. Recently, the beneficial effects of 8-PN as an herbal alternative for menopausal vasomotor complaints have been described [14]. Of all of the phytoestrogens, 8-PN has the highest affinity for the ER-α receptor [12] and is therefore an attractive molecule for osteoporosis research. Previous evidence has demonstrated the osteoprotective effects of 8-PN [4, 15, 16]. Unfortunately, in all of these studies, only the tibiae were investigated. There are no data on the osteoprotective effects of 8-PN in vertebrae or femora, which have a main role in osteoporosis. Furthermore, in these former studies, the dosage of 8-PN varied considerably. In some, high dosages were used to demonstrate osteoprotective effects [4, 16]. This dosage is not advisable because the risk for endometrial cancer is increased via an ER-α receptor-driven mechanism [17, 18]. From the oncological point of view, the dosage should be as low as possible even for the phytoestrogen 8-PN. An osteoprotective treatment with 8-PN is only reasonable if there is no increased risk of cancer. Thus, we wanted to investigate the effects of 8-PN on femora and vertebrae at a safe dose [16].

In addition to pharmaceutical therapy, mechanical stimulation is an alternative treatment for osteoporosis. Whole-body vibration is a safe and successful treatment [19]. According to the mechanostat model described by Frost, mechanical stimulation induces bone growth as a result of local elastic bone deformation [20]. Mechanical stress stimulates osteocytes, osteoblasts, and other cells of the bone lining to produce bone matrix via multiple pathways [21]. The use of vibration has been shown to increase both cortical and trabecular bone in animals [22].

The aim of this study was to evaluate the combined effects of 8-PN, an ER-α agonist, and whole-body vertical vibration (WBVV) as an osteoporosis treatment for the first time. We used the ovariectomized rat, which is a standard animal model for osteoporosis research [23]. Osteoporosis predominantly affects trabecular bones, such as the distal radius, femoral neck, and vertebral body. Because vertebral and femoral fractures are an important indicator of the progression of osteoporosis, lumbar vertebrae and femora were analyzed.

2. Materials and Methods

2.1. Animals and Substances. All of the procedures were approved by the local Institutional Animal Care and Use Committee (permission number 33.9-42502-04-12/0854, district authorities of Oldenburg, Germany). Consistent with recommendations in previous studies, experiments were performed on 64 three-month-old female Sprague-Dawley rats weighing 230–290 g (Fa. Winkelmann, Borchen, Germany) [23]. The rats were maintained according to German animal protection laws and fed a soy-free diet (ssniff Special Diet, Soest, Germany). A total of 52 rats were ovariectomized

($n = 52$) at the age of three months. The other 12 rats were not subjected to surgery (non-OVX, $n = 12$). After surgical treatment, the rats were divided into different groups.

The rats not operated on were placed in Group 1 (non-OVX). Ovariectomized rats receiving no treatment comprised Group 2 (OVX). Group 3 (OVX + VIB) contained ovariectomized rats treated with low-magnitude, high-frequency WBVV. Groups 4 (OVX-8-PN) and 5 (OVX-8-PN + VIB) received injections of 8-PN (Orgentis Chemicals GmbH, Gatersleben, Germany) five weeks after ovariectomy. Rats in Group 5 (OVX-8-PN + VIB) were treated with both 8-PN and WBVV.

For WBVV treatment, the rats were placed on a vibration platform twice daily for 15 min each, 5 days per week for 10 weeks beginning five weeks after ovariectomy. The vibration motor was constructed by Schultheis (Vibra Drehstrom-Vibrationsmotor Typ HVL/HVE, Offenbach, Germany), and it vibrated at a frequency of 35 Hz with a mean amplitude of 0.47 mm. Transmitted acceleration rate measured at the back of the rat was 0.2 g.

Rats treated with 8-PN received daily subcutaneous injections of 8-PN at a concentration of 1.77 mg/kg for ten weeks beginning five weeks after ovariectomy. 8-PN (purity > 98% by HPLC) was diluted in 30% hydroxypropyl-β-cyclodextrin (AppliChem GmbH, Darmstadt, Germany).

All animals were sacrificed under CO_2-anesthesia at 15 weeks after ovariectomy. The lumbar vertebrae were removed for ashing for mineral content analysis (second lumbar vertebrae), biomechanical testing (third lumbar vertebrae), microcomputed tomography (fourth lumbar vertebrae), and gene expression analysis (sixth lumbar vertebrae). The vertebrae were stored in tubes at −20°C until the analyses were performed. For gene expression analysis, the samples were stored at −80°C.

2.2. Compression Test. Biomechanical tests were performed according to the protocol standardized by Sehmisch et al. [24]. A mechanical testing machine (Zwick, type 145 660 Z020/TND, Ulm, Germany) was used to measure the resistance of the lumbar vertebrae to force. The thawed vertebrae were fixed to the aluminum base (Figure 1(a)), and the stamp was lowered at a speed of 50 mm/min with a primary force of 1 N to fix the upper body plate. Measurements were obtained with a relative accuracy of 0.2–0.4% over the range of 2–500 N. The measurements were automatically stopped when the linear increase of the curve declined more than 10 N. Strength admission was recorded using testXpert software (Zwick, Ulm, Germany). The actual strength was measured in increments of 0.1 mm. We tested the femora in a similar way as described by Tezval et al. [25] (Figure 1(b)).

We quantified the maximum load (F_{max}), yield load (yL), and stiffness (S) as described by Sehmisch et al. [10, 26]. The maximum load (F_{max}) is the most force that the ground plate can withstand. The yield load (yL) is the inflection point from elastic deformation to plastic deformation. The stiffness measures the elasticity of the bone.

2.3. Microcomputed Tomography. An eXplore Locus SP microcomputed tomography scanner (GE Healthcare, Chalfont

(a)   (b)

FIGURE 1: The thawed vertebrae (a) and femora (b) were fixed to the aluminum base, and the stamp was lowered with a primary force of 1 N to fix the bones at 50 mm/min. The range of the testing machine is 2–500 N.

St Giles, UK) was used to analyze bone mineral density and other structural bone properties. Each scan included six vertebrae simultaneously. To compare the different scans, a test block was integrated into every scan. The test block consisted of five different materials with known mineral densities. To generate 3D models, GEHC Micro View v. 2.1.2 (GE Healthcare, Chalfont St Giles, UK) was used. We measured the following properties consistent with ASBMR nomenclature: trabecular thickness (Tb.Th), trabecular number (Tb.N), cortical thickness (Ct.Th), number of trabecular nodes (N.Nd), and bone volume fraction (BV/TV) [27, 28]. The vertebral body volume was calculated using the formula for a cylinder. For this calculation, the 2 cranial and 2 caudal perpendicular diameters and the dorsal and ventral heights were measured on the 3D images.

2.4. Microradiographic Analysis of Femora. We used microradiographic analysis to obtain more information about structural properties. For these tests, 150 μm thick sagittal sections of femoral heads were used. The sections were cut out between the epiphyseal and intertrochanteric line. A Leica microscope (Leica-Systems MZ 7.5, Wetzlar, Germany) was used to measure the parameters. The pictures were digitalized by Qwin software (Leica, Wetzlar, Germany). We measured trabecular nodes (N.Nd), trabecular connectivity [N.Nd/mm^2], trabecular bone area, trabecular thickness (Tb.Wi), trabecular density, and cortical density.

2.5. Ashing. The mass of mineralized bone from vertebrae was measured by ashing. The second lumbar vertebrae were heated in a muffle oven at 750°C for 30 min, and the bones were weighed to the nearest 0.00001 g before and after ashing. The mineral content (ash weight) is expressed relative to the wet weight of each vertebra (%).

   The calcium content was assessed using an atomic absorption spectrometer (4100, PerkinElmer, Waltham, USA)

according to CEN. The orthophosphate content was measured using a colorimetric method (ZeissDM4 spectrophotometer, Oberkochen, Germany) according to CEN.

2.6. Serum Analysis. Alkaline phosphatase (ALP) activity was measured in blood samples using an electrochemiluminescence immunoassay (Roche Diagnostics, Mannheim, Germany). The immunoassay was performed according to the manufacturer's instructions (Roche Diagnostics, Mannheim, Germany).

2.7. Gene Expression Analysis. For gene expression analyses, the sixth lumbar vertebrae were homogenized using a micro-dismembrator S (Sartorius, Göttingen, Germany). The RNeasy Mini Kit (Qiagen, Hilden, Germany) was used to extract the RNA, and the RNA was reverse-transcribed using Superscript RNase H-reverse transcriptase (Promega, Mannheim, Germany). The expression levels of alkaline phosphatase (ALP), receptor activator of nuclear factor κB ligand (RANKL), osteocalcin, tartrate-resistant acid phosphatase (TRAP), and osteoprotegerin (OPG) were measured using quantitative real-time polymerase chain reaction (qRT-PCR) based on SYBR green detection (QuantiTect SYBR Green PCR Kit, Qiagen) in an iCycler (CFX96, Bio-Rad Laboratories, Munich, Germany). Primers from Qiagen (QuantiTect Primer Assays, Qiagen) were used, and quantitative real-time PCR was performed according to the manufacturer's instructions. Gene expression was calculated using the $2^{\Delta\Delta CT}$ method [29], and the results shown are normalized to the gene expression in untreated female rats (non-OVX). The reference gene was β2-microglobulin.

2.8. Statistical Analysis. Data are shown as the means and standard deviation (SD). Significant differences were analyzed by one-way ANOVA with a Tukey-Kramer post hoc

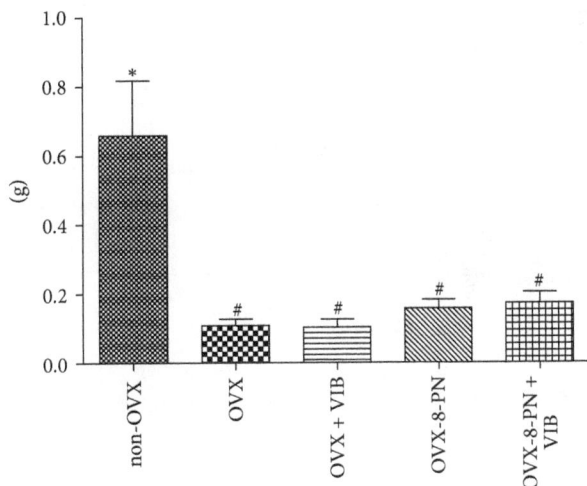

FIGURE 2: The uterine wet weight of ovariectomized rats treated with 8-PN increased nonsignificantly compared with ovariectomized rats that received no treatment. $^*p < 0.05$ versus OVX, $^{\#}p < 0.05$ versus non-OVX.

test (Graph Pad Prism, San Diego, USA). p values < 0.05 were considered significant.

3. Results

At the beginning of the study, the rats had approximately the same weights ($260 \pm 12.2\,\text{g}$). After the ovariectomy, typical changes in metabolism induced increases in body weight. The non-OVX rats also increased in body weight by the end of the evaluation period (15 weeks), which is consistent with normal growth. However, the non-OVX rats gained significantly less weight than the ovariectomized rats (Table 1). Only ovariectomized rats treated with 8-PN and WBVV (OVX-8-PN + VIB) demonstrated no significant increase compared with non-OVX rats. As expected, the uterine wet weight was highest in the non-OVX rats. Ovariectomized rats treated with 8-PN tended to have higher uterine wet weights than those of the other ovariectomized rats (Figure 2).

3.1. Biomechanical Assessment of Vertebrae.
To exclude the effects of different vertebral body sizes and volumes, all F_{max}, yL, and stiffness measurements were normalized to the bone volume determined in the micro-CT analysis [24]. In our study, treatment with 8-PN did not improve the biomechanical properties of vertebrae (Figure 3). In contrast, single therapy using 8-PN significantly worsened the biomechanical properties compared with those of non-OVX rats and tended to worsen these properties compared with untreated ovariectomized rats. Adjunctive treatment of WBVV and 8-PN caused no significant changes in the biomechanical properties. Single therapy with WBVV did not significantly affect the F_{max}, yL, or stiffness compared with those of ovariectomized rats that received no treatment (Figure 3). Non-OVX rats demonstrated the best biomechanical results.

3.2. Biomechanical Assessment of Femora.
For femora, absolute values were measured. In contrast to the results for the vertebrae, treatment with 8-PN as a single therapy caused no significant decrease in the biomechanical properties in the femora compared with those of non-OVX rats and a slight but nonsignificant increase compared with those of untreated ovariectomized rats (Figure 4). Dual therapy with WBVV and 8-PN caused no significant improvements in ovariectomized rats. There were no significant effects attributable to WBVV as a single or adjunctive therapy. Altogether, the results in the femora are consistent with the results shown in the vertebrae.

3.3. Microcomputed Tomography of Vertebrae.
The bone mineral density (BMD) significantly decreased after treatment with WBVV compared with that in untreated ovariectomized rats (Figure 5). Neither treatment with 8-PN alone nor dual therapy with WBVV and 8-PN showed improving effects on the BMD of ovariectomized rats. Non-OVX rats had significantly higher BMD than that of all ovariectomized rats irrespective of any therapy. In the BV/TV of trabecular bone, treatment with 8-PN alone and as adjunctive therapy showed a slight increase but with no statistical effect. For the trabecular thickness (Tb.th), trabecular number (Tb.N), and cortical thickness (Ct.Th), no improving effects of 8-PN or WBVV were observed compared with the values in untreated ovariectomized rats.

3.4. Microradiographic Analysis of Femora.
Neither vibration therapy nor treatment with 8-PN or adjunctive therapy had a significant effect on the structural bone properties in the femoral neck. Non-OVX rats demonstrated the best results.

3.5. Ashing of Vertebrae.
The non-OVX rats had significantly higher mineral content than that of the ovariectomized rats. Compared with untreated rats, rats treated with WBVV (OVX + VIB, OVX-8-PN + VIB) had lower mineral contents (Table 1).

The Ca^{2+}/PO_4^{3-} ratios did not differ from ovariectomized rats.

3.6. Serum Analysis.
Ovariectomized rats treated with WBVV alone had significantly higher concentrations of alkaline phosphatase (ALP) (OVX + VIB 149.6 ± 33.8 U/I) than that of ovariectomized rats that received no treatment (OVX 113.4 ± 18.3 U/I) (Table 1). Treatment with 8-PN alone (OVX-8-PN 137.4 ± 28.1 U/I) and dual therapy with WBVV (OVX-PN + WBVV 135.9 ± 8.4) also caused significantly increased ALP levels compared with that in non-OVX rats but with no statistical effects compared with untreated ovariectomized rats. Non-OVX rats had the lowest levels of ALP.

3.7. Gene Expression Analysis.
The mRNA-expression of the bone-resorptive enzyme ALP significantly increased in the rats treated with WBVV and 8-PN as dual therapy compared with untreated ovariectomized rats (Table 1). Single treatment with 8-PN and WBVV resulted in a nonsignificant increase in ALP-mRNA. The non-OVX rats had the lowest expression of RANKL-mRNA. A nonsignificant increase in OPG

TABLE 1: Body weight, μCT, protein expression, and serum analysis data after treatment with WBVV, 8-PN, or combined therapy.

Sample size	non-OVX 11		OVX 11		OVX + VIB 11		OVX-8-PN 11		OVX-8-PN + VIB 12	
	Mean	STD	Mean	STD	Mean	STD	Mean	STD	Mean	STD
Bodyweight										
Pre-OVX [g]	258.0	8.9	264.3	13.8	258.3	11.8	264.1	10.8	257.5	14.8
After 15 weeks [g]	346.6[a]	19.1	400.7[c]	47.4	392.5[c]	24.0	388.3[c]	21.3	367.0	22.1
μCT vertebrae										
Trabecular number (Tb.N) [n]	111.5	26.98	117.5	25.56	105.4	27.27	117.6	25.45	113.8	27.91
Number of trabecular nodes N.Nd [n]	133.8	36.76	140.8	34.79	122.8	36.33	143.6	36.05	137.1	37.56
Mean trabecular junctions at one node (Tb.N/Nd) [n]	2.33	0.14	2.33	0.16	2.25	0.14	2.37	0.14	2.36	0.17
Trabecular thickness (Tb.Th) [mm]	0.05669	0.02143	0.04956	0.02006	0.04975	0.0138	0.04884	0.01932	0.04827	0.02725
BV/TV whole body of vertebrae [%]	43.82	3.22	39.84	1.94	42.66	4.46	41.68	5.82	43.32	2.92
BV/TV trabecular bone [%]	52.55	6.75	48.53	7.04	47.30	8.53	50.13	9.24	58.03	15.85
Cortical thickness (CT.Th) [mm]	0.2367	0.0212	0.1977	0.0393	0.2116	0.0412	0.2042	0.0356	0.2021	0.0408
Microradiographic analysis femora										
Trabecular nodes (N.Nd) [n]	109.1	21.7	87.5	19.2	78.4[c]	23.1	87.0	8.7	91.6	9.9
Trabecular connectivity [N.Nd/mm^2]	21.85[a]	3.93	16.97[c]	3.29	14.09[c]	4.13	16.32[c]	1.46	17.21[c]	2.87
Trabecular bone area [mm^2]	4.99	0.45	5.16	0.70	5.62	0.69	5.33	0.28	5.39	0.64
Trabecular thickness (Tb.Wi) [mm]	0.00698	0.00171	0.00597	0.00079	0.00564[c]	0.00056	0.00553[c]	0.00071	0.00572[c]	0.00071
Trabecular density [%]	75.50[a]	11.77	58.50[c]	10.29	50.80[c]	10.75	53.80[c]	7.15	56.48[c]	9.99
Cortical density [%]	97.44	1.41	96.62	1.49	97.03	0.96	96.52	0.88	96.14	1.22
Ashing										
Mineral content (%)	43.00[a]	7.59	37.82[c]	1.66	35.97[c]	1.54	37.46	1.68	36.00[c]	2.10
Ca^{2+}/PO$_4^{3-}$	1.541	0.065	1.500	0.220	1.562	0.156	1.609	0.065	1.569	0.031
Gene expression										
ALP	1.056	0.325	0.825	0.190	1.376	0.435	1.300	0.584	1.665[a]	0.238
Osteocalcin	1.114	0.574	1.631	0.549	2.066	1.085	1.142	0.471	1.306	0.537
RANKL	1.199	0.556	1.956	0.310	2.938[c]	0.874	2.054	0.633	1.661	0.816
OPG	1.218	0.968	1.296	0.732	1.595	1.066	1.472	1.756	2.187	1.801
TRAP	1.069	0.494	1.152	0.248	1.136	0.569	0.867	0.471	0.841	0.384
Serum analysis										
ALP [U/I] serum	91.4	15.37	113.4[b]	18.34	149.6[ac]	33.8	137.4[c]	28.07	135.9[c]	8.39

[a] $p < 0.05$ versus OVX.
[b] $p < 0.05$ versus adjunctive VIB.
[c] $p < 0.05$ versus non-OVX.

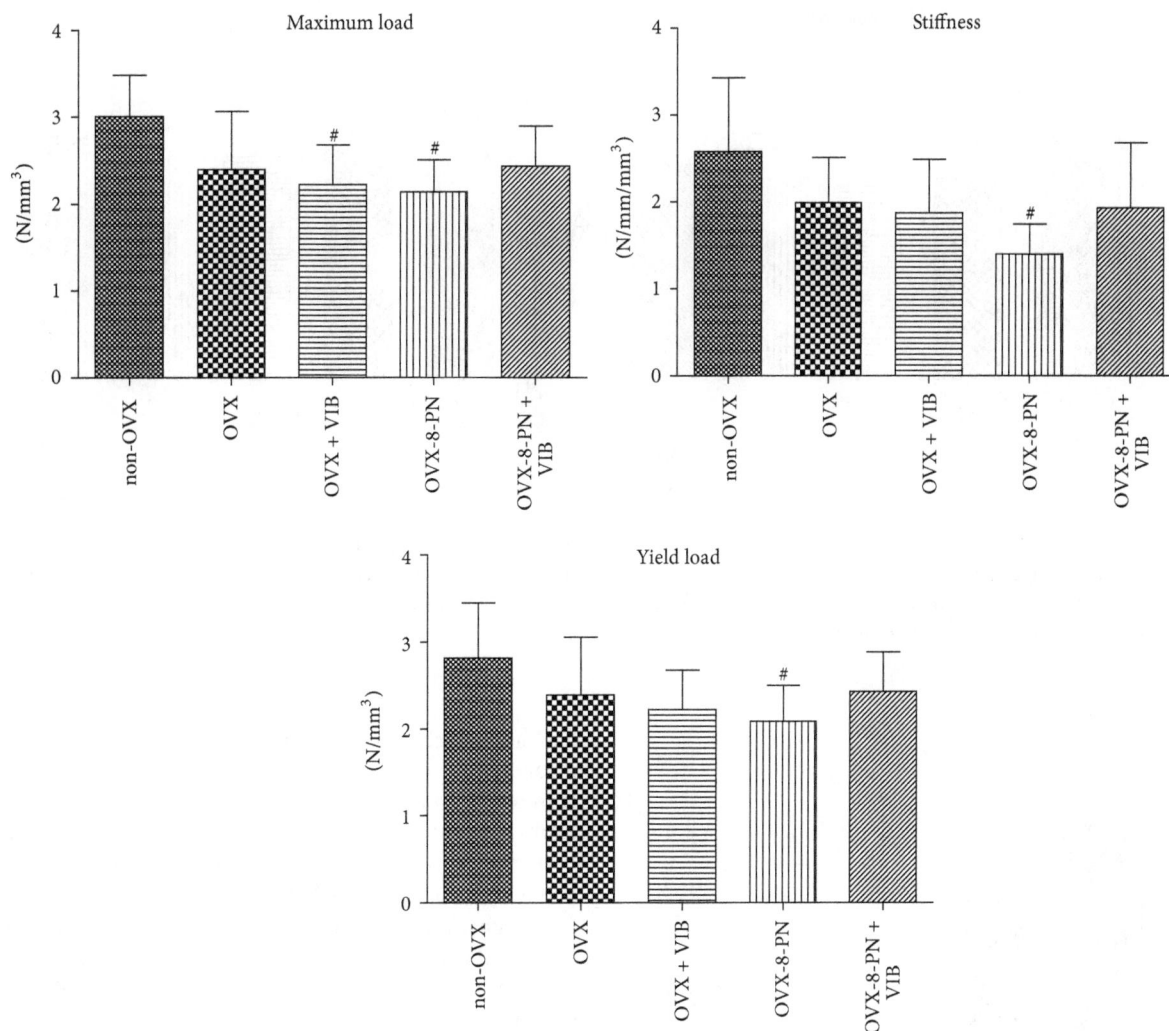

FIGURE 3: Biomechanical assessment of vertebrae (measurements normalized to the bone volume). Single therapy with 8-PN or WBVV induced in vertebrae a significant decrease in biomechanical properties compared with non-OVX rats. Compared with untreated ovariectomized rats, treatment with 8-PN tended to worsen biomechanical properties. Adjunctive therapy using 8-PN with WBVV produced no significant improvements. Non-OVX rats had the highest values for all of the biomechanical properties. Vertebrae measurements were normalized to the bone volume. $^{\#}p < 0.05$ versus non-OVX.

expression was observed in ovariectomized rats following WBVV treatment (Table 1).

4. Discussion

Several recent studies have investigated vibration or phytoestrogen treatment as potential new therapies for osteoporosis [4, 9, 19, 30, 31]. Almost all phytoestrogens tested in previous studies predominately acted via the estrogen receptor ER-β [4, 10, 11, 32]. Unfortunately, this receptor only minimally affects bones. In contrast to ER-β, the estrogen receptor ER-α exerts crucial effects on trabecular and cortical bone [6]. Of all of the phytoestrogens, 8-PN has the highest affinity for ER-α [4, 12]. This property makes 8-PN unique and interesting. However, to date, there has only been limited research into the effects of 8-PN on osteoporosis and on bones in general [4, 15, 16]. Osteoprotective effects were only shown in tibia and not in femora or spine, which are predominately affected

by osteoporosis. From the authors' point of view, conclusive data supporting the benefits of 8-PN as an osteoprotective drug in the case of osteoporosis are still lacking.

In the present study, we could not demonstrate fundamental improvements in osteoporotic vertebrae and femora after treatment with 8-PN. Neither the biomechanical nor the morphological properties improved significantly in our study. Instead, we could demonstrate that 8-PN nonsignificantly worsened the biomechanical properties and BMD in the vertebrae. In contrast to the bone data, a slight increase in uterine weight confirmed the systemic estrogen-like effects. Our results for the biomechanical and structural bone parameters in femora and vertebrae differ from those of previous studies. In 2008, Sehmisch et al. reported that 8-PN significantly improved the biomechanical properties of bone in tibiae. However, only minimal improvements in bone structure were observed [4]. In 1998, Miyamoto et al. showed an increase in BMD after the administration of 8-PN [15].

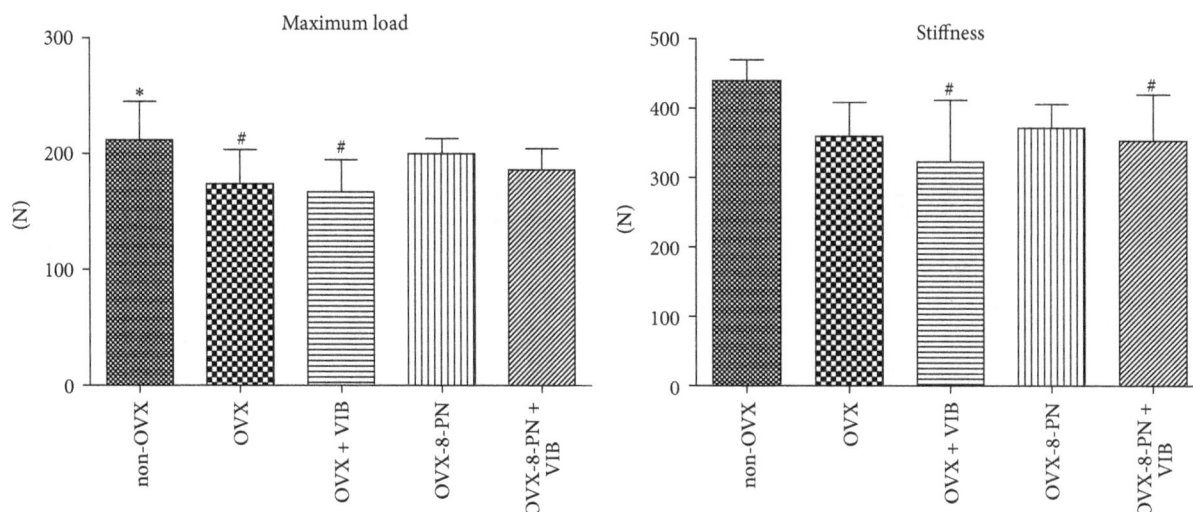

Maximum load

Stiffness

FIGURE 4: Biomechanical assessment of femora (absolute values): treatment with 8-PN as single therapy caused no significant decrease in the biomechanical properties compared with those of non-OVX rats. Compared with untreated ovariectomized rats, treatment with 8-PN showed only a nonsignificant improvement. There were no significant effects attributable to WBVV as a single or adjunctive therapy with 8-PN. $^*p < 0.05$ versus OVX, $^\#p < 0.05$ versus non-OVX.

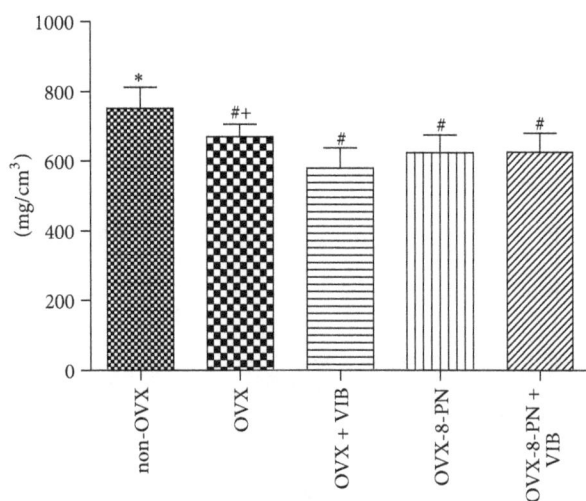

FIGURE 5: The BMD significantly decreased after treatment with WBVV compared with that in untreated ovariectomized rats. Neither treatment with 8-PN alone nor dual therapy with WBVV and 8-PN showed improving effects on the BMD of ovariectomized rats. Non-OVX rats had significantly higher BMD than that of all ovariectomized rats irrespective of any therapy. $^*p < 0.05$ versus OVX, $^\#p < 0.05$ versus non-OVX, and $^+p < 0.05$ versus adjunctive VIB.

Similar results were shown by Hümpel et al. [16]. However, none of these studies tested femora or spine parameters. Additionally, the rates and methods of 8-PN administration differed considerably. The dosages in these previous studies differed from 1.77 mg/kg per day to 68 mg/kg per day [4, 15, 16]. This fact is important from a safety point of view. Even the 1.77 mg/kg dosage demonstrated a weak stimulation of endometrial luminal epithelial cells due to an ER-α-driven mechanism [16]. In the present study, we administered 8-PN

subcutaneously at a dose of 1.77 mg/kg to determine whether a cancer-safe dose has beneficial effects on an osteoporotic spine and femora neck. We were unable to confirm this effect. From the authors' point of view, the administration of a higher 8-PN concentration to improve osteoprotective effects is not advisable due to safety reasons.

Mechanical stimulation by vibration therapy has shown beneficial effects on the structure and biomechanical properties of bones, including vertebrae, in several animal studies of osteoporosis [19, 22, 30, 33, 34]. A systemic meta-analysis in 2010 showed significant but small effects in postmenopausal women [33].

In the present study, we could not demonstrate significant improvements in the biomechanical properties and bone structure after WBVV. The effects of WBVV as a single or adjunctive therapy were more pronounced in previous studies [19, 22, 30, 31]. However, all of these studies were performed on different bones (tibia, femur, and lumbar vertebrae) and at varying frequencies. We used a frequency of 35 Hz and amplitude of 0.47 mm in our study based on our own previous and external studies [34–36]. Compared with ovariectomized rats that received no treatment, rats that received WBVV treatment showed higher bone turnover, as demonstrated by increased RANKL, ALP, and osteocalcin expression. However, overall, we could not confirm that beneficial effects were exerted on bones in our setting. In our opinion, the optimal setting including frequency, amplitude, and acceleration for WBVV has not yet been determined, and data in the literature for the rat osteopenia model are contradictory [31, 34, 35]. In the present study, only bone parameters were investigated. It is reasonable that muscle status can be improved by WBVV. Increased muscle strength is beneficial for preventing falls and maintaining bone mass. Further studies are needed to study the integral effects of vibration therapy in the case of osteoporosis.

Adjunctive therapy using WBVV and the ER-α agonist 8-PN has not previously been investigated. The present study is the first to investigate a dual treatment using WBVV and 8-PN. No significant improvements in bone structure were observed. There were no effects on biomechanical properties. According to our results, adjunctive therapy with 8-PN and WBVV at 35 Hz has no effects on bones, which are predominately affected by osteoporosis.

5. Conclusion

In conclusion, we cannot confirm the osteoprotective effects of 8-PN at a cancer-safe dose in primary affected osteoporotic bones. In our opinion, higher concentrations of 8-PN are not advisable due to safety reasons. Adjunctive therapy with WBVV at 35 Hz has no significant effects on bones. Further studies are needed to investigate the integral effects and best setting of WBVV in the case of osteoporosis.

Abbreviations

8-PN:	8-Prenylnaringenin
ALP:	Alkaline phosphatase
BV/TV:	Bone volume fraction
BMD:	Bone mineral density
Ct.Th:	Cortical thickness
ER:	Estrogen receptor
F_{max}:	Maximum load
HRT:	Hormone replacement therapy
N.Nd:	Number of trabecular nodes
N.Nd/mm^2:	Trabecular connectivity
OPG:	Osteoprotegerin
RANKL:	Receptor activator of nuclear factor κB ligand
S:	Stiffness
Tb.N:	Trabecular number
Tb.Th:	Trabecular thickness
yL:	Yield load.

Conflict of Interests

The authors declare that there is no conflict of interests regarding the publication of this paper.

Acknowledgments

The present study was funded by the German Research Foundation (DFG SE 1966/5-1). The authors are grateful to R. Castro-Machguth and A. Witt for technical support.

References

[1] R. Osterkamp, "Population developments in Germany until 2050. Demographic and economic consequences for geriatric medicine and surgery," Chirurg, vol. 76, no. 1, pp. 10–18, 2005.

[2] R. Pacifi, "Cytokines, estrogen, and postmenopausal osteoporosis—the second decade," Endocrinology, vol. 139, no. 6, pp. 2659–2661, 1998.

[3] J. E. Rossouw, G. L. Anderson, R. L. Prentice et al., "Risks and benefits of estrogen plus progestin in healthy postmenopausal women: principal results from the women's health initiative randomized controlled trial," The Journal of the American Medical Association, vol. 288, no. 3, pp. 321–333, 2002.

[4] S. Sehmisch, F. Hammer, J. Christoffel et al., "Comparison of the phytohormones genistein, resveratrol and 8-prenylnaringenin as agents for preventing osteoporosis," Planta Medica, vol. 74, no. 8, pp. 794–801, 2008.

[5] G. Zaman, H. L. Jessop, M. Muzylak et al., "Osteocytes use estrogen receptor α to respond to strain but their ERα content is regulated by estrogen," Journal of Bone and Mineral Research, vol. 21, no. 8, pp. 1297–1306, 2006.

[6] A. E. Börjesson, M. K. Lagerquist, S. H. Windahl, and C. Ohlsson, "The role of estrogen receptor α in the regulation of bone and growth plate cartilage," Cellular and Molecular Life Sciences, vol. 70, no. 21, pp. 4023–4037, 2013.

[7] A. E. Börjesson, S. H. Windahl, M. K. Lagerquist et al., "Roles of transactivating functions 1 and 2 of estrogen receptor-α In bone," Proceedings of the National Academy of Sciences of the United States of America, vol. 108, no. 15, pp. 6288–6293, 2011.

[8] S. H. Windahl, O. Vidal, G. Andersson, J. A. Gustafsson, and C. Ohlsson, "Increased cortical bone mineral content but unchanged trabecular bone mineral density in female ER$\beta^{-/-}$ mice," The Journal of Clinical Investigation, vol. 104, no. 7, pp. 895–901, 1999.

[9] C. R. Sirtori, "Risks and benefits of soy phytoestrogens in cardiovascular diseases, cancer, climacteric symptoms and osteoporosis," Drug Safety, vol. 24, no. 9, pp. 665–682, 2001.

[10] S. Sehmisch, M. Erren, L. Kolios et al., "Effects of isoflavones equol and genistein on bone quality in a rat osteopenia model," Phytotherapy Research, vol. 24, supplement 2, pp. S168–S174, 2010.

[11] J. V. Turner, S. Agatonovic-Kustrin, and B. D. Glass, "Molecular aspects of phytoestrogen selective binding at estrogen receptors," Journal of Pharmaceutical Sciences, vol. 96, no. 8, pp. 1879–1885, 2007.

[12] T. F. H. Bovee, R. J. R. Helsdingen, I. M. C. M. Rietjens, J. Keijer, and R. L. A. P. Hoogenboom, "Rapid yeast estrogen bioassays stably expressing human estrogen receptors α and β, and green fluorescent protein: a comparison of different compounds with both receptor types," Journal of Steroid Biochemistry and Molecular Biology, vol. 91, no. 3, pp. 99–109, 2004.

[13] S. R. Milligan, J. C. Kalita, A. Heyerick, H. Rong, L. De Cooman, and D. De Keukeleire, "Identification of a potent phytoestrogen in hops (Humulus lupulus L.) and beer," The Journal of Clinical Endocrinology & Metabolism, vol. 84, no. 6, pp. 2249–2252, 1999.

[14] H. T. Depypere and F. H. Comhaire, "Herbal preparations for the menopause: beyond isoflavones and black cohosh," Maturitas, vol. 77, no. 2, pp. 191–194, 2014.

[15] M. Miyamoto, Y. Matsushita, A. Kiyokawa et al., "Prenylflavonoids: a new class of non-steroidal phytoestrogen (Part 2). Estrogenic effects of 8-isopentenylnaringenin on bone metabolism," Planta Medica, vol. 64, no. 6, pp. 516–519, 1998.

[16] M. Hümpel, W.-D. Schleuning, O. Schaefer, P. Isaksson, and R. Bohlmann, "Use of 8-prenylnaringenin for hormone replacement therapy," European Patent Application, Bulletin 2005/16, 2005.

[17] W. Zhou and J. M. Slingerland, "Links between oestrogen receptor activation and proteolysis: relevance to hormone-regulated cancer therapy," Nature Reviews Cancer, vol. 14, no. 1, pp. 26–38, 2014.

[18] P. Zhang, K. Gao, X. Jin et al., "Endometrial cancer-associated mutants of SPOP are defective in regulating estrogen receptor-α

protein turnover," *Cell Death and Disease*, vol. 6, no. 3, Article ID e1687, 2015.

[19] M. Tezval, M. Biblis, S. Sehmisch et al., "Improvement of femoral bone quality after low-magnitude, high-frequency mechanical stimulation in the ovariectomized rat as an osteopenia model," *Calcified Tissue International*, vol. 88, no. 1, pp. 33–40, 2011.

[20] H. M. Frost, "The Utah paradigm of skeletal physiology: an overview of its insights for bone, cartilage and collagenous tissue organs," *Journal of Bone and Mineral Metabolism*, vol. 18, no. 6, pp. 305–316, 2000.

[21] C. H. Turner, Y. Takano, and I. Owan, "Aging changes mechanical loading thresholds for bone formation in rats," *Journal of Bone and Mineral Research*, vol. 10, no. 10, pp. 1544–1549, 1995.

[22] C. Rubin, A. S. Turner, C. Mallinckrodt, C. Jerome, K. Mcleod, and S. Bain, "Mechanical strain, induced noninvasively in the high-frequency domain, is anabolic to cancellous bone, but not cortical bone," *Bone*, vol. 30, no. 3, pp. 445–452, 2002.

[23] D. N. Kalu, "The ovariectomized rat model of postmenopausal bone loss," *Bone and Mineral*, vol. 15, no. 3, pp. 175–191, 1991.

[24] S. Sehmisch, M. Erren, T. Rack et al., "Short-term effects of parathyroid hormone on rat lumbar vertebrae," *Spine*, vol. 34, no. 19, pp. 2014–2021, 2009.

[25] M. Tezval, E. K. Stuermer, S. Sehmisch et al., "Improvement of trochanteric bone quality in an osteoporosis model after short-term treatment with parathyroid hormone: a new mechanical test for trochanteric region of rat femur," *Osteoporosis International*, vol. 21, no. 2, pp. 251–261, 2010.

[26] E. K. Stürmer, D. Seidlová-Wuttke, S. Sehmisch et al., "Standardized bending and breaking test for the normal and osteoporotic metaphyseal tibias of the rat: effect of estradiol, testosterone, and raloxifene," *Journal of Bone and Mineral Research*, vol. 21, no. 1, pp. 89–96, 2006.

[27] D. W. Dempster, J. E. Compston, M. K. Drezner et al., "Standardized nomenclature, symbols, and units for bone histomorphometry: a 2012 update of the report of the ASBMR Histomorphometry Nomenclature Committee," *Journal of Bone and Mineral Research*, vol. 28, no. 1, pp. 2–17, 2013.

[28] A. M. Parfitt, M. K. Drezner, F. H. Glorieux et al., "Bone histomorphometry: standardization of nomenclature, symbols, and units. Report of the ASBMR Histomorphometry Nomenclature Committee," *Journal of Bone and Mineral Research*, vol. 2, pp. 595–610, 1987.

[29] K. J. Livak and T. D. Schmittgen, "Analysis of relative gene expression data using real-time quantitative PCR and the $2^{-\Delta\Delta C_T}$ method," *Methods*, vol. 25, no. 4, pp. 402–408, 2001.

[30] S. Sehmisch, R. Galal, L. Kolios et al., "Effects of low-magnitude, high-frequency mechanical stimulation in the rat osteopenia model," *Osteoporosis International*, vol. 20, no. 12, pp. 1999–2008, 2009.

[31] S. Judex, X. Lei, D. Han, and C. Rubin, "Low-magnitude mechanical signals that stimulate bone formation in the ovariectomized rat are dependent on the applied frequency but not on the strain magnitude," *Journal of Biomechanics*, vol. 40, no. 6, pp. 1333–1339, 2007.

[32] T. Hertrampf, G. H. Degen, A. A. Kaid et al., "Combined effects of physical activity, dietary isoflavones and 17β-estradiol on movement drive, body weight and bone mineral density in ovariectomized female rats," *Planta Medica*, vol. 72, no. 6, pp. 484–487, 2006.

[33] L. Slatkovska, S. M. H. Alibhai, J. Beyene, and A. M. Cheung, "Effect of whole-body vibration on BMD: a systematic review and meta-analysis," *Osteoporosis International*, vol. 21, no. 12, pp. 1969–1980, 2010.

[34] M. Komrakova, S. Sehmisch, M. Tezval et al., "Identification of a vibration regime favorable for bone healing and muscle in estrogen-deficient rats," *Calcified Tissue International*, vol. 92, no. 6, pp. 509–520, 2013.

[35] K. S. Leung, H. F. Shi, W. H. Cheung et al., "Low-magnitude high-frequency vibration accelerates callus formation, mineralization, and fracture healing in rats," *Journal of Orthopaedic Research*, vol. 27, no. 4, pp. 458–465, 2009.

[36] M. Komrakova, E. K. Stuermer, M. Tezval et al., "Evaluation of twelve vibration regimes applied to improve spine properties in ovariectomized rats," *Bone Reports*, 2014.

Factors Associated with Anemia among Children Aged 6–23 Months Attending Growth Monitoring at Tsitsika Health Center, Wag-Himra Zone, Northeast Ethiopia

Haile Woldie,[1] Yigzaw Kebede,[2] and Amare Tariku[1]

[1]*Department of Human Nutrition, Institute of Public Health, College of Medicine and Health Science, University of Gondar, Gondar, Ethiopia*
[2]*Department of Epidemiology and Biostatistics, Institute of Public Health, College of Medicine and Health Science, University of Gondar, Gondar, Ethiopia*

Correspondence should be addressed to Amare Tariku; amaretariku15@yahoo.com

Academic Editor: Christel Lamberg-Allardt

Background. Globally, about 47.4% of children under five are suffering from anemia. In Ethiopia, 60.9% of children under two years are suffering from anemia. Anemia during infancy and young childhood period is associated with poor health and impaired cognitive development, leading to reduced academic achievement and earnings potential in their adulthood life. However, there is scarcity of information showing the magnitude of iron deficiency anemia among young children in Ethiopia. Therefore, this study aimed at assessing prevalence and associated factors of iron deficiency anemia among children under two (6–23 months). *Methods.* Institution based cross-sectional study was carried out from March to May, 2014, at Tsitsika Health Center in Wag-Himra Zone, Northeast Ethiopia. Systematic random sampling technique was employed. Automated hemoglobin machine was used to determine the hemoglobin level. Socioeconomic and demographic data were collected by using a pretested and structured questionnaire. Binary logistic regression analysis was used to identify associated factors and odds ratio with 95% CI was computed to assess the strength of association. *Results.* Total of 347 children participated in this study. The overall prevalence of anemia was 66.6%. In multivariate logistic regression analysis, male sex (AOR = 3.1 (95% CI: 1.60–5.81)), 9–11 months of age (AOR = 9.6 (95% CI: 3.61–25.47)), poor dietary diversity (AOR = 3.2 (95% CI: 1.35–7.38)), stunting (AOR = 2.7 (95% CI: 1.20–6.05)), diarrhea (AOR = 4.9 (1.63–14.59)), no formal education (AOR = 2.6 (95% CI: 1.26–5.27)), early initiation of complementary food (AOR = 11.1 (95% CI: 4.08–30.31)), and lowest wealth quintile (AOR = 3.0 (95% CI: 1.01–8.88)) were significantly associated with anemia. *Conclusion.* The overall prevalence of anemia among children who aged 6–23 months has sever public health importance in the study area. Integrated efforts need to be prioritized to improve health as well as appropriate infant and young child feeding practice among children under.

1. Background

Currently, micronutrient deficiencies are coming to be the most prevalent nutritional deficiencies causing serious developmental problems in the world [1]. Anemia is one of micronutrient deficiencies which have serious public health significance in the world. It is second leading nutritional cause of disability [2–4]. Anemia is an important outcome indicator of poor nutrition and health with its major consequences on socioeconomic development of a population [5]. Anemia can occur at any time and at all stages of the life cycle [3] but young children and pregnant women are the most at risk segment of the community [1].

Globally, about 47.4% of children under five are suffering from anemia [6]. In developing countries, it affects 46–66% of children aged under five years [7]. Africa and Asia are found with sever public health importance of anemia [8]. About 67.6% of children under five in Africa are suffering from anemia while they are 65.5% in Southeast Asia [6]. The highest overall prevalence of anemia in children aged under 5 years is recorded in the Western and Central African Region as 75% [9]. According to the Ethiopian 2011 DHS report, the overall prevalence of anemia among children under two years (6–23 months) was 60.9% [10].

Among children, iron deficiency anemia is consequence of complex interaction of several factors. As different studies claimed that iron deficiency anemia is significantly associated with low birth weight [11], sex [11–14], age [11, 12, 14–16], rural residence, infectious disease (malaria, tuberculosis, intestinal parasitic infestation, and HIV/AIDS) [15, 17], undernutrition (stunting, wasting, and underweight) [13, 15, 18], poor socioeconomic status [12, 14, 16, 18], household food insecurity [12], duration of lactation [18] and poor dietary iron intake [18], poor maternal educational status [11, 14, 18], and maternal anemia [12].

Anemia during childhood period is strongly associated with poor health and physical development [3, 19], mild and moderate mental retardation [3, 20], and poor motor development and control [16, 21] leading to reduced academic achievement and work capacity thereby reducing earning potential and damaging national economic growth in the future [19]. Iron deficiency anemia also increases risk of mortality and morbidity from infectious disease [3, 11, 22].

Still anemia with its devastating implication has public health importance among children under five in Africa [23]. In Ethiopia, as part of other African countries, burden of anemia also proves public health significance among children [10]. However, there is scarcity of information showing the burden of anemia and its risk factor among children under two. Therefore, this study was aimed to assess prevalence of anemia and its associated factors among children aged 6–23 months in Wag-Himra zone, Northeast Ethiopia.

2. Methods

2.1. Study Area. This study was conducted at Tsitsika Health Center, Ziquala Woreda, Amhara regional state, Northeast Ethiopia. Tsitsika (Woreda capital of Ziquala Woreda) is 730 km far from Addis Ababa, the capital city of Ethiopia, and founded at 900 meters above sea level. According to the regional finance and economic development projection, the total population of the study area for the year 2014 was 49671. The total numbers of children aged under five and under two years were 6987 and 2473, respectively. Residents of this Woreda largely depend on subsistence farming and producing cereals as main agricultural product. Since 2008, the study area has been implementing community based nutrition intervention program (growth monitoring, nutritional screening, vitamin-A supplementation, outpatient therapeutic feeding program, etc.). At the time of the study, Ziquala Woreda has six health centers and 15 health posts. Growth monitoring targeting children under two years has been carried out by using Weight-for-Age anthropometric index as routine procedure in all health institutions of current study area.

2.2. Study Population and Sampling Techniques. All children aged 6–23 months with mothers/caretakers attending growth monitoring clinic during the study period were included in the study. Those children who did not register from the Health Center logbook for their health status were excluded

from the study. Systematic random sampling technique was employed to recruit a total of 347 children. The total sample size was determined by using the formula to estimate a single population proportion with the following assumptions: population proportion (P), that is, 60.91% taken from Ethiopian 2011 DHS report as prevalence of anemia among children aged 6–23 months, 95% confidence level, and 5% margin of error. Then, the final sample size 366 was obtained. The sampling interval was calculated from the 2013 growth monitoring registration book of the Health Center. The number of children aged 6–23 months and participated for growth monitoring during calendar year of March to May/2013 was taken as 763 from the logbook. Children who have no record regarding their morbidity status on the registration book were not included in the study. Sampling interval was determined by 763/366 = 2.01 ≈ 2; then, lottery method was employed to select starting or first sample in the study. The procedure continued until the required sample size was obtained.

2.3. Data Collection Tools and Procedures. Socioeconomic and demographic characteristics of the family and child, feeding practice, health care utilization, and child morbidity status (within two weeks before data collection) were collected by using a pretested and structured questionnaire through interviewing of the mother/caretakers of the child. Hemoglobin level of the child was measured from capillary blood and one drop of capillary blood was carefully collected from the middle finger of the child by finger prick. Strict aseptic technique and a separate lancet for each child were employed. Automated hemoglobin machine made from Germany with model kx-21 and serial number b-0839Model was used to determine the hemoglobin concentration and the result was expressed in g/dL and the presence and severity of anemia were determined according to age based criteria of WHO cut-off point. For children aged 6–24 months who have hemoglobin level >11 g/dL, they were considered as nonanemic, 10.0–10.9 g/dL as mildly anemic, 7–9.9 g/dL moderately anemic, and <7 g/dL as severely anemic [24].

Nutritional status of the child was assessed by taking anthropometric body measurement of the child. Length of a child was measured in a recumbent position to the nearest 0.1 cm by using a board with an upright wooden base and a movable headpiece, on a flat surface. Weight measurement of a child was taken by a Salter scale (model-2356S) with the calibration of 100 g unit. It is designed and manufactured under the authority and recognition of the United Nations Children's Fund (UNICEF). The scale was adjusted to zero before weighing every child and measurement was recorded to the nearest 0.1 kg. Each measurement was repeated and the mean value was calculated and recorded on the questionnaire. All children were without any shoes during the measurement.

Age of child was determined by two methods: for 81 children aged between 12–23 months, their birth date was extracted from the immunization card; while for 266 children without immunization card and who aged under 12 months, their age was determined by using information given by the mother/female caretaker of the child. The three standard indices (Length-for-Age, Weight-for-Length, and Weight-for-Age) were analyzed by ANTHRO software and used to determine the nutritional status of children. Each of the three measurements was expressed in standard deviation units of Z-score from the median of WHO-2006, standard population [25]. Children with a measurement <-2 of Z-score were determined as stunted for Length-for-Age, wasted for Weight-for-Length, and underweight for Weight-for-Age. Information related to morbidity status of the child (intestinal parasite, diarrhea, malaria, and upper respiratory tract infection) before two weeks was captured by looking up the health centre logbook.

2.4. Measurement of Dietary Diversity Score. Dietary diversity scores of a child were determined by using WHO and "indicators for assessing infant and young child feeding practices" minimum dietary diversity for children age 6–23 months and by employing 24 hrs recall method. Mothers or female care takers were asked to report all food items and beverages given to the child during the previous day of the survey. Then, all food items and beverages consumed by the child were categorized into seven food groups as (1) grains, roots, and tubers, (2) legumes and nuts, (3) dairy products, (4) flesh foods, (5) eggs, (6) vitamin-A rich fruits and vegetables, and (7) other fruits and vegetables [26]. Using dietary diversity score 4 (minimum dietary diversity score) as cut-off point, a child was defined as having "poor dietary diversity" if he/she consumed less than 4 food groups while having "good dietary diversity" if he/she had 4 or more food groups.

2.5. Determination of Wealth Index. The wealth index was used in the study and constructed from the data collected in the household questionnaire. The standardized tool for measurement of wealth index was adopted from Ethiopian DHS-2011 [10]. This index consist of seven selected household asset data, that is, availability of electric city, television, refrigerator, mobile telephone, nonmobile telephone, a bed with cotton/sponge/spring mattress, and electric mittade (local name for electric stove or oven), and via a principal components analysis. The wealth index was divided into five categories (lowest, second, middle, fourth, and highest).

2.6. Data Quality Control. Two-day intensive training was given for data collectors and supervisors regarding study objective, interview techniques, anthropometric measurements, and ethical issues during data collection. Pretest without hemoglobin level determination was done among 5% of the total sample size in the nearest health postproviding growth monitoring service before three days of the actual data collection in order to sort out language barriers and contextual difference on the structured questionnaires. Questionnaire was checked daily for accuracy, consistency, and completeness by supervisor. Furthermore, the supervisors and the principal investigator give feedback and correction regarding the collected data on daily basis to the data collectors.

2.7. Data Processing and Analysis. Data was cleaned, coded, and entered using EPI-INFO version 3.5.3 and exported to SPSS version 16 for analysis. Bivariate analysis was done to see the association of each independent variable with the outcome variable (anemia status). Those independent variables having P value less than 0.2 in the bivariate analysis were entered into the multivariate analysis to determine the effect of each explanatory variable on outcome variable and to control the possible effect of confounders. Odds ratio with 95% confidence level was used to determine the strength of association. In the multivariate analysis, independent variables with P value ≤ 0.05 were considered as significant.

2.8. Ethical Consideration. Ethical approval was obtained from Institutional Review Board of University of Gondar. Each mother/caretaker was informed about the objective of the study and written informed consent was secured before questionnaire administered. A child with a confirmed anemia was referred to the concerned body in the Health Center.

3. Results

3.1. Socioeconomic and Demographic Characteristics of a Child and Family. A total of 347 children aged 6–23 months with their mothers/caretakers were included in the study giving response rate of 97%. Fourteen percent of mothers/caretakers of children had no formal education. Eighty percent of children were living with both parents and 55.9% had one sibling aged under five years. Nineteen percent of families were at the lowest level of wealth quintile range and 10.4% were at the highest one (Table 1).

3.2. Feeding Practice and Nutritional Status of Children. Only one child was found without history of ever breast feed and 88.2% of children were found with history of current breast feeding status during the interview. About 20.5% of children had early introduction of complementary foods while 25.1% were found with history of cow's milk consumption before 12 months of their age. Eighty five percent of children had poor dietary diversity scores. Nearly 24% of children were stunted (Table 2).

3.3. Morbidity and Health Care Related Characteristics of the Child. About 59.4% of children were born at home. Regarding the morbidity status, 14.4% were with malaria infection and 14.7% of children had diarrhea in the last two weeks (Table 3).

3.4. Prevalence of Anemia among Children Aged 6–23 Months. The overall prevalence of anemia was 66.6%. Burden was higher among males with the magnitude of 55.4%. Among the four age groups, the highest prevalence was recorded in the age group of 9–11 months (79.6%), followed by 6–8 months (69.2%).

3.5. Factors Associated with Anemia among Children Aged 6–23 Months. The result of both bi- and multivariate analyses revealed that sex of the child, age, history of diarrhea before

TABLE 1: Socioeconomic and demographic characteristics of the family and children aged 6–23 months attending growth monitoring at Tsitsika Health Center, Northeast Ethiopia ($n = 347$).

Background characteristics	Frequencies	Percent (%)
Sex of the child		
Male	171	49.3
Female	176	50.7
Age of the child (in months)		
6–8 months	91	26.2
9–11 months	93	26.8
12–17 months	100	28.8
18–23 months	63	18.2
Respondent relation to the child		
Mother	335	96.5
Other	12	3.5
Marital status of the mother/caretaker		
Single	36	10.4
Married	306	88.2
Other*	5	1.4
Educational status of the mother/caretaker		
No formal education	140	40.3
Primary education	114	32.9
≥Secondary education	93	26.8
Employment status of the mother/caretaker		
Housewife	266	76.7
Civil servant	27	7.8
Farmer	27	7.8
Merchant	11	3.2
Other**	16	4.6
Father educational status		
No formal education	78	22.7
Primary education (1–8)	147	42.4
≥Secondary education and above	121	34.9
Father employment status		
Farmer	149	42.9
Civil servant	74	21.3
Merchant	59	17.0
Private employed	47	13.5
Other	18	5.2
Birth order of the child		
1st	65	18.7
2nd	74	21.3
3rd	69	19.9
4th and above	139	40.1
Living arrangement of the child		
Living with both parents	305	87.9
Living with mother only	38	10.9
Living with grandparents	4	1.2

TABLE 1: Continued.

Background characteristics	Frequencies	Percent (%)
Number of siblings aged <5 years		
0	150	43.2
1	194	55.9
≥2	3	0.9
Number of children in the family		
2-3	96	27.7
4–6	235	67.7
≥7	16	4.6
Household wealth index		
Lowest	67	19.3
Second	78	22.5
Middle	108	31.1
Fourth	58	16.7
Highest	36	10.4

*divorced, widowed, and separated.
**Student and house servant.

two weeks, maternal educational status, dietary diversity, introduction of complementary foods, stunting, and household wealth quintile were significantly associated with the anemia (Figure 2 and Table 4).

4. Discussion

The result of this study revealed that 66.6% of children were anemic (95% CI: 0.619–0.713). The result is slightly higher than 2011 Ethiopian DHS report, 60.9% [10], and Bangladesh, 60% [27]. But the finding is lower than study report in Nepal, 69% [28], and Ghana, 84.3% [29]. This could be because, in developing countries, complementary foods for children are mostly porridges made of locally available staple cereals [30]. Cereals are known to be rich in phytates, which are nutrients causing poor bioavailability of iron. Similarly, staple food in the study area is cereal based given that children in the current study area share the same risk with other developing countries. Approximately 36% of children were with history of early and late introduction of complementary foods. Both carries risks contributing to persistent young child malnutrition [30]. Both practices are well known to cause anemia among young children. The current study area (Figure 1) is one of chronical foods in secured area in the region. This household food insecurity might hinder child from obtaining adequate and appropriate complementary food due to poor household food purchasing power.

Male children were 3.1 times more likely to be anemic as compared to females (AOR = 3.1 (95% CI: 1.60–5.81)). This finding is similar with study reports in Ghana [29] and Bangladesh [27]. Other studies conducted in Tanzania [31] and Brazil [32] found that sex difference did not show association with anemia. The possible explanation for this discrepancy could be due to sate of rapid growth of male children in the first months of life which increases their micronutrient requirement including iron [33]. If this physiological state is not compensated with appropriate and iron

TABLE 2: Feeding practice and nutritional status of children aged 6–23 months attending growth monitoring clinic at Tsitsika Health Center, Northeast Ethiopia (*n* = 347).

Background characteristics	Frequencies	Percent (%)
Ever breast feed		
Yes	346	99.7
No	1	0.3
Current breast feeding status		
Yes	306	88.2
No	41	11.8
Introduction of complementary foods		
Early (<6 months)	71	20.5
Timely (6–8 months)	223	64.3
Late (≥9 months)	53	15.3
History of pica consumption		
Yes	87	25.0
No	254	75.0
History of cow's milk consumption		
Yes	88	25.1
No	259	74.9
Meat consumption/week		
Yes	89	25.6
No	258	74.4
Fruit consumption per week		
Yes	101	29.1
No	246	70.9
Dietary diversity score		
Poor*	296	85.3
Good**	51	14.7
Length-for-Age		
Stunted (< −2-Z-score)	82	23.6
Not Stunted (> −2-Z-score)	265	76.4
Weight-for-Length		
Wasted (< −2-Z-score)	54	15.6
Not Wasted (> −2-Z-score)	293	84.4
Weight-for-Age		
Underweight (< −2-Z-score)	61	17.5
Not underweight (> −2-Z-score)	286	82.5

Note: * child received foods from <3 food groups in the previous 24 hrs.
** Child who received foods from ≥4 food groups in the previous 24 hrs.

TABLE 3: Morbidity and health care related characteristics of children aged 6–23 months attending growth monitoring clinic at Tsitsika Health Center, Northeast Ethiopia, 2014 (*n* = 347).

Background characteristics	Frequencies	Percent (%)
Birth place of the child		
Home	206	59.4
Health institution	141	40.6
Immunization status of the child		
Partial immunization	246	70.9
Full immunization	101	29.1
ITN* utilization		
Yes	139	40.1
No	208	59.9
History of malaria infection		
Yes	50	14.4
No	297	85.6
History of intestinal parasite in the past 2 weeks		
Yes	11	3.2
No	336	96.8
History of diarrheal in the past 2 weeks		
Yes	51	14.7
No	296	85.3
History of URTIs** in the past 2 weeks		
Yes	7	2.0
No	340	98.0

Note: *ITN: insecticide threatened bed nets.
**URTIs: upper respiratory tract infections.

the anemic pregnant mothers are more likely to give birth of child with poor iron stores [12].

Children with early (<6 months) and late (≥9 months) introduction of complementary foods were 11.1 times (AOR = 11.1 (95% CI: 4.08–30.31)) and 4.3 times (AOR = 4.3 (95% CI: 1.78–10.18)) more likely to be anemic than children with timely initiation of complementary food, respectively. It is evident that most digestive enzymes are inadequate until the first six months of age [35] and introducing liquid or solid food during this time causes interference with the absorption of iron in the breast milk [36]. Early exposure of infants (before six months of age) to microbial pathogens due to complementary foods increases the risk of infection for diarrheal disease, thereby malabsorption [37]. Breast milk has minimal iron to fulfill nutritional requirement of growing infant [38], given that providing breast milk alone coupled with rapid iron depletion beyond six months also increases risk of anemia for younger infant.

Those children with poor dietary diversity score were near to three times more likely to be anemic than children with good dietary diversity scores (AOR = 3.2 (95% CI: 1.35–7.38)). Cereal based monotonous diets (undiversified diet) are known to cause micronutrient deficiency including anemia [39]. It is also evidenced that dietary diversity is proxy indicator for micronutrient adequacy of diet [40].

rich complementary foods at this critical stage, risk of iron deficiency anemia will be higher among male children as compared to their counterpart.

Children in the age group of 6–8, 9–11, and 12–17 months were 3.5 times (AOR = 3.5 (95% CI: 1.46–8.26)), 9.6 times (AOR = 9.6 (95% CI: 3.61–25.47)), and 2.9 times (AOR = 2.9 (95% CI: 1.23–6.75)) more likely to be anemic than children in the age range of 18–23 months, respectively. This could be because prenatal iron store depletion is highest starting at six months of age [34]. In addition, it may be due to poor maternal iron reserve during pregnancy. It is known that

TABLE 4: Factors associated with anemia among children aged 6–23 months attending growth monitoring clinic at Tsitsika Health Center, Northeast Ethiopia, (n = 347).

Background characteristics	Anemia status of children		COR (95%: CI)	AOR (95%: CI)
	Yes	No		
Sex				
Male	131	40	2.5 (1.57–3.95)*	3.1 (1.60–5.81)*
Female	100	76	1.00	1.00
Age				
6–8 months	63	28	2.1 (1.05–3.98)**	3.5 (1.46–8.26)*
9–11 months	74	19	3.5 (1.75–7.17)*	9.6 (3.61–25.47)*
12–17 months	61	39	1.4 (0.75–2.69)	2.9 (1.23–6.75)**
18–23 months	33	30	1.00	1.00
Introduction of complementary foods				
<6 months	65	6	8.5 (3.53–20.42)*	11.1 (4.08–30.31)*
≥9 months	41	12	2.7 (1.34–5.37)*	4.3 (1.78–10.18)*
6–8 months	125	98	1.00	1.00
Dietary diversity				
Poor	208	88	2.9 (1.57–5.27)*	3.2 (1.35–7.38)*
Good	23	28	1.00	1.00
Length-for-Age				
Stunted	70	12	3.8 (1.95–7.29)*	2.7 (1.20–6.05)**
Not stunted	161	104	1.00	1.00
History of diarrhoea before 2 weeks				
Yes	47	5	5.5 (2.13–14.31)*	4.9 (1.63–14.59)*
No	185	111	1.00	1.00
Educational level of the mother				
No formal education	109	31	3.4 (1.947–6.083)*	2.6 (1.26–5.27)**
Primary education	75	39	1.9 (1.07–3.30)**	1.8 (0.83–3.71)
Secondary education and above	47	46	1.00	1.00
HH wealth index				
Lowest	53	14	3.8 (1.57–9.12)*	3.0 (1.01–8.88)**
Second	55	23	2.4 (1.06–5.40)**	2.8 (1.02–7.81)**
Middle	73	35	2.1 (0.97–4.50)	1.2 (0.46–3.28)
Fourth	32	26	1.2 (0.54–2.83)	0.9 (0.33–2.52)
Highest	18	18	1.00	1.00

Note: P value * < 0.01, and ** = 0.01–0.05.
HH: household.

Children with history of diarrhea before two weeks of the study were 4.9 times more likely to be anemic than children without diarrhea (AOR = 4.9 (1.63–14.59)). This finding is consistent with study reports in Indonesia [41, 42] and Brazil [32]. This could mainly operate through loss of appetite and malabsorption from diarrhea which in turn increases likelihood of developing anemia.

Stunted children were 2.7 times more likely to be anemic than their counterpart (AOR = 2.7 (95% CI: 1.20–6.05)). This finding is similar to studies conducted in Bangladesh [27], Brazil [43], and Burma [44]. This is could be because undernourished children are often anemic [43], low hemoglobin level has compromising effect of the linear growth [45], and coexisting of other micronutrient deficiencies and stunting

may increase the development of anemia by a synergism association. But the current study cannot provide cause and effect relationship between stunting and iron deficiency anemia.

Children of mothers with no formal education were 2.6 times more likely to be anemic than children of mother with secondary and above education level (AOR = 2.6 (95% CI: 1.26–5.27)). This finding is similar to study conducted in Kenya [14], Ghana [39], and Bangladesh [28]. But a study conducted in Timor-Lest [46] reported that maternal educational status was inversely associated with their children's nutritional status. Children of mothers with secondary education had significantly lower mean hemoglobin concentration than mothers with primary and no education. Moreover, mothers'

FIGURE 1: Map of Ziquala Woreda (study area), Northeast Ethiopia.

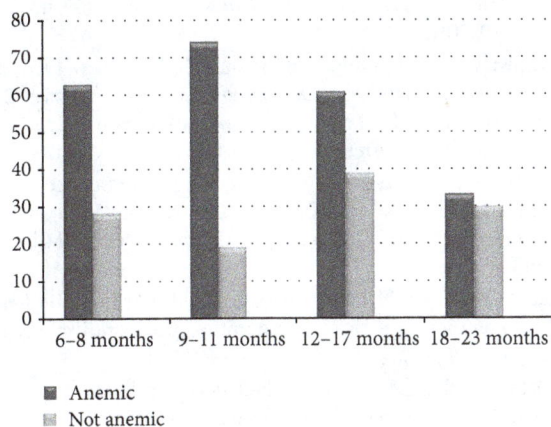

Anemic
Not anemic

FIGURE 2: Number of children (6–23 months) with anemia by age category.

level of education may positively influence practices related to the health care and feeding practice of their children. Educated mothers are more conscious of their children's health and introducing scientifically proved feeding practices, which help to improve their children nutritional status [28]. It is also confirmed that, maternal education is strong predictor for nutritional outcomes of children [39].

Children from families of lowest wealth quintiles were three times more likely to be anemic than children from highest wealth quintile (AOR = 3.0 (95% CI: 1.01–8.88)). The finding of the current study is in line with studies conducted in Ghana [39], Brazil [33], and Bangladesh [28]. Poor household economic status might result in loss of power to purchase diversified and nutrient rich food and secure the household per capita food availability. Other studies also reveal that poor household economic status is associated with household food insecurity [22]. In turn, household food insecurity is strong determinant factor for undernutrition including iron deficiency anemia [12, 47].

Some limitations of the study should be noted and taken into consideration. Cross-sectional nature of this study did not reveal causal links between independent variables and iron deficiency anemia. Anemia status from the study group was determined by using the hemoglobin level but not by using the latest test indicators like serum ferritin.

5. Conclusion

Burden of anemia among children aged 6–23 months in the study area is higher and it has severe public health significance according to the WHO cut-off points. Stunting, mother with no formal education, lowest wealth quintiles, having diarrhea before two weeks, poor dietary diversity scores, early and late introduction of complementary foods, and sex and age of the child were significantly associated with anemia. Well integrated interventions to improve the health status and infant and young child feeding practices need to be prioritized to prevent deficiency of anemia targeting children aged under two years of age.

Conflict of Interests

The authors declare that they have no competing interests.

Authors' Contribution

Haile Woldie designed the study, carried out statistical analysis, and thesis writes up process and manuscript preparation. Yigzaw Kebede and Amare Tariku participated in proposal writing, reviewing, and approval of the thesis and manuscript writing. Haile Woldie, Yigzaw Kebede, and Amare Tariku contributed equally to this work.

Acknowledgments

The authors would like to express their sincere gratitude to those children and their mother/caretaker for their willingness and positive cooperation for being part of the study. The authors' heartfelt gratitude will also go to the ENGIEN/JHPIEGO, Ethiopia, for the financial support of this study.

References

[1] R. Aikawa, N. C. Khan, S. Sasaki, and C. W. Binns, "Risk factors for iron-deficiency anaemia among pregnant women living in rural Vietnam," *Public Health Nutrition*, vol. 9, no. 4, pp. 443–448, 2006.

[2] World Health Organization, *WHO Vitamin and Mineral Nutrition/Anemia*, 2011.

[3] E. McLean, M. Cogswell, I. Egli, D. Wojdyla, and B. De Benoist, "Worldwide prevalence of anaemia, WHO Vitamin and Mineral Nutrition Information System, 1993–2005," *Public Health Nutrition*, vol. 12, no. 4, pp. 444–454, 2009.

[4] S. Awasthi, R. Das, T. Verma, and S. Vir, "Anemia and undernutrition among preschool children in Uttar Pradesh, India," *Indian Pediatrics*, vol. 40, no. 10, pp. 985–990, 2003.

[5] World Health Organization, *Global Burden of Diseases 2004 Update*, World Health Organization, Geneva, Switzerland, 2008.

[6] B. D. Benoist, E. McLean, I. Egll, M. Cogswell, and D. Wojdyla, *Worldwide Prevalence of Anemia 1993—2005: WHO Global Database on Anemia*, World Health Organization, 2008.

[7] B. Lozoff and M. K. Georgieff, "Iron deficiency and brain development," *Seminars in Pediatric Neurology*, vol. 13, no. 3, pp. 158–165, 2006.

[8] WHO, *Worldwide Prevalence of Anemia 1993-2005: WHO Global Database on Anemia*, World Health Organization, Geneva, Switzerland, 2008.

[9] S. Villalpando, T. Shamah-Levy, C. I. Ramírez-Silva, F. Mejía-Rodríguez, and J. A. Rivera, "Prevalence of anemia in children 1 to 12 years of age: results from a nationwide probabilistic survey in Mexico," *Salud Pública de México*, vol. 45, no. 4, pp. S490–S498, 2003.

[10] Central Statistical Authority (Ethiopia) and ORC Macro, *Ethiopia Demographic and Health Survey 2011*, Central Statistical Authority, Addis Ababa, Ethiopia; ORC Macro, Calverton, Md, USA, 2011.

[11] E. Pollitt, "Early iron deficiency anemia and later mental retardation," *The American Journal of Clinical Nutrition*, vol. 69, no. 1, pp. 4–5, 1999.

[12] S.-R. Pasricha, J. Black, S. Muthayya et al., "Determinants of anemia among young children in rural India," *Pediatrics*, vol. 126, no. 1, pp. e140–e149, 2010.

[13] G. Egbi, M. Steiner-Asiedu, F. S. Kwesi et al., "Anaemia among school children older than five years in the Volta Region of Ghana," *Pan African Medical Journal*, vol. 17, no. 1, article 10, 2014.

[14] O. Ngesa, H. Mwambi, and J. A. Stoute, "Prevalence and risk factors of anaemia among children aged between 6 months and 14 years in Kenya," *PLoS ONE*, vol. 9, no. 11, Article ID e113756, 2014.

[15] A. Shet, S. Mehta, N. Rajagopalan et al., "nemia and growth failure among HIV-infected children in India: a retrospective analysis," *BMC Pediatrics*, vol. 9, article 37, 2009.

[16] M. A. Cardoso, K. K. G. Scopel, P. T. Muniz, E. Villamor, and M. U. Ferreira, "Underlying factors associated with anemia in amazonian children: a population-based, cross-sectional study," *PLoS ONE*, vol. 7, no. 5, Article ID e36341, 2012.

[17] A. Desalegn, A. Mossie, L. Gedefaw, and C. M. Schooling, "Nutritional iron deficiency anemia: magnitude and its predictors among school age children, southwest ethiopia: a community based cross-sectional study," *PLoS ONE*, vol. 9, no. 12, Article ID e114059, 2014.

[18] S. Villalpando, T. Shamah-Levy, C. I. Ramírez-Silva, F. Mejía-Rodríguez, and J. A. Rivera, "Prevalence of anemia in children 1 to 12 years of age. Results from a nationwide probabilistic survey in Mexico," *Salud Pública de México*, vol. 45, no. 4, pp. S490–S498, 2003.

[19] C. G. Victora, L. Adair, C. Fall et al., "Maternal and child undernutrition: consequences for adult health and human capital," *The Lancet*, vol. 371, no. 9609, pp. 340–357, 2008.

[20] T. Shafir, R. Angulo-Barroso, J. Su, S. W. Jacobson, and B. Lozoff, "Iron deficiency anemia in infancy and reach and grasp development," *Infant Behavior and Development*, vol. 32, no. 4, pp. 366–375, 2009.

[21] L. Manning, M. Laman, A. Rosanas-Urgell et al., "Severe anemia in papua new guinean children from a malaria-endemic area:

[22] T. Birhane, S. Shiferaw, S. Hagos, and K. S. Mohindra, "Urban food insecurity in the context of high food prices: a community based cross sectional study in Addis Ababa, Ethiopia," *BMC Public Health*, vol. 14, no. 1, article 680, 2014.

[23] R. E. Black, C. G. Victora, S. P. Walker et al., "Maternal and child undernutrition and overweight in low-income and middle-income countries," *The Lancet*, vol. 382, no. 9890, pp. 427–451, 2013.

[24] World Health Organization, *Iron Deficiency Anemia Assessment, Prevention and Control: A Guide for Programme Managers*, WHO, Geneva, Switzerland, 2001.

[25] World Health Organization, *World Health Organization Child Growth Standards: Methods and Development*, World Health Organization, Geneva, Switzerland, 2006.

[26] World Health Organization, *Indicators for Assessing Infant and Young Child Feeding Practices*, WHO Press, Geneva, Switzerland, 2010.

[27] M. K. Uddin, M. H. Sardar, M. Z. Hossain et al., "Prevalence of anaemia in children of 6 months to 59 months in Narayanganj, Bangladesh," *Journal of Dhaka Medical College*, vol. 19, no. 2, pp. 126–130, 2011.

[28] Ministry of Health and Population (MOHP), New ERA, and ICF International, *Nepal Demographic and Health Survey 2011*, Ministry of Health and Population, Kathmandu, Nepal; New ERA, and ICF International, Calverton, Md, USA, 2012.

[29] Ghana Statistical Service (GSS), Ghana Health Service (GHS), and ICF Macro, *Ghana Demographic and Health Survey 2013: Key Findings*, GSS, GHS, and ICF Macro, Calverton, Md, USA, 2013.

[30] A. W. Onyango, "Dietary diversity, child nutrition and health in contemporary African communities," *Comparative Biochemistry and Physiology*, vol. 136, pp. 61–69, 2003.

[31] D. Schellenberg, J. R. M. A. Schellenberg, A. Mushi et al., "The silent burden of anaemia in Tanzanian children: a community-based study," *Bulletin of the World Health Organization*, vol. 81, no. 8, pp. 581–590, 2003.

[32] L. P. Leal, M. B. Filho, P. I. C. de Lira, J. N. Figueiroa, and M. M. Osório, "Prevalence of anemia and associated factors in children aged 6–59 months in Pernambuco, Northeastern Brazil," *Revista de Saúde Pública*, vol. 45, no. 3, pp. 457–466, 2011.

[33] C. M. Chaparro, "Setting the stage for child health and development: prevention of iron deficiency in early infancy," *Journal of Nutrition*, vol. 138, no. 12, pp. 2529–2533, 2008.

[34] M. M. Black, A. M. Quigg, K. M. Hurley, and M. R. Pepper, "Iron deficiency and iron-deficiency anemia in the first two years of life: strategies to prevent loss of developmental potential," *Nutrition Reviews*, vol. 69, supplement 1, pp. S64–S70, 2011.

[35] S. R. D. M. Saldiva, S. I. Venancio, A. G. C. Gouveia, A. L. D. S. Castro, M. M. L. Escuder, and E. R. J. Giugliani, "Regional influence on early consumption of foods other than breast milk in infants less than 6 months of age in Brazilian State capitals and the Federal District," *Cadernos de Saude Publica*, vol. 27, no. 11, pp. 2253–2262, 2011.

[36] F. A. Oski and S. A. Landaw, "Inhibition of iron absorption from human milk by baby food," *The American Journal of Diseases of Children*, vol. 134, no. 5, pp. 459–460, 1980.

[37] J. K. Meinzen-Derr, M. L. Guerrero, M. Altaye, H. Ortega-Gallegos, G. M. Ruiz-Palacios, and A. L. Morrow, "Risk of infant

a case-control etiologic study," *PLoS Neglected Tropical Diseases*, vol. 6, no. 12, Article ID e1972, 2012.

anemia is associated with exclusive breast-feeding and maternal anemia in a Mexican cohort," *Journal of Nutrition*, vol. 136, no. 2, pp. 452–458, 2006.

[38] G. Wiseman, *Nutrition and Health*, Taylor & Francis e-Library, New York, NY, USA, 2004.

[39] P. S. Mamiro, P. Kolsteren, D. Roberfroid, S. Tatala, A. S. Opsomer, and J. H. van Camp, "Feeding practices and factors contributing to wasting, stunting, and iron-deficiency anaemia among 3–23-month old children in Kilosa district, rural Tanzania," *Journal of Health, Population and Nutrition*, vol. 23, no. 3, pp. 222–230, 2005.

[40] World Health Organization and Working Group on Infant and Young Child Feeding Indicators, *Developing and Validating Simple Indicators of Dietary Quality and Energy Intake of Infants and Young Child in Developing Countries*, World Health Organization, Washington, DC, USA, 2007.

[41] R. D. Semba, S. de Pee, M. O. Ricks, M. Sari, and M. W. Bloem, "Diarrhea and fever as risk factors for anemia among children under age five living in urban slum areas of Indonesia," *International Journal of Infectious Diseases*, vol. 12, no. 1, pp. 62–70, 2008.

[42] C. T. Howard, S. de Pee, M. Sari, M. W. Bloem, and R. D. Semba, "Association of diarrhea with anemia among children under age five living in rural areas of Indonesia," *Journal of Tropical Pediatrics*, vol. 53, no. 4, pp. 238–244, 2007.

[43] M. M. Osório, P. I. C. Lira, M. Batista-Filho, and A. Ashworth, "Prevalence of anemia in children 6–59 months old in the state of Pernambuco, Brazil," *Revista Panamericana de Salud Pública*, vol. 10, no. 2, pp. 101–107, 2001.

[44] A. Zhao, Y. Zhang, Y. Peng et al., "Prevalence of anemia and its risk factors among children 6–36 months old in Burma," *The American Journal of Tropical Medicine and Hygiene*, vol. 87, no. 2, pp. 306–311, 2012.

[45] A. T. Soliman, M. M. Al dabbagh, A. H. Habboub, A. Adel, N. A. Humaidy, and A. Abushahin, "Linear growth in children with iron deficiency anemia before and after treatment," *Journal of Tropical Pediatrics*, vol. 55, no. 5, Article ID fmp011, pp. 324–327, 2009.

[46] Factors associated with Hemoglobin concentration among Timor-Leste children Aged 6–59 months, Australia, 2009.

[47] D. Ali, K. K. Saha, P. H. Nguyen et al., "Household food insecurity is associated with higher child undernutrition in Bangladesh, Ethiopia, and Vietnam, but the effect is not mediated by child dietary diversity," *Journal of Nutrition*, vol. 143, no. 12, pp. 2015–2021, 2013.

14

Fructose Metabolism and Relation to Atherosclerosis, Type 2 Diabetes, and Obesity

Astrid Kolderup[1] **and Birger Svihus**[2]

[1]*Faculty of Public Health, Hedmark University College, P.O. Box 400, 2418 Elverum, Norway*
[2]*Norwegian University of Life Sciences, P.O. Box 5003, 1432 Aas, Norway*

Correspondence should be addressed to Astrid Kolderup; astrid.kolderup@hihm.no

Academic Editor: Michael B. Zemel

A high intake of sugars has been linked to diet-induced health problems. The fructose content in sugars consumed may also affect health, although the extent to which fructose has a particularly significant negative impact on health remains controversial. The aim of this narrative review is to describe the body's fructose management and to discuss the role of fructose as a risk factor for atherosclerosis, type 2 diabetes, and obesity. Despite some positive effects of fructose, such as high relative sweetness, high thermogenic effect, and low glycaemic index, a high intake of fructose, particularly when combined with glucose, can, to a larger extent than a similar glucose intake, lead to metabolic changes in the liver. Increased *de novo* lipogenesis (DNL), and thus altered blood lipid profile, seems to be the most prominent change. More studies with realistic consumption levels of fructose are needed, but current literature does not indicate that a normal consumption of fructose (approximately 50–60 g/day) increases the risk of atherosclerosis, type 2 diabetes, or obesity more than consumption of other sugars. However, a high intake of fructose, particularly if combined with a high energy intake in the form of glucose/starch, may have negative health effects via DNL.

1. Introduction

Sugars are important sources of energy in our diet, and a high intake of sugars has increasingly been identified as a considerable cause of major diet-induced health problems, namely, atherosclerosis, type 2 diabetes, and obesity [1–4]. An increased intake of sugar sweetened beverages in particular has been associated with these health problems in many epidemiological studies [5–7]. The main constituents of sugars are fructose and glucose, which can be present either alone or in combination, although most commonly fructose is mixed 50 : 50 with glucose. The ratio of glucose : fructose in the diet has been argued to be important for some of the effects of sugars [8, 9]. Thus, it is important to study the effects of fructose and glucose together, but also it is important to study the effects of fructose alone. Fructose has been claimed to be of benefit because it may aid glycemic control [10, 11], but also it has been claimed to be more harmful than other sugars,

especially with regard to the development of atherosclerosis, type 2 diabetes, and obesity [8].

Fructose is found in a variety of foods. In table sugar, it is bound to glucose to form the disaccharide sucrose, whereas in honey it occurs in monosaccharide form. In fruit, berries, and vegetables, fructose occurs in both monosaccharide and disaccharide forms. Measured as intake from caloric sweeteners in USA, fructose intake was rather stable throughout the fifties and sixties but increased from the seventies until the end of the nineties, after which intake has declined [12]. The average American fructose intake was estimated to be 49 g per day in 2004 [13]. In Norway, the average daily intake of fructose can be estimated to be approximately 56 g/day, based on data from the Norwegian Directorate of Health's survey of consumption patterns [14] and composition data [15]. Globally, the main source of fructose is sucrose, which constitutes >90% of the energizing sweeteners used in the world [16]. In some countries, such as USA and Japan, high-fructose corn syrup (HFCS) is also an important source of fructose. HFCS is a mixture of fructose and glucose in different concentrations. It can contain up to 90% fructose [17], but the dominating concentration of fructose is 42% or 55% in commercial products [9]. In USA, consumption of HFCS increased sharply from 1970 until 1999, but it has declined since then [12, 18].

This review paper aims to describe how fructose, compared with other sugars, is managed by the body, thus clarifying the impact of fructose on atherosclerosis, type 2 diabetes, and obesity.

2. Materials and Methods

The review is based on scientific peer-reviewed papers obtained using a nonsystematic search of the databases PubMed and Web of Science. The first step in the selection of literature was to identify the relevant keywords to search for in these databases. Various combinations of the following keywords were used: "fructose," "glucose," "sucrose," and "sugar" combined with "metabolism," "insulin resistance," "overweight," "obesity," "relative sweetness," "absorption capacity," "glycemic index," "de novo lipogenesis," "thermogenesis," "type 2 diabetes," and "appetite regulation."

The next step in the selection process consisted of inclusion or exclusion of papers based on the relevance to the study's aim. Through this initial selection process, both original and review papers were included. The reference lists of the papers were thoroughly studied to discover any possibly relevant papers that were not included. As far as possible, only original papers and reviews based on controlled trials and mechanistic studies were included. Unless otherwise specified, the results presented and discussed in this paper are statistically significant, and the experiments were conducted with a control group. Although animals may metabolize fructose differently from humans, animal studies have been included when found relevant, for example, due to lack of human studies. One possible limitation in this study lies in the nonsystematic nature of the search strategy, resulting in the possibility that relevant papers were not considered.

3. The Body's Fructose Management

3.1. Absorption. Although there is some uncertainty about the mechanisms of fructose absorption, most of fructose seems to be absorbed by facilitated transport in the jejunum by the fructose transporter GLUT5 [18–20]. The body has limited capacity to absorb pure fructose, and intake of fructose can therefore lead to malabsorption [20–24]. Malabsorption of fructose results in bacterial fermentation, which leads to formation of short-chain fatty acids (acetate, propionate, and butyrate) and gases (hydrogen, methane, and carbon dioxide) [25, 26]. These processes can affect the motility of the intestine and cause various symptoms such as abdominal pain, bloating, and altered stool [27]. A large individual variation in absorption capacity for fructose has been observed. When fructose is consumed as a single oral dose, the maximum absorption capacity has been shown to vary between 5 and 50 g [28]. Several factors seem to affect

this capacity, such as age and health [27, 29, 30], but of the dietary factors the presence of glucose is the most important [20, 23]. A significant increase in fructose absorption has been shown when fructose is coingested with equal amounts of glucose [21, 31]. There is still uncertainty about how glucose increases absorption capacity for fructose, but it may amongst others be due to an effect of glucose on the presence of fructose transporters [20]. Fructose does not appear in any fruit or vegetable without glucose [15]. Evolutionarily, this may explain why humans do not have the ability to absorb large amounts of pure fructose.

3.2. Metabolism. After absorption, fructose is transported by the portal vein to the liver, where it is effectively absorbed by liver cells [32], resulting in only small amounts entering the systemic circulation. The concentration of fructose in the blood is therefore only about 0.01 mmol/L, unlike that of glucose which is approximately 5.5 mmol/L [33]. Metabolism of fructose thus occurs primarily in the liver, but fructose may also be metabolized by enterocytes. In a study of pigs, it was shown that intestinal lactate production from fructose could account for 12% of absorbed fructose [34], but the functional significance of intestinal metabolism of fructose in humans remains unknown [18]. Bjorkman and Felig [35] also showed that infused fructose (48.6 g fructose in liquid form infused intravenously) was metabolized in the kidneys in humans who had fasted for 60 hours. Despite the artificially high blood fructose level, that study still shows that the kidneys have a relatively large capacity to metabolize fructose. It has been shown that GLUT5 is expressed in the membrane of fat, kidney, muscle, and brain cells [32, 36], but, due to very low levels of fructose in the blood, negligible amounts of fructose are probably metabolized in these tissues [9, 18].

As discussed above, the liver will metabolize a large majority of the ingested fructose, compared to only about 15–30% of ingested glucose [37, 38]. Most of the reactions in liver fructolysis are the same as those occurring in glycolysis, but fructose enters at a later stage in the glycolytic reaction chain than glucose [39]. Thus, fructose avoids the main control step in glycolysis, the phosphofructokinase step, which is tightly regulated by the energy status of the cell. The first step in fructolysis is the phosphorylation of fructose by fructokinase to fructose 1-phosphate. Unlike the phosphofructokinase in glycolysis, this enzyme is not inhibited by ATP [40]. The enzyme is considered virtually unregulated and, even when the liver's energy status (ATP) is high, the fructokinase will metabolize fructose to fructose 1-phosphate. Furthermore, fructose 1-phosphate is cleaved to dihydroxyacetone phosphate and glyceraldehyde by the enzyme aldolase B. Glyceraldehyde may be phosphorylated by triose kinase, with ATP as the phosphate donor, to form the glycolytic intermediate glyceraldehyde 3-phosphate [41]. After these steps, the carbon atoms from fructose follow the glycolytic steps.

Bypassing the phosphofructokinase step makes the flow of fructose carbon atoms through the biochemical pathways less controlled than for glucose. In this way, the liver will metabolize fructose in an unlimited way, as opposed to the case of glucose. This will influence the type and amount of metabolic products produced by the liver and is the main reason why fructose and glucose have different metabolic effects.

In the liver, fructose can enter metabolic pathways: it can be oxidized, converted to glucose (and glycogen), or converted to lactic acid, or enter *de novo* lipogenesis (DNL). After an overnight fast, approximately 50% of the fructose eaten as an oral dose of approximately 30–70 g is converted to glucose via gluconeogenesis [42]. Others have shown that after a similar fructose intake about 45% is oxidized; however, this includes both direct fructose oxidation and oxidation of glucose and lactic acid formed from fructose [43]. Another metabolic fate of fructose is to form lactic acid, but this seems to occur only at high fructose intakes [44–48]. The ingestion of 72 g fructose is the lowest amount that has been experimentally demonstrated to result in lactic acid formation [49]. Intake of fructose may also lead to formation of fatty acids via DNL. The diet composition may influence the distribution of fructose. Theytaz et al. [50] showed that coingestion of glucose with fructose decreased oxidation and gluconeogenesis from fructose carbon atoms. Intake of fructose together with glucose thus seems to affect the metabolic fate of fructose. To some degree, this effect may be due to higher insulin secretion after intake of glucose compared to fructose [51]. Insulin will, amongst others, decrease glucose production from fructose [52], and insulin will also stimulate DNL [53]. The extent to which fructose enters DNL is central to the health effects of fructose.

3.3. De Novo Lipogenesis (DNL). DNL is the metabolic pathway that transforms surplus nonfat energy into fat by synthesis of fatty acids from acetyl-CoA [54]. The liver is the primary site of DNL, but DNL can also occur in lactating mammary glands and adipose tissue. Key lipogenic enzymes are present in adipose tissue, but to what degree adipose tissue contributes to the total DNL seems to be unclear. Although some data indicate that DNL in adipose tissue is of little importance [55–57], it has also been shown that it may play a more significant role [58], especially after ingestion of large amounts of carbohydrates [59, 60]. Some controversy exists regarding the capacity of the liver to carry out *de novo* lipogenesis in humans [60, 61], although a high capacity has been demonstrated in several experiments [62, 63].

As a result of the metabolic difference between glucose and fructose, a higher percentage of fructose compared to glucose can be converted to fat in the liver via DNL [63]. This has been shown in a number of animal and human studies, in which these sugars have been consumed in equal quantities under similar experimental conditions [63–66]. The greater potential of fructose to stimulate DNL is the main reason why fructose has been portrayed as particularly harmful. As described earlier, the liver absorbs most of the ingested fructose, and fructose metabolism bypasses the main control step in glycolysis, which means that a greater proportion of fructose, compared with glucose, is available for DNL. A normal diet, however, provides 3 to 5 times more glucose than fructose [12, 67], and this may influence the practical relevance of this metabolic difference. If, for example, glucose intake is very high, the liver may need to

handle larger amounts of glucose than fructose. At two points in liver fructolysis, intermediates can enter lipid synthesis [9]. Fructose is converted to dihydroxyacetone phosphate, an intermediate in equilibrium with glycerol 3-phosphate, which forms the basis for the glycerol in triglycerides and phospholipids. Meanwhile, a large proportion of fructose carbons are metabolized directly to pyruvic acid and then acetyl-CoA. On metabolism of a large amount of fructose to acetyl-CoA, the amount of acetyl-CoA may exceed the citric acid cycle capacity of mitochondria. High levels of fructose may thus act as a nonregulated source of hepatic acetyl-CoA, which is a substrate that can enter DNL [9]. In addition, a high intake of fructose seems to stimulate gene expression and activity of lipogenic enzymes in the liver [40, 41, 66, 68]. Recently, fructose has also been shown to give a higher increase in fibroblast growth factor 21 (FGF21) than glucose. FGF21 is a hormone involved in glucose and lipid homeostasis [69], and high levels of FGF21 seem to be associated with metabolic disease [70].

It is clear that a high intake of fructose will cause a significant increase in DNL activity [71]. However, there is a paucity of knowledge about the effect of a normal fructose intake on this activity [72]. No research seems to have been conducted to assess the minimum amount or individual range of fructose that must be eaten to obtain a significant increase in DNL, which among others may be due to difficulties in quantifying DNL. In fact, measurement of hepatic DNL must be considered semiquantitative [54, 73]. Chong et al. [73] found that ^{13}C-labelled fructose contributed only 0.4% of the triglycerides in very-low-density lipoproteins (VLDLs) in men. This was measured 6 hours after eating a meal of 0.5 g fat/kg body weight and 0.75 g fructose/kg body weight after an overnight fast. The small increase in triglyceride levels in the blood after a fructose intake may be a result of delayed triglyceride secretion. Incorporation of carbon from ^{13}C-labelled fructose in VLDLs will not always reflect DNL, because some of the newly synthesized fatty acids will have delayed secretion [74, 75]. The level of DNL activity after fructose intake seems to vary significantly among individuals [76–78], and it also seems to vary during the day [79, 80].

Excessive intake of fructose, and hence increased DNL, may increase the risk of disease, because it may potentially cause both increased cholesterol levels in the blood and accumulation of fat in the liver [81]. In several animal studies, fat accumulation in the liver has been demonstrated after a high intake of fructose [82, 83], although most human studies have failed to demonstrate such an effect [84–86]. In fact, Chiu et al. [87] concluded, based on a systematic review with meta-analysis of controlled feeding trials, that fructose does not increase lipid content in liver when isocalorically exchanged for other carbohydrates. However, they pointed out that fructose ingested in large doses can raise liver lipid content, an effect that may be due to excess energy rather than fructose per se. Bravo et al. [88] conducted a study where participants consumed sucrose or HFCS for ten weeks at different levels of intake. In the study, fructose consumed as a part of a normal diet did not promote increased liver lipid content even at high intakes (90th percentile consumption level). In a study referred to by Rippe and Angelopoulos [89], no increase in liver fat was observed after consumption of HFCS or sucrose at levels up to 30% of energy for ten weeks. However, Maersk et al. [90] found that intake of 1 L sucrose-sweetened soft drink per day for six months increased liver lipids in overweight participants compared with intake of same amount of milk, diet cola, and water. This may illustrate that the combination of fructose and glucose, as in sucrose, can lead to increased level of liver lipids. However, it is not possible to conclude that this is an effect of fructose, since there was no glucose control group in this study. Johnston et al. [91] found no difference in liver triglyceride level between overweight participants that ate a hypercaloric high-fructose or high-glucose diet. The authors concluded that this result indicates that the hypercaloric state rather than macronutrient composition is important for the accumulation of liver lipids. Carbohydrate-induced accumulation of fat in the liver can lead to nonalcoholic fatty liver disease (NAFLD) [92]. However, it is still unclear if fructose via increased DNL is particularly conducive to NAFLD [93]. The effect of fructose on lipid accumulation is thus unclear, but the effect of fructose on the blood lipid profile seems to be better documented.

4. Atherosclerosis

It appears that high fructose intake can create an unfavorable lipid profile in blood via DNL [94]. The main product of DNL is palmitic acid [95], a fatty acid specifically shown to increase the risk of atherosclerosis [96]. Fatty acids formed by DNL will mainly be packed in VLDLs delivered into the bloodstream. This may, in turn, increase the level of low-density lipoproteins (LDLs) in the blood. In several studies, fructose has to a greater extent than glucose increased blood levels of triglycerides [51, 65, 97, 98] and LDLs [65, 99–103]. Aeberli et al. [104] showed that fructose increased the small dense LDLs, the type of LDLs that may in particular be linked to cardiovascular risk [105]. The level of high-density lipoproteins (HDLs) in blood does not seem to be affected by fructose [100, 101]. In most studies, an intake of fructose >100 g/day has been necessary to observe the adverse effects on lipid profiles [51, 65, 85, 98, 100, 106–108]. However, in a recent study by Aeberli et al. [109], a daily intake of about 77 g fructose and 34 g glucose for 3 weeks resulted in increased levels of total cholesterol and LDLs in the blood of healthy young men, compared with a daily intake of approximately 109 g glucose and 28 g fructose over the same time period. The fact that both groups also ingested an unknown amount of starch and the fact that food intake was not controlled reduce the solidity of these results. Maersk et al. [90] found that intake of 50 g fructose per day together with 50 g glucose could have a negative effect on blood triglyceride level. The lack of a control group ingesting glucose makes it difficult to conclude that this is an effect of fructose. Conversely, Lowndes et al. [110] found no negative effect on lipid profile in overweight or obese individuals consuming HFCS or sucrose incorporated in a eucaloric diet for ten weeks at levels corresponding to the 25th and 50th percentiles of adult

fructose consumption. Using the current knowledge, it does not appear that the consumption of moderate amounts of fructose (<50 g/day) alone will result in an unfavorable blood lipid profile [86, 111].

Due to the insignificant levels of fructose in peripheral blood, as described above, only glucose has the potential to be a substrate for DNL in adipose tissue. Although DNL in adipose tissue seems to be small as earlier discussed, glucose will, because of its presence in blood and by raising blood insulin level, probably to a higher extent than fructose, stimulate DNL in adipose tissue. Intake of glucose, in amounts that exceed the total capacity for glycogen storage and glucose oxidation, may thus increase DNL in adipose tissue more than the same amount of fructose. While fat formed in the liver has to be transported as lipoproteins in blood, this is avoided if the fat is formed directly in adipose tissue. Considering known negative health effects of lipoprotein residues, DNL occurring in adipose tissue may be preferable compared with DNL in the liver. This may illustrate a metabolic difference between glucose and fructose when consuming large amounts of sugars.

Another possible difference between fructose and glucose on risk factors for atherosclerosis is the effect of these sugars on blood uric acid level. Increased uric acid level has been associated with atherosclerosis in epidemiological studies, but the causality is uncertain [112–114]. Fructose appears to increase uric acid levels in the blood to a higher extent than glucose, especially at high intakes and when consumed as excess energy [86, 115, 116]. Intake of 0.5 g fructose/kg body weight is the lowest quantity shown to result in uric acid formation [117]. An increased blood level of uric acid can theoretically lead to elevated blood pressure because uric acid inhibits an enzyme in the endothelial cells of the arteries called endothelial nitric oxide synthase (eNOS). Activated eNOS leads to increased production of nitric oxide (NO), an important vasodilator. Thus, inhibition of eNOS may lead to vasoconstriction. Although only 0.5 g fructose/kg body weight has been shown to give uric acid formation and increased uric acid level theoretically could increase blood pressure, results from studies of the effect of fructose on blood pressure are very inconsistent [65, 72, 118, 119]. An average intake of fructose does not seem to lead to increased blood pressure [111, 120, 121]. The lack of causal link between uric acid level and atherosclerosis makes it difficult to draw conclusions on this effect of fructose.

5. Type 2 Diabetes

A high intake of sugar-sweetened beverages, with fructose as one of the major types of monosaccharides, has been associated with development of type 2 diabetes [5, 122]. Although this association does not prove causation, it is important to study the role of fructose in the development of type 2 diabetes. Central to the understanding of type 2 diabetes is the effect of nutrients on blood glucose homeostasis. Fructose must be converted to glucose in the liver to cause an increase in blood glucose level. As the conversion takes time and only a portion of the fructose will form glucose, fructose increases blood glucose less than similar levels of glucose [51]. Thus, the glycemic index for fructose is only 23 [10]. This, together with lack of stimulation of the pancreatic β cells [123], gives lower insulin secretion after intake of fructose compared with glucose [51, 124, 125]. These effects are positive because they contribute to blood glucose homeostasis. Additionally, moderate amounts of fructose have been shown to have positive effects on glycemic control [86, 126, 127]. However, it is claimed that fructose may also contribute negatively to blood glucose homeostasis by causing insulin resistance in the liver [9]. There is evidence that a high intake of fructose can cause insulin resistance in animals [128, 129], but several human studies have failed to demonstrate such an association [103, 130–132]. The operational definition of insulin resistance or sensitivity seems unclear, and many different methods have been used to measure it [133]. Thus, it is difficult to compare studies of insulin resistance. In human studies, in which fructose has been reported to cause insulin resistance, the daily intake of fructose has been as high as 110 g [109], approximately 250 g [134], 80 g [135], and 138 g [65]. This may indicate that the fructose intake must be high to potentially cause insulin resistance [86]. In the studies by Aeberli et al. [109], Stanhope et al. [65], and Beck-Nielsen et al. [134], total food intake was not controlled. Thus, the observed effect of fructose may also have been caused by differences in food intake between the control and experimental groups. In all the studies, in which insulin resistance has been shown, fructose was eaten together with glucose or starch, so the observations could also be the result of a combination of fructose and glucose. A number of hypotheses on how fructose can cause insulin resistance in the liver have been proposed. Lipid accumulation in the liver [136–138], metainflammation [83], and oxidative stress [139] are, either via inhibitory phosphorylation of the insulin receptor or the signaling molecules involved in insulin signaling, possible mechanisms for fructose-induced insulin resistance [9]. However, there are too few studies in humans and these are too divergent to be able to conclude firmly that there is a link between the consumption of fructose and insulin resistance. More long-term studies in which the daily intake of fructose is moderate are needed.

6. Obesity

It is debatable whether fructose is less satiating than other sugars and thus can contribute to obesity through a high food intake. In a study by Page et al. [92], magnetic resonance (MR) images were taken of human brains after giving 75 g of a fructose or glucose drink. Glucose, but not fructose, reduced activity (regional cerebral blood flow) in the hypothalamus in the areas involved in energy regulation and reward systems; this is probably an indicator of satiety and may indicate that fructose is less satiating than glucose. Fructose will also to a lesser extent than glucose increase blood levels of insulin [51], leptin [51, 140], gastric inhibitory polypeptide [141], and glucagon-like peptide-1 [92], while at the same time it will attenuate levels of ghrelin less [51]. Although these hormonal effects may indicate that fructose is less satiating than glucose, this has not been confirmed in studies of the ability of fructose to satiate. In such studies, it has been shown that fructose

has a greater appetite-reducing effect than glucose, when intake occurs before a meal [142, 143], or that there is no difference in the effects on appetite between fructose and glucose [144, 145]. Therefore, the effect of fructose on appetite remains unclear.

Although it is conceivable that fructose, via lack of stimulation of satiety signals, could contribute to obesity, fructose has several properties that act against obesity. As previously mentioned, the small intestine has a limited capacity to absorb fructose. This can lead to malabsorption at least if large amounts are consumed and consumption occurs without glucose-providing nutrients. Malabsorption of fructose will make less fructose enter the bloodstream and thus less energy will be available to the cells. In this way the malabsorption will act against obesity. It has also been shown in numerous studies that fructose has a greater thermogenic effect than glucose [46, 146–148]. This means that the body uses more energy after eating fructose rather than glucose, so less energy will be available to be stored as fat. The relative sweetness of fructose is also greater than for glucose and sucrose [149, 150]. Although this will decrease with increasing temperature [151, 152], the high relative sweetness allows smaller amounts of fructose than glucose and sucrose to be used to achieve a particular sweetness in most applications. On the basis of these properties, it does not appear that fructose is more fattening than other sugars. This also agrees with experimental studies of the relationship between fructose intake and obesity in animals [153, 154] and humans [65, 86, 106, 155].

7. Substrate Oxidation

It has been proposed that fructose can inhibit lipid oxidation [146]. For liver lipid oxidation this is logical, because the liver acquires energy from fructose and thus does not need to oxidize fat. Fructose can also increase DNL. It would not be expedient to form and break down fat simultaneously in the liver, and increased levels of malonyl-CoA (due to active DNL) will inhibit β-oxidation [41, 156]. It would also be logical for fructose to increase total body lipid oxidation more than glucose due to fructose's smaller contribution as an extrahepatic energy source. In some studies, however, fructose has been shown to increase the respiration quotient (RQ), the ratio of CO_2 exhaled to O_2 consumed, [157] more than glucose. This indicates that fructose to a higher degree than glucose reduces total body lipid oxidation and increases total body carbohydrate oxidation. Blaak and Saris [158] conducted a study in which participants ate 75 g fructose, starch, or glucose after a 12-hour fast in a crossover study. Fructose resulted in a significantly larger increase in the RQ, measured 6 hours after ingestion, than both glucose and starch. Tappy et al. [146] also showed a greater increase in RQ 4 hours after ingestion of 75 g fructose, compared with similar healthy participants eating 75 g glucose. Schwarz et al. [46] conducted a similar study (75 g fructose/glucose) with the same result. These results may be explained by the fact that fructose, more than glucose, enters DNL under specific conditions. Due to the fact that RQ varies for different substrates (e.g., 1 for carbohydrates and 0.7 for fats), RQ is used to determine the source of substrate oxidation [159]. However, during active DNL, CO_2 is produced without using O_2. Simultaneous occurrence of DNL and carbohydrate oxidation can lead to RQ values greater than 1 [159]. An increased RQ caused by DNL can, therefore, be misinterpreted as reduced lipid oxidation and increased carbohydrate oxidation. Such a misinterpretation may have occurred in the studies described above. Thus the effect of fructose on total body substrate oxidation remains unclear.

8. Discussion

The distribution of fructose into metabolic pathways, especially DNL, is of key importance to the health effects of fructose. The distribution varies with the amount of fructose consumed, the duration of fructose exposure, the composition of diet/meal, and whether the measurement took place postprandially, after absorption, or under fasting conditions. Individual physiological, enzymatic, and endocrine factors are also important. Diet composition and the amount of fructose eaten and absorbed will be the focus of this discussion.

Malabsorption of fructose will affect the amount of fructose absorbed and can thus be an important confounding factor in studies in which factors that affect absorption capacity have not been taken into account [39]. Truswell et al. [21] showed that intake of 50 g fructose led to malabsorption in over 50% of the study participants. The results of studies in which fructose is malabsorbed can thus be inaccurate due to individual differences in the absorption capacity of fructose. As the small intestine has a large absorption capacity for glucose and a limited one for fructose, it is problematic to compare fructose with glucose as the sole carbohydrate source. In future studies, this should be controlled for, for example, by using the hydrogen breath test to assess fructose malabsorption.

The composition of diet and especially the amount of glucose/starch may have influence on the health effects of fructose. As fructose is present with glucose in most food products, it is more practical and relevant to look at the effects of fructose and glucose together than the effects of fructose alone. A larger increase in DNL after eating fructose and glucose together (50 : 50 glucose : fructose) rather than the same amount as pure glucose has been shown [63]. Eating glucose with fructose is likely to affect fructose's health effects by stimulating the flow of fructose to DNL [50]. This effect could be due to both increased absorption capacity for fructose when coingested with glucose and therefore greater availability of fructose carbon atoms going towards DNL and increased blood insulin levels when glucose is present in the diet. Insulin stimulates DNL directly and indirectly by inhibiting other important metabolic pathways for fructose, such as gluconeogenesis. It is also plausible that glucose will compete with fructose as an energy source for enterocytes and hepatic cells, thus making more fructose available for DNL. Coingestion of glucose with fructose has been shown to decrease oxidation and gluconeogenesis from fructose carbon atoms [50]. Thus, it appears that the combination of fructose and glucose is particularly unfortunate, although fructose is not a prerequisite for DNL. It is unclear how much glucose/starch must be included with fructose for this effect to be significant.

The amount of fructose consumed seems to be of great importance to the effects of fructose on health. The negative health effects of fructose have mainly been demonstrated at high intakes, and several studies have found an average intake of fructose to cause no health problems [12, 110, 111, 160]. Although the average daily intake of fructose is 50–60 g/day, some of the population will consume larger amounts [161, 162], so the negative health effects of fructose may be relevant at least for a proportion of the population. Many of the current studies are poorly suited to determine health risks of fructose because (a) the fructose intakes are unrealistically high, (b) fructose is given in isolation and not mixed with other carbohydrates as in practice, and/or (c) the studies are conducted on animals. Differences between human and animal physiology limit their applicability to humans. There is a need for more human studies under conditions more similar to the way fructose is normally consumed. Such studies, particularly related to DNL, will be necessary to understand the effects of a normal fructose intake. The effects of fructose on triglyceride and cholesterol levels in the blood, fat accumulation in liver, and insulin signaling all seem to be linked to the extent to which fructose enters DNL. Surprisingly, there have been no satisfactory studies to assess the proportion of fructose that enters DNL at different levels of intake. Such studies should be carried out and should include intake of both pure fructose and fructose together with glucose.

It is also important to note that, despite the metabolic difference between glucose and fructose, glucose consumption far exceeds fructose consumption in the human diet [12]. This quantitative aspect must be considered when comparing the health effects of glucose and fructose.

9. Conclusion

Although there is a paucity of published literature regarding physiological effects of fructose in humans, current literature does not indicate that a normal consumption of fructose (approximately 50–60 g/day) increases the risk of atherosclerosis, type 2 diabetes, or obesity more than consumption of other sugars. However, a high intake of fructose, particularly if combined with a high energy intake in the form of glucose/starch, may have negative health effects via DNL. More studies are clearly needed, particularly studies under more realistic consumption levels of fructose.

Conflict of Interests

The authors declare that there is no conflict of interests regarding the publication of this paper.

References

[1] A. M. Cohen, A. Teitelbaum, M. Balogh, and J. J. Groen, "Effect of interchanging bread and sucrose as main source of carbohydrate in a low fat diet on the glucose tolerance curve

of healthy volunteer subjects," *The American Journal of Clinical Nutrition*, vol. 19, no. 1, pp. 59–62, 1966.

[2] L. H. Storlien, E. W. Kraegen, A. B. Jenkins, and D. J. Chisholm, "Effects of sucrose vs starch diets on *in vivo* insulin action, thermogenesis, and obesity in rats," *The American Journal of Clinical Nutrition*, vol. 47, no. 3, pp. 420–427, 1988.

[3] A. Giaccari, G. Sorice, and G. Muscogiuri, "Glucose toxicity: the leading actor in the pathogenesis and clinical history of type 2 diabetes—mechanisms and potentials for treatment," *Nutrition, Metabolism and Cardiovascular Diseases*, vol. 19, no. 5, pp. 365–377, 2009.

[4] M. R. Laughlin, "Normal roles for dietary fructose in carbohydrate metabolism," *Nutrients*, vol. 6, no. 8, pp. 3117–3129, 2014.

[5] V. S. Malik, B. M. Popkin, G. A. Bray, J.-P. Després, W. C. Willett, and F. B. Hu, "Sugar-sweetened beverages and risk of metabolic syndrome and type 2 diabetes: a meta-analysis," *Diabetes Care*, vol. 33, no. 11, pp. 2477–2483, 2010.

[6] R. Dhingra, L. Sullivan, P. F. Jacques et al., "Soft drink consumption and risk of developing cardiometabolic risk factors and the metabolic syndrome in middle-aged adults in the community," *Circulation*, vol. 116, no. 5, pp. 480–488, 2007.

[7] A. T. Høstmark, "The Oslo Health Study: soft drink intake is associated with the metabolic syndrome," *Applied Physiology, Nutrition and Metabolism*, vol. 35, no. 5, pp. 635–642, 2010.

[8] R. H. Lustig, "Fructose: metabolic, hedonic, and societal parallels with ethanol," *Journal of the American Dietetic Association*, vol. 110, no. 9, pp. 1307–1321, 2010.

[9] P. J. Havel, "Dietary fructose: implications for dysregulation of energy homeostasis and lipid/carbohydrate metabolism," *Nutrition Reviews*, vol. 63, no. 5, pp. 133–137, 2005.

[10] M. S. Segal, E. Gollub, and R. J. Johnson, "Is the fructose index more relevant with regards to cardiovascular disease than the glycemic index?" *European Journal of Nutrition*, vol. 46, no. 7, pp. 406–417, 2007.

[11] L. G. Sánchez-Lozada, M. Le, M. Segal, and R. J. Johnson, "How safe is fructose for persons with or without diabetes?" *The American Journal of Clinical Nutrition*, vol. 88, no. 5, pp. 1189–1190, 2008.

[12] J. S. White, "Challenging the fructose hypothesis: new perspectives on fructose consumption and metabolism," *Advances in Nutrition*, vol. 4, no. 2, pp. 246–256, 2013.

[13] B. P. Marriott, N. Cole, and E. Lee, "National estimates of dietary fructose intake increased from 1977 to 2004 in the United States," *The Journal of Nutrition*, vol. 139, no. 6, pp. S1228–S1235, 2009.

[14] Helsedirektoratet, "Utviklingen i norsk kosthold. Matforsyningsstatistikk," 2013, (Norwegian), https://helsedirektoratet.no/Lists/Publikasjoner/Attachments/370/Utviklingen-i-norsk-kosthold-2013-matforsyningsstatistikk-IS-2116.pdf.

[15] United States Department of Argiculture, *National Nutrient Database for Standard Reference*, United States Department of Argiculture, 2012, http://ndb.nal.usda.gov/ndb/search/list?fg=&man=&lfacet=&count=&max=&sort=&qlookup=&offset=&format=Full&new=.

[16] J. S. White, "Straight talk about high-fructose corn syrup: what it is and what it ain't," *The American Journal of Clinical Nutrition*, vol. 88, no. 6, pp. 1716S–1721S, 2008.

[17] L. Ferder, M. D. Ferder, and F. Inserra, "The role of high-fructose corn syrup in metabolic syndrome and hypertension," *Current Hypertension Reports*, vol. 12, no. 2, pp. 105–112, 2010.

[18] L. Tappy and K.-A. Le, "Metabolic effects of fructose and the worldwide increase in obesity," *Physiological Reviews*, vol. 90, no. 1, pp. 23–46, 2010.

[19] M. Madero, S. E. Perez-Pozo, D. Jalal, R. J. Johnson, and L. G. Sánchez-Lozada, "Dietary fructose and hypertension," *Current Hypertension Reports*, vol. 13, no. 1, pp. 29–35, 2011.

[20] H. F. Jones, R. N. Butler, and D. A. Brooks, "Intestinal fructose transport and malabsorption in humans," *The American Journal of Physiology—Gastrointestinal and Liver Physiology*, vol. 300, no. 2, pp. G202–G206, 2011.

[21] A. S. Truswell, J. M. Seach, and A. W. Thorburn, "Incomplete absorption of pure fructose in healthy subjects and the facilitating effect of glucose," *The American Journal of Clinical Nutrition*, vol. 48, no. 6, pp. 1424–1430, 1988.

[22] J. E. Riby, T. Fujisawa, and N. Kretchmer, "Fructose absorption," *American Journal of Clinical Nutrition*, vol. 58, no. 5, pp. S748–S753, 1993.

[23] C. P. Corpe, C. F. Burant, and J. H. Hoekstra, "Intestinal fructose absorption: clinical and molecular aspects," *Journal of Pediatric Gastroenterology and Nutrition*, vol. 28, no. 4, pp. 364–374, 1999.

[24] P. L. Beyer, E. M. Caviar, and R. W. McCallum, "Fructose intake at current levels in the United States may cause gastrointestinal distress in normal adults," *Journal of the American Dietetic Association*, vol. 105, no. 10, pp. 1559–1566, 2005.

[25] M. Pimentel, H. C. Lin, P. Enayati et al., "Methane, a gas produced by enteric bacteria, slows intestinal transit and augments small intestinal contractile activity," *American Journal of Physiology—Gastrointestinal and Liver Physiology*, vol. 290, no. 6, pp. G1089–G1095, 2006.

[26] J. M. W. Wong, R. de Souza, C. W. C. Kendall, A. Emam, and D. J. A. Jenkins, "Colonic health: fermentation and short chain fatty acids," *Journal of Clinical Gastroenterology*, vol. 40, no. 3, pp. 235–243, 2006.

[27] P. R. Gibson, E. Newnham, J. S. Barrett, S. J. Shepherd, and J. G. Muir, "Review article: fructose malabsorption and the bigger picture," *Alimentary Pharmacology & Therapeutics*, vol. 25, no. 4, pp. 349–363, 2007.

[28] J. L. Madsen, J. Linnet, and J. J. Rumessen, "Effect of nonabsorbed amounts of a fructose-sorbitol mixture on small intestinal transit in healthy volunteers," *Digestive Diseases and Sciences*, vol. 51, no. 1, pp. 147–153, 2006.

[29] J. Dyer, I. S. Wood, A. Palejwala, A. Ellis, and S. P. Shirazi-Beechey, "Expression of monosaccharide transporters in intestine of diabetic humans," *American Journal of Physiology: Gastrointestinal and Liver Physiology*, vol. 282, no. 2, pp. G241–G248, 2002.

[30] G. L. Kellett and E. Brot-Laroche, "Apical GLUT2—a major pathway of intestinal sugar absorption," *Diabetes*, vol. 54, no. 10, pp. 3056–3062, 2005.

[31] C. M. F. Kneepkens, R. J. Vonk, and J. Fernandes, "Incomplete intestinal absorption of fructose," *Archives of Disease in Childhood*, vol. 59, no. 8, pp. 735–738, 1984.

[32] V. Douard and R. P. Ferraris, "Regulation of the fructose transporter GLUT5 in health and disease," *American Journal of Physiology—Endocrinology and Metabolism*, vol. 295, no. 2, pp. E227–E237, 2008.

[33] G. A. Bray, "How bad is fructose?" *The American Journal of Clinical Nutrition*, vol. 86, no. 4, pp. 895–896, 2007.

[34] O. Bjorkman, M. Crump, and R. W. Phillips, "Intestinal metabolism of orally administered glucose and fructose in Yucatan miniature swine," *The Journal of Nutrition*, vol. 114, no. 8, pp. 1413–1420, 1984.

[35] O. Bjorkman and P. Felig, "Role of the kidney in the metabolism of fructose in 60-hour fasted humans," *Diabetes*, vol. 31, no. 6, pp. 516–520, 1982.

[36] G. J. Litherland, E. Hajduch, G. W. Gould, and H. S. Hundal, "Fructose transport and metabolism in adipose tissue of Zucker rats: diminished GLUT5 activity during obesity and insulin resistance," *Molecular and Cellular Biochemistry*, vol. 261, no. 1, pp. 23–33, 2004.

[37] P. Lam, K. Ng, K. L. Stanhope et al., "Effects of consuming dietary fructose versus glucose on de novo lipogenesis in overweight and obese human subjects," *Berkeley Scientific Journal*, vol. 15, no. 2, 2011.

[38] L. Tappy and K.-A. Lê, "Does fructose consumption contribute to non-alcoholic fatty liver disease?" *Clinics and Research in Hepatology and Gastroenterology*, vol. 36, no. 6, pp. 554–560, 2012.

[39] K. L. Stanhope, "Role of fructose-containing sugars in the epidemics of obesity and metabolic syndrome," *Annual Review of Medicine*, vol. 63, pp. 329–343, 2012.

[40] V. T. Samuel, "Fructose induced lipogenesis: from sugar to fat to insulin resistance," *Trends in Endocrinology and Metabolism*, vol. 22, no. 2, pp. 60–65, 2011.

[41] P. A. Mayes, "Intermediary metabolism of fructose," *American Journal of Clinical Nutrition*, vol. 58, no. 5, pp. S754–S765, 1993.

[42] J. Delarue, S. Normand, C. Pachiaudi, M. Beylot, F. Lamisse, and J. P. Riou, "The contribution of naturally labelled 13C fructose to glucose appearance in humans," *Diabetologia*, vol. 36, no. 4, pp. 338–345, 1993.

[43] S. Z. Sun and M. W. Empie, "Fructose metabolism in humans—what isotopic tracer studies tell us," *Nutrition& Metabolism*, vol. 9, article 89, 2012.

[44] L. H. Smith Jr., R. H. Ettinger, and D. Seligson, "A comparison of the metabolism of fructose and glucose in hepatic," *The Journal of Clinical Investigation*, vol. 32, no. 4, pp. 273–282, 1953.

[45] H. Sahebjami and R. Scalettar, "Effects of fructose infusion on lactate and uric acid metabolism," *The Lancet*, vol. 297, no. 7695, pp. 366–369, 1971.

[46] J.-M. Schwarz, Y. Schutz, F. Froidevaux et al., "Thermogenesis in men and women induced by fructose vs glucose added to a meal," *American Journal of Clinical Nutrition*, vol. 49, no. 4, pp. 667–674, 1989.

[47] C. Couchepin, L. Ê. Kim-Anne, M. Bortolotti et al., "Markedly blunted metabolic effects of fructose in healthy young female subjects compared with male subjects," *Diabetes Care*, vol. 31, no. 6, pp. 1254–1256, 2008.

[48] V. Lecoultre, R. Benoit, G. Carrel et al., "Fructose and glucose co-ingestion during prolonged exercise increases lactate and glucose fluxes and oxidation compared with an equimolar intake of glucose," *The American Journal of Clinical Nutrition*, vol. 92, no. 5, pp. 1071–1079, 2010.

[49] J. L. Kelsay, K. M. Behall, and W. M. Clark, "Glucose, fructose, lactate and pyruvate in blood, and lactate and pyruvate in parotid saliva in response to sugars with and without other foods," *American Journal of Clinical Nutrition*, vol. 27, no. 8, pp. 819–825, 1974.

[50] F. Theytaz, S. de Giorgi, L. Hodson et al., "Metabolic fate of fructose ingested with and without glucose in a mixed meal," *Nutrients*, vol. 6, no. 7, pp. 2632–2649, 2014.

[51] K. L. Teff, S. S. Elliott, M. Tschöp et al., "Dietary fructose reduces circulating insulin and leptin, attenuates postprandial suppression of ghrelin, and increases triglycerides in women,"

The Journal of Clinical Endocrinology & Metabolism, vol. 89, no. 6, pp. 2963–2972, 2004.

[52] J. Girard, "The inhibitory effects of insulin on hepatic glucose production are both direct and indirect," *Diabetes*, vol. 55, no. 2, pp. S65–S69, 2006.

[53] G. Boden, S. Salehi, P. Cheung et al., "Comparison of in vivo effects of insulin on SREBP-1c activation and INSIG-1/2 in rat liver and human and rat adipose tissue," *Obesity*, vol. 21, no. 6, pp. 1208–1214, 2013.

[54] M. K. Hellerstein, J.-M. Schwarz, and R. A. Neese, "Regulation of hepatic de novo lipogenesis in humans," *Annual Review of Nutrition*, vol. 16, pp. 523–557, 1996.

[55] E. Shrago, J. A. Glennon, and E. S. Gordon, "Comparative aspects of lipogenesis in mammalian tissues," *Metabolism: Clinical and Experimental*, vol. 20, no. 1, pp. 54–62, 1971.

[56] Z. K. Guo, L. K. Cella, C. Baum, E. Ravussin, and D. A. Schoeller, "De nova lipogenesis in adipose tissue of lean and obese women: application of deuterated water and isotope ratio mass spectrometry," *International Journal of Obesity*, vol. 24, no. 7, pp. 932–937, 2000.

[57] F. Diraison, V. Yankah, D. Letexier, E. Dusserre, P. Jones, and M. Beylot, "Differences in the regulation of adipose tissue and liver lipogenesis by carbohydrates in humans," *Journal of Lipid Research*, vol. 44, no. 4, pp. 846–853, 2003.

[58] A. Strawford, F. Antelo, M. Christiansen, and M. K. Hellerstein, "Adipose tissue triglyceride turnover, de novo lipogenesis, and cell proliferation in humans measured with ^2H_2O," *American Journal of Physiology—Endocrinology & Metabolism*, vol. 286, no. 4, pp. E577–E588, 2004.

[59] C. Chascione, D. H. Elwyn, M. Davila, K. M. Gil, J. Askanazi, and J. M. Kinney, "Effect of carbohydrate intake on de novo lipogenesis in human adipose tissue," *American Journal of Physiology: Endocrinology and Metabolism*, vol. 253, no. 6, pp. E664–E669, 1987.

[60] A. Aarsland, D. Chinkes, and R. R. Wolfe, "Hepatic and whole-body fat synthesis in humans during carbohydrate overfeeding," *The American Journal of Clinical Nutrition*, vol. 65, no. 6, pp. 1774–1782, 1997.

[61] R. J. Stubbs, N. Mazlan, and S. Whybrow, "Carbohydrates, appetite and feeding behavior in humans," *The Journal of Nutrition*, vol. 131, no. 10, pp. 2775S–2781S, 2001.

[62] K. J. Acheson, Y. Schutz, T. Bessard, K. Anantharaman, J.-P. Flatt, and E. Jequier, "Glycoprotein storage capacity and de novo lipogenesis during massive carbohydrate overfeeding in man," *American Journal of Clinical Nutrition*, vol. 48, no. 2, pp. 240–247, 1988.

[63] E. J. Parks, L. E. Skokan, M. T. Timlin, and C. S. Dingfelder, "Dietary sugars stimulate fatty acid synthesis in adults," *Journal of Nutrition*, vol. 138, no. 6, pp. 1039–1046, 2008.

[64] O.-J. Park, D. Cesar, D. Faix, K. Wu, C. H. L. Shackleton, and M. K. Hellerstein, "Mechanisms of fructose-induced hyper-triglyceridaemia in the rat. Activation of hepatic pyruvate dehydrogenase through inhibition of pyruvate dehydrogenase kinase," *Biochemical Journal*, vol. 282, no. 3, pp. 753–757, 1992.

[65] K. L. Stanhope, J. M. Schwarz, N. L. Keim et al., "Consuming fructose-sweetened, not glucose-sweetened, beverages increases visceral adiposity and lipids and decreases insulin sensitivity in overweight/obese humans," *The Journal of Clinical Investigation*, vol. 119, no. 5, pp. 1322–1334, 2009.

[66] R. Crescenzo, F. Bianco, I. Falcone, P. Coppola, G. Liverini, and S. Iossa, "Increased hepatic de novo lipogenesis and

mitochondrial efficiency in a model of obesity induced by diets rich in fructose," *European Journal of Nutrition*, vol. 52, no. 2, pp. 537–545, 2013.

[67] T. J. Carden and T. P. Carr, "Food availability of glucose and fat, but not fructose, increased in the US between 1970 and 2009: analysis of the USDA food availability data system," *Nutrition Journal*, vol. 12, article 130, 2013.

[68] K. M. Hirahatake, J. K. Meissen, O. Fiehn, and S. H. Adams, "Comparative effects of fructose and glucose on lipogenic gene expression and intermediary metabolism in HepG2 liver cells," *PLoS ONE*, vol. 6, no. 11, Article ID e26583, 2011.

[69] J. R. Dushay, E. Toschi, E. K. Mitten, F. M. Fisher, M. A. Herman, and E. Maratos-Flier, "Fructose ingestion acutely stimulates circulating FGF21 levels in humans," *Molecular Metabolism*, vol. 4, no. 1, pp. 51–57, 2014.

[70] T. Bobbert, F. Schwarz, A. Fischer-Rosinsky et al., "Fibroblast growth factor 21 predicts the metabolic syndrome and type 2 diabetes in Caucasians," *Diabetes Care*, vol. 36, no. 1, pp. 145–149, 2013.

[71] D. Faeh, K. Minehira, J.-M. Schwarz, R. Periasami, P. Seongsu, and L. Tappy, "Effect of fructose overfeeding and fish oil administration on hepatic de novo lipogenesis and insulin sensitivity in healthy men," *Diabetes*, vol. 54, no. 7, pp. 1907–1913, 2005.

[72] V. Ha, J. L. Sievenpiper, R. J. De Souza et al., "Effect of fructose on blood pressure: a systematic review and meta-analysis of controlled feeding trials," *Hypertension*, vol. 59, no. 4, pp. 787–795, 2012.

[73] M. F.-F. Chong, B. A. Fielding, and K. N. Frayn, "Mechanisms for the acute effect of fructose on postprandial lipemia," *The American Journal of Clinical Nutrition*, vol. 85, no. 6, pp. 1511–1520, 2007.

[74] A. Vedala, W. Wang, R. A. Neese, M. P. Christiansen, and M. K. Hellerstein, "Delayed secretory pathway contributions to VLDL-triglycerides from plasma NEFA, diet, and de novo lipogenesis in humans," *Journal of Lipid Research*, vol. 47, no. 11, pp. 2562–2574, 2006.

[75] K. L. Stanhope and P. J. Havel, "Fructose consumption: potential mechanisms for its effects to increase visceral adiposity and induce dyslipidemia and insulin resistance," *Current Opinion in Lipidology*, vol. 19, no. 1, pp. 16–24, 2008.

[76] I. Marques-Lopes, D. Ansorena, I. Astiasaran, L. Forga, and J. A. Martínez, "Postprandial de novo lipogenesis and metabolic changes induced by a high-carbohydrate, low-fat meal in lean and overweight men," *The American Journal of Clinical Nutrition*, vol. 73, no. 2, pp. 253–261, 2001.

[77] J. D. Horton, J. L. Goldstein, and M. S. Brown, "SREBPs: activators of the complete program of cholesterol and fatty acid synthesis in the liver," *Journal of Clinical Investigation*, vol. 109, no. 9, pp. 1125–1131, 2002.

[78] C. Tran, D. Jacot-Descombes, V. Lecoultre et al., "Sex differences in lipid and glucose kinetics after ingestion of an acute oral fructose load," *British Journal of Nutrition*, vol. 104, no. 8, pp. 1139–1147, 2010.

[79] L. C. Hudgins, M. K. Hellerstein, C. E. Seidman, R. A. Neese, J. D. Tremaroli, and J. Hirsch, "Relationship between carbohydrate-induced hypertriglyceridemia and fatty acid synthesis in lean and obese subjects," *Journal of Lipid Research*, vol. 41, no. 4, pp. 595–604, 2000.

[80] M. T. Timlin and E. J. Parks, "Temporal pattern of de novo lipogenesis in the postprandial state in healthy men," *American Journal of Clinical Nutrition*, vol. 81, no. 1, pp. 35–42, 2005.

[81] K.-A. Le, M. Ith, R. Kreis et al., "Fructose overconsumption causes dyslipidemia and ectopic lipid deposition in healthy subjects with and without a family history of type 2 diabetes," *The American Journal of Clinical Nutrition*, vol. 89, no. 6, pp. 1760–1765, 2009.

[82] R. J. L. Allen and J. S. Leahy, "Some effects of dietary dextrose, fructose, liquid glucose and sucrose in the adult male rat," *British Journal of Nutrition*, vol. 20, no. 2, pp. 339–347, 1966.

[83] I. Bergheim, S. Weber, M. Vos et al., "Antibiotics protect against fructose-induced hepatic lipid accumulation in mice: role of endotoxin," *Journal of Hepatology*, vol. 48, no. 6, pp. 983–992, 2008.

[84] K.-A. Lê, D. Faeh, R. Stettler et al., "A 4-wk high-fructose diet alters lipid metabolism without affecting insulin sensitivity or ectopic lipids in healthy humans," *American Journal of Clinical Nutrition*, vol. 84, no. 6, pp. 1374–1379, 2006.

[85] G. Silbernagel, J. MacHann, S. Unmuth et al., "Effects of 4-week very-high-fructose/glucose diets on insulin sensitivity, visceral fat and intrahepatic lipids: an exploratory trial," *British Journal of Nutrition*, vol. 106, no. 1, pp. 79–86, 2011.

[86] J. L. Sievenpiper, R. J. de Souza, A. I. Cozma, L. Chiavaroli, V. Ha, and A. Mirrahimi, "Fructose vs. glucose and metabolism: do the metabolic differences matter?" *Current Opinion in Lipidology*, vol. 25, no. 1, pp. 8–19, 2014.

[87] S. Chiu, J. L. Sievenpiper, R. J. de Souza et al., "Effect of fructose on markers of non-alcoholic fatty liver disease (NAFLD): a systematic review and meta-analysis of controlled feeding trials," *European Journal of Clinical Nutrition*, vol. 68, no. 4, pp. 416–423, 2014.

[88] S. Bravo, J. Lowndes, S. Sinnett, Z. Yu, and J. Rippe, "Consumption of sucrose and high-fructose corn syrup does not increase liver fat or ectopic fat deposition in muscles," *Applied Physiology, Nutrition and Metabolism*, vol. 38, no. 6, pp. 681–688, 2013.

[89] J. M. Rippe and T. J. Angelopoulos, "Sucrose, high-fructose corn syrup, and fructose, their metabolism and potential health effects: what do we really know?" *Advances in Nutrition*, vol. 4, no. 2, pp. 236–245, 2013.

[90] M. Maersk, A. Belza, H. Stødkilde-Jørgensen et al., "Sucrose-sweetened beverages increase fat storage in the liver, muscle, and visceral fat depot: a 6-mo randomized intervention study," *The American Journal of Clinical Nutrition*, vol. 95, no. 2, pp. 283–289, 2012.

[91] R. D. Johnston, M. C. Stephenson, H. Crossland et al., "No difference between high-fructose and high-glucose diets on liver triacylglycerol or biochemistry in healthy overweight men," *Gastroenterology*, vol. 145, no. 5, pp. 1016–1025, 2013.

[92] K. A. Page, O. Chan, J. Arora et al., "Effects of fructose vs glucose on regional cerebral blood flow in brain regions involved with appetite and reward pathways," *The Journal of the American Medical Association*, vol. 309, no. 1, pp. 63–70, 2013.

[93] J. B. Moore, P. J. Gunn, and B. A. Fielding, "The role of dietary sugars and de novo lipogenesis in non-alcoholic fatty liver disease," *Nutrients*, vol. 6, no. 12, pp. 5679–5703, 2014.

[94] S. W. Rizkalla, "Health implications of fructose consumption: a review of recent data," *Nutrition & Metabolism*, vol. 7, article 82, 2010.

[95] A. Aarsland and R. R. Wolfe, "Hepatic secretion of VLDL fatty acids during stimulated lipogenesis in men," *Journal of Lipid Research*, vol. 39, no. 6, pp. 1280–1286, 1998.

[96] W. E. Connor, "Harbingers of coronary heart disease: dietary saturated fatty acids and cholesterol. Is chocolate benign

because of its stearic acid content?" *The American Journal of Clinical Nutrition*, vol. 70, no. 6, pp. 951–952, 1999.

[97] J. C. Cohen and R. Schall, "Reassessing the effects of simple carbohydrates on the serum triglyceride responses to fat meals," *American Journal of Clinical Nutrition*, vol. 48, no. 4, pp. 1031–1034, 1988.

[98] K. L. Teff, J. Grudziak, R. R. Townsend et al., "Endocrine and metabolic effects of consuming fructose- and glucose-sweetened beverages with meals in obese men and women: influence of insulin resistance on plasma triglyceride responses," *The Journal of Clinical Endocrinology & Metabolism*, vol. 94, no. 5, pp. 1562–1569, 2009.

[99] J. Hallfrisch, S. Reiser, and E. S. Prather, "Blood lipid distribution of hyperinsulinemic men consuming three levels of fructose," *The American Journal of Clinical Nutrition*, vol. 37, no. 5, pp. 740–748, 1983.

[100] S. Reiser, A. S. Powell, D. J. Scholfield, P. Panda, K. C. Ellwood, and J. J. Canary, "Blood lipids, lipoproteins, apoproteins, and uric acid in men fed diets containing fructose or high-amylose cornstarch," *The American Journal of Clinical Nutrition*, vol. 49, no. 5, pp. 832–839, 1989.

[101] J. P. Bantle, J. E. Swanson, W. Thomas, and D. C. Laine, "Metabolic effects of dietary fructose in diabetic subjects," *Diabetes Care*, vol. 15, no. 11, pp. 1468–1476, 1992.

[102] J. E. Swanson, D. C. Laine, W. Thomas, and J. P. Bantle, "Metabolic effects of dietary fructose in healthy subjects," *American Journal of Clinical Nutrition*, vol. 55, no. 4, pp. 851–856, 1992.

[103] K. L. Stanhope, A. A. Bremer, V. Medici et al., "Consumption of fructose and high fructose corn syrup increase postprandial triglycerides, LDL-cholesterol, and apolipoprotein-B in young men and women," *Journal of Clinical Endocrinology and Metabolism*, vol. 96, no. 10, pp. E1596–E1605, 2011.

[104] I. Aeberli, P. A. Gerber, M. Hochuli et al., "Low to moderate sugar-sweetened beverage consumption impairs glucose and lipid metabolism and promotes inflammation in healthy young men: a randomized controlled trial," *The American Journal of Clinical Nutrition*, vol. 94, no. 2, pp. 479–485, 2011.

[105] M. R. Diffenderfer and E. J. Schaefer, "The composition and metabolism of large and small LDL," *Current Opinion in Lipidology*, vol. 25, no. 3, pp. 221–226, 2014.

[106] G. Livesey and R. Taylor, "Fructose consumption and consequences for glycation, plasma triacylglycerol, and body weight: meta-analyses and meta-regression models of intervention studies," *The American Journal of Clinical Nutrition*, vol. 88, no. 5, pp. 1419–1437, 2008.

[107] M. M. Swarbrick, K. L. Stanhope, S. S. Elliott et al., "Consumption of fructose-sweetened beverages for 10 weeks increases postprandial triacylglycerol and apolipoprotein-B concentrations in overweight and obese women," *British Journal of Nutrition*, vol. 100, no. 5, pp. 947–952, 2008.

[108] D. David Wang, J. L. Sievenpiper, R. J. De Souza et al., "Effect of fructose on postprandial triglycerides: a systematic review and meta-analysis of controlled feeding trials," *Atherosclerosis*, vol. 232, no. 1, pp. 125–133, 2014.

[109] I. Aeberli, M. Hochuli, P. A. Gerber et al., "Moderate amounts of fructose consumption impair insulin sensitivity in healthy young men: a randomized controlled trial," *Diabetes Care*, vol. 36, no. 1, pp. 150–156, 2013.

[110] J. Lowndes, S. Sinnett, S. Pardo et al., "The effect of normally consumed amounts of sucrose or high fructose corn syrup on lipid profiles, body composition and related parameters in overweight/obese subjects," *Nutrients*, vol. 6, no. 3, pp. 1128–1144, 2014.

[111] J. M. Rippe, "The metabolic and endocrine response and health implications of consuming sugar-sweetened beverages: findings from recent randomized controlled trials," *Advances in Nutrition*, vol. 4, no. 6, pp. 677–686, 2013.

[112] M. Hashemi, M. Yavari, N. Amiri et al., "Uric acid: a risk factor for coronary atherosclerosis?" *Cardiovascular Journal of South Africa*, vol. 18, no. 1, pp. 16–19, 2007.

[113] T. C. Rodrigues, D. M. Maahs, R. J. Johnson et al., "Serum uric acid predicts progression of subclinical coronary atherosclerosis in individuals without renal disease," *Diabetes Care*, vol. 33, no. 11, pp. 2471–2473, 2010.

[114] D. I. Feig, D.-H. Kang, and R. J. Johnson, "Uric acid and cardiovascular risk," *The New England Journal of Medicine*, vol. 359, no. 17, pp. 1811–1821, 2008.

[115] C. L. Cox, K. L. Stanhope, J. M. Schwarz et al., "Consumption of fructose- but not glucose-sweetened beverages for 10 weeks increases circulating concentrations of uric acid, retinol binding protein-4, and gamma-glutamyl transferase activity in overweight/obese humans," *Nutrition & Metabolism*, vol. 9, article 68, 2012.

[116] D. D. Wang, J. L. Sievenpiper, R. J. de Souza et al., "The effects of fructose intake on serum uric acid vary among controlled dietary trials," *The Journal of Nutrition*, vol. 142, no. 5, pp. 916–923, 2012.

[117] J. Perheentupa and K. Raivio, "Fructose-induced hyperuricaemia," *The Lancet*, vol. 2, no. 7515, pp. 528–531, 1967.

[118] C. M. Brown, A. G. Dulloo, G. Yepuri, and J.-P. Montani, "Fructose ingestion acutely elevates blood pressure in healthy young humans," *The American Journal of Physiology—Regulatory Integrative and Comparative Physiology*, vol. 294, no. 3, pp. R730–R737, 2008.

[119] S. E. Perez-Pozo, J. Schold, T. Nakagawa, L. G. Sánchez-Lozada, R. J. Johnson, and J. L. Lillo, "Excessive fructose intake induces the features of metabolic syndrome in healthy adult men: role of uric acid in the hypertensive response," *International Journal of Obesity*, vol. 34, no. 3, pp. 454–461, 2010.

[120] V. H. Jayalath, J. L. Sievenpiper, R. J. de Souza et al., "Total fructose intake and risk of hypertension: a systematic review and meta-analysis of prospective cohorts," *Journal of the American College of Nutrition*, vol. 33, no. 4, pp. 328–339, 2014.

[121] V. Ha, V. H. Jayalath, A. I. Cozma, A. Mirrahimi, R. J. de Souza, and J. L. Sievenpiper, "Fructose-containing sugars, blood pressure, and cardiometabolic risk: a critical review," *Current Hypertension Reports*, vol. 15, no. 4, pp. 281–297, 2013.

[122] M. Wang, M. Yu, L. Fang, and R. Hu, "Association between sugar-sweetened beverages and type 2 diabetes: a meta-analysis," *Journal of Diabetes Investigation*, vol. 6, no. 3, pp. 360–366, 2015.

[123] M.-F. Kong, I. Chapman, E. Goble et al., "Effects of oral fructose and glucose on plasma GLP-1 and appetite in normal subjects," *Peptides*, vol. 20, no. 5, pp. 545–551, 1999.

[124] A. M. Grant, M. R. Christie, and S. J. H. Ashcroft, "Insulin release from human pancreatic islets in vitro," *Diabetologia*, vol. 19, no. 2, pp. 114–117, 1980.

[125] D. L. Curry, "Effects of mannose and fructose on the synthesis and secretion of insulin," *Pancreas*, vol. 4, no. 1, pp. 2–9, 1989.

[126] A. I. Cozma, J. L. Sievenpiper, R. J. de Souza et al., "Effect of fructose on glycemic control in diabetes: a systematic review

and meta-analysis of controlled feeding trials," *Diabetes Care*, vol. 35, no. 7, pp. 1611–1620, 2012.

[127] J. L. Sievenpiper, L. Chiavaroli, R. J. De Souza et al., "'Catalytic' doses of fructose may benefit glycaemic control without harming cardiometabolic risk factors: A small meta-analysis of randomised controlled feeding trials," *British Journal of Nutrition*, vol. 108, no. 3, pp. 418–423, 2012.

[128] I. S. Hwang, H. Ho, B. B. Hoffman, and G. M. Reaven, "Fructose-induced insulin resistance and hypertension in rats," *Hypertension*, vol. 10, no. 5, pp. 512–516, 1987.

[129] Y.-J. Huang, V. S. Fang, C.-C. Juan, Y.-C. Chou, C.-F. Kwok, and L.-T. Ho, "Amelioration of insulin resistance and hypertension in a fructose-fed rat model with fish oil supplementation," *Metabolism: Clinical and Experimental*, vol. 46, no. 11, pp. 1252–1258, 1997.

[130] P. A. Crapo and O. G. Kolterman, "The metabolic effects of 2-week fructose feeding in normal subjects," *The American Journal of Clinical Nutrition*, vol. 39, no. 4, pp. 525–534, 1984.

[131] G. Grigoresco, S. W. Rizkalla, P. Halfon et al., "Lack of detectable deleterious effects on metabolic control of daily fructose ingestion for 2-mo in NIDDM patients," *Diabetes Care*, vol. 11, no. 7, pp. 546–550, 1988.

[132] A. L. Sunehag, G. Toffolo, M. S. Treuth et al., "Effects of dietary macronutrient content on glucose metabolism in children," *Journal of Clinical Endocrinology & Metabolism*, vol. 87, no. 11, pp. 5168–5178, 2002.

[133] R. D. Feinman and E. J. Fine, "Fructose in perspective," *Nutrition & Metabolism*, vol. 10, no. 1, article 45, 2013.

[134] H. Beck-Nielsen, O. Pedersen, and H. O. Lindskov, "Impaired cellular insulin binding and insulin sensitivity induced by high-fructose feeding in normal subjects," *The American Journal of Clinical Nutrition*, vol. 33, no. 2, pp. 273–278, 1980.

[135] J. Hallfrisch, K. C. Ellwood, O. E. Michaelis, S. Reiser, T. M. O'Dorisio, and E. S. Prather, "Effects of dietary fructose on plasma-glucose and hormone responses in normal and hyperinsulinemic men," *The Journal of Nutrition*, vol. 113, no. 9, pp. 1819–1826, 1983.

[136] K. C. Eiffert, R. B. McDonald, and J. S. Stern, "High sucrose diet and exercise: effects on insulin-receptor function of 12- and 24-mo-old Sprague-Dawley rats," *Journal of Nutrition*, vol. 121, no. 7, pp. 1081–1089, 1991.

[137] Y. Wei and M. J. Pagliassotti, "Hepatospecific effects of fructose on c-jun NH2-terminal kinase: implications for hepatic insulin resistance," *American Journal of Physiology—Endocrinology and Metabolism*, vol. 287, no. 5, pp. E926–E933, 2004.

[138] Y. Nagai, S. Yonemitsu, D. M. Erion et al., "The role of peroxisome proliferator-activated receptor gamma coactivator-1 β in the pathogenesis of fructose-induced insulin resistance," *Cell Metabolism*, vol. 9, no. 3, pp. 252–264, 2009.

[139] S. Delbosc, E. Paizanis, R. Magous et al., "Involvement of oxidative stress and NADPH oxidase activation in the development of cardiovascular complications in a model of insulin resistance, the fructose-fed rat," *Atherosclerosis*, vol. 179, no. 1, pp. 43–49, 2005.

[140] K. L. Stanhope, S. C. Griffen, B. R. Bair, M. M. Swarbrick, N. L. Keim, and P. J. Havel, "Twenty-four-hour endocrine and metabolic profiles following consumption of high-fructose corn syrup-, sucrose-, fructose-, and glucose-sweetened beverages with meals," *The American Journal of Clinical Nutrition*, vol. 87, no. 5, pp. 1194–1203, 2008.

[141] L. T. Tran, V. G. Yuen, and J. H. McNeill, "The fructose-fed rat: a review on the mechanisms of fructose-induced

insulin resistance and hypertension," *Molecular and Cellular Biochemistry*, vol. 332, no. 1-2, pp. 145–159, 2009.

[142] J. Rodin, D. Reed, and L. Jamner, "Metabolic effects of fructose and glucose: implications for food intake," *American Journal of Clinical Nutrition*, vol. 47, no. 4, pp. 683–689, 1988.

[143] J. Rodin, "Comparative effects of fructose, aspartame, glucose, and water preloads on calorie and macronutrient intake," *The American Journal of Clinical Nutrition*, vol. 51, no. 3, pp. 428–435, 1990.

[144] L. Spitzer and J. Rodin, "Effects of fructose and glucose preloads on subsequent food-intake," *Appetite*, vol. 8, no. 2, pp. 135–145, 1987.

[145] Z. S. Warwick and H. P. Weingarten, "Dynamics of intake suppression after a preload: role of calories, volume, and macronutrients," *The American Journal of Physiology—Regulatory Integrative and Comparative Physiology*, vol. 266, no. 4, pp. R1314–R1318, 1994.

[146] L. Tappy, J.-P. Randin, J.-P. Felber et al., "Comparison of thermogenic effect of fructose and glucose in normal humans," *The American Journal of Physiology—Endocrinology and Metabolism*, vol. 250, no. 6, pp. E718–E724, 1986.

[147] D. C. Simonson, L. Tappy, E. Jequier, J.-P. Felber, and R. A. DeFronzo, "Normalization of carbohydrate-induced thermogenesis by fructose in insulin-resistant states," *The American Journal of Physiology—Endocrinology and Metabolism*, vol. 254, no. 2, pp. E201–E207, 1988.

[148] J.-M. Schwarz, K. J. Acheson, L. Tappy et al., "Thermogenesis and fructose metabolism in humans," *American Journal of Physiology: Endocrinology and Metabolism*, vol. 262, no. 5, pp. E591–E598, 1992.

[149] J. Hugenholtz, "The lactic acid bacterium as a cell factory for food ingredient production," *International Dairy Journal*, vol. 18, no. 5, pp. 466–475, 2008.

[150] M.-J. Gwak, S.-J. Chung, Y. J. Kim, and C. S. Lim, "Relative sweetness and sensory characteristics of bulk and intense sweeteners," *Food Science and Biotechnology*, vol. 21, no. 3, pp. 889–894, 2012.

[151] R. S. Shallenberger, "Intrinsic chemistry of fructose," *Pure and Applied Chemistry*, vol. 50, no. 11-12, pp. 1409–1420, 1978.

[152] R. E. Wrolstad, *Food Carbohydrate Chemistry*, John Wiley & Sons, New York, NY, USA, 1st edition, 2012.

[153] S. R. Blakely, J. Hallfrisch, S. Reiser, and E. S. Prather, "Long-term effects of moderate fructose feeding on glucose-tolerance parameters in rats," *The Journal of Nutrition*, vol. 111, no. 2, pp. 307–314, 1981.

[154] K. L. Stanhope and P. J. Havel, "Endocrine and metabolic effects of consuming beverages sweetened with fructose, glucose, sucrose, or high-fructose corn syrup," *The American Journal of Clinical Nutrition*, vol. 88, no. 6, pp. 1733S–1737S, 2008.

[155] J. L. Sievenpiper, R. J. de Souza, A. Mirrahimi et al., "Effect of fructose on body weight in controlled feeding trials: a systematic review and meta-analysis," *Annals of Internal Medicine*, vol. 156, no. 4, pp. 291–304, 2012.

[156] J.-M. Schwarz, P. Linfoot, D. Dare, and K. Aghajanian, "Hepatic de novo lipogenesis in normoinsulinemic and hyperinsulinemic subjects consuming high-fat, low-carbohydrate and low-fat, high-carbohydrate isoenergetic diets," *The American Journal of Clinical Nutrition*, vol. 77, no. 1, pp. 43–50, 2003.

[157] K. N. Frayn, P. Lund, and M. Walker, "Interpretation of oxygen and carbon dioxide exchange across tissue beds in vivo," *Clinical Science*, vol. 85, no. 4, pp. 373–384, 1993.

[158] E. E. Blaak and W. H. M. Saris, "Postprandial thermogenesis and substrate utilization after ingestion of different dietary carbohydrates," *Metabolism: Clinical and Experimental*, vol. 45, no. 10, pp. 1235–1242, 1996.

[159] E. Ferrannini, "The theoretical bases of indirect calorimetry: a review," *Metabolism: Clinical and Experimental*, vol. 37, no. 3, pp. 287–301, 1988.

[160] L. Tappy and B. Mittendorfer, "Fructose toxicity: is the science ready for public health actions?" *Current Opinion in Clinical Nutrition and Metabolic Care*, vol. 15, no. 4, pp. 357–361, 2012.

[161] M. B. Vos, J. E. Kimmons, C. Gillespie, J. Welsh, and H. M. Blank, "Dietary fructose consumption among US children and adults: the Third National Health and Nutrition Examination Survey," *The Medscape Journal of Medicine*, vol. 10, no. 7, article 160, 2008.

[162] M. Sland, M. Haugen, F.-L. Eriksen et al., "High sugar consumption and poor nutrient intake among drug addicts in Oslo, Norway," *British Journal of Nutrition*, vol. 105, no. 4, pp. 618–624, 2011.

15

Breakfast Protein Source Does Not Influence Postprandial Appetite Response and Food Intake in Normal Weight and Overweight Young Women

Christina M. Crowder, Brianna L. Neumann, and Jamie I. Baum

Department of Food Science, University of Arkansas, 2650 North Young Avenue, Fayetteville, AR 72704, USA

Correspondence should be addressed to Jamie I. Baum; baum@uark.edu

Academic Editor: Pedro Moreira

Breakfasts higher in protein lead to a greater reduction in hunger compared to breakfasts higher in carbohydrate. However, few studies have examined the impact of higher protein breakfasts with differing protein sources. Our objective was to determine if protein source (animal protein (AP) versus plant protein (PP)) influences postprandial metabolic response in participants consuming a high protein breakfast (~30% energy from protein). Normal weight (NW; $n = 12$) and overweight women (OW; $n = 8$) aging 18–36 were recruited to participate. Participants completed two visits in a randomized, cross-over design with one week between visits. Subjects had 15 minutes to consume each breakfast. Blood glucose and appetite were assessed at baseline, 15, 30, 45, 60, and 120 minutes postprandial. Participants kept a 24-hour dietary record for the duration of each test day. No difference was found between NW and OW participants or breakfasts for postprandial appetite responses. AP had a significantly lower glucose response at 30 minutes compared with PP (-11.6%; 127 ± 4 versus 112 ± 4 mg/dL; $P < 0.05$) and a slower return to baseline. There was no difference in daily energy intake between breakfasts. These data suggest that protein source may influence postprandial glucose response without significantly impacting appetite response in breakfast consumers.

1. Introduction

Early adulthood is a vulnerable life stage for weight gain, especially among women. The average weight gain for women between the ages of twenty and thirty is 12–25 lbs [1]. Weight gain during early adulthood increases the risk of developing a number of chronic health conditions such as type 2 diabetes mellitus, osteoarthritis, and some cancers [2, 3]. For example, after the age of eighteen years, women are 1.9 times more likely to develop type 2 diabetes if body weight increased 10–16 pounds and were 2.7 times more likely to develop type 2 diabetes if body weight increased 16–22 pounds [1].

Breakfast is often cited as the most important meal of the day for children, but this is also true for adults. There are many benefits associated with eating a healthy breakfast including improved micronutrient intake, decreased incidence of overweight and obesity, and lower cholesterol levels [4–7]. Several studies, in both adults and children, have shown that individuals who eat breakfast tend to weigh less than those who omit breakfast as eating a healthy breakfast can reduce hunger throughout the day [8, 9]. Consuming more protein (20–30 g) at breakfast than found in the standard cereal-based breakfast (10–15 g) may increase subjective feeling of fullness and satiety throughout the day [10, 11] and decrease calorie intake at lunch [11]. In addition, overweight women consuming sources of protein for breakfast five times a week for eight weeks lost 65% more weight and reduced their waist circumference by 83% more than those participants eating a carbohydrate-based breakfast [10].

The use of high protein diets to reduce the amount of food consumed at the next meal is a strategy used to help maintain negative energy balance during weight loss or to maintain weight equilibrium [12]. Protein-based breakfasts positively affect postprandial blood glucose homeostasis, of which tighter control is strongly associated with a lower risk of type 2 diabetes, hypertension, and cardiovascular disease.

Healthy participants as well as metabolically compromised individuals with type 2 diabetes both respond positively to high protein breakfasts, resulting in favorably altered biomarkers including reduced HbA1C%, postprandial glucose, postprandial insulin, and lower systolic blood pressure [13, 14].

Although several studies demonstrate positive effects of protein consumption at breakfast, very few have focused on the source or quality of the protein. Protein quality is important because although equal quantities of plant or animal protein may have the same caloric content, the digestibility and content of amino acids impact blood glucose regulation differently [15]. Therefore, the objective of this study is to determine if protein source (animal protein versus plant protein) at breakfast influences satiety and glucose response and decreases daily food intake.

2. Methods and Materials

2.1. Participants. Female participants ($n = 20$; ages 18–36) were recruited using the university daily newsletter, social media, and word of mouth. Participants who were underweight (BMI $\leq$ 18.4), were smokers, were taking medication (with the exception of hormonal birth control), had food allergies and/or dietary restrictions (e.g., weight loss, vegetarian), disliked the foods served during the study, and/or had any known existing medical conditions that prevented them from eating the breakfasts were excluded from the study. Participants were recruited on a rolling basis and grouped based on their BMI score into normal weight (NW; BMI < 25; $n = 12$) or overweight (OW; BMI $\geq$ 25; $n = 8$) groups (Figure 1). A total of forty-seven women were screened and twenty-five participants started the study. Twenty-two of the women screened did not meet the study criteria. Twenty participants completed the study and were used in data analysis. Refer to Table 1 for participant characteristics. Females aged 18–36 were the focus of this study since this population is at a higher risk for weight gain [1] and there have been several papers published using the population that are focused on breakfast [16, 17]. Ethical approval for the study was obtained from the Office of Research Compliance Institutional Review Board of the University of Arkansas (Fayetteville, AR). Written consent was obtained from all participants prior to beginning the study.

2.2. Study Design. The study was conducted using a randomized, crossover design in which each subject received two different breakfasts, animal protein-based (AP) and plant protein-based (PP), with at least a one-week washout period between each test day and no more than 14 days between testing days. Participants were instructed to fast overnight and limit their physical activity prior to each study day. Upon arrival, baseline measurements of blood glucose and appetite were collected. Food items for each breakfast were portioned, weighed, and labeled appropriately for each subject. Participants were then given 15 minutes to consume the test breakfast. Participants were asked to rate the appearance and taste of the breakfast using a visual analog scale (VAS) [18]. Blood glucose and appetite were analyzed at 0,

TABLE 1: Participant characteristics[1].

Characteristics	NW	OW
Participants (n)	8	12
Age (y)	25 ± 1^a	25 ± 1^a
Weight (kg)	61.3 ± 2.1^a	87.8 ± 7.8^b
Height (m)	1.66 ± 1.2^a	1.65 ± 1.8^a
BMI (kg/m^2)	22.2 ± 0.6^a	31.9 ± 2.7^b
Ethnicity		
Asian	2	0
Caucasian	7	6
Indian	2	1
Latina	1	1

[1] Age, weight, height, and BMI are expressed as means ± SEM. NW: normal weight participants; OW: overweight participants. Means in a row without a common letter are significantly different ($P < 0.05$).

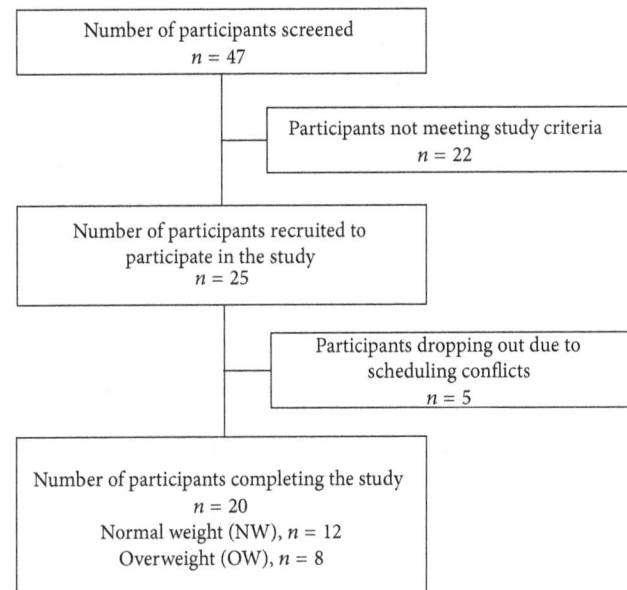

FIGURE 1: Flow diagram of the participant screening and selection process.

15, 30, 45, 60, 90, and 120 minutes postprandial. In addition, participants were instructed to keep a 24-hour dietary food record for the remainder of each test day.

2.3. Test Breakfasts and Dietary Assessment. The nutritional composition for the test breakfasts is described in Table 2. The AP had 29% protein, 29% fat, and 42% carbohydrates. The PP breakfast consisted of 27% protein, 26% fat, and 47% carbohydrates. The AP breakfast consisted of one commercially available breakfast sandwich (Jimmy Dean Delights Turkey Sausage, Egg White, Cheese and English Muffin Breakfast Sandwich), 85 g plain, nonfat Greek yogurt, 6 almonds, and 85 g fresh blueberries. The PP breakfast contained 2 vegan sausage patties (76 g; MorningStar Farms, Kellogg's), 32.3 g of vegan country white bread (Rudi's), 1 slice of vegan American cheese (19 g; Go Veggie, Galaxy Nutritional Products), 85 g of blueberry soy yogurt (WholeSoy & Co.), and 28 g of fresh

TABLE 2: Dietary characteristics of test breakfasts.

	Animal protein (AP) breakfast	Plant protein (PP) breakfast
Total kcal	368	387
Protein (g)	27	26
Fat (g)	12	11
Carbohydrate (g)	38	46
Fiber (g)	4	5
Breakfast appearance, mm[1]	74.8 ± 3.6[a]	63.6 ± 3.5[b]
Breakfast palatability, mm[1]	73.1 ± 3.5[a]	65.9 ± 3.8[a]

[1]Values are expressed as means ± SEM, $n = 20$. CHO: carbohydrate-based breakfast; PRO: protein-based breakfast. Means in a row without a common letter are significantly different ($P < 0.05$).

blueberries. Since we used commercially prepared products, we do not know the exact contribution of each protein source from each product. Participants were asked to record their food intake for the remainder of the test day using 24-hour dietary intake records. The participants were provided with detailed instructions and examples for completing the dietary intake records. The test breakfast composition and 24-hour dietary intake records were analyzed using the Genesis R&D diet analysis software package (Salem, OR).

2.4. Anthropometric Measurements. Body height was measured to the nearest 0.01 cm using a stadiometer (Detecto, St. Louis, MO) with participants barefoot, in the freestanding position. Body weight was measured in the fasting state with participants barefoot to the nearest 0.01 kg using calibrated balance scale (Detecto, St. Louis, MO). BMI was calculated as weight (kg) divided by height (m) squared.

2.5. Blood Glucose Measurement. Blood glucose samples were measured using the fingerstick method at 0, 15, 30, 45, 60, 90, and 120 minutes postprandial using a Lifescan One Touch UltraSmart System (New Brunswick, NJ). One blood sample per time point was collected in a capillary tube (Health Management Systems, Corp; Plano, TX). Samples were measured in duplicate from the sample collected in the capillary tube and the average was used in analysis [19, 20].

2.6. Appetite and Palatability Ratings. Participants were asked to rate their perceived hunger, fullness, desire for food, prospective food consumption, desire for something sweet, and desire for something savory using a 100 mm visual analog scale (VAS) [18]. The VAS is a validated questionnaire incorporating a 100 mm horizontal line scale with questions worded as "how strong is your feeling of" and end anchors of "not at all" to "extremely." Taste and appearance of test breakfasts were collected using the same method.

2.7. Statistical Analysis. Summary statistics were calculated for all data (sample means and sample standard deviations). Net incremental area under the curve (niAUC) was calculated for appetite ratings and glucose values and was used in analyses [21]. Two-sample independent t-tests were used to

determine initial differences between NW and OW participants and to analyze participant characteristics, breakfast appearance and palatability, and comparisons of niAUC between test breakfasts (AP versus PP). Twenty-four-hour energy and macronutrient intake were analyzed using one-factor analysis of variance (ANOVA). Two-factor, crossover, repeated measures analysis of variance (ANOVA) was used to examine significant differences between breakfast and weight groups over time for blood glucose and appetite ratings. The Bonferroni correction for multiple comparisons was applied when significance was observed within the analyses. Results are reported as means ± SEMs. All analyses were conducted using Prism GraphPad Software Version 6.0 (La Jolla, CA). $P < 0.05$ was considered statistically significant.

3. Results

3.1. Participant Characteristics and Compliance. The physical characteristics of the participants are presented in Table 1. There was no difference in age or height between the NW and OW groups. Body weight and BMI were higher in the OW group ($P < 0.05$).

3.2. Appetite and Palatability Responses. The results for perceived hunger, fullness, desire to eat, prospective food consumption, and food cravings are presented in Figure 2. There was no difference in appetite ratings or food cravings between NW and OW groups or between AP and PP breakfasts. However, there was an effect of time on both appetite and food cravings for both group and breakfast ($P < 0.0001$).

The perceived taste and appearance responses to each breakfast were measured immediately following breakfast consumption. There was no difference in taste between AP or PP breakfasts (Table 2). Participants preferred the appearance of the AP versus the PP breakfast ($P < 0.05$).

3.3. Blood Glucose Response. The results for postprandial glucose response are presented in the line graphs (individual time points) and bar graphs (niAUC) in Figure 3. Overall, there was an effect of time on postprandial blood glucose response ($P < 0.0001$), with no effect of diet or weight group over time. Postprandial blood glucose was higher at 30 min following with PP breakfast compared to the AP breakfast, 126.8 ± 4.4 mg/dL versus 112.1 ± 3.9 mg/dL, respectively ($P < 0.05$). Participants had a lower percent change in blood glucose response from the postprandial peak at 30 min to 120 min postprandial following the AP breakfast versus the PP breakfast ($-26.9 ± 4.3\%$ and $-46.5 ± 4.9\%$, resp.; $P < 0.01$).

3.4. 24-Hour Food Intake Assessment. Nutrient composition of the 24-hour food intake records is shown in Table 3. Overall, there was no difference in 24-hour nutritional intake between weight groups or breakfast type. However, there was a trend for participants to have a higher caloric intake following the AP breakfast compared to the PP breakfast ($P = 0.09$). In general, the OW group ate an additional 133 kcal more than NW group. The OW group consumed on average 44% of kcals from carbohydrate, 38% of kcals from

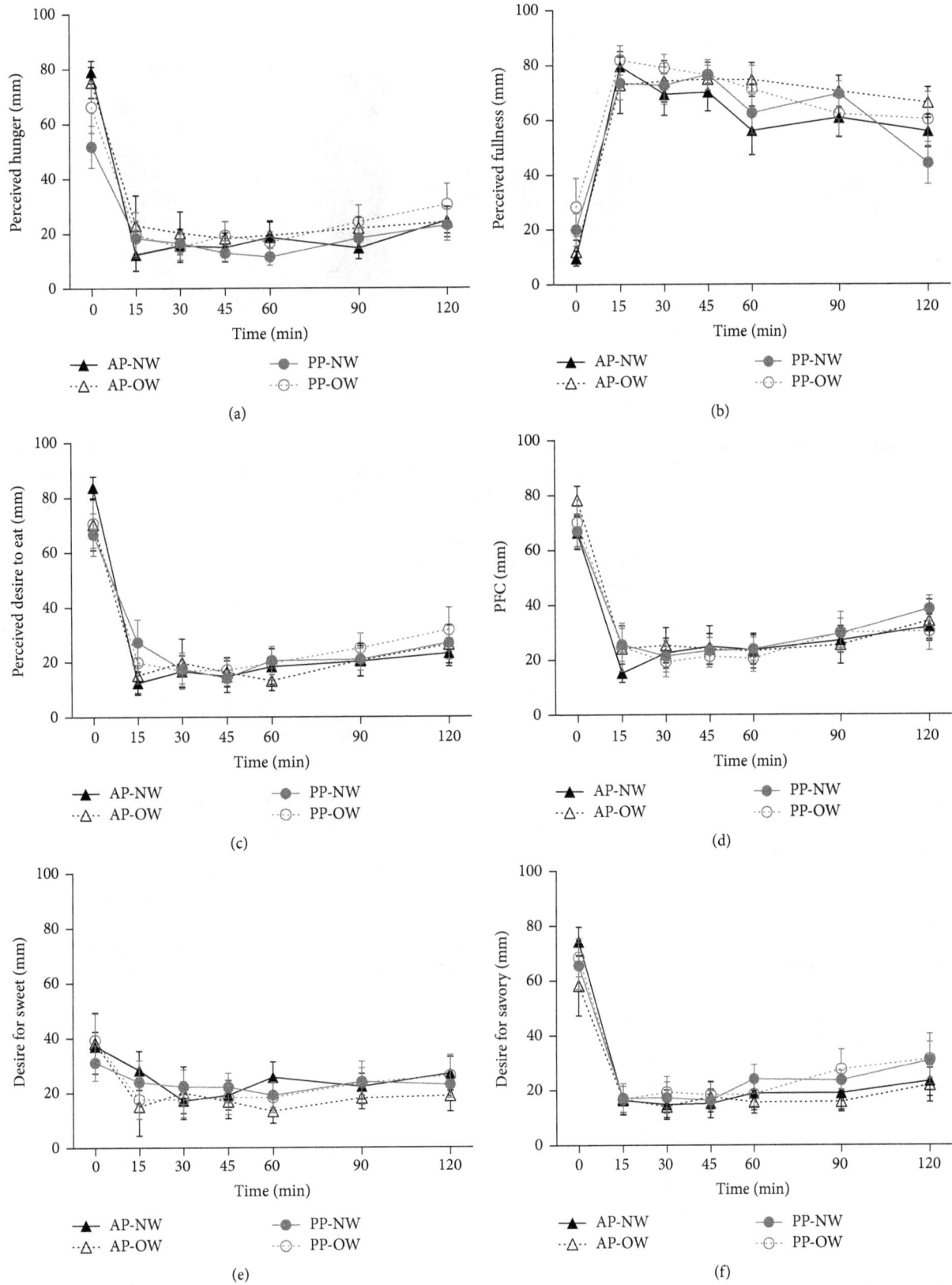

FIGURE 2: Appetite responses following test breakfasts. Values expressed as means ± SEM. Data are depicted as appetite rating over time per weight group and breakfast type and net incremental area under the curve (niAUC). (a) Perceived hunger. (b) Perceived fullness. (c) Perceived desire to eat. (d) Prospective food consumption. (e) Desire for something sweet. (f) Desire for something savory. AP: animal protein; NW: normal weight; OW: overweight; PP: plant protein.

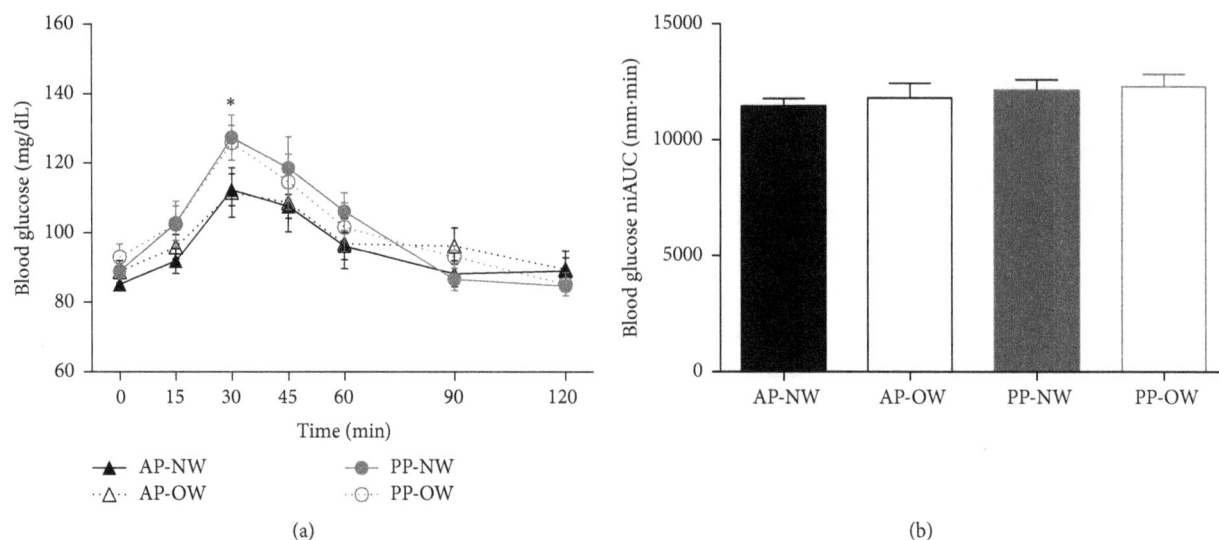

(a)

(b)

FIGURE 3: Glucose response to the test breakfasts. (a) Glucose response to the test breakfasts over time. (b) Glucose net incremental area under the curve (niAUC). Values expressed as means ± SEM. ∗ indicates that blood glucose values for AP were significantly different than PP ($P < 0.05$). AP: animal protein; NW: normal weight; OW: overweight; PP: plant protein.

TABLE 3: Energy and macronutrient content of 24-hour food intake.

	AP-NW	AP-OW	PP-NW	PP-OW
Energy (kcal)	2327 ± 141	2417 ± 251	2041 ± 161	2218 ± 269
Carbohydrate (g)	271 ± 13.3	275.6 ± 22.9	308.18 ± 55.6	237.6 ± 35.3
Fat (g)	93.5 ± 11.4	100.4 ± 13.7	83.1 ± 19.8	95.6 ± 13.7
Protein (g)	123.1 ± 20.9	107.3 ± 20	107.4 ± 10.5	93.4 ± 14.1

[1] Values are expressed as means ± SEM. AP: animal protein; NW: normal weight; OW: overweight; PP: plant protein.

fat, and 17% of kcals from protein after each test breakfast, while the NW group consumed on average 53% of kcals from carbohydrate, 36% of kcals from fat, and 21% of kcals from protein.

4. Discussion

This is one of the first studies to examine the effect of complete meals comparing plant protein and animal protein sources, on postprandial appetite and glucose response in NW and OW females. The present study suggests protein source within the context of a higher protein meal exhibits no difference in appetite response or total nutritional intake; however, protein source could play a role in regulating postprandial blood glucose levels by decreasing the postprandial peak in blood glucose levels.

No difference in postprandial appetite response between AP or PP was detected; however, these results are consistent with several studies in current literature that have tested isolated proteins that were not part of a complete meal. Several studies have compared the effect of protein source on appetite within a mixed meal [22–24], demonstrating equal appetite responses to plant and animal proteins within higher protein meals (>22% protein). When whey protein was compared to casein and soy at 10% energy of a test breakfast, whey exhibited a greater satiating response; however, this difference

diminished when the protein level was increased to 25% energy of a test breakfast, which is similar to the higher protein breakfast composition used in this study [22]. Another study examined beef versus soy within a mixed meal and found no difference in hunger or fullness responses over seven hours [24]. The similar effect of protein sources on appetite response within a high protein diet may be attributed to an overall increased consumption of amino acids [25, 26].

Furthermore, fiber is known to influence appetite response [27]. Although PP breakfast had a slightly higher fiber content (1 g) compared to AP, there is evidence that fiber quantity may have little impact on satiety within a high protein diet. One study demonstrated that when mixed meals, matching in protein content with differing fiber amounts, were ingested, there was no difference found hunger or fullness area under the curve analysis [28] suggesting that protein quantity may influence satiety to a greater extent than fiber content. However, additional research needs to be explored comparing high protein/fiber diets and their effect on appetite.

An increase in protein intake throughout the day, starting with breakfast, may help an individual to feel more satisfied and respond to neural signals of satiety and blood glucose regulation [29]. Though not significant, OW participants consumed fewer calories following the AP breakfast. In general, OW participants consumed less protein and consumed more calories compared to NW participants over the 24-hour test period. The underlying mechanism is still unknown, but high protein diets appear to spontaneously reduce food intake in individuals which could be attributed to satiating effect of protein [30].

Despite there being no significant differences in glucose response between breakfasts or weight groups over the 120 min postprandial period (niAUC), there was a trend for a more stable postprandial glucose response following AP breakfast for both NW and OW groups. The control of postprandial glucose levels is important for HbA1C% levels and diabetes risk [31, 32]. Both eucaloric and hypocaloric diets with increased protein lead to more stable postprandial glucose levels with lesser peak excursions and incremental area under the curve [33–36]. The higher postprandial glucose levels for both NW and OW following the PP breakfast could be attributed to the disparity in breakfast carbohydrate content or differing amino acid profiles of the test breakfasts. It has been observed that healthy individuals and those with higher postprandial glucose levels may do better with a high animal protein-based breakfast compared to a lower protein, carbohydrate-based breakfast [17]. Another possibility is that the lower blood glucose observed, following the AP breakfast, could be due to an increase in insulin production; however, insulin response was not measured in this study and needs to be further explored.

4.1. Limitations. The first limitation of this study is the short postprandial data collection period following breakfast consumption. Two hours postprandial may not be enough time to fully capture the postprandial appetite and glucose response, as meals are generally four to five hours apart and initiated by habit or hunger [37]. Many studies take postprandial measurements for four hours or longer following the test meal to ensure that appetite responses and metabolic measurements (e.g., glucose) return to baseline [16, 24]. Therefore, we may not have captured the entire postprandial breakfast response. Since there were no differences in postprandial appetite responses niAUC, we do not think measuring over a longer period would change our results. Additionally, the discrepancy in caloric and carbohydrate values and fiber content of the test breakfasts may have contributed to the differences observed in postprandial glucose response. The AP breakfast had lower postprandial glucose response at 30 min, which could be due to the lower carbohydrate and fiber content of this breakfast. However, since our conclusions are consistent with current literature, they do not warrant dismissal [22, 26, 38]. Finally, blood glucose was measured via fingerstick, not via intravenous blood draw, which limited the number of postprandial analyses conducted.

5. Conclusions

There was no difference in postprandial appetite response or 24-hour food intake after consumption of breakfasts higher in protein with differing protein sources, AP versus PP, in either NW or OW women. However, consumption of PP generated a higher postprandial glucose peak compared to AP. Taken together, these data suggest that protein source, as part of

breakfast higher in protein, does not differentially affect appetite response but may differentially affect postprandial metabolism.

Abbreviations

PRO: Protein
CHO: Carbohydrate
NW: Normal weight
OW: Overweight
AP: Animal protein-based breakfast
PP: Plant protein-based breakfast
VAS: Visual analog scale
BMI: Body mass index.

Conflict of Interests

The authors declare that there is no conflict of interests regarding the publication of this paper.

Authors' Contribution

Christina M. Crowder and Brianna L. Neumann contributed equally to the paper.

Acknowledgments

The authors would like to thank The University of Arkansas Undergraduate Honors Program and University of Arkansas Dale Bumpers College of Agriculture, Food, and Life Sciences Undergraduate Program for funding of this project.

References

[1] M. J. Hutchesson, J. Hulst, and C. E. Collins, "Weight management interventions targeting young women: a systematic review," *Journal of the Academy of Nutrition and Dietetics*, vol. 113, no. 6, pp. 795–802, 2013.

[2] Y. Wang, A. E. Wluka, J. A. Simpson et al., "Body weight at early and middle adulthood, weight gain and persistent overweight from early adulthood are predictors of the risk of total knee and hip replacement for osteoarthritis," *Rheumatology*, vol. 52, no. 6, pp. 1033–1041, 2013.

[3] L. Lu, H. Risch, M. L. Irwin et al., "Long-term overweight and weight gain in early adulthood in association with risk of endometrial cancer," *International Journal of Cancer*, vol. 129, no. 5, pp. 1237–1243, 2011.

[4] C. H. S. Ruxton and T. R. Kirk, "Breakfast: a review of associations with measures of dietary intake, physiology and biochemistry," *The British Journal of Nutrition*, vol. 78, no. 2, pp. 199–213, 1997.

[5] E. Pollitt and R. Mathews, "Breakfast and cognition: an integrative summary," *The American Journal of Clinical Nutrition*, vol. 67, no. 4, pp. 804S–813S, 1998.

[6] J. L. Stanton Jr. and D. R. Keast, "Serum cholesterol, fat intake, and breakfast consumption in the United States adult population," *Journal of the American College of Nutrition*, vol. 8, no. 6, pp. 567–572, 1989.

[7] A. Keski-Rahkonen, J. Kaprio, A. Rissanen, M. Virkkunen, and R. J. Rose, "Breakfast skipping and health-compromising

behaviors in adolescents and adults," *European Journal of Clinical Nutrition*, vol. 57, no. 7, pp. 842–853, 2003.

[8] P. Deshmukh-Taskar, T. A. Nicklas, J. D. Radcliffe, C. E. O'Neil, and Y. Liu, "The relationship of breakfast skipping and type of breakfast consumed with overweight/obesity, abdominal obesity, other cardiometabolic risk factors and the metabolic syndrome in young adults. The National Health and Nutrition Examination Survey (NHANES): 1999–2006," *Public Health Nutrition*, vol. 16, no. 11, pp. 2073–2082, 2013.

[9] P. R. Deshmukh-Taskar, T. A. Nicklas, C. E. O'Neil, D. R. Keast, J. D. Radcliffe, and S. Cho, "The relationship of breakfast skipping and type of breakfast consumption with nutrient intake and weight status in children and adolescents: the National Health and Nutrition Examination Survey 1999–2006," *Journal of the American Dietetic Association*, vol. 110, no. 6, pp. 869–878, 2010.

[10] J. S. Vander Wal, A. Gupta, P. Khosla, and N. V. Dhurandhar, "Egg breakfast enhances weight loss," *International Journal of Obesity*, vol. 32, no. 10, pp. 1545–1551, 2008.

[11] J. Ratliff, J. O. Leite, R. de Ogburn, M. J. Puglisi, J. VanHeest, and M. L. Fernandez, "Consuming eggs for breakfast influences plasma glucose and ghrelin, while reducing energy intake during the next 24 hours in adult men," *Nutrition Research*, vol. 30, no. 2, pp. 96–103, 2010.

[12] M. P. G. M. Lejeune, E. M. R. Kovacs, and M. S. Westerterp-Plantenga, "Additional protein intake limits weight regain after weight loss in humans," *The British Journal of Nutrition*, vol. 93, no. 2, pp. 281–289, 2005.

[13] M. C. Gannon, F. Q. Nuttall, A. Saeed, K. Jordan, and H. Hoover, "An increase in dietary protein improves the blood glucose response in persons with type 2 diabetes," *The American Journal of Clinical Nutrition*, vol. 78, no. 4, pp. 734–741, 2003.

[14] H. R. Rabinovitz, M. Boaz, T. Ganz et al., "Big breakfast rich in protein and fat improves glycemic control in type 2 diabetics," *Obesity*, vol. 22, no. 5, pp. E46–E54, 2014.

[15] D. J. Millward, D. K. Layman, D. Tomé, and G. Schaafsma, "Protein quality assessment: impact of expanding understanding of protein and amino acid needs for optimal health," *The American Journal of Clinical Nutrition*, vol. 87, no. 5, pp. 1576S–1581S, 2008.

[16] H. J. Leidy and E. M. Racki, "The addition of a protein-rich breakfast and its effects on acute appetite control and food intake in 'breakfast-skipping' adolescents," *International Journal of Obesity*, vol. 34, no. 7, pp. 1125–1133, 2010.

[17] T. M. Rains, H. J. Leidy, K. D. Sanoshy, A. L. Lawless, and K. C. Maki, "A randomized, controlled, crossover trial to assess the acute appetitive and metabolic effects of sausage and egg-based convenience breakfast meals in overweight premenopausal women," *Nutrition Journal*, vol. 14, no. 1, p. 17, 2015.

[18] A. Flint, A. Raben, J. E. Blundell, and A. Astrup, "Reproducibility, power and validity of visual analogue scales in assessment of appetite sensations in single test meal studies," *International Journal of Obesity*, vol. 24, no. 1, pp. 38–48, 2000.

[19] J. I. Baum, M. Gray, and A. Binns, "Breakfasts higher in protein increase postprandial energy expenditure, increase fat oxidation, and reduce hunger in overweight children from 8 to 12 years of age," *The Journal of Nutrition*, vol. 145, no. 10, pp. 2229–2235, 2015.

[20] N. M. Poquette, X. Gu, and S.-O. Lee, "Grain sorghum muffin reduces glucose and insulin responses in men," *Food and Function*, vol. 5, no. 5, pp. 894–899, 2014.

[21] H. J. Leidy, L. C. Ortinau, S. M. Douglas, and H. A. Hoertel, "Beneficial effects of a higher-protein breakfast on the appetitive, hormonal, and neural signals controlling energy intake regulation in overweight/obese, 'breakfast-skipping,' late-adolescent girls," *The American Journal of Clinical Nutrition*, vol. 97, no. 4, pp. 677–688, 2013.

[22] M. A. B. Veldhorst, A. G. Nieuwenhuizen, A. Hochstenbach-Waelen et al., "Effects of high and normal soyprotein breakfasts on satiety and subsequent energy intake, including amino acid and 'satiety' hormone responses," *European Journal of Nutrition*, vol. 48, no. 2, pp. 92–100, 2009.

[23] V. Lang, F. Bellisle, J.-M. Oppert et al., "Satiating effect of proteins in healthy subjects: a comparison of egg albumin, casein, gelatin, soy protein, pea protein, and wheat gluten," *The American Journal of Clinical Nutrition*, vol. 67, no. 6, pp. 1197–1204, 1998.

[24] S. M. Douglas, T. R. Lasley, and H. J. Leidy, "Consuming beef vs. soy protein has little effect on appetite, satiety, and food intake in healthy adults," *The Journal of Nutrition*, vol. 145, no. 5, pp. 1010–1016, 2015.

[25] M. Veldhorst, A. Smeets, S. Soenen et al., "Protein-induced satiety: effects and mechanisms of different proteins," *Physiology and Behavior*, vol. 94, no. 2, pp. 300–307, 2008.

[26] M. A. B. Veldhorst, A. G. Nieuwenhuizen, A. Hochstenbach-Waelen et al., "Comparison of the effects of a high- and normal-casein breakfast on satiety, 'satiety' hormones, plasma amino acids and subsequent energy intake," *The British Journal of Nutrition*, vol. 101, no. 2, pp. 295–303, 2009.

[27] X. Hu, J. Gao, Q. Zhang et al., "Soy fiber improves weight loss and lipid profile in overweight and obese adults: a randomized controlled trial," *Molecular Nutrition and Food Research*, vol. 57, no. 12, pp. 2147–2154, 2013.

[28] L. J. Karhunen, K. R. Juvonen, S. M. Flander et al., "A psyllium fiber-enriched meal strongly attenuates postprandial gastrointestinal peptide release in healthy young adults," *The Journal of Nutrition*, vol. 140, no. 4, pp. 737–744, 2010.

[29] S. C. Woods, "The control of food intake: behavioral versus molecular perspectives," *Cell Metabolism*, vol. 9, no. 6, pp. 489–498, 2009.

[30] G. H. Anderson and S. E. Moore, "Dietary proteins in the regulation of food intake and body weight in humans," *The Journal of Nutrition*, vol. 134, no. 4, pp. 974S–979S, 2004.

[31] L. A. Leiter, A. Ceriello, J. A. Davidson et al., "Postprandial glucose regulation: new data and new implications," *Clinical Therapeutics*, vol. 27, supplement 2, pp. S42–S56, 2006.

[32] G. Boden, K. Sargrad, C. Homko, M. Mozzoli, and T. P. Stein, "Effect of a low-carbohydrate diet on appetite, blood glucose levels, and insulin resistance in obese patients with type 2 diabetes," *Annals of Internal Medicine*, vol. 142, no. 6, pp. 403–411, 2005.

[33] J. H. O'Keefe, N. M. Gheewala, and J. O. O'Keefe, "Dietary strategies for improving post-prandial glucose, lipids, inflammation, and cardiovascular health," *Journal of the American College of Cardiology*, vol. 51, no. 3, pp. 249–255, 2008.

[34] E. Farnsworth, N. D. Luscombe, M. Noakes, G. Wittert, E. Argyiou, and P. M. Clifton, "Effect of a high-protein, energy-restricted diet on body composition, glycemic control, and lipid concentrations in overweight and obese hyperinsulinemic men and women," *The American Journal of Clinical Nutrition*, vol. 78, no. 1, pp. 31–39, 2003.

[35] D. K. Layman, H. Shiue, C. Sather, D. J. Erickson, and J. Baum, "Increased dietary protein modifies glucose and insulin homeostasis in adult women during weight loss," *The Journal of Nutrition*, vol. 133, no. 2, pp. 405–410, 2003.

[36] M. C. Gannon and F. Q. Nuttall, "Control of blood glucose in type 2 diabetes without weight loss by modification of diet composition," *Nutrition and Metabolism*, vol. 3, article 16, 2006.

[37] S. C. Woods, "The eating paradox: how we tolerate food," *Psychological Review*, vol. 98, no. 4, pp. 488–505, 1991.

[38] R. Abou-Samra, L. Keersmaekers, D. Brienza, R. Mukherjee, and K. Macé, "Effect of different protein sources on satiation and short-term satiety when consumed as a starter," *Nutrition Journal*, vol. 10, article 139, 2011.

Short-Term High Fat Intake Does Not Significantly Alter Markers of Renal Function or Inflammation in Young Male Sprague-Dawley Rats

Catherine Crinigan, Matthew Calhoun, and Karen L. Sweazea

School of Nutrition and Health Promotion, School of Life Sciences, Arizona State University, Tempe, AZ 85287, USA

Correspondence should be addressed to Karen L. Sweazea; karen.sweazea@asu.edu

Academic Editor: Jonathan M. Hodgson

Chronic high fat feeding is correlated with diabetes and kidney disease. However, the impact of short-term high fat diets (HFD) is not well-understood. Six weeks of HFD result in indices of metabolic syndrome (increased adiposity, hyperglycemia, hyperinsulinemia, hyperlipidemia, hyperleptinemia, and impaired endothelium-dependent vasodilation) compared to rats fed on standard chow. The hypothesis was that short-term HFD would induce early signs of renal disease. Young male Sprague-Dawley rats were fed either HFD (60% fat) or standard chow (5% fat) for six weeks. Morphology was determined by measuring changes in renal mass and microstructure. Kidney function was measured by analyzing urinary protein, creatinine, and hydrogen peroxide (H_2O_2) concentrations, as well as plasma cystatin C concentrations. Renal damage was measured through assessment of urinary oxDNA/RNA concentrations as well as renal lipid peroxidation, tumor necrosis factor alpha (TNFα), and interleukin 6 (IL-6). Despite HFD significantly increasing adiposity and renal mass, there was no evidence of early stage kidney disease as measured by changes in urinary and plasma biomarkers as well as histology. These findings suggest that moderate hyperglycemia and inflammation produced by short-term HFD are not sufficient to damage kidneys or that the ketogenic HFD may have protective effects within the kidneys.

1. Introduction

Diabetes is currently the leading cause of 44% of newly diagnosed cases of renal failure [1] as sustained hyperglycemia damages the filtration membranes of the kidneys resulting in albuminuria and proteinuria. In fact, proteinuria is often used as a surrogate marker for renal disease [2–4]. As the incidence of diabetes continues to increase throughout the world, rates of albuminuria and diabetic kidney disease (DKD) are likewise expected to increase [2].

The mechanisms leading to the development of chronic kidney disease (CKD) in subjects with metabolic syndrome, however, are not as well-understood [5]. According to the World Health Organization, the criteria for diagnosing metabolic syndrome indicate that an individual must have insulin resistance, type 2 diabetes, impaired glucose tolerance, or impaired fasting glucose along with two additional criteria: abdominal or overall obesity, dyslipidemia, hypertension, or microalbuminuria [6]. These characteristics of metabolic syndrome also increase the risk of developing cardiovascular disease and diabetes, which subsequently increase the risk of renal dysfunction and failure [6]. Foster et al. [7] examined data from the Framingham Heart Study that explored the relationship between renal fat accumulation, hypertension, and CKD. Participants were assessed for renal fat by computed tomography. Positive associations were identified between fatty kidney and hypertension and between fatty kidney and CKD. After adjusting for visceral adipose tissue, no association was found between fat infiltration in the kidney and diabetes. These data indicate that there may be an independent association between hypertension, renal fat accumulation, and CKD [7].

Deji et al. [5] showed that feeding 6-week-old mice a high fat diet (60% fat) for 12 weeks increased body mass, plasma glucose, insulin, and triglyceride concentrations and induced hypertension and kidney disease as evidenced by increased albuminuria, alterations in renal morphology, and renal lipid accumulation. Similarly, Altunkaynak et al. [8] demonstrated that feeding adult female Sprague-Dawley rats a moderate fat diet (30% fat) for 12 weeks caused the animals to become overweight and to develop increased kidney masses and volumes with significant morphological changes indicative of renal disease. Moreover, renal fat accumulation is evident with even moderate fat intake in rabbits (10% corn oil + 5% lard for 8–12 weeks) [9]. Stemmer et al. [10] compared lean chow-fed Wistar rats with HFD-fed rats that were either sensitive or partially resistant to diet-induced obesity after eleven months of HFD (40% butter fat). A positive correlation was established between levels of adiposity and the severity of renal damage. Their results indicate that lipotoxicity is not a strong contributor to renal dysfunction as they found that plasma triglyceride and free fatty acid as well as renal triglyceride levels did not differ significantly between the treatment and control groups [10].

While long-term feeding protocols are found in the literature, the effects of shorter-term feeding protocols to mimic early onset pathological changes within the kidney following high fat feeding are limited. A variety of rodent models have been used to investigate the effects of consuming HFD over the course of eight weeks to 11 months [5, 8, 10, 11], and although the effects of HFD on kidney morphology have been studied at four weeks [12], to our knowledge there are currently no studies that have examined the effects of shorter-term intake of high fat (as opposed to combined high carbohydrate, high fat) diets on renal function. For this reason, the present study was designed to examine the short-term effects of feeding rats a HFD (60% fat) for six weeks on renal morphology and biomarkers of function. Prior studies have shown that young (1.5-month-old) male Sprague-Dawley rats fed on a HFD (60% kcal from fat) for as few as six weeks develop indices of metabolic syndrome and cardiovascular disease: increased visceral adiposity, hyperglycemia, impaired glucose tolerance, hyperleptinemia, systemic inflammation (plasma TNFα), lipid peroxidation (TBARS), and impaired endothelium-dependent vasodilation [13, 14]. Other studies have shown that increased visceral adiposity, endothelial dysfunction, hypertension, inflammation, and oxidative stress (characteristics shared by these animals) increase blood pressure in the kidneys and promote renal damage, although this has yet to be examined in the 6-week HFD rat model. Therefore, exploring the effect of short-term HFD on the kidneys will fill a gap in the literature and allow for a better understanding of early pathological changes that occur due to consuming a HFD. It was hypothesized that six weeks of high fat intake (60% fat) would lead to early stages of renal disease in 1.5-month-old male Sprague-Dawley rats, as evidenced by morphological and functional changes in the kidney, compared to control rats on a standard rodent chow diet (5% fat).

2. Materials and Methods

2.1. Animal Models. Prior studies in our laboratory have shown that feeding young male (1.5-month-old) Sprague-Dawley rats (140–160 g body weight, Harlan Teklad Industries (Madison, WI, USA)) a HFD (60% kcal from fat; Cat. number D12492, Research Diets Inc., New Brunswick, NJ, USA) for six weeks results in symptoms associated with metabolic syndrome including increased adiposity, hyperleptinemia, hyperglycemia, endothelial dysfunction, and hypertension in comparison to animals fed on standard rodent chow [13, 14]. The current study examined isolated kidneys, plasma, and urine samples from a subset of animals examined in these prior IACUC-approved studies from September 2009 to May 2010. Blood samples were obtained via cardiac puncture and centrifuged at 13,000 rpm at 4°C for 15 minutes to separate formed elements from the plasma, which was then stored at −80°C until analyses. Urine was collected with a 25-gauge needle inserted directly into the bladder. The urine was then centrifuged at 13,000 rpm at 4°C for 10 minutes to remove debris and the supernatant was then stored at −80°C until analyses. Following a midline laparotomy, kidneys were removed and stored at −80°C until analyses while a separate subset was embedded in Optimal Cutting Temperature (OCT; Cat. number 4583; Sakura Finetek USA, Inc., Torrance, CA, USA) compound prior to freezing in isopentane cooled by liquid nitrogen and storing at −80°C.

2.2. Morphometrics. Animal morphometrics (body mass, epididymal fat pad mass, waist circumference, and tail length) and renal tissue masses were measured. Tail length is used as marker of overall growth in rats [15]. Epididymal fat pad mass was used to determine variations in adiposity between HFD and chow-fed rats because the epididymal fat pad can be easily removed from the animal objectively. Moreover, recent studies have shown that this fat pad is highly correlated (R^2 = 0.94) with total fat volume from the base of the skull to the distal tibia as measured by *in vivo* microCT scans [16].

A subset of frozen kidneys (n = 4 chow and n = 5 HFD rats) embedded in OCT compound were sectioned using a cryostat (Leica Biosystems CM1950; Buffalo Grove, IL, USA) and sections (14 μm) were collected onto (+)-glass microscope slides (Histobond© adhesive slides, VWR VistaVision, Radnor, PA, USA). Sections were allowed to air-dry for 10 mins and were then fixed with 10% formalin for an additional 10 mins. Following 3-4 washes with tap water, sections were stained with Mayer's hematoxylin (Cat. number 6194A16; Hardy Diagnostics, Santa Maria, CA, USA) for 2 mins and then rinsed with warm tap water. The slides were then dipped in acid alcohol (0.5% HCl in 70% ethanol) and washed in Scotts Tap Water Substitute (Cat. number 26070-06; Electron Microscopy Sciences, Hatfield, PA, USA) for 5 mins followed by a rinse with deionized water. The sections were then counterstained with Eosin Y (Cat. number IS4054; Aldon Corporation, Avon, NY, USA) for 3 mins followed by a rinse with tap water and coverslips were mounted onto the glass microscope slides using SHUR/Mount (Cat. number LC-W; Triangle Biomedical Sciences, Durham, NC, USA). Sections were viewed using a light microscope (Olympus

BX50) and images collected with an Olympus DP70 camera (Melville, NY, USA) to assess morphology.

2.3. Biomarkers of Renal Function. Measuring the excretion of total proteins in the urine, proteinuria, is a marker of early renal disease [3, 4]. In fact, multiple studies have found a stronger association between proteinuria and renal disease outcomes than any other tested factors [4]. Moreover, clinical studies on humans have identified proteinuria as among the first clinical symptoms of kidney damage [17, 18]. Total urinary protein concentrations were measured on a random set (n = 8/group) using the Bradford technique (Cat. number 500-0006; Bio-Rad, Hercules, CA, USA). Urinary creatinine concentrations were measured on the same set of urine samples (n = 8/group) using an available kit (Cat. number CR01; Oxford Biomedical Research, Rochester Hills, MI) and values were used to calculate the protein : creatinine ratio. Interestingly, creatinine clearance is also correlated with risk of renal failure [19].

Cystatin C produced by nucleated cells is filtered by the glomerulus and then catabolized by tubular cells in the kidney [20]. For this reason, plasma concentrations of cystatin C can be used as a biomarker of glomerular filtration rate [20]. Urinary changes in cystatin C can be used as markers of acute renal injury, although not often used in CKD assessment [21]. Serum creatinine concentrations are likewise often assessed as a marker of renal function; however, more recent studies indicate that serum or plasma cystatin C is a stronger, and more consistent, surrogate indicator of glomerular filtration rate and renal function [19–27]. Concentrations of cystatin C in the serum increase as much as 1-2 days prior to serum creatinine in conditions of acute kidney injuries [21] as well as in individuals with type 2 diabetes who have normoalbuminuria [28]. While serum creatinine levels can be affected by sex, age, diet, muscle mass, and body mass, cystatin C levels are independent of gender, muscle mass, and malignancy [21, 27]. Plasma cystatin C was therefore measured on a subset of samples (n = 10 per group) using an available ELISA kit (Cat. number MSCTC0; R&D Systems, Minneapolis, MN, USA) according to the manufacturer's protocol.

Urinary H_2O_2, a biomarker of inflammation, oxidative stress, and renal function, was likewise measured on a subset of samples (n = 6 per group) using an available kit (Cat. number ab102500; Abcam, Cambridge, MA, USA). Creatinine concentrations were measured from the same urine samples (Cat. number CR01; Oxford Biomedical Research, Rochester Hills, MI, USA) and the urinary H_2O_2 : creatinine ratio was calculated. Urinary and tissue H_2O_2 levels are often examined in studies of renal function. The increased renal perfusion pressure that is associated with hypertension has been found to elevate excretion of H_2O_2 [29]. Since H_2O_2 can stimulate proteinuria, it is sometimes used as an indicator of the initiation of renal pathology [30].

2.4. Inflammatory Markers. Chronic inflammation is often associated with multiple disease states and has been hypothesized to contribute to both morbidity and mortality in CKD patients [31]. Studies on animals fed on HFD have found lipid accumulation in glomeruli and proximal tubules within the kidneys along with increased expression of inflammatory markers, particularly tumor necrosis factor alpha (TNFα) and interleukin-6 (IL-6) [5, 32]. TNFα production is elevated with high fat intake and may in part cause insulin resistance [33] and both inflammatory cytokines can predict risk for chronic kidney disease [32]. Briefly, a subset of kidneys from both chow and HFD rats were transferred to a ground glass homogenizer containing Tris-HCl buffer (10 mM Tris (pH 7.6; Cat. number 161-0716; BioRad, Hercules, CA, USA), 1 mM EDTA, 1% triton X-100, 0.1% sodium deoxycholate, 0.03% protease inhibitor cocktail (Cat. number P2714; Sigma-Aldrich, St. Louis, MO, USA), and 1 mM phenylmethanesulfonyl fluoride (PMSF)). Samples were then centrifuged at 14,000 rpm for 10 minutes at 4°C to remove insoluble debris. Total protein concentrations of the supernatants were determined using the Bradford Technique (Cat. number 500-0006; Bio-Rad). Each sample (100 μg total protein) was mixed with 6 μL 5x sample buffer (0.6 mL 1 M Tris-HCl, pH 6.8; 5 mL 50% glycerol; 2 mL 10% sodium dodecyl sulfate; 1 mL 1% bromophenol blue; 0.9 mL deionized water and 2% β-mercaptoethanol as a reducing agent) and then boiled for 3 minutes. The mixture was then resolved using 4–15% gradient Tris-HCl SDS-PAGE gels (Cat. number 456-1083; Bio-Rad) and transferred to PVDF membranes (Cat. number 152-0176; Bio-Rad) for 90 minutes at 200 V. Membranes were then blocked for two hours in Tris buffered saline containing 0.05% Tween 20, 3% BSA fraction V, and 5% nonfat milk followed by an overnight incubation at 4°C in primary antibodies for IL-6 (1 : 250; Cat. number ab6672; Abcam, Cambridge, MA, USA), TNFα (1 : 500; Cat. number 11948; Cell Signaling, Danvers, MA, USA), or the loading control beta-actin (1 : 2000; Cat. number ab8227; Abcam, Cambridge, MA, USA). Primary antibodies were prepared in Tris buffered saline with 0.05% Tween 20 (TTBS). Membranes were then exposed to anti-rabbit IgG secondary antibody (1 : 1000 for IL-6 and TNFα and 1 : 2000 for beta actin; Cat number 7074S; Cell Signaling Technology, Danvers, MA) in TTBS for one hour at room temperature followed by exposure to Pierce enhanced chemiluminescence western blotting substrate (Cat. number 32279; Thermo Scientific, Rockford, IL, USA) for 1 min. Immunoreactive bands were visualized by exposure to X-ray film (Cat. number 34090; Thermo Scientific, Rockford, IL, USA) and analyzed using NIH ImageJ software. Protein expression of IL-6 and TNFα were normalized to beta-actin to determine the level of IL-6 (n = 5 chow and 6 HFD animals) or TNFα (n = 8/group) in each kidney sample.

2.5. Oxidative Stress. Kidneys are especially susceptible to oxidative stress due to a high concentration of long-chain polyunsaturated fatty acids, which easily undergo lipid peroxidation when exposed to reactive oxygen species [34]. Lipid peroxidation forms malondialdehyde as an end product, which can be detected via a thiobarbituric acid reactive substance (TBARS) assay, a marker of oxidative damage that is often examined. A subset of samples (n = 5/group) were used to evaluate renal lipid peroxidation using a commercially

TABLE 1: Morphometric measurements.

Measure	Chow (n)	HFD (n)	p value
Body mass (g)	349.8 ± 7.98 (9)	376.4 ± 10.3 (8)	0.056
Epididymal fat pad mass (g)	3.49 ± 0.18 (9)	5.60 ± 0.38 (8)	**<0.001**
Waist circumference (cm)	16.59 ± 0.30 (9)	17.96 ± 0.27 (8)	**0.004**
Tail length (cm)	20.1 ± 0.30 (9)	20.6 ± 0.32 (8)	0.276
Renal mass (g)	1.05 ± 0.04 (9)	1.19 ± 0.04 (8)	**0.019**

Data expressed as mean ± SEM and analyzed by Student's t-tests.

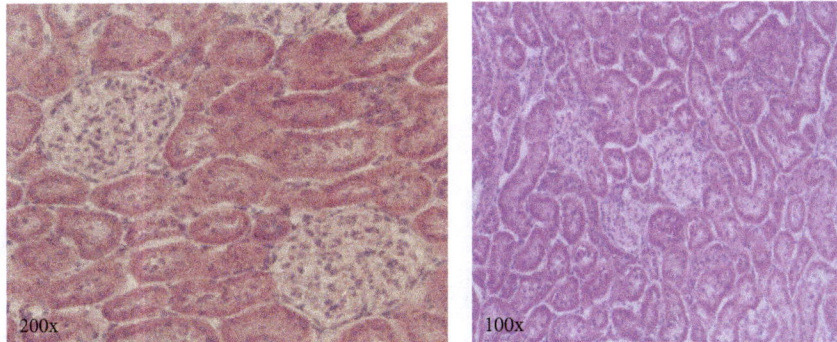

FIGURE 1: Representative images of hematoxylin and eosin stained renal tissue sections from HFD animals. Hematoxylin and eosin staining shows no signs of morphological damage.

available TBARS assay kit (Cat. number 0801192; ZeptoMetrix Corporation, Buffalo, NY, USA).

8-Hydroxy-2′-deoxyguanosine (8-OHdG) is a sensitive biomarker of oxidative stress in tissue and bodily fluids and is a major product of damage caused by oxidative stress to DNA. 8-OHdG is produced by enzymatic cleavage of the guanine base, the base most prone to oxidation. Plasma samples (n = 10 chow; n = 11 HFD) were filtered with a 30 kDa ultrafilter (Cat. number UFC503096; EMD Millipore, Billerica, MA, USA) and 8-hydroxyguanosine, 8-OHdG, and 8-hydroxyguanine were examined as biomarkers of oxidative DNA and RNA damage using a commercially available kit (Cat. number 589320; Cayman Chemical, Ann Arbor, MI, USA), as oxidative stress has been shown to be elevated during early renal failure.

2.6. Statistical Analyses. Data are expressed as mean ± SEM. Statistical analyses were computed using SigmaPlot (Systat Software Inc., Version 13.0; San Jose, CA, USA). Data were tested for normality and then analyzed using Student's t-tests or the Mann-Whitney U tests, as appropriate. p values of ≤ 0.05 were considered significant.

3. Results

3.1. Morphometrics. Although the HFD rats tended to weigh more than the chow-fed rats, this difference was not statistically significant for the animals examined in this study (Table 1). Rats fed on a HFD demonstrated significantly increased epididymal fat pad mass compared to chow-fed animals, establishing that adiposity was increased by the HFD (Table 1). The waist circumferences of the HFD animals were significantly increased when compared to chow-fed controls, indicating that the HFD increased abdominal adiposity (Table 1). Tail lengths, a marker of overall growth, were not significantly different (Table 1). The renal masses of the HFD rats were significantly greater compared to the renal masses of animals in the chow group (Table 1).

Morphological analyses using hematoxylin and eosin-stained tissue sections showed no structural differences between the chow and HFD groups, indicating that although the mass of the HFD kidneys was increased, damage to the microstructure of the kidneys was not evident (Figure 1).

3.2. Biomarkers of Renal Function. The urine protein : creatinine ratios of the chow and HFD rats did not differ significantly, indicating that the filtering ability of the HFD rat kidneys was not damaged (Table 2). There was also no difference in the urine creatinine concentrations of the two groups (Table 2). Although HFD plasma cystatin C concentrations tended to be higher than chow rats, the difference was not sufficient to merit statistical significance (Table 2). Urinary hydrogen peroxide concentrations were significantly increased in the HFD animals. However, the H_2O_2 : creatinine ratios were not significantly different between the chow and HFD groups (Table 2).

3.3. Inflammatory Markers. Western blot analyses of renal tissue established no significant difference in the TNFα protein expression of the chow and HFD rat kidneys (Figure 2). Quantification of western blots of IL-6 expression in renal tissues likewise showed no difference between the kidneys from chow and HFD rats (Figure 3).

TABLE 2: Plasma and urinary markers of renal functionand damage.

Measure	Chow (n)	HFD (n)	p value
Urine creatinine (mg/L)	765.0 ± 110.6 (8)	900.3 ± 137.4 (8)	0.456
Urine protein : creatinine ratio (AU)	6.01 ± 2.86 (8)	3.89 ± 1.31 (8)	0.574
Urine H_2O_2 (pM/μL)	0.023 ± 0.003 (6)	0.048 ± 0.005 (6)	**0.002**
Urine H_2O_2 : creatinine ratio (AU)	0.014 ± 0.002 (6)	0.026 ± 0.007 (6)	0.142
Plasma cystatin C (μg/mL)	14.35 ± 0.99 (10)	16.2 ± 1.12 (10)	0.233
Renal TBARS (mM/L)	26.9 ± 2.02 (5)	28.08 ± 0.80 (5)	0.600
Plasma oxDNA/RNA (pg/mL)	887 ± 145 (10)	1003 ± 188 (11)	0.634

Data expressed as mean ± SEM and analyzed by Student's t-tests with the exception of urine protein : creatinine ratio, which was analyzed by the Mann-Whitney U test.

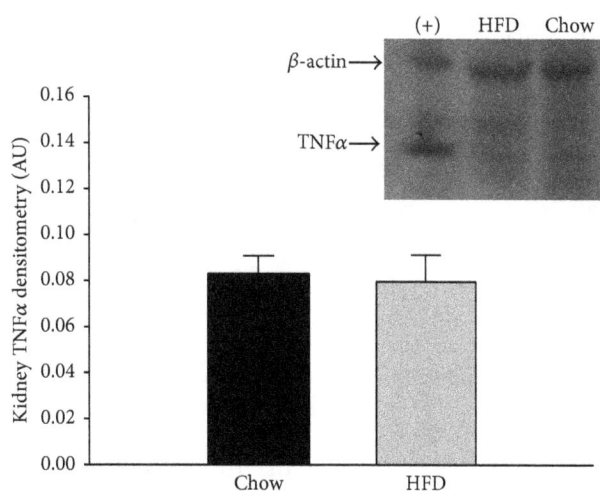

FIGURE 2: Renal tissue TNFα protein expression in chow and HFD rats. There was no difference in the TNFα expression of the two diet groups (n = 8/group). Retroperitoneal adipose tissue from a HFD rat was used as a positive control (+) and is shown in the first column. Data were analyzed by Student's t-tests and are expressed as means ± SEM. p = 0.803.

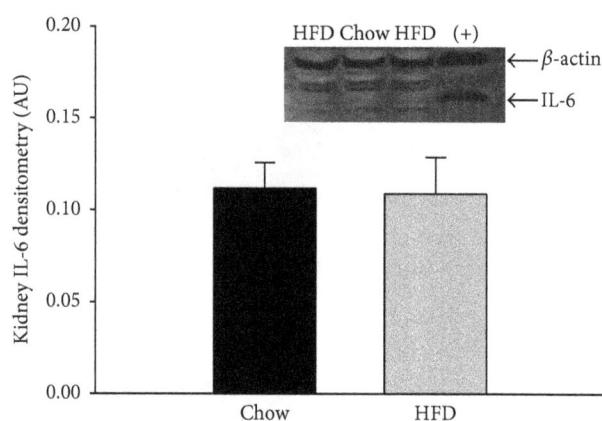

FIGURE 3: Renal tissue IL-6 protein expression in chow and HFD rats. There was no difference in the IL-6 expression of the two diet groups (n = 5 chow and 6 HFD rats). Retroperitoneal adipose tissue from a HFD rat was used as a positive control (+) and is shown in the last column. Data were analyzed by Student's t-tests and are expressed as means ± SEM. p = 0.903.

3.4. Oxidative Stress Markers. Renal TBARS, a measure of tissue oxidative stress via lipid peroxidation, was not significantly elevated in the HFD rats (Table 2). Plasma levels of oxidized DNA and RNA, quantified as plasma levels of multiple oxidative stress markers including 8-hydroxyguanosine from RNA, 8-OHdG from DNA, and 8-hydroxyguanine, were examined. There were no significant differences between the chow and HFD rat groups (Table 2).

4. Discussion

Young (1.5-month-old) male Sprague-Dawley rats fed on HFD (60% fat) for six weeks developed significant increases in waist circumference and epididymal fat pad mass along with a trend towards increased body mass (Table 1). There was no significant difference in the tail lengths of the animals on the HFD and chow diets (Table 1), indicating that the animals were of similar body size. These data also indicate that the rats

used for the present study developed increased adiposity. The use of an animal model that mimics the effects of metabolic syndrome and prediabetes is consistent with prior studies of HFD intake by both mice and rats [5, 8, 33, 35–39]. Previous studies from our laboratory also demonstrate that 6 weeks of HFD results in hyperglycemia, impaired glucose tolerance, hypertension, endothelial dysfunction, oxidative stress (plasma TBARS), and inflammation (plasma TNFα) in these animals, indicating that the HFD rats are also models of metabolic syndrome [13, 14].

The renal masses of the HFD rats examined in the present study were significantly higher than the chow rats (Table 1). The lack of difference in tail lengths (Table 1) signifies that the increased mass was not caused by overall growth of the animals and instead suggests structural damage to the kidneys. Altunkaynak et al. [8] found similar increases in body and renal masses of rats fed on HFD (30% fat) for 12 weeks. The increased renal mass may be attributed to hyperglycemia as proliferation and hypertrophy of mesangial cells and thickening of the glomerular basement membrane promote increased renal mass [40] along with vasodilation, inflammation, and increased connective tissue [8]. Despite the increased overall mass of the kidneys, hematoxylin and eosin staining of the HFD kidneys showed no evidence of morphological damage (Figure 1). This is in contrast to long-term studies of HFD that have been shown to cause morphological renal changes such as glomerular capillary dilation, enlarged lumens in the tubules and Bowman's capsule, amassing of extracellular proteins, nephron degradation, glomerular membrane thickening, glomerulosclerosis, renal and tubular interstitial cell necrosis, and shortened tubular epitheliums [5, 8, 41].

Proteinuria indicates that the filtering ability of the glomerulus is compromised and allowing proteins, which normally remain in the blood, to be passed through the filtration membrane and excreted. In evaluating renal function, albumin is sometimes specifically examined; however, by studying total protein excretion, a wider picture of filtration ability can be ascertained. Proteinuria, specifically, can be used as a marker of endothelial dysfunction as well as a surrogate outcome for renal disease progression. Urinary protein concentrations were normalized to urinary creatinine concentrations in order to account for variations in urine output [42]. Creatinine is also filtered by the kidneys and was therefore evaluated for differences between the HFD and chow-fed groups. Ruggiero et al. [36] found no significant changes in creatinine levels after C57BL mice consumed a 45% fat diet for sixteen weeks; the authors mentioned this indicated early stages of renal damage without severe functional damage (Table 2). In the present study, no statistically significant differences were observed between the urinary protein to creatinine ratio for animals ingesting the HFD as compared to the chow-fed animals (Table 2), indicating that the filtration abilities of the glomeruli were not damaged by the HFD or at least not to the extent of allowing large proteins to be excreted. With a longer duration feeding protocol, research suggests that the HFD may result in sustained elevations in blood glucose concentrations further leading to increased stress on the kidneys.

Plasma cystatin C is considered a marker of early renal dysfunction and the lack of difference between the cystatin C concentrations of the HFD and chow-fed rats indicates that the six-week feeding protocol may not have been long enough to initiate functional renal damage (Table 2). The HFD group did develop higher concentrations of plasma cystatin C; however, this difference was not statistically significant (Table 2). Although cystatin C levels in plasma have been shown to be an early sensitive marker of renal dysfunction, these results indicate that the kidneys of the HFD-fed rats were not damaged enough so as to prevent cystatin C from being filtered [22–24, 43]. Muntner et al. [44] demonstrated that serum cystatin C can be elevated in subjects who do not exhibit signs of micro- or macroalbuminuria or kidney disease. Rather, the authors concluded that serum cystatin C was correlated with cardiovascular disease in overweight subjects [44]. Therefore, plasma cystatin C alone may not be an accurate estimation of GFR.

H_2O_2 is an early marker of oxidative stress and inflammation. Despite the increased renal perfusion pressure that is associated with hypertension and elevated H_2O_2

excretion, there was no significant difference in urinary H_2O_2 : creatinine ratio between the two groups (Table 2). Elevated H_2O_2 in the HFD group would have indicated early renal damage, as H_2O_2 has been found to be elevated prior to GFR decline, proteinuria, and more robust inflammatory and fibrotic responses following high fat feedings [30, 37, 38]. Overall, these results demonstrate that although prior studies indicate that the HFD rats develop plasma and vascular oxidative stress and inflammation in addition to endothelial dysfunction [13, 14], these pathologies were not sufficient to impair renal function.

Western blot analyses established no significant difference between the TNFα concentrations of renal tissue isolated from HFD and chow-fed rats (Figure 2). Although increased TNFα expression in adipose tissue has been shown to promote to insulin resistance in overweight and obese rodent models [45, 46], there was no indication that the HFD increased renal tissue expression of this inflammatory cytokine (Figure 2). There was also no increase in IL-6 expression (Figure 3), a proinflammatory cytokine. Other experimental models of renal injury indicate that increases in IL-6 secretion are not necessarily an integral step in progressive renal failure [47]. Overall, the lack of significant findings in markers of early renal dysfunction, such as proteinuria and urinary H_2O_2, supports the lack of increases in cytokine expression. Chronic increased IL-6 levels are an indication of continued TNFα production and activation of inflammatory pathways. Despite the controversy regarding the differing effects of IL-6 as a cytokine and myokine, it continues to be studied as a marker of inflammation in HFD-induced obesity. Studies on both overweight and obese women have found that serum TNFα and IL-6 levels were positively correlated [48] while Ozay et al. [35] found that HFD-induced obesity in male Wistar albino rats (22% fat for 12 weeks) increased plasma TNFα without altering plasma IL-6. Conversely, Kaya et al. [49] found that IL-6 levels in obese human subjects were higher in both plasma and adipose tissue compared to TNFα. Stemmer et al. [10] examined the effect of HFD (40% of calories from butter fat for 11 months) on male Wistar rats and the creation of an inflammatory renal environment. Their results indicate an increase of IL-6 and TNFα in retroperitoneal fat and in the kidneys, while there was no increase in circulating IL-6.

Previous studies of the 6-week HFD rats showed that plasma TBARS, a measure of whole body oxidative stress, was increased in comparison to chow-fed controls [13]. Despite the increased plasma TBARS, renal TBARS were not significantly different between chow and HFD-fed rats (Table 2). These results indicate that although the HFD rats exhibited systemic oxidative stress, oxidative stress within the kidneys is not yet evident. Moreover, no significant difference between oxidative DNA/RNA damage between the experimental and control groups was observed (Table 2). Oxidative stress, examined via TBARS and DNA/RNA oxidation, is positively correlated [50]. The lack of significant TNFα and IL-6 expression in renal tissue examined in the present study is consistent with the observed lack of oxidative stress within the kidneys as inflammation and oxidative stress are typically correlated [51].

5. Conclusions

In conclusion, aside from morphological changes, the only major statistically significant findings in this study were an increase in visceral adiposity and renal mass following a short-term HFD. This study is novel in that it is the first to examine a HFD for only six weeks; prior to this study, the shortest feeding protocol that is evident in the literature is four weeks at which point only morphological changes were examined as opposed to measures of renal function. Previous research conducted on the six-week HFD rat model showed evidence of metabolic syndrome including significant vascular oxidative stress and inflammation, endothelial dysfunction, hypertension, hyperglycemia, hyperleptinemia, and impaired glucose tolerance [13, 14].

It is possible that the methods used in this study might not have been sensitive enough to detect subtle changes in renal function. In prior studies by our laboratory, 6 weeks of HFD significantly increased plasma concentrations of the ketone beta-hydroxybutyrate from 4.72 ± 0.31 in chow rats to 7.18 ± 0.77 mg/dL in the HFD animals [14]. Considering that ketogenic diets can reduce renal responses to glucose, it is also possible that the HFD exerts a protective effect in the kidney [51]. This was shown in a prior study of mouse models of type 2 diabetes with nephropathy in which 8 weeks of consuming a ketogenic diet fully normalized albumin : creatinine ratios and the expression of stress-related genes [51]. Therefore, it is apparent from the current study that prolonged hyperglycemia and inflammation are likely necessary to induce significant changes within the kidneys or that HFD may exert some protective effects through increased generation of ketones.

Conflict of Interests

The authors declare that there is no conflict of interests regarding the publication of this paper.

Acknowledgments

This study was funded by a Graduate Research Support Program Award from ASU (Catherine Crinigan) as well as a Sigma Xi Grant-in-Aid of Research (Catherine Crinigan). The authors would also like to thank Kristin Ricklefs (ASU), Marc Girard (University of Poitiers, France), and Zoha Ahmed (ASU) for technical assistance with the study.

References

[1] Centers for Disease Control and Prevention, *National Diabetes Statistics Report: Estimates of Diabetes and Its Burden in the United States, 2014*, U.S. Department of Health and Human Services, Atlanta, Ga, USA, 2014.

[2] A. T. Reutens, "Epidemiology of diabetic kidney disease," *Medical Clinics of North America*, vol. 97, no. 1, pp. 1–18, 2013.

[3] W. F. Keane and G. Eknoyan, "Proteinuria, albuminuria, risk, assessment, detection, elimination (PARADE): a position paper of the National Kidney Foundation," *The American Journal of Kidney Diseases*, vol. 33, no. 5, pp. 1004–1010, 1999.

[4] A. S. Levey, D. Cattran, A. Friedman et al., "Proteinuria as a surrogate outcome in CKD: report of a scientific workshop sponsored by the National Kidney Foundation and the US Food and Drug Administration," *American Journal of Kidney Diseases*, vol. 54, no. 2, pp. 205–226, 2009.

[5] N. Deji, S. Kume, S.-I. Araki et al., "Structural and functional changes in the kidneys of high-fat diet-induced obese mice," *American Journal of Physiology: Renal Physiology*, vol. 296, no. 1, pp. F118–F126, 2009.

[6] G. M. Reaven, "The metabolic syndrome: time to get off the merry-go-round?" *Journal of Internal Medicine*, vol. 269, no. 2, pp. 127–136, 2011.

[7] M. C. Foster, S.-J. Hwang, S. A. Porter, J. M. Massaro, U. Hoffmann, and C. S. Fox, "Fatty kidney, hypertension, and chronic kidney disease: the framingham heart study," *Hypertension*, vol. 58, no. 5, pp. 784–790, 2011.

[8] M. E. Altunkaynak, E. Özbek, B. Z. Altunkaynak, I. Can, D. Unal, and B. Unal, "The effects of high-fat diet on the renal structure and morphometric parametric of kidneys in rats," *Journal of Anatomy*, vol. 212, no. 6, pp. 845–852, 2008.

[9] T. M. Dwyer, H. L. Mizelle, K. Cockrell, and P. Buhner, "Renal sinus lipomatosis and body composition in hypertensive, obese rabbits," *International Journal of Obesity and Related Metabolic Disorders*, vol. 19, no. 12, pp. 869–874, 1995.

[10] K. Stemmer, D. Perez-Tilve, G. Ananthakrishnan et al., "High-fat-diet-induced obesity causes an inflammatory and tumor-promoting microenvironment in the rat kidney," *Disease Models & Mechanisms*, vol. 5, no. 5, pp. 627–635, 2012.

[11] R. Buettner, K. G. Parhofer, M. Woenckhaus et al., "Defining high-fat-diet rat models: metabolic and molecular effects of different fat types," *Journal of Molecular Endocrinology*, vol. 36, no. 3, pp. 485–501, 2006.

[12] A. Pranprawit, F. M. Wolber, J. A. Heyes, A. L. Molan, and M. C. Kruger, "Short-term and long-term effects of excessive consumption of saturated fats and/or sucrose on metabolic variables in Sprague Dawley rats: a pilot study," *Journal of the Science of Food and Agriculture*, vol. 93, no. 13, pp. 3191–3197, 2013.

[13] K. L. Sweazea, M. Lekic, and B. R. Walker, "Comparison of mechanisms involved in impaired vascular reactivity between high sucrose and high fat diets in rats," *Nutrition and Metabolism*, vol. 7, article 48, 2010.

[14] K. L. Sweazea and B. R. Walker, "High fat feeding impairs endothelin-1 mediated vasoconstriction through increased iNOS-derived nitric oxide," *Hormone and Metabolic Research*, vol. 43, no. 7, pp. 470–476, 2011.

[15] L. González Bosc, M. L. Kurnjek, A. Müller, and N. Basso, "Effect of chronic angiotensin II inhibition on the cardiovascular system of the normal rat," *The American Journal of Hypertension*, vol. 13, no. 12, pp. 1301–1307, 2000.

[16] Y. K. Luu, S. Lublinsky, E. Ozcivici et al., "*In vivo* quantification of subcutaneous and visceral adiposity by micro-computed tomography in a small animal model," *Medical Engineering & Physics*, vol. 31, no. 1, pp. 34–41, 2009.

[17] N. Kambham, G. S. Markowitz, A. M. Valeri, J. Lin, and V. D. D'Agati, "Obesity-related glomerulopathy: an emerging epidemic," *Kidney International*, vol. 59, no. 4, pp. 1498–1509, 2001.

[18] R. D. Adelman, I. G. Restaino, U. S. Alon, and D. L. Blowey, "Proteinuria and focal segmental glomerulosclerosis in severely obese adolescents," *The Journal of Pediatrics*, vol. 138, no. 4, pp. 481–485, 2001.

[19] M. Praga, E. Hernández, E. Morales et al., "Clinical features and long-term outcome of obesity-associated focal segmental glomerulosclerosis," *Nephrology Dialysis Transplantation*, vol. 16, no. 9, pp. 1790–1798, 2001.

[20] A. Grubb, "Diagnostic value of analysis of cystatin C and protein HC in biological fluids," *Clinical Nephrology*, vol. 38, supplement 1, pp. S20–S27, 1992.

[21] J. C. Sirota, J. Klawitter, and C. L. Edelstein, "Biomarkers of acute kidney injury," *Journal of Toxicology*, vol. 2011, Article ID 328120, 10 pages, 2011.

[22] J. S. C. Chew, M. Saleem, C. M. Florkowski, and P. M. George, "Cystatin C: a paradigm of evidence based laboratory medicine," *The Clinical Biochemist Reviews*, vol. 29, no. 2, pp. 47–62, 2008.

[23] J. S. C. Chew-Harris, C. M. Florkowski, P. M. George, J. L. Elmslie, and Z. H. Endre, "The relative effects of fat versus muscle mass on cystatin C and estimates of renal function in healthy young men," *Annals of Clinical Biochemistry*, vol. 50, no. 1, pp. 39–46, 2013.

[24] R. Hojs, S. Bevc, R. Ekart, M. Gorenjak, and L. Puklavec, "Serum cystatin C as an endogenous marker of renal function in patients with chronic kidney disease," *Renal Failure*, vol. 30, no. 2, pp. 181–186, 2008.

[25] N. V. McNamara, R. Chen, M. R. Janu, P. Bwititi, G. Car, and M. Seibel, "Early renal failure detection by cystatin C in Type 2 diabetes mellitus: varying patterns of renal analyte expression," *Pathology*, vol. 41, no. 3, pp. 269–275, 2009.

[26] V. R. Dharnidharka, C. Kwon, and G. Stevens, "Serum cystatin C is superior to serum creatinine as a marker of kidney function: a meta-analysis," *American Journal of Kidney Diseases*, vol. 40, no. 2, pp. 221–226, 2002.

[27] O. Schück, V. Teplan, M. Stollová, and J. Skibová, "Estimation of glomerular filtration rate in obese patients with chronic renal impairment based on serum cystatin C levels," *Clinical Nephrology*, vol. 62, no. 2, pp. 92–96, 2004.

[28] Y. K. Jeon, M. R. Kim, J. E. Huh et al., "Cystatin C as an early biomarker of nephropathy in patients with type 2 diabetes," *Journal of Korean Medical Science*, vol. 26, no. 2, pp. 258–263, 2011.

[29] C. Jin, C. Hu, A. Polichnowski et al., "Effects of renal perfusion pressure on renal medullary hydrogen peroxide and nitric oxide production," *Hypertension*, vol. 53, no. 6, pp. 1048–1053, 2009.

[30] T. Yoshioka, I. Ichikawa, and A. Fogo, "Reactive oxygen metabolites cause massive, reversible proteinuria and glomerular sieving defect without apparent ultrastructural abnormality," *Journal of the American Society of Nephrology*, vol. 2, no. 4, pp. 902–912, 1991.

[31] H. F. Tbahriti, D. Meknassi, R. Moussaoui et al., "Inflammatory status in chronic renal failure: the role of homocysteinemia and pro-inflammatory cytokines," *World Journal of Nephrology*, vol. 2, pp. 31–37, 2013.

[32] C. Rüster and G. Wolf, "Adipokines promote chronic kidney disease," *Nephrology Dialysis Transplantation*, vol. 28, no. 4, pp. 8–14, 2013.

[33] S. E. Borst and C. F. Conover, "High-fat diet induces increased tissue expression of TNF-α," *Life Sciences*, vol. 77, no. 17, pp. 2156–2165, 2005.

[34] E. Ozbek, "Induction of oxidative stress in kidney," *International Journal of Nephrology*, vol. 2012, Article ID 465897, 9 pages, 2012.

[35] R. Ozay, E. Uzar, A. Aktas et al., "The role of oxidative stress and inflammatory response in high-fat diet induced peripheral neuropathy," *Journal of Chemical Neuroanatomy*, vol. 55, pp. 51–57, 2014.

[36] C. Ruggiero, M. Ehrenshaft, E. Cleland, and K. Stadler, "High-fat diet induces an initial adaptation of mitochondrial bioenergetics in the kidney despite evident oxidative stress and mitochondrial ROS production," *American Journal of Physiology: Endocrinology and Metabolism*, vol. 300, no. 6, pp. E1047–E1058, 2011.

[37] A.-E. Declèves, A. V. Mathew, R. Cunard, and K. Sharma, "AMPK mediates the initiation of kidney disease induced by a high-fat diet," *Journal of the American Society of Nephrology*, vol. 22, no. 10, pp. 1846–1855, 2011.

[38] A. E. Declèves, J. J. Rychak, D. J. Smith, and K. Sharma, "Effects of high-fat diet and losartan on renal cortical blood flow using contrast ultrasound imaging," *The American Journal of Physiology—Renal Physiology*, vol. 305, no. 9, pp. F1343–F1351, 2013.

[39] A. H. Stark, B. Timar, and Z. Madar, "Adaptation of Sprague Dawley rats to long-term feeding of high fat or high fructose diets," *European Journal of Nutrition*, vol. 39, no. 5, pp. 229–234, 2000.

[40] E. N. Obineche, E. Mensah-Brown, S. I. Chandranath, I. Ahmed, O. Naseer, and A. Adem, "Morphological changes in the rat kidney following long-term diabetes," *Archives of Physiology and Biochemistry*, vol. 109, no. 3, pp. 241–245, 2001.

[41] T. Thethi, M. Kamiyama, and H. Kobori, "The link between the renin-angiotensin-aldosterone system and renal injury in obesity and the metabolic syndrome," *Current Hypertension Reports*, vol. 14, no. 2, pp. 160–169, 2012.

[42] S. S. Waikar, V. S. Sabbisetti, and J. V. Bonventre, "Normalization of urinary biomarkers to creatinine during changes in glomerular filtration rate," *Kidney International*, vol. 78, no. 5, pp. 486–494, 2010.

[43] G. S. Hotamisligil, N. S. Shargill, and B. M. Spiegelman, "Adipose expression of tumor necrosis factor-α: direct role in obesity-linked insulin resistance," *Science*, vol. 259, no. 5091, pp. 87–91, 1993.

[44] P. Muntner, D. Mann, J. Winston, S. Bansilal, and M. E. Farkouh, "Serum cystatin C and increased coronary heart disease prevalence in US adults without chronic kidney disease," *The American Journal of Cardiology*, vol. 102, no. 1, pp. 54–57, 2008.

[45] W. P. Cawthorn and J. K. Sethi, "TNF-α and adipocyte biology," *FEBS Letters*, vol. 582, no. 1, pp. 117–131, 2008.

[46] S. A. Jones, D. J. Fraser, C. A. Fielding, and G. W. Jones, "Interleukin-6 in renal disease and therapy," *Nephrology Dialysis Transplantation*, vol. 30, no. 4, pp. 564–574, 2015.

[47] S. Fenkci, S. Rota, N. Sabir, Y. Sermez, A. Guclu, and B. Akdag, "Relationship of serum interleukin-6 and tumor necrosis factor alpha levels with abdominal fat distribution evaluated by ultrasonography in overweight or obese postmenopausal women," *Journal of Investigative Medicine*, vol. 54, no. 8, pp. 455–460, 2006.

[48] P. A. Kern, S. Ranganathan, C. Li, L. Wood, and G. Ranganathan, "Adipose tissue tumor necrosis factor and interleukin-6 expression in human obesity and insulin resistance," *American Journal of Physiology—Endocrinology and Metabolism*, vol. 280, no. 5, pp. E745–E751, 2001.

[49] Y. Kaya, A. Çeb, N. Söylemez, H. Dem, H. H. Alp, and E. Bakan, "Correlations between oxidative DNA damage, oxidative stress and coenzyme Q10 in patients with coronary artery disease,"

International Journal of Medical Sciences, vol. 9, no. 8, pp. 621–626, 2012.

[50] H. K. Vincent and A. G. Taylor, "Biomarkers and potential mechanisms of obesity-induced oxidant stress in humans," *International Journal of Obesity*, vol. 30, no. 3, pp. 400–418, 2006.

[51] M. M. Poplawski, J. W. Mastaitis, F. Isoda, F. Grosjean, F. Zheng, and C. V. Mobbs, "Reversal of diabetic nephropathy by a ketogenic diet," *PLoS ONE*, vol. 6, no. 4, Article ID e18604, 2011.

Awareness and Perception of Plant-Based Diets for the Treatment and Management of Type 2 Diabetes in a Community Education Clinic: A Pilot Study

Vincent Lee,[1] Taylor McKay,[2] and Chris I. Ardern[1,3]

[1]*Diabetes Education Centre, Southlake Regional Health Centre, Newmarket, ON, Canada L3Y 2B1*
[2]*Department of Human Health and Nutritional Science, University of Guelph, Guelph, ON, Canada N1G 2W1*
[3]*School of Kinesiology and Health Science, York University, Toronto, ON, Canada M3J 1P3*

Correspondence should be addressed to Vincent Lee; vlee@southlakeregional.org

Academic Editor: C. S. Johnston

Objective. To assess awareness, barriers, and promoters of plant-based diet use for management of type 2 diabetes (T2D) for the development of an appropriate educational program. *Design.* Cross-sectional study of patients and healthcare providers. *Setting.* Regional Diabetes Education Centre in ON, Canada. *Participants.* n = 98 patients attending the Diabetes Education Centre and n = 25 healthcare providers. *Variables Measures.* Patient questionnaires addressed demographics, health history, and eating patterns, as well as current knowledge, confidence levels, barriers to, promoters of, and interests in plant-based diets. Staff questionnaires addressed attitudes and current practice with respect to plant-based diets. *Analysis.* Mean values, frequency counts, and logistic regression (alpha = 0.05). *Results.* Few respondents (9%) currently followed a plant-based diet, but 66% indicated willingness to follow one for 3 weeks. Family eating preferences and meal planning skills were common barriers to diet change. 72% of healthcare providers reported knowledge of plant-based diets for diabetes management but low levels of practice. *Conclusions and Implications.* Patient awareness of the benefits of a plant-based diet for the management of diabetes remains suboptimal and may be influenced by perception of diabetes educators and clinicians. Given the reported willingness to try (but low current use of) plant-based diets, educational interventions targeting patient and provider level knowledge are warranted.

1. Introduction

Diabetes has become a global epidemic affecting an estimated 371 million people (in 2012), a number that is expected to reach 552 million by 2030 [1]. With healthcare costs approaching $490 billion for the treatment of diabetes [2], alternative (patient-centered) lifestyle management approaches and cost-effective dietary interventions such as plant-based diets are a focus of increasing attention [3].

Recent research has revealed that 58% of type 2 diabetes (T2DM) cases can be prevented or delayed through lifestyle changes such as increased physical activity, healthy eating, and weight loss [3]. Other large cohort studies have shown that the prevalence of T2DM is significantly lower amongst people following a range of plant-based diets [4–6] and that

those with greater adherence to plant-based foods, such as a low-fat vegan diet, experience the greatest benefit. Tuso et al. (2013) define a plant-based diet as a regimen that encourages whole, plant-based foods and discourages meats, dairy products, and eggs as well as all refined and processed foods [7]. (The definition of other variants of plant-based diets is included in the questionnaire.) Various studies suggest that plant-based diets can be an effective Medical Nutrition Therapy (MNT) for the treatment and management of T2DM [8], specifically by improving body weight, cardiovascular risk factors, and insulin sensitivity [9–11] and reducing the need for diabetic medications [12–14]. Providing MNT to people with diabetes demonstrates effectiveness in reducing hospitalization and physician services by 9.5% and 23.5%, respectively, which, in turn, reduces healthcare costs in the long-term [15]. Studies show that a plant-based diet is as effective, if not more effective than an ADA-recommended diabetes diet at reducing certain clinical markers such as HbA1c levels [14]. With the growing body of evidence, the new 2013 *Canadian Diabetes Association Clinical Practice Guidelines* (CDACPG 2013) recommend the use of plant-based diets for management of T2DM [16]. However, this dietary pattern is often perceived to be extreme and difficult to follow, and this perception may be influenced by the healthcare providers that diabetic patients encounter. Despite a strong understanding of the health benefits of a plant-based diet, healthcare providers commonly cite low patient interest and difficulties in facilitating patient adoption as reasons for not promoting plant-based diets.

In order to provide insight into the justification for (and development of) an effective and patient-focused education program, a survey of patients and clinicians was undertaken to assess the awareness, confidence, perceived barriers and promoters, and educational needs for using a plant-based diet in the prevention and management of T2DM.

2. Methods

2.1. Participants and Study Design. The Diabetes Education Centre (DEC) at Southlake Regional Health Centre (SRHC) provides assessment and therapeutic and self-management education for adults with type 1, type 2, gestational diabetes, and prediabetes in York Region, ON, Canada. With the mission of providing a broad-based education on the prevention and management of diabetes, the DEC has approximately 12,500 patient visits annually. This pilot study was approved by the Research Ethics Board prior to patient enrolment and study commencement. Participants from the DEC community were subsequently recruited for one of two surveys: a patient survey or a health professional survey. All patients visiting the clinic for an appointment during the survey period were approached in the waiting room and given the option to complete the patient survey. The patient survey ran from April 22, 2013, to June 5, 2013, and 100 individuals agreed to participate. Inclusion criteria for patient participation included being a patient of the DEC diagnosed with prediabetes, type 1 diabetes, or type 2 diabetes. Patients with gestational diabetes and type 1 diabetes on insulin pump therapy and nonpatients were excluded from the study. The final analytic sample included 98 individuals (prediabetes: $n = 14$; type 1 diabetes: $n = 17$; and T2DM: $n = 62$).

2.2. Patient Questionnaire. Survey data was derived from dichotomous (yes/no) and Likert-type scale closed-ended questions. Additional open-ended questions were used to acquire more specific demographics, health history, and behavioural information (e.g., height, weight, and opinions about diabetes education needs). Since a validated questionnaire in this particular topic was not available, questions were carefully designed to address the following areas: (1) present knowledge, (2) confidence level, (3) potential barriers/promoters, and (4) interests and needs for establishing a future education program.

2.3. Staff Questionnaire. Staff members of the diabetes team were also asked to provide responses to a brief questionnaire on their attitudes and practices regarding plant-based diets. The health professional survey was offered to all staff members working at the DEC and included registered nurses (RN), endocrinologists, and registered dietitians (RD). The survey ran from March 25, 2013, to April 12, 2013, and was completed by 25 staff members: 11 RN, 1 endocrinologist, and 13 RD.

2.4. Data Analysis. Mean values (μ) and frequency counts (n, %) were used to describe the demographics, health history, and dietary practices and perceptions of participants for continuous and categorical variables, respectively. After developing a character profile of participants (e.g., Body Mass Index (BMI), diabetes type, duration of diabetes, new versus continuing patient, weight management strategies, etc.), logistic regression was used to explore the relationship between clinical and education-related factors on the willingness of patients to change to a vegetarian diet. All analyses were conducted using SPSS (v 19), with significance set at alpha = 0.05.

3. Results

3.1. Patient Knowledge and Perceived Barriers to Uptake of a Vegetarian Diet. In general, study participants tended to be male (55%), over age 50 (71%), be overweight or obese (73%), have T2DM (68%), be diagnosed in the last 10 years (65%), and be returning patients (55%) (Table 1). The majority of patients (89%) had not heard of using a plant-based diet to treat or manage T2DM. Furthermore, only 8 (9%) participants reported adherence to a plant-based diet of any type, 3 of whom had followed the diet for less than 1 year. Given the appropriate support, 66% of nonvegetarians were willing to follow a trial plant-based diet for 3 weeks. Nonetheless, almost half of participants cited concerns regarding "family eating habits" (48%), a lack of "meal planning skills" (45%), and a "preference to eat meat" (45%) as primary deterrents to following a plant-based diet. Other factors such as "food cost" (22%), "ease of cooking" (19%), "time constraints" (19%), and "other" factors (6%) were also common (results not shown). Few respondents were confident in their ability to follow a vegetarian (vegan, pesco-, or lacto-ovo) diet, with 17–28% of participants indicating that they were "not confident at all" (results not shown).

Overall, less than half of all participants were aware of the benefits of a plant-based diet to improve diabetes, weight, heart disease, high blood pressure, or high cholesterol. Awareness also varied according to DEC attendance (Figure 1(a)) and willingness to try a plant-based diet (Figure 1(b)) While there was a trend for higher awareness within those who were willing to try plant-based diets and those who were returning patients in the DEC, these differences did not reach statistical significance (Figure 2).

When asked what supports would benefit dietary change, 22% of participants indicated that they did not intend on making a change. Stratified by time since diagnosis, more longer-term than newly diagnosed diabetics (30% versus 10%,

TABLE 1: Knowledge and perception of plant-based diets in patients attending a Diabetes Education Centre.

Age	
<50 y	28 (28.6%)
≥50 y	70 (71.4%)
Sex (% male)	54 (55.1%)
Body Mass Index	
Normal weight (18.5–24.9 kg/m^2)	26 (26.5%)
Overweight (25.0–29.9 kg/m^2)	26 (26.5%)
Obese (≥30.0 kg/m^2)	46 (46.9%)
Diabetes type	
Prediabetes	14 (15.1%)
Type 1 diabetes	17 (18.7%)
Type 2 diabetes	62 (68.1%)
Time since diagnosis of diabetes*	
0–10 years	57 (65.5%)
10+ years	30 (34.5%)
Dietary practices	
Not on plant-based diet	85 (91.4%)
Semivegetarian	6 (6.5%)
Pesco-vegetarian	2 (2.2%)
Patient history in diabetes clinic	
New patient	41 (44.6%)
Returning patient	52 (55.4%)

Note: values may not add up to 100% due to missing responses and rounding.
*excludes $N = 11$ prediabetics.
Values for continuous measures are μ. Categorical measures are N (%).

(a)

(b)

FIGURE 1: Percentage of patients who are aware of the benefits of a plant-based diet on various chronic conditions. Chi-square analysis comparing willingness to change diet and status of patient, all nonsignificant.

$P < 0.05$) were unwilling to consider a plant-based diet, despite greater awareness of alternative dietary treatments to diabetes. At the bivariate level, patients interested in educational resources (OR = 42.9, 95% CI: 12.9–142.4) and those who were motivated by potential health (13.0, 4.9–34.2) or weight loss (4.0, 1.6–9.6) benefits of a plant-based diet had higher odds of being willing to make the necessary change (Table 2). Further adjustment for age and sex only served to strengthen the association between motivation and willingness to change [health: 17.0 (5.9–48.8); weight loss: 4.8 (1.9–12.0)].

3.2. Staff Perception and Use of Plant-Based Diets. A majority of staff (72%) were aware of the use of plant-based diets for treatment of T2DM, but only 32% are currently recommending this dietary pattern to patients (Table 3). While the reasons are likely to vary by clinician and individual patient risk profile, the three most commonly cited reasons were as follows: (1) this eating pattern is not realistic and too difficult to adhere to (and could lead to meal imbalance); (2) there is low perceived acceptance by patients; (3) there are lack of clear clinical practice guidelines and diet-specific educational support.

4. Discussion and Conclusion

4.1. Discussion of Patient Questionnaire Results. Study results reflect that approximately 89% of patients were not aware

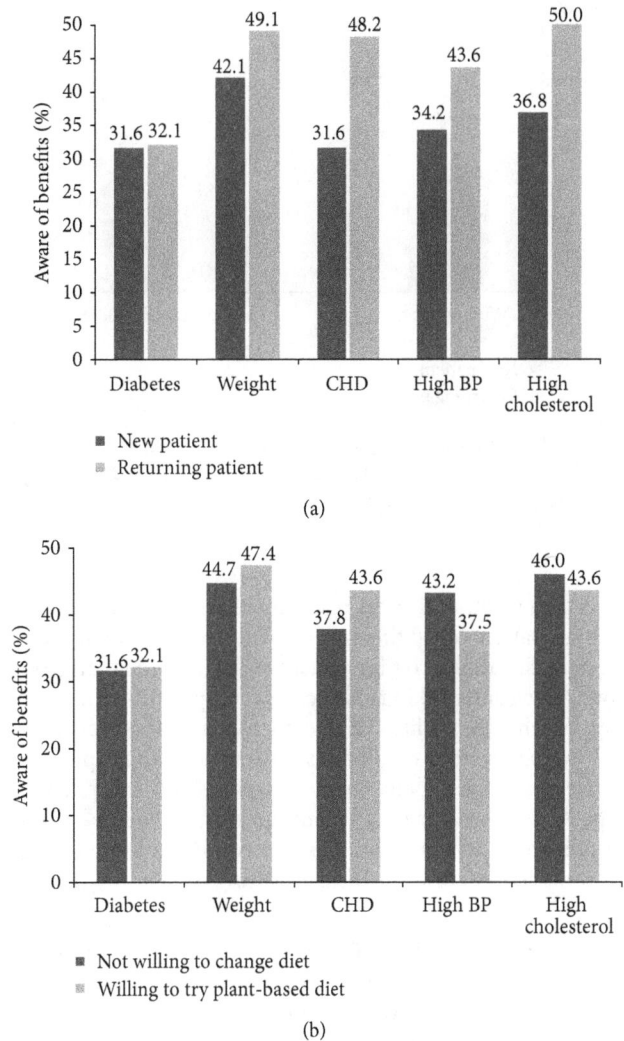

of using an alternate diet such as a plant-based diet for the prevention and management of T2DM and many of them cited low confidence in adopting this eating pattern. However, two-thirds of the patients showed willingness to follow a plant-based diet for the short-term and expressed interest in attending a vegetarian education program. Patients' low awareness and confidence level on the use of plant-based diets for managing T2DM can be partially attributed to the fact that, despite the growing interest in the health benefits of a plant-based diet, the vegetarian population remains relatively small in Canada (4%) [17]. Another plausible reason is that the patients were not well informed about this eating pattern, as less than half of the diabetes team recommended this dietary pattern to patients, potentially influencing their awareness and confidence level.

The top three barriers for making dietary changes towards a plant-based diet included family's influence, preference

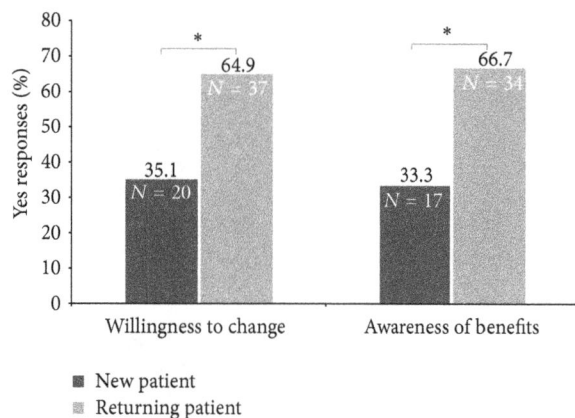

FIGURE 2: Awareness of the benefits and willingness to try a plant-based diet in new and returning patients. Chi-square analysis comparing new and returning patients; $^{*}P < 0.05$.

of eating meat, and meal planning skills. To promote this change, patients cited their top educational needs to be a vegetarian education program consisting of individual or group counselling and cooking instructions components. This result suggests that the traditional theory-based nutrition education at the DEC setting may be insufficient to address patients' barriers; a bigger focus should be placed on the practical aspects, such as teaching patients and family members how to prepare appetizing plant-based meals, in order to change their perception towards this new eating pattern.

4.2. Discussion of Staff Perception. One of the common reasons for diabetes educators not to be recommending this diet to patients was that this dietary approach is too difficult to follow with low perceived approval (i.e., patients are unlikely to accept it). This notion is contrary to the patient survey results that almost two-thirds of patients were willing to follow this dietary pattern at least for short-term when educational support is provided. Katcher et al. (2010) also indicate in a workplace study that a vegan diet is well accepted with over 95% adherence rate, and subjects report increased energy level, better digestion, better sleep, and increased satisfaction when compared with the control group [18]. Participants recruited for the Katcher et al. study that had a BMI ≥ 25 and/or previous diagnosis with type 2 diabetes were randomized into a low fat vegan diet group or a placebo group for 22 weeks. The treatment group received weekly instructions but no meals were provided [18]. Previous studies also demonstrate that the adherence and acceptability of a vegetarian diet are comparable to those of other therapeutic diets [19–21]. A number of staff members also expressed their second reason for not recommending this diet as the clinical practice guidelines and scientific evidence regarding this dietary pattern are not clear. Although randomized controlled intervention studies regarding the use of plant-based diets for the treatment of diabetes have been rather limited until recently, a number of high profile studies show that this diet is not only nutritionally adequate for long-term use [14, 22] but also effective in promoting weight loss, reducing

TABLE 2: Unadjusted logistic regression between clinical and patient-education factors on willingness to change to a vegetarian diet*.

	Odds ratio (95% CI)
Patient interest in education on vegetarian diets	
No	1.0 (referent)
Yes	**42.9 (12.9–142.4)**
Interest in plant-based diet is to improve health	
No	1.0 (referent)
Yes	**13.0 (4.9–34.2)**
Interest in plant-based diet is to lose weight	
No	1.0 (referent)
Yes	**4.0 (1.6–9.6)**
Age	
18–29 y	1.0 (referent)
30–49 y	4.0 (0.6–27.4)
50–65 y	3.5 (0.6–20.1)
65+ y	1.60 (0.2–11.1)
Sex	
Female	1.0 (referent)
Male	1.3 (0.6–2.9)
Demographic and clinical characteristics	
Prediabetes	1.0 (referent)
Type 1 diabetes	1.4 (0.3–5.9)
Type 2 diabetes	1.0 (0.3–3.1)
Time since diabetes diagnosis	
0–10 y	1.0 (referent)
10+ y	1.3 (0.5–3.1)
Body Mass Index	
Normal weight (18.5–24.9 kg/m^2)	1.0 (referent)
Overweight (25.0–29.9 kg/m^2)	0.9 (0.3–2.6)
Obese (≥30.0 kg/m^2)	1.1 (0.4–3.0)
Confidence in becoming vegetarian	
Somewhat confident or confident	1.0 (referent)
Not at all confident	1.2 (0.5–2.9)
Heard of a plant-based diet	
No	1.0 (referent)
Yes	2.1 (0.5–8.3)
Aware of benefits of a plant-based diet	
No	1.0 (referent)
Yes	1.3 (0.6–2.8)
Patient history in diabetes clinic	
First visit	1.0 (referent)
Returning patient	1.3 (0.5–3.1)

*Sample includes only participants who are not currently on a plant-based diet ($N = 85$). Significant associations are presented in bold.

insulin resistance, reducing diabetes medications (43% versus 5%) [10], and improving plasma lipids levels and overall glycemic control [12, 13]. Studies show that a plant-based vegan diet may be as effective as an ADA-recommended diet at causing weight loss [12, 13] and decreasing fat intake [22].

TABLE 3: Staff perception and recommendation for patient use of plant-based diets.

Heard of using a plant-based diet to treat diabetes	
Yes	18 (72.0%)
No	6 (24.0%)
No response	1 (4.0%)
Perceived confidence planning a plant-based diet	
Confident	8 (32.0%)
Somewhat confident	3 (12.0%)
Not confident	10 (40.0%)
No response	4 (16.0%)
Current practice regarding plant-based diets	
Currently recommending	8 (32.0%)
Not recommending	14 (56.0%)
No response	3 (12.0%)

Note: values may not add up to 100% due to missing responses and rounding. Numbers are N (%).

It may be more effective than an ADA-recommended diet at reducing the use of diabetes medication, HbA1c levels, and plasma lipids [13]. It is therefore possible that our survey simply captured a lag-time in dissemination of this new information from current research findings to clinician to patient. Alternatively, because there are a number of vegetarian food guides and practical guidelines that can be used by nutritional professionals [14, 23, 24], the disparity in recommendations and the varying effect of plant-based diets may in themselves be a challenge for nutrition counselling.

4.3. Limitations. As with any study, the results of this preliminary survey must be interpreted with caution. First, the small sample size (n = 98) and selective nature of participant recruitment limit the applicability of the results to a larger and more general population. Second, the patients who completed the surveys may be individuals who were already more interested in vegetarian diets and have a healthier risk profile overall. It is also possible that not all patients understood the terminology being used. When designing the patient and staff questionnaires, it was intended to compare the acceptance and confidence levels of using different types of vegetarian diets such as lacto-ovo vegetarian, pesco-vegetarian, semivegetarian, and vegan diets. However, as many patients had limited exposure to these diets, many of these questions were left unanswered or a single response was selected as an overall rating for all diet types. The questions used in both surveys were not validated due to the preliminary nature of the study as well as a lack of validated questionnaires available for use in this particular subject area. This may limit the ability to compare the current study to other similar studies. Although more than half (66%) of patients expressed an interest in following a plant-based diet, this interest was expressed only for the three-week period; therefore it does not guarantee long-term interest in or attendance in education programs or the longer-term change in eating habits that would be required for effective diabetes management. Finally, the heterogeneity of

the surveyed clinicians limits the ability of the current study to ascertain a certain opinion of a given professional group.

4.4. Conclusion. Patient awareness of (and interest in) the benefits of a plant-based diet for the management of diabetes remains suboptimal and may be influenced by the perception of diabetes educators and clinicians. To provide assurance of the acceptability and efficacy of plant-based diets to patients, offering diet-specific education programs by nutrition professionals in community-based diabetes centres is warranted. Developing these programs in partnership with local nutrition service providers such as community kitchens, grocery stores, and local food network could foster exchange of teaching experience and new perspectives amongst educators and enable sharing of important teaching resources such as a demonstration kitchen. As such, additional training on plant-based diets may require the development of a more standardized and user-friendly practice guideline on plant-based diets to facilitate patient education. With its proven multiple health benefits, a plant-based diet has clearly shown to be beneficial in improving clinical outcomes, and also it has great potential to alleviate healthcare cost in the prevention and management of diabetes as well as other chronic diseases. The current study provides support for the need to further investigate the cost-effectiveness of this dietary pattern in a clinical setting.

4.5. Implications for Future Practice. There is now considerable evidence to support the use of plant-based diets as an effective Medical Nutrition Therapy for chronic diseases such as T2DM [8, 10, 12–14]. Nonetheless, results from this pilot study suggest low awareness and confidence, but a willingness to try a plant-based eating pattern, which supports the need for a patient-focused vegetarian education program. For diabetes educators and registered dietitians, developing diet-specific education programs (such as the Mediterranean and vegetarian diets) are now supported within best practice guidelines such as CDACPG 2013. Depending on available resources, the nutrition program should consist of one-on-one or group counseling sessions and cooking instructions components, and educators should address the patient's barriers to change (e.g., family's eating preference and meal planning skills) to increase a patient's likelihood of making long-term lifestyle changes.

Disclosure

The authors confirm that all patient/personal identifiers have been removed or disguised so the patient/person(s) described are not identifiable and cannot be identified through the details of the story.

Conflict of Interests

The authors declare that there is no conflict of interests regarding the publication of this paper.

References

[1] Clinical Practice Guidelines Expert Committee, "Clinical practice guideline—introduction," *Canadian Journal of Diabetes*, vol. 37, pp. S1–S3, 2013.

[2] World Health Organization, "World Health Statistics 2008," http://www.who.int/whosis/whostat/2008/en/index.html.

[3] Diabetes Prevention Program Research Group, "Reduction in the incidence of type 2 diabetes with lifestyle intervention or metformin," *The New England Journal of Medicine*, vol. 346, no. 6, pp. 393–40, 2002.

[4] S. Tonstad, T. Butler, R. Yan, and G. E. Fraser, "Type of vegetarian diet, body weight, and prevalence of type 2 diabetes," *Diabetes Care*, vol. 32, no. 5, pp. 791–796, 2009.

[5] G. E. Fraser, "Associations between diet and cancer, ischemic heart disease, and all-cause mortality in non-Hispanic white California Seventh-day Adventists," *The American Journal of Clinical Nutrition*, vol. 70, no. 3, supplement, pp. 532S–538S, 1999.

[6] D. A. Snowdon and R. L. Phillips, "Does a vegeterian diet reduce the occurrence of diabetes?" *American Journal of Public Health*, vol. 75, no. 5, pp. 507–512, 1985.

[7] P. J. Tuso, M. H. Ismail, B. P. Ha, and C. Bartolotto, "Nutritional update for physicians: plant-based diets," *The Permanente Journal*, vol. 17, no. 2, pp. 61–66, 2013.

[8] J. G. Pastors, H. Warshaw, A. Daly, M. Franz, and K. Kulkarni, "The evidence for the effectiveness of medical nutrition therapy in diabetes management.," *Diabetes Care*, vol. 25, no. 3, pp. 608–613, 2002.

[9] C.-J. Hung, P.-C. Huang, Y.-H. Li, S.-C. Lu, L.-T. Ho, and H.-F. Chou, "Taiwanese vegetarians have higher insulin sensitivity than omnivores," *British Journal of Nutrition*, vol. 95, no. 1, pp. 129–135, 2006.

[10] H. Kahleova, M. Matoulek, H. Malinska et al., "Vegetarian diet improves insulin resistance and oxidative stress markers more than conventional diet in subjects with type2 diabetes," *Diabetic Medicine*, vol. 28, no. 5, pp. 549–559, 2011.

[11] L. M. Goff, J. D. Bell, P. W. So, A. Dornhorst, and G. S. Frost, "Veganism and its relationship with insulin resistance and intramyocellular lipid," *European Journal of Clinical Nutrition*, vol. 59, no. 2, pp. 291–298, 2005.

[12] N. D. Barnard, J. Cohen, D. J. A. Jenkins et al., "A low-fat vegan diet improves glycemic control and cardiovascular risk factors in a randomized clinical trial in individuals with type 2 diabetes," *Diabetes Care*, vol. 29, no. 8, pp. 1777–1783, 2006.

[13] N. D. Barnard, J. Cohen, D. J. A. Jenkins et al., "A low-fat vegan diet and a conventional diabetes diet in the treatment of type 2 diabetes: a randomized, controlled, 74-wk clinical trial," *The American Journal of Clinical Nutrition*, vol. 89, no. 5, 2009.

[14] American Dietetic Association, "Position of the American Dietetic Association: vegetarian diets," *Journal of the American Dietetic Association*, vol. 109, no. 7, pp. 1266–1282, 2009.

[15] J. F. Sheils, R. Rubin, and D. C. Stapleton, "The estimated costs and savings of medical nutrition therapy: the medicare population," *Journal of the American Dietetic Association*, vol. 99, no. 4, pp. 428–435, 1999.

[16] P. D. Dworatzek, K. Arcudi, R. Gougeon, N. Husein, J. L. Sievenpiper, and S. L. Williams, "Nutrition therapy," *Canadian Journal of Diabetes*, vol. 37, no. 1, pp. S45–S55, 2013.

[17] American Dietetic Association, "Position of the American dietetic association and dietitians of Canada: vegetarian diets," *Journal of the American Dietetic Association*, vol. 103, no. 6, pp. 748–765, 2003.

[18] H. I. Katcher, H. R. Ferdowsian, V. J. Hoover, J. L. Cohen, and N. D. Barnard, "A worksite vegan nutrition program is well-accepted and improves health-related quality of life and work productivity," *Annals of Nutrition and Metabolism*, vol. 56, no. 4, pp. 245–252, 2010.

[19] N. D. Barnard, L. W. Scherwitz, and D. Ornish, "Adherence and acceptability of a low-fat, vegetarian diet among patients with cardiac disease," *Journal of Cardiopulmonary Rehabilitation*, vol. 12, no. 6, pp. 423–431, 1992.

[20] N. Barnard, A. R. Scialli, P. Bertron, D. Hurlock, and K. Edmunds, "Acceptability of a therapeutic low-fat, vegan diet in premenopausal women," *Journal of Nutrition Education and Behavior*, vol. 32, no. 6, pp. 314–319, 2000.

[21] N. D. Barnard, A. R. Scialli, G. Turner-McGrievy, and A. J. Lanou, "Acceptability of a very-low-fat, vegan diet compares favorably to a more moderate low-fat diet in a randomized, controlled trial," *Journal of Cardiopulmonary Rehabilitation*, vol. 24, pp. 229–235, 2004.

[22] G. M. Turner-McGrievy, N. D. Barnard, J. Cohen, D. J. A. Jenkins, L. Gloede, and A. A. Green, "Changes in nutrient intake and dietary quality among participants with type 2 diabetes following a low-fat vegan diet or a conventional diabetes diet for 22 weeks," *Journal of the American Dietetic Association*, vol. 108, no. 10, pp. 1636–1645, 2008.

[23] V. Messina, V. Melina, and A. R. Mangels, "A new food guide for North American vegetarians," *Journal of the American Dietetic Association*, vol. 103, no. 6, pp. 771–775, 2003.

[24] *General Conference Nutrition Council. My Vegetarian Food Pyramid*, Loma Linda University, 2014, http://www.llu.edu/llu/nutrition/vegfoodpyramid.pdf.

Evaluation of Antioxidant Capacity of
Solanum sessiliflorum (Cubiu) Extract: An *In Vitro* Assay

Diego Rocha de Lucena Herrera Mascato,[1] **Janice B. Monteiro,**[2]
Michele M. Passarinho,[1] **Denise Morais Lopes Galeno,**[1] **Rubén J. Cruz,**[3]
Carmen Ortiz,[4] **Luisa Morales,**[5] **Emerson Silva Lima,**[6] **and Rosany Piccolotto Carvalho**[1]

[1]*Programa Multi-Institucional de Pós-Graduação em Biotecnologia, Universidade Federal do Amazonas,*
69077-000 Manaus, AM, Brazil
[2]*Division of Biochemistry, Ponce Health Sciences University, School of Medicine, Ponce Research Institute, Ponce, PR 00732-7004, USA*
[3]*Biology Department, University of Puerto Rico at Ponce, Ponce, PR 00734, USA*
[4]*Divisions of Pharmacology & Toxicology & Cancer Biology, Ponce Health Sciences University, School of Medicine,*
Ponce Research Institute, Ponce, PR 00732-7004, USA
[5]*Public Health Program, Ponce Health Sciences University, Ponce, PR 00732-7004, USA*
[6]*Faculdade de Ciências Farmacêuticas, Universidade Federal do Amazonas, 69010-300 Manaus, AM, Brazil*

Correspondence should be addressed to Carmen Ortiz; caortiz@stu.psm.edu
and Rosany Piccolotto Carvalho; prosany@ufam.edu.br

Academic Editor: C. S. Johnston

Cubiu is a vegetable of Solanaceae family, native to the Amazon, which is widely distributed through Brazil, Peru, and Colombia. It is used in food, medicine, and cosmetics by native populations. Research has shown that cubiu extracts have antioxidant activities with great biological relevance. We performed a phytochemical screening to identify the main chemical groups that could confer antioxidant activity to this extract. Several tests and qualitative precipitation specific staining for major classes of secondary metabolites were used. Antioxidant capacity *in vitro* tests (DPPH and ABTS) were also used to assess the extract's ability to sequester free radicals of 70% hydroethanolic and aqueous extracts of cubiu flour. Alkaloids, organic acids, phenols, flavonoid glycosides, and coumarins were found in the hydroethanolic extract while the aqueous extract presented anthocyanins, gums, tannins and mucilage, amino groups, and volatile and fixed acids. For *in vitro* tests, the IC_{50} value obtained in the DPPH assay was $606.3 \pm 3.5\,\mu g/mL$ while that for the ABTS assay was $290.3 \pm 10.7\,\mu g/mL$. Although cubiu extracts present chemical compounds directly related to antioxidant activity, our results show that it has a low antioxidant activity. Additional studies will be needed to isolate and characterize specific compounds to further assess antioxidant activity.

1. Introduction

Cubiu (*Solanum sessiliflorum* Dunal), also known in Spanish-speaking countries as "tupiro," "topiro," and "cocona" and known as "Orinoco apple" and "peach tomato" in English-speaking countries, is a vegetable of the family Solanaceae which is native to the Amazon [6] and widely distributed through the humid equatorial regions of Brazil, Peru, and Colombia. Its domestication was made by the Indians of Western Amazonia [7]. The fruit can weigh between 20 and 450 grams and contains between 200 and 500 oval flat seeds. The fruits have the most diverse forms. Those cylindrical in shape have generally 4 loci, while the heart-shaped, round, and flattened ones have between 6 and 8 loci. The fruit color changes as it matures, ranging from green when it is not ripe to yellow-orange when it is mature and finally to red when

TABLE 1: Chemical composition of cubiu (*Solanum sessiliflorum*) in 100 g of whole pulp [1–5].

Component	Villachica (1996) [4]	Pahlen (1977) [3]	Andrade et al. (1997) [1]	Yuyama et al. (1997, 1998) [2, 5]
Units (g)	89	91	93	90
Energy (kcal)	41	33	31	45
Proteins (g)	0.9	0.6	—	0.9
Lipids (g)	—	1.4	—	1.9
N-free extract (g)	—	5.7	—	4.7
Fiber (g)	0.2	0.4	—	0.9
Ashes (g)	0.7	0.9	—	0.9
Total sugars (%)	—	—	4.6	—
Reducing sugars (%)	—	—	3.9	1
Nonreducing sugars (%)	—	—	1.8	1
Soluble solids (°Brix) %	—	5	8	—
Citric acid (%)	—	—	0.8	—
Brix/acidity	—	—	5.9	—
Phenolics (mg)	—	—	14.4	—
Tannins (mg)	—	—	142	—

coffee is ready to be consumed. The fruit has a pulp color ranging from light yellow to yellowish cream, measuring between 0.2 and 2.5 cm thick [6].

The pulp is the edible part of the fruit. As food, it is consumed "naturally" as strip-like alcoholic beverages, jellies, jams, juices, and sauces [1, 8]. The macerated leaves of the plant are used by Peru-Brazil indigenous people to avoid the formation of blisters in the event of skin burns, while the juice from the fruit's locular cavity is used to alleviate the itching of spider bites [9]. The pure juice is used for the control of cholesterol, diabetes, excess uric acid, and other affections caused by kidney and liver failure, besides being used to eliminate lice [9]. Previous studies have shown the composition of cubiu as presented in Table 1.

Cubiu is a juicy fruit with humidity between 88 and 93%. It has a variable soluble solids content (°Brix) between 5 and 8, of which the majority are reducing sugars [1]. Since the ratio of °Brix and acidity of the fruit is low, it is considered acidic. This ensures a high dilution factor (great for juice production) and little preference for fresh consumption (due to sourness of the fruit). The composition of phenolic compounds is also low [1]. Taking into consideration the nutritional value of cubiu (Table 2), it presents a high dietary potential due to its low caloric and high fiber content. This supports its indication in the diet of hypercholesterolemic and hyperglycemic patients [2].

In a study conducted in Norbert Wiener University in Lima (Peru), it was found that administration of cubiu extract (40 mL/day) for three days decreased the level of LDL, triglycerides, serum cholesterol, and glucose. However, it increased serum HDL levels in 100 volunteers of both sexes [10]. Furthermore, it was found that the fruit has a phenolic compound called tannin. Phenolic compounds are products of secondary plant metabolism which are essential for their growth and reproduction. Tannins are formed under conditions of stress such as infection, injury, and UV radiation [11]. These also act as natural

second class antioxidants [12]. According to their mode of action, antioxidants can be classified into primary and secondary. Primary antioxidants act by donating hydrogen or electrons to free radicals, therefore, converting them into stable products and/or reacting with free radicals, forming lipid-antioxidant complexes that can react with another free radical. Secondary antioxidants act by delaying the initiation step of autooxidation by different mechanisms including metal complexation, sequestration of oxygen, decomposition of hydroperoxides to form nonradical species, absorption of UV light, or deactivation of singlet oxygen species [13].

The phenolic antioxidants interact preferentially with the peroxyl radical due to its prevalence in the autooxidation step [15]. This mechanism of action present in plant extracts plays an important role in reducing lipid oxidation in tissues, plants, and animals because it preserves food quality and reduces the risk of developing diseases such as arteriosclerosis and cancer [16, 17]. Since antioxidants are of great relevance for the field of medicine today, there is a need to investigate the antioxidant potential of cubiu and to elucidate the role of this fruit in the diet of the population.

2. Material and Methods

2.1. Feedstock. For this study, cubiu fruits used were obtained from the Horticultural Experimental Station Alejo von der Pahlen from the National Institute of Amazonian Research (NIAR) located in Manaus, AM. The fruits were transported in coolers to the Laboratory of Food and Nutrition at the NIAR where they were processed to obtain cubiu flour as described by da Silva Filho et al. (2005) [18].

2.2. Preparation of Cubiu Flour. After the fruits were received, the ripe intact fruits (free of cuts, holes, and other defects) were selected for processing. Fruits were washed, brushed, sanitized using a 200 ppm hypochlorite solution for

TABLE 2: Vitamin and mineral composition of *Solanum sessiliflorum* in 100 g of cubiu whole pulp according to several studies and percentage of the daily recommendation of the National Research Council (1989) [14].

Component	Villachica (1996) [4]	Pahlen (1977) [3]	Andrade et al. (1997) [1]	Yuyama et al. (1997, 1998) [2, 5]	%NRC
Ascorbic acid (mg)	4.5	—	13.9	—	15.3
Niacin (mg)	2.3	2.5	—	—	14.1
Carotene (mg)	0.2	0.2	—	—	—
Thiamine (mg)	0.1	0.3	—	—	15.4
Riboflavin (mg)	0.1	—	—	—	6.6
Calcium (mg)	16	12	—	—	1.2
Magnesium (mg)	—	—	—	23.7	7.5
Phosphorus (mg)	30	14	—	—	1.8
Potassium (mg)	—	—	—	385.4	19.3
Sodium (mg)	—	—	—	371	74.2
Copper (mg)	—	—	—	329	14.6
Iron (mg)	—	—	—	324	2.6
Zinc (mg)	—	—	—	157	1.1
Manganese (mg)	—	—	—	97	2.8

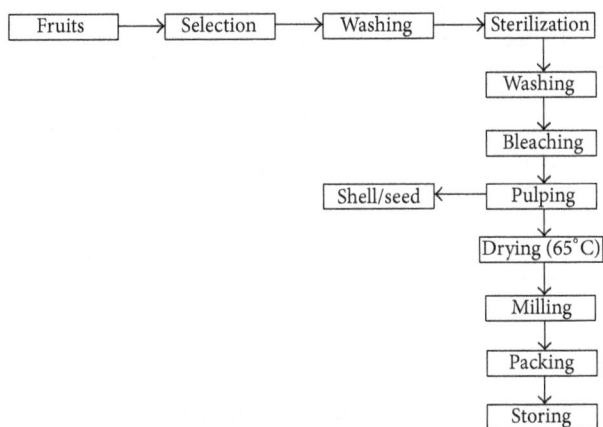

FIGURE 1: Processing of cubiu fruits to obtain the flour as described by da Silva Filho et al. [18].

30 minutes, and rinsed with running water. After this process, the fruits were stripped of their pulp and had the placenta (inner part of the fruit that contains the seeds and the juice of the locular cavity) and the epicarp being removed with a stainless steel knife. Thermal bleaching was then performed at 90°C for 3 minutes, followed by a heat shock in an ice bath for 3 minutes. Bleached fruits were packed in 2 kg polyethylene bags and frozen at −15°C. The fruits were then dried in an oven with forced air circulation at 60°C for 48 hours. After drying, the fruits were liquified, homogenized, packed, and frozen at −15°C until they were grinded to obtain the flour. The flour was packed in plastic bags and stored in a freezer with temperature −15°C until the extract was prepared (Figure 1).

2.3. Preparations of Cubiu Flour Extracts.

Cubiu extracts were prepared by dissolving 2 kg of cubiu flour in 2 L of ethanolic solvent. The solvent was prepared using 500 mL of distilled water and 1500 mL of ethyl alcohol (Sigma,

MO). Mixtures were then placed on the Ultrasonic Cleaner (UNIQUE, Brazil) for 15 minutes to complete homogenization and filtered using filter paper and a disposable glass funnel three times. Then, the extracts were vacuum filtered using a Buchner funnel and an Erlenmeyer filtration flask. Subsequently, the extracts were placed in round bottom flasks for rotoevaporation with a 25°C water bath. The extracts were returned to their viscous form and then freeze-dried using a lyophilizer. After this, they were kept in a freezer at −11°F for later use as described by Rogério (2006) [19].

2.4. Assessment of Antioxidant Capacity

2.4.1. DPPH Method.

The DPPH assay is based on the reduction of 2,2-diphenyl-1-picrylhydrazyl radical (DPPH) through the donation of a hydrogen atom of the test compound to the radical of the molecule [20]. Measurement of DPPH radical scavenging activity was performed according to the methodology described by Molyneux (2004) [21] with some modifications. Initially, 2 mg of DPPH was dissolved in 12 mL of absolute ethanol. In a flat bottom 96-well plate, 270 μL of DPPH solution was mixed with 30 μL ethanol. Thereafter, the test consisted of 270 μL of the solution of DPPH mixed with 30 μL of 70% hydroethanolic cubiu extract. The reaction was incubated for 30 minutes at room temperature in the absence of light. After incubation, the absorbance was read at 517 nm. To express antioxidant activity IC_{50} was calculated, that is, the minimum concentration required to reduce the initial DPPH reaction by 50%. Ascorbic acid was used as standard. The antioxidant activity was calculated using the following formula:

$$\%\text{Inhibition} = 100 - \left(\frac{\text{Abs sample}}{\text{Abs control}} \right) \times 100. \tag{1}$$

2.4.2. ABTS Method.

This method is based on the reduction of ABTS (2,2′azinobis-3-ethylbenzothiazoline-6-sulfonic acid) and was performed as described by Re et al. (1999) [22]

with modifications. ABTS was dissolved in MiliQ water to a 7 mM concentration. ABTS$^{\bullet+}$ was produced by reacting ABTS stock solution with 5 mM potassium persulfate and kept in the dark at room temperature for 12–16 h. Once ABTS$^{\bullet+}$ is formed, MiliQ water was added to the solution (dilution 1 : 7). A 96-well plate flat-bottomed was added by 270 μL of ABTS solution with 30 μL of water. This solution was monitored by reading at 714 nm in a microplate reader (DTX 800, Beckman, CA, USA) to obtain absorbance of approximately 1.00 (control). Then, 30 μL of cubiu extract from different concentrations was added to 270 μL of ABTS$^{\bullet+}$ and the reaction was incubated for 15 min in the dark at room temperature. After incubation, the absorbance at 714 nm was measured. Gallic acid in the same cubiu extract concentrations was measured following the same procedures described above and was used as positive controls. The antioxidant activity was calculated using the following equation:

$$\%\text{Inhibition} = 100 - \left(\frac{\text{Abs sample}}{\text{Abs control}} \right) \times 100. \qquad (2)$$

2.5. Systematic Analysis in Phytochemistry. This assay analyzes all the qualitative features of the principal chemical groups active in natural products, by means of staining or precipitation reactions. The systematic testing for phytochemical analysis was performed according to Moreira (1979) using maceration, aqueous extraction, and hydroethanolic extraction [23].

2.6. Preparation of Aqueous and Hydroethanolic Cubiu Extract. The extracts were performed by dissolving 40 g of the powder of dried cubiu fruits in 200 mL of 70% ethyl alcohol or water. The mixture was placed in an ultrasound at 50°C for 10 min. The mash was filtered through filter paper and supplemented with the same solvent volume to 200 mL. The extract was kept refrigerated until completion of the phytochemical assays.

2.7. Assessment for the Presence of Alkaloids, Organic Acids, and Phenols. Experiments were conducted using general alkaloid reagents (Mayer, Dragendorff, Bouchardat, and Bertrand as described in Table 4) as follows: 50 mL of the hydroethanolic extract was evaporated to dryness in a water bath at 70°C, followed by dissolution of the residue in 1 mL of ethanol and 20 mL of 1% HCl. Extracts were transferred to test tubes and the reagent was added on each one. The appearance of precipitate indicated a positive reaction. To retest, 15 mL of the aqueous extract was transferred to a separating funnel and alkalized with ammonium hydroxide (pH 10). An ether/chloroform (3 : 1) extraction was performed and the extract obtained was also tested for alkaloids. The remaining solution prepared for the study of alkaloids was dried and redissolved in 5 mL of distilled water. The pH of the solution was measured, where an acidic pH indicated the presence of organic acids. Two drops of 1% $FeCl_2$ aqueous solution were added to the solution obtained in the study of organic acids, to confirm the presence of phenols.

2.8. Assessment for the Presence of Flavonoid Glycosides: Oxalic-Boric Complex (R. Wilson-Tauböck), Pacheco, and Shinoda Tests. 10 mL of each of the extractions was dried using a water bath. Five drops of acetone and 30 mg of a mixture of boric acid and oxalic acid at a ratio of 1 : 1 were added to the dried extracts. After stirring and drying the mixture, 5 mL of ethyl ether was added and the solution was transferred to test tubes. Solutions were seen under UV light. The reaction is considered positive when the result has a greenish yellow fluorescent appearance. For the Pacheco test, a few crystals of NaOAc and 100 μL of acetic anhydride were added to the dried extract. The mixture was heated in a water bath and 100 μL of concentrated HCl was added. The result of the reaction is considered positive when there was purple coloring. For the Shinoda test, 5 mL of the hydroethanolic extract was mixed with 200 mg of magnesium turnings and 1 mL of HCl. The formation of orange color indicates the presence of flavonoids.

2.9. Assessment for the Presence of Coumarins. Thirty milliliters of hydroethanolic extract was acidified to pH 1 and concentrated to 10 mL in a water bath at 60°C. The residual extract was mixed with 5 mL of deionized water and transferred into a separation funnel with ethyl ether in 3 portions of 10 mL. The volume of the organic extract was reduced to 5 mL in a water bath at 60°C. Three drops of hydroethanolic extract were placed in two points of a premarked filter paper and left to dry and 1 drop of 1 N NaOH was added on each spot. The stains were covered with a coin and observed under UV light. A blue or yellow-green fluorescence indicated a positive reaction.

2.10. Assessment for the Presence of Anthraquinones. 20 mL of hydroethanolic extract was brought to a boil under reflux for 15 minutes by adding 3 mL of 10% H_2SO_4. After cooling, the mixture was transferred to a separatory funnel along with 30 mL of distilled water to perform an extraction with 10 mL of toluene. The extract was concentrated to 10 mL and mixed with 10 mL of NaOH. The appearance of pink or red color indicated the presence of hydroxy-anthraquinones and naphthoquinones.

2.11. Assessment for the Presence of Sterols and Triterpenes. 20 mL of hydroethanolic extract was evaporated to perform an extraction with dichloromethane. Extracts were concentrated (3 mL) and mixed with 2 mL of acetic anhydride and 3 drops of H_2SO_4. The development of blue-green color demonstrated the presence of steroids and/or triterpenes.

2.12. Assessment for the Presence of Anthocyanin and Saponin Glycosides. Three portions (5 mL each) of aqueous extract were neutralized with 5% KOH until a pH of 5.5 was reached. Changes in color indicated the presence of the neutralized portions of anthocyanin glycosides. To test for the presence of saponin glycosides, the remaining solution was stirred vigorously for 5 min. The appearance of persistent foam in the sample indicated the presence of saponins. This was further confirmed by adding 1% HCl to the sample.

2.13. Assessment for the Presence of Cyanogenic Glycosides. 15 mL of the aqueous extract was transferred to a test tube. After adding 1 mL of 1 N H_2SO_4, a strip of picrosodium paper was secured inside the tube while it was in a water bath at 60°C for 30 minutes. The appearance of red color on the paper indicated the presence of cyanogenic glycosides.

2.14. Assessment for the Presence of Gums, Tannins, and Mucilage. Five drops of 10% basic acetate and neutral lead acetate were added to 2 portions of 5 mL of aqueous extract. The precipitate formation is indicative of the presence of gums, mucilages, and tannins. To assess the presence of tannins, 5 drops of 1% $FeCl_2$ were added to 5 mL of the aqueous extract. Upon dark precipitate formation, 5 mL of the aqueous extract was transferred to a flat bottom flask to add 5 drops of 37% formaldehyde and 4 mL HCl. The mixture was left to reflux for 1 hour and when it cooled, it was filtered and washed with distilled water and alcohol. A few drops of 5% KOH were added to the material retained on the filter. The appearance of color indicated the presence of condensed tannins. The filtrate was then mixed with 10 drops of 1% $FeCl_3$, the formation of a dark blue precipitate confirmed the presence of hydrolyzable tannins.

2.15. Assessment for the Presence of Amino Groups. After concentrating 10 mL of the aqueous extract at a temperature of 50°C, 5 drops of concentrated extract were dropped on a filter paper. After drying, the filter paper was sprayed with a solution of ninhydrin in butanol and heated at 90–100°C for 15 min. The appearance of blue-violet color indicated the presence of amino groups.

2.16. Assessment for the Presence of Volatile and Fixed Acids. 10 mL of aqueous extract was acidified with 1 N H_2SO_4 and brought to a boil in a water bath. A pH strip was used to measure the acidity of the vapors produced during boiling. The acidic color indicates the presence of volatile acids. 20 mL of aqueous extract was transferred to a distillation flask along with 2 mL of 1 N NaOH. The mixture was left to reflux for 30 minutes and after cooling and acidifying with 1 N H_2SO_4 an extraction with ethyl ether was performed. The extracts obtained were filtered and evaporated to dryness. After heating the residue for 10 min at 100°C and adding 5 mL of 1 N NH_4OH, it was filtered again. Three drops were transferred to a filter paper to obtain a spot of 1 cm in diameter. The paper was dried at 100°C for 10 minutes and treated with Nessler's reagent. The color development indicated the presence of fixed acids.

3. Results and Discussion

The results for the systematic phytochemistry analysis can be seen in Tables 3 and 5.

In this first set of assays, we obtained positive results for the presence of alkaloids and most of them precipitated in neutral or slightly acidic medium. It should be emphasized that these precipitates can also be caused by proteins, purines, alpha pyrones, certain coumarins, lignans, and phenolic

TABLE 3: Tests for the 70% hydroethanolic extract of the fruit *Solanum sessiliflorum* Dunal.

Compound classes assessed	Test results
Alkaloids	
Mayer reagent	−
Dragendorff reagent	++
Bouchardat reagent	−
Bertrand reagent	++
Reaction of confirmation	Dragendorff and Bertrand
Organic acids	++ pH = 5.5–6
Phenols	+++
Heterogeneous flavonoids	
Tauböck or Oxalo-boric reaction	+++
Pacheco reaction	−
Shinoda reaction	+
Coumarins	+
Anthraquinones	−
Sterols and Triterpenes	−

+++: strongly positive/++: positive/+: traits/−: negative.

TABLE 4: Reagent composition and color of precipitate for the different tests used to determine the presence of alkaloids on the 70% hydroethanolic extract of the *Solanum sessiliflorum* Dunal fruit.

RGA	Composition	Color of precipitate
(1) Mayer	Potassium-mercuric iodide	Orange
(2) Dragendorff	Bismuth nitrate	White
(3) Bertrand	Silicotungstic acid	White
(4) Bouchard/Wagner	Potassium triiodide	Brown

TABLE 5: Tests for the aqueous fruit extract of *Solanum sessiliflorum* Dunal.

Presence of chemical groups	Test results
Anthocyanin heterosides	++
Saponin glycosides	−
Cyanogenic glycosides	−
Gums, tannins, and mucilage	+++
Tannins	+
Amine groups	+++
Volatile acids	++
Fixed acids	+

+++: strongly positive/++: positive/+: traits/−: negative.

hydroxyl compounds. Negative results with these reagents are indicative of the absence of alkaloids; however, the formation of precipitates can only be considered as probable presence thereof [24].

The acidity of the fruit was confirmed through the measure of pH of 5.5–6. This correlated with the results shown by Andrade et al. 1997 where it was calculated using the ratio of Brix (soluble solids content) and the acidity of the fruit yielding a low ratio, therefore showing its acidic potential.

It also presented a strong positive result for the presence of phenols, in contrast with the findings reported in a study by Andrade et al. (1997) in which a relatively low amount of phenolic compounds (14.4 mg) in 100 g of cubiu whole pulp was found [1]. According to Moreira and Mancini-Filho (2004), phenolic compounds are natural second class or secondary antioxidants [12]. This class of antioxidants acts by delaying the initiation of autooxidation by different mechanisms, for example, hydroperoxide decomposition to form nonradical species [13]. This finding confirms the potential antioxidant activity of cubiu.

In our study, we also found evidence of the presence of flavonic heterosides. Previous studies have shown that flavonoids exert several activities over different biological systems showing antimicrobial, antiviral, antiulcerogenic, cytotoxic, antineoplastic, antioxidant, antihepatotoxic, antihypertensive, hypolipidemic, anti-inflammatory, and antiplatelet activities. Flavonoids have also been found to increase capillary permeability and inhibit protein exudation and leukocyte migration [25]. These effects may be related to the inhibitory effect that flavonoids exert over different enzyme systems including hydrolases, isomerases, oxygenases, oxidoreductases, polymerases, phosphatases, proteins, and amino acid oxidases phosphokinases [26]. The presence of coumarins can be related to antioxidant activities, along with other possible mechanisms, such as anti-inflammatory, antispasmodic, antitumor properties, and interaction with several enzymes.

In the second set of assays (Table 5), the presence of anthocyanin heterosides, acids (volatile and fixed) and, in greater quantity, amino groups and gums, mucilages, and tannins was noted.

The anthocyanin or anthocyanin heterosides are substances belonging to the flavonoid family. These are pigments that give color to flowers, fruits, leaves, stems, and roots of plants [27]. In previous studies, anthocyanins were found to have an inhibitory action on lipid peroxidation [28] and a protective action over vascular epithelial cells against reactive oxygen species, principally to H_2O_2-induced loss of cell viability [29].

Gums are generally considered a pathological result due to physical injury suffered by plant tissues by the action of microorganisms or due to unfavorable conditions such as drought. Mucilage gums were detected by precipitation in tube assays. However, neither of both has antioxidant characteristics [30]. The amino groups were detected by the presence of purple coloring. They are compounds bound to an NH_2 radical, structurally similar to the alkaloids but with no oxidative activity due to the gradual reduction of the radicals present in their structures [24].

The antioxidant activity of the 70% hydroethanolic extract of cubiu was evaluated using in vitro assays: DPPH and ABTS. In this study, the sequestering capacity of the hydroethanolic extract of cubiu was expressed as the final concentration of extract required to inhibit oxidation of DPPH by 50% (Figure 2). The antioxidants in the extract react with DPPH, which is a stable free radical, and convert it into 2,2-diphenyl-1-picryl hydrazine. The degree of discoloration indicates the antioxidant potential of the extract. An extract that has a high antioxidant potential has a low IC_{50} [31].

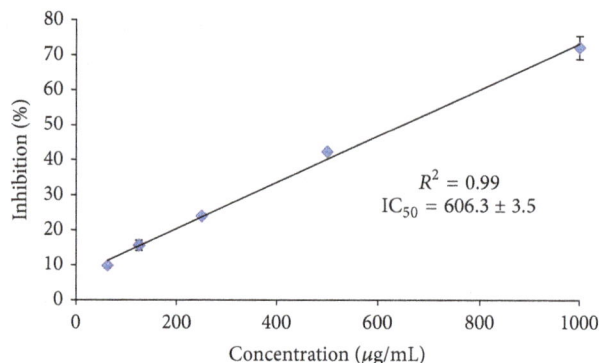

FIGURE 2: Colorimetric assessment of antioxidant capacity using in the DPPH assay. Values represent mean ± standard deviation of the mean of replicate readings ($n = 3$). The IC_{50} values denote the concentration of the sample, which is required to scavenge 50% of free radicals.

Our results show that IC_{50} of cubiu was 606.3 ± 3.5 µg/mL, a high value when compared with ascorbic acid (IC_{50} of 2.74 ± 0.3 µg/mL), which used as a reference substance.

In a study comparing the antioxidant activity of native fruits, the lower IC_{50} values were obtained by gallic acid (1.38 mg/mL), ethanolic and aqueous pequi bark extracts (9.44 and 17.98 g/mL, resp.), ethanolic cagaita seed extract (14.15 mg/mL), and ethanolic extract of araticum seed and peel (30.97 and 49.18 mg/mL, resp.) [32]. In a study that evaluated the antioxidant activity of several brands of orange juice using the DPPH assay, it was found that the ones with activity were the orange file (66.24%) and the orange bahia (60.32%) [33]. However, it is difficult to compare the antioxidant activity of different samples because the authors used different dilutions of the samples to conduct the analysis, since each sample has a different antioxidant power. Furthermore, the ways to analyze and display the results are different. Stratil et al. (2007) reported that the comparison of results regarding the antioxidant capacity published in individual methods and among groups that use the same method is often problematic [34].

Regarding the results obtained by the ABTS method (Figure 3), the values obtained were 290.3 ± 10.7 µg/mL. In a study performed in the Singapore market, the ABTS method was used to measure the Equivalent Antioxidant Capacity L-Ascorbic Acid (EACA) and the results were expressed in mg of ascorbic acid (AA) per 100 g of the homogenate containing the fruit extract. In this study, the fruit with the highest antioxidant capacity was sapodilla (3396 mg/100 g), followed by strawberry (472 mg/100 g), plum (312 mg/100 g), and carambola (278 mg/100 g) [35]. Dos Santos et al. (2008) evaluated the antioxidant capacity of commercial açaí pulp using the ABTS method and expressing the results as TEAC (Trolox Equivalent Antioxidant Capacity). The TEAC of açaí pulp ranged from 10.21 to 52.47 mM of Trolox/g sample. This confirms the difficulty of comparing the antioxidant activity reported in different studies since each one uses a different method of analysis [36].

The results obtained in this study were conflicting when compared to those presented in other studies. Nascimento

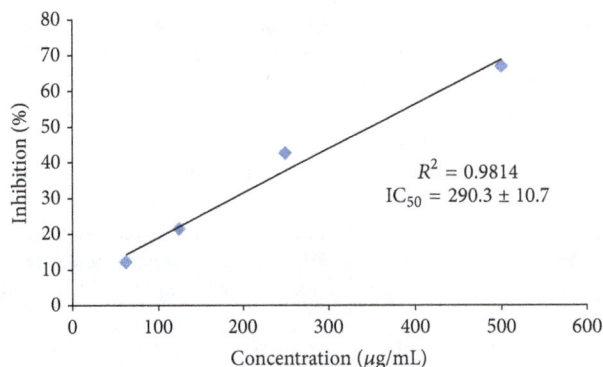

FIGURE 3: Results from the ABTS test. Values represent mean ± standard deviation of the mean of replicate readings (n = 3). The IC_{50} values denote the concentration of the sample, which is required to scavenge 50% of free radicals.

and Pereira (2011) used the DPPH method and stated that cubiu extracts obtained from methanol fraction contained the highest antioxidant activity (between 100 and 200 mE BHT/g of freeze-dried material) [37], which explains the fact that in this study the antioxidant activity was found to be low since the hydroethanolic extract was used. Another interesting finding of the study is that, of all the extracts obtained from different parts of the fruit, the bark extract stood out with higher antioxidant activity, which is usually discarded in food preparations. Ledur et al. (2012) used the DPPH method to evaluate the antioxidant capacity of the 70% hydroethanolic extract of cubiu. The bark extract showed the lowest IC_{50} when compared to the seed and pulp extracts (278.7 mg/mL and 309.98 mg/mL, resp.) [38]. These values are lower than the results obtained in our study; however, they confirm the fact that cubiu has little antioxidant capacity.

4. Conclusion

To achieve the objective of the study, flour of cubiu (*Solanum sessiliflorum* Dunal) fruit was prepared and used for the preparation of the 70% hydroethanolic extract that was further analyzed using different phytochemical assays and antioxidant capacity tests. From the phytochemicals assays, the presence of phenolic compounds and flavonoids was noted. Both of these substances are recognized in the literature for their great antioxidant potential. However, despite confirmation of the presence of these compounds, cubiu extract showed a low antioxidant activity when compared to other fruits such as sapodilla, strawberry, orange, cherry [35], and many others known for their great antioxidant effect.

Conflict of Interests

The authors declare no conflict of interests.

Acknowledgments

This research was supported by the RCMI program G12MD007579 (Janice B. Monteiro) and the MBRS-RISE program funded by RISE Grant R25GM082406 from the National Institutes of Health (Carmen Ortiz) at Ponce School of Medicine and Health Sciences and Puerto Rico Clinical and Translational Research Consortium Grant U54MD007587. The authors would like to acknowledge the support of the Conselho Nacional de Desenvolvimento Científico e Tecnológico (CNPq) and the Fundação de Amparo à Pesquisa do Estado do Amazônas (FAPEAM).

References

[1] J. S. Andrade, I. M. A. Rocha, and D. F. Silva Filho, "Características físicas e composição química de frutos de populações naturais de cocona (*Solanum sessiliflorum* Dunal) encontradas na Amazônia Central," *Revista Brasileira de Fruticultura*, 1997.

[2] L. K. O. Yuyama, J. P. Aguiar, S. H. M. Macedo, T. Gioia, and D. F. Silva Filho, "Composição centesimal de diversas populações de cubiu (*Solanum sessiliflorum* Dunal) da Estação Experimental do Instituto Nacional de Pesquisas da Amazônia, INPA," in *Proceedings of the Anais do II Simpósio Latino Americano de Ciências de Alimentos*, Campinas, Brazil, 1997.

[3] A. V. D. Pahlen, "Cubiu (*Solanum topiro* (Humb. & Bonpl.)), uma fruteira da Amazonia," *Acta Amazonica*, vol. 7, no. 3, pp. 301–307, 1977.

[4] H. Villachica, "Cocona (*Solanum sessiliflorum* Dunal)," in *Frutales y Hortalizas Promisorios de la Amazonía*, pp. 98–102, Secretaria Pro-Tempore, Lima, Peru, 1996.

[5] L. K. O. Yuyama, "Conteúdos minerais em algumas populações de cocona (*Solanum sessiliflorum* Dunal): dados preliminares," in *Anuais do Congresso Brasileiro de Ciêcia e Tecnologia*, Rio de Janeiro, Brazil, 1998.

[6] D. F. da Silva Filho, J. S. de Andrade, C. R. Clement, F. M. Machado, and H. Noda, "Correlações fenotípicas, genéticas e ambientais entre descritores morfológicos e quimicos em frutos de cubiu (*Solanum sessiliflorum* Dunal) da amazônia," *Acta Amazónica*, vol. 29, no. 4, pp. 503–511, 1999.

[7] R. E. Schultes, "Amazonian cultigens and their northward and westward migrations in pre-columbian times," in *Pre-Columbian Plant Migration*, D. Stone, Ed., pp. 19–38, Harvard University Press, Cambridge, Mass, USA, 1984.

[8] D. F. de Silva Filho, H. Noda, and C. R. Clement, "Genetic variability of economic characters in 30 accessions of cubiu *Solanum sessiliflorum* Dunal, Solanaceae evaluated in Central Amazonia," *Revista Brasileira de Genetica*, vol. 16, no. 2, pp. 409–417, 1993.

[9] J. Salick, "Cocona (*Solanum sessiliflorum* Dunal), an overview of productions and breeding potentials," in *Proceedings of the International Symposium on New Crops for Food Industry*, University Southampton, Southampton, UK, 1989.

[10] M. A. Pardo S, "Efecto de *Solanum sessiliflorum* Dunal sobre el metabolismo lipídico y de la glucosa," *Ciencia e Investigación*, vol. 7, no. 2, pp. 44–48, 2004.

[11] M. Naczk and F. Shahidi, "Extraction and analysis of phenolics in food," *Journal of Chromatography A*, vol. 1054, no. 1-2, pp. 95–111, 2004.

[12] A. V. B. Moreira and J. Mancini-Filho, "Influência dos compostos fenólicos de especiarias sobre a lipoperoxidação e o perfil lipídico de tecidos de ratos," *Revista de Nutrição*, vol. 17, no. 4, pp. 411–424, 2004.

[13] G. O. Adegoke, M. Vijay Kumar, A. G. Gopala Krishna, M. C. Varadaraj, K. Sambaiah, and B. R. Lokesh, "Antioxidants and

lipid oxidation in foods—a critical appraisal," *Journal of Food Science and Technology*, vol. 35, no. 4, pp. 283–298, 1998.

[14] National Research Council, *Recomended Dietary Allowances*, M.A. Press, 1989.

[15] E. A. Decker, "Strategies for manipulating the prooxidative/antioxidative balance of foods to maximize oxidative stability," *Trends in Food Science & Technology*, vol. 9, no. 6, pp. 241–248, 1998.

[16] N. Ramarathnam, T. Osawa, H. Ochi, and S. Kawakishi, "The contribution of plant food antioxidants to human health," *Trends in Food Science & Technology*, vol. 6, no. 3, pp. 75–82, 1995.

[17] M. Namiki, "Antioxidants/antimutagens in food," *Critical Reviews in Food Science and Nutrition*, vol. 29, no. 4, pp. 273–300, 1990.

[18] D. F. da Silva Filho, L. K. Yuyama, J. P. Aguiar, M. C. Oliveira, and L. H. Martins, "Caracterização e avaliação do potencial agronômico e nutricional de etnovariedades de cubiu (*Solanum sessiliflorum* Dunal) da Amazônia," *Acta Amazonica*, vol. 35, no. 4, pp. 399–405, 2005.

[19] A. P. Rogério, *Estudo da atividade anti-inflamatória, analgésica, anti-edematogênica e antipirética do extrato de Lafoensia pacari e do ácido elágico [Ph.D. thesis]*, Faculdade de Ciências Farmacêuticas de Ribeirão Preto, Universidade de São Paulo, São Paulo, Brazil, 2006.

[20] O. I. Aruoma, J. P. E. Spencer, D. Warren, P. Jenner, J. Butler, and B. Halliwell, "Characterization of food antioxidants, illustrated using commercial garlic and ginger preparations," *Food Chemistry*, vol. 60, no. 2, pp. 149–156, 1997.

[21] P. Molyneux, "The use of the stable free radical diphenylpicrylhydrazyl (DPPH) for estimating antioxidant activity," *Songklanakarin Journal of Science and Technology*, vol. 26, no. 2, pp. 211–219, 2004.

[22] R. Re, N. Pellegrini, A. Proteggente, A. Pannala, M. Yang, and C. Rice-Evans, "Antioxidant activity applying an improved ABTS radical cation decolorization assay," *Free Radical Biology and Medicine*, vol. 26, no. 9-10, pp. 1231–1237, 1999.

[23] E. A. Moreira, "Marcha sistemática de análise em fitoquímica," *Tribuna Farmacêutica*, vol. 47, no. 1, pp. 1–19, 1979.

[24] C. M. O. Simões, E. P. Schenkel, G. Gosmann, J. C. P. Mello, L. A. Mentz, and P. R. Petrovick, *Farmacognosia: Da Planta ao Medicamento*, Florianópolis, Porto Alegre, Brazil, 2001.

[25] L. E. Pelzer, T. Guardia, A. O. Juarez, and E. Guerreiro, "Acute and chronic antiinflammatory effects of plant flavonoids," *Farmaco*, vol. 53, no. 6, pp. 421–424, 1998.

[26] L. R. Ferguson, "Role of plant polyphenols in genomic stability," *Mutation Research/Fundamental and Molecular Mechanisms of Mutagenesis*, vol. 475, no. 1-2, pp. 89–111, 2001.

[27] P. Markakis, "Stability of anthocyanins in foods," in *Anthocyanins As Food Colors*, P. Markakis, Ed., chapter 6, pp. 163–180, Academic Press, 1982.

[28] M. S. Narayan, K. A. Naidu, G. A. Ravishankar, L. Srinivas, and L. V. Venkataraman, "Antioxidant effect of anthocyanin on enzymatic and non-enzymatic lipid peroxidation," *Prostaglandins Leukotrienes and Essential Fatty Acids*, vol. 60, no. 1, pp. 1–4, 1999.

[29] K. A. Youdim, A. Martin, and J. A. Joseph, "Incorporation of the elderberry anthocyanins by endothelial cells increases protection against oxidative stress," *Free Radical Biology and Medicine*, vol. 29, no. 1, pp. 51–60, 2000.

[30] B. Nair, "Sustainable utilization of gum and resin by improved tapping technique in some species," in *Harvesting of Non-Wood Forest Products*, International Agro-Hydrology Research and Training Center, Menemen, Turkey, 2000.

[31] G. K. Jayaprakasha and B. S. Patil, "In vitro evaluation of the antioxidant activities in fruit extracts from citron and blood orange," *Food Chemistry*, vol. 101, no. 1, pp. 410–418, 2006.

[32] R. Roesler, L. G. Malta, L. C. Carrasco, R. B. Holanda, C. A. S. Sousa, and G. M. Pastore, "Atividade antioxidante de frutas do cerrado," *Food Science and Technology (Campinas)*, vol. 27, no. 1, pp. 53–60, 2007.

[33] M. A. L. Couto and S. G. Canniatti-Brazaca, "Quantification of vitamin C and antioxidant capacity of citrus varieties," *Ciência e Tecnologia de Alimentos*, vol. 30, supplement 1, pp. 15–19, 2010.

[34] P. Stratil, B. Klejdus, and V. Kubáň, "Determination of phenolic compounds and their antioxidant activity in fruits and cereals," *Talanta*, vol. 71, no. 4, pp. 1741–1751, 2007.

[35] L. P. Leong and G. Shui, "An investigation of antioxidant capacity of fruits in Singapore markets," *Food Chemistry*, vol. 76, no. 1, pp. 69–75, 2002.

[36] G. M. Dos Santos, G. A. Maia, P. H. M. de Sousa, J. M. C. da Costa, R. W. de Figueiredo, and G. M. do Prado, "Correlação entre atividade antioxidante e compostos bioativos de polpas comerciais de açaí (*Euterpe oleracea* Mart)," *Archivos Latinoamericanos de Nutrición*, vol. 58, no. 2, pp. 187–192, 2008.

[37] G. S. Nascimento and I. R. O. Pereira, "Avaliação da atividade antioxidante das partes da fruta Cubiu (*Solanum sessiliflorum*)," in *Jornada de Iniciação Científica*, Universidade Presbiteriana Mackenzie, São Paulo, Brazil, 2011.

[38] P. C. Ledur, F. Rogalski, R. M. Oliveira, C. P. Mostardeiro, G. F. F. S. Montagner, and I. B. M. Cruz, *Determinação da Capacidade Antioxidante do Extrato Bruto de Solanum sessiliflorum (Cubiu)*, UNIFRA, 2012.

Absorption, Metabolism, and Excretion by Freely Moving Rats of 3,4-DHPEA-EDA and Related Polyphenols from Olive Fruits (*Olea europaea*)

Shunsuke Kano,[1] Haruna Komada,[2] Lina Yonekura,[1,2] Akihiko Sato,[2] Hisashi Nishiwaki,[3] and Hirotoshi Tamura[1,2]

[1]*The United Graduate School of Agricultural Sciences, Ehime University, 3-5-7 Tarumi, Matsuyama 790-8566, Japan*
[2]*The Graduate School of Agriculture, Kagawa University, 2393 Ikenobe, Miki-cho, Kagawa 761-0795, Japan*
[3]*Faculty of Agriculture, Ehime University, 3-5-7 Tarumi, Matsuyama 790-8566, Japan*

Correspondence should be addressed to Hirotoshi Tamura; tamura@ag.kagawa-u.ac.jp

Academic Editor: Stan Kubow

Absorption, metabolism, and excretion of 3,4-DHPEA-EDA, oleuropein, and hydroxytyrosol isolated from olive fruits were newly evaluated after oral and intravenous administration in freely moving rats cannulated in the portal vein, jugular vein, and bile duct. Orally administered 3,4-DHPEA-EDA, an important bioactive compound in olive pomace, was readily absorbed and metabolized to hydroxytyrosol, homovanillic acid, and homovanillyl alcohol, as shown by dose-normalized 4 h area under the curve ($AUC_{0 \to 4h}$/Dose) values of 27.7, 4.5, and 4.2 μM·min·kg/μmol, respectively, in portal plasma after oral administration. The parent compound 3,4-DHPEA-EDA was not observed in the portal plasma, urine, and bile after oral and intravenous administration. Additionally, hydroxytyrosol, homovanillic acid, and homovanillyl alcohol in the portal plasma after oral administration of hydroxytyrosol showed 51.1, 22.8, and 7.1 μM·min·kg/μmol $AUC_{0 \to 4h}$/Dose, respectively. When oleuropein, a polar glucoside, was injected orally, oleuropein in the portal plasma showed 0.9 μM·min·kg/μmol $AUC_{0 \to 4h}$/Dose. However, homovanillic acid was detected from oleuropein in only a small amount in the portal plasma. Moreover, the bioavailability of hydroxytyrosol and oleuropein for 4 hours was 13.1% and 0.5%, respectively. Because the amount of 3,4-DHPEA-EDA in olive fruits is about 2-3 times greater than that of hydroxytyrosol, the metabolites of 3,4-DHPEA-EDA will influence biological activities.

1. Introduction

Olive oil and table olives (*Olea europaea L.*) are major products in food markets all over the world and are important as dietary sources of many polyphenols such as oleuropein, oleuropein aglycones, oleocanthal, hydroxytyrosol, and luteolin, which exert biological effects such as inhibiting LDL oxidation [1–4] and atherosclerosis [5, 6], reducing postprandial blood glucose levels and the risk of diabetes [7–9], and inhibiting inflammation [10, 11]. Therefore, studies on the absorption and metabolism of olive bioactive compounds are essential to determine which compounds are biologically important in the human body after the consumption of olive products. Research works addressing the oral absorption and metabolism of olive bioactives in humans are often done

by the detection of the polyphenols and related metabolites in the urine and the systemic venous blood [12–18]. Cell-based assays and the rat everted gut sac have been used to overcome the limitations of studies done in humans but would not provide data on first-pass and systemic metabolism and excretion, which require the use of a living animals. The use of small laboratory animals such as the rat is convenient, especially when measuring the absorption of compounds that are not commercially available or isolated in house, but in most cases the analysis of circulating metabolites can only be done in sacrificed animals [19–21]. Recently, the use of jugular cannulated rats enabled the study of the kinetics of absorption, metabolism, and excretion anthocyanins in freely moving rats, providing bioavailability data under more natural conditions for the animal [22]. However, to get a

TABLE 1: Identification of metabolites from plasma, urine, and bile after administration of 3,4-DHPEA-EDA, hydroxytyrosol, and oleuropein.

Rt (min)	λ_{max} (nm)	Formula	Calc. mass [M−H]⁻	Observed [M−H]⁻	Mass error (mDa)	Compound identification
3.96	230, 278	$C_8H_{10}O_3$	153.0552	153.0554	0.2	Hydroxytyrosol
8.95	228, 278	$C_8H_{12}O_3$	167.0708	167.0731	2.3	Homovanillyl alcohol
9.67	230, 278	$C_9H_{10}O_4$	181.0501	181.0533	3.2	Homovanillic acid
20.10	230, 280	$C_{25}H_{32}O_8$	539.1765	539.1776	1.1	Oleuropein

Data was obtained from HT metabolites isolated from urine.
Data was obtained from oleuropein isolated from plasma.

more accurate evaluation of absorption, the portal vein cannulation has the great advantage of allowing blood sampling immediately after intestinal absorption, avoiding dilution in the systemic circulation and further metabolic changes of the absorbed phytochemicals. Therefore, we used portal vein cannulated rats to evaluate the intestinal bioavailability and absorption kinetics after an oral dose and jugular cannulated rats to evaluate the systemic metabolism and plasma levels of the same compounds after an intravenous dose. In both cases the rats were moving freely and unanesthetized during the experiment.

The biokinetics of olive polyphenols, oleuropein [23–27], and hydroxytyrosol [24, 28–31] has been investigated in rats. However, the biokinetics of oleuropein aglycones in living rats has never been studied in spite of their significant biological activities [32]. As oleuropein aglycones are one of the most abundant phytochemicals in the olive fruit, monitoring its absorption, metabolism, and excretion would be indispensable for understanding physiological functions of olive fruit.

In this experiment, we monitored the oleuropein aglycone 3,4-DHPEA-EDA and related polyphenols in the portal plasma and urine after oral administration, and in the systemic venous plasma and urine after intravenous administration; both procedures were done in freely moving unanesthetized rats. In addition, biliary excretion was also monitored in bile cannulated rats after intravenous administration of the phytochemicals. Together, those three procedures enabled a comprehensive view of the bioavailability and biokinetics of 3,4-DHPEA-EDA, including intestinal absorption and first-pass metabolism, systemic metabolism, and urinary and biliary excretion. This approach provides a more accurate indication of the biologically important compounds from olive fruits.

2. Materials and Methods

2.1. Chemicals. Oleuropein (Ole) was obtained from Nacalai Tesque, Inc. (Kyoto, Japan). Hydroxytyrosol (HT, 2-(3,4-dihydroxyphenyl) ethyl alcohol) was purchased from Tokyo Chemical Co. (Tokyo, Japan). Oleuropein aglycone (3,4-DHPEA-EDA, 3,4-dihydroxyphenylethyl elenolate dialdehydic form) was purified using ODS-preparative HPLC according to a previous study [32]. Homovanillic acid (HVA, 4-hydroxy-3-methoxyphenylacetic acid), homovanillyl alcohol (HVAOH, 4-hydroxy-3-methoxyphenethyl alcohol), MOPS [3-(N-morpholino)propanesulfonic acid], β-glucuronidase

Type VII-A, and sulfatase Type H-1 were obtained from Sigma-Aldrich (St. Louis, MO, USA). Trifluoroacetic acid (TFA) was obtained from Wako Pure Chemical Co. (Tokyo, Japan).

2.2. Animals and Diets. SPF Wistar ST rats (6 weeks of age, male, 180 ± 20 g of body weight) were purchased from Japan SLC Inc. (Hamamatsu, Japan). The rats were housed in an air-conditioned room at 22 ± 2°C under cycles of 12 hours dark and 12 hours light and given free access to a commercial diet and water for a week. The rats were then fasted for over 16 hours before the experiments. The present study was approved by the Animal Care and Use Committee at Kagawa University, and the rat treatment conformed to the Rules of Animal Experiments at Kagawa University.

2.3. Animal Treatment: Jugular Vein, Portal Vein, and Bile Duct Cannulation. After fasting, the rats were cannulated with a polyethylene tube (PE-50; 0.58 mm I.D., 0.965 mm O.D.) into the jugular vein, portal vein, or bile duct under anesthesia with isoflurane inhalation using a WP-SAA01 (LMS Co., Ltd., Tokyo, Japan). A small hole was made in the vein or duct using pointed scissors and cannulated with the polyethylene tube. After inserting the tube, it was fixed at the edge of the vein or duct. The jugular vein and portal vein cannulas penetrated under the skin and were then tied at the back of the rats, and the bile duct cannula was tied at the abdomen of the rats. Bile duct-cannulated rats were then restrained in a Bollman cage (Natsume Seisakusho Co., Ltd., Tokyo, Japan). Before further experiments, the rats were allowed to recover from anesthesia.

2.4. Identification of Metabolites. Metabolites from 3,4-DHPEA-EDA, HT, and Ole were monitored using a UPLC/Xevo Q-TOF MS (Waters Corp., Milford, MA, USA). Retention time and UV λ_{max} from a diode array detector UFLC, and molecular ions of Xevo Q-TOF MS of the metabolites were compared with those of authentic compounds (Table 1).

2.5. Collection of Blood, Urine, and Bile Samples. 3,4-DHPEA-EDA (300 mg/kg body weight; 936.51 μmol/kg) dissolved in 50% polyethylene glycol aqueous solution was administered orally via a stomach tube to the portal vein cannulated rats. Blood samples (450 μL) were then collected using a heparinized syringe attached to the cannula before administration and 5, 15, 30, 60, 120, and 240 min after administration. Urine was also collected before

TABLE 2: Pharmacokinetic parameters after oral administration of 3,4-DHPEA-EDA, hydroxytyrosol, and oleuropein.

Oral administration	Dose (μmol)	Metabolites	AUC (μM·min)	AUC/Dose (μM·min·kg/μmol)	C_{max} (μM)	C_{max}/Dose (μM·kg/μmol)	T_{max} (min)
3,4-DHPEA-EDA	936.51	HT	25935.0 ± 3103.1	27.7 ± 3.3	178.87 ± 45.3	0.22 ± 0.05	30
		HVA	4180.0 ± 1646.5	4.5 ± 1.8	21.5 ± 6.9	0.03 ± 0.007	60
		HVAOH	3934.5 ± 789.7	4.2 ± 0.8	22.2 ± 2.0	0.03 ± 0.002	30
		Total	**34049.5 ± 5539.3**	**36.4 ± 5.9**			
Hydroxytyrosol	648.66	HT	33160 ± 5541.5	51.1 ± 8.5	394.7 ± 58.0	0.61 ± 0.089	15
		HVA	14794.1 ± 510.1	22.8 ± 0.8	150.6 ± 9.7	0.23 ± 0.015	30
		HVAOH	4604.4 ± 885.2	7.1 ± 1.4	26.6 ± 5.2	0.04 ± 0.008	60
		Total	**52585 ± 6936.8**	**81.0 ± 10.7**			
Oleuropein	555.02	Ole	1015.8 ± 222.6	1.8 ± 0.4	11.9 ± 3.2	0.021 ± 0.006	5
		HVA	515.7 ± 112.4	0.9 ± 0.2	4.1 ± 1.4	0.007 ± 0.003	60
		Total	**1531.5 ± 335**	**2.7 ± 0.6**			

Values of means and ± SEM are in triplicate.

administration and 0 to 2 h and 2 to 4 h after administration. For intravenous administration, 3,4-DHPEA-EDA (10 mg/kg; 31.22 μmol/kg) was injected into the jugular vein, and blood (450 μL) was collected using a heparinized syringe before administration and 5, 15, 30, 60, 120, and 240 min after administration via the jugular vein cannula. The blood (450 μL) collected was immediately cooled on ice and centrifuged at 15,000 rpm for 5 min at 4°C to obtain over 200 μL plasma. Urine was also collected before administration and 0 to 2 h and 2 to 4 h after administration. Bile was collected via the bile duct cannula before administration and 0 to 2 h and 2 to 4 h after intravenous administration. The collected plasma, urine, and bile were immediately stored in an ice box until further analyses.

HT (oral administration: 100 mg/kg; 648.66 μmol/kg, intravenous administration: 10 mg/kg; 64.87 μmol/kg) and Ole (oral administration: 300 mg/kg; 555.02 μmol/kg, intravenous administration: 10 mg/kg; 18.50 μmol/kg) were investigated using the same method as described above.

All animal studies were performed in triplicate per compound.

2.6. Determination of 3,4-DHPEA-EDA, HT, and Ole and Their Metabolites in Plasma, Urine, and Bile. Metabolites were deconjugated by using β-glucuronidase (500 U) and sulfatase (50 U) in 50 μL 625 mM MOPS (pH 6.8), which was added to 200 μL of plasma, urine, or bile. The mixture was incubated for 45 min at 37°C and subjected to extraction using QuEChERS [33]. After that, each sample was evaporated to dryness *in vacuo*, dissolved in 200 μL 0.5% TFA-1% acetonitrile aqueous solution, and then analyzed using a UFLC system (Prominence UFLC; Shimadzu, Kyoto, Japan) equipped with a Shim-pack XR-ODS column (3.0 mm I.D. × 100 mm; Shimadzu, Kyoto, Japan) at 40°C with a mobile phase of 0.5% TFA-1% acetonitrile aqueous solution (A) and 0.5% TFA-75% acetonitrile aqueous solution (B). The elution conditions are described as follows: isocratic elution of 0% B was maintained for 4.0 min; linear gradient 0 to

4% B, 4.0–4.5 min; linear gradient 4.0–16.0% B, 4.5–16.0 min; linear gradient 16.0–100% B, 16.0–25.0 min; isocratic elution of 100% B, 25.0–26.0 min; and linear gradient 100 to 0% B, 26.0–26.1 min, with a flow rate of 0.5 mL/min. Injection volume was 50 μL. All eluted peaks were detected at 279 nm with a diode array detector (SPD-M20A; Shimadzu, Kyoto, Japan).

2.7. Calculation of Pharmacokinetic Parameters after Administration. The absorption of 3,4-DHPEA-EDA, HT, and Ole was evaluated by dose-normalized 4 h area under the curve ($AUC_{0 \to 4h}$/Dose) and C_{max}/Dose values (Table 2). $Dose_{i.g.}$ and $Dose_{i.v.}$ are the amount of oral (i.g.) and intravenous (i.v.) administration, respectively. C_{max} is the maximum concentration, and T_{max} is the time at which C_{max} is observed. AUC (μM·min) is the area under the plasma concentration curve.

3. Results

3.1. Absorption and Metabolism of 3,4-DHPEA-EDA, HT, and Ole after Oral Administration. The metabolites HT, HVA, and HVAOH were observed when 3,4-DHPEA-EDA was administrated orally (Figure 1(a)). 3,4-DHPEA-EDA was not detected in the portal plasma after oral administration (300 mg/kg; 936.51 μmol/kg), even though HT as a metabolite was detected at a fairly high concentration, with a dose-normalized C_{max} (C_{max}/$Dose_{i.g.}$) of 0.22 μM·kg/μmol and AUC ($AUC_{0 \to 4h-i.g.}$/$Dose_{i.g.}$) of 27.7 μM·min·kg/μmol (Table 2). The other metabolites of 3,4-DHPEA-EDA were HVA (C_{max}/$Dose_{i.g.}$ 0.03 μM·kg/μmol; $AUC_{0 \to 4h-i.g.}$/$Dose_{i.g.}$ 4.5 μM·min·kg/μmol) and HVAOH (C_{max}/$Dose_{i.g.}$ 0.03 μM·kg/μmol; $AUC_{0 \to 4h-i.g.}$/$Dose_{i.g.}$ 4.2 μM·min·kg/μmol) (Table 2). T_{max} (time at maximum concentration) for HT in the portal plasma was 30 min after oral administration of 3,4-DHPEA-EDA (Figure 1(a)). HT in the portal plasma subsequently decreased gradually in concentration until 240 min. In addition, HVA and HVAOH showed T_{max} at 60 min and 30 min, respectively, even though those detected

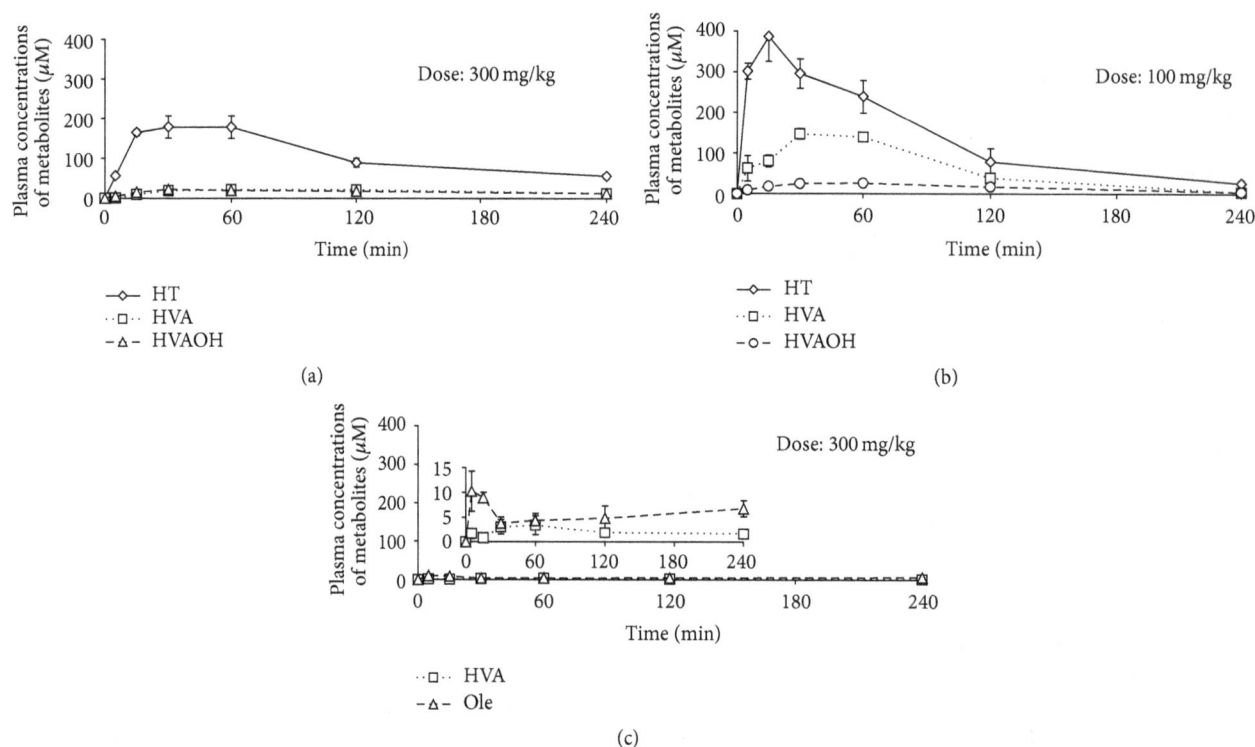

FIGURE 1: Metabolites from 3,4-DHPEA-EDA (a), hydroxytyrosol (b), and oleuropein (c) in plasma from the portal vein over time after oral administration (300 mg/kg for 3,4-DHPEA-EDA, 100 mg/kg for HT, and 300 mg/kg for Ole). Values of means and SEM are from measurements performed in triplicate. Sampling times are described in the text. HT: hydroxytyrosol, Ole: oleuropein, HVA: homovanillic acid, and HVAOH: homovanillyl alcohol.

amounts in the portal plasma were very low. Observation of HVAOH in the portal plasma was the first from 3,4-DHPEA-EDA.

The plasma concentration of HT (Figure 1(b)) immediately increased just after HT oral administration (100 mg/kg; 648.66 μmol/kg) and showed $AUC_{0 \to 4\,h\text{-i.g.}}/Dose_{\text{i.g.}}$ of 51.1 μM·min·kg/μmol (Table 2). The $AUC_{0 \to 4\,h\text{-i.g.}}/Dose_{\text{i.g.}}$ values for HVA and HVAOH were 22.8 and 7.1 μM·min·kg/μmol, respectively (Table 2). Moreover, the $C_{\max}/Dose_{\text{i.g.}}$ of HT was 0.61 μM·kg/μmol at $T_{\max}$ 15 min. Those of HVA and HVAOH were 22.8 μM·kg/μmol at 30 min and 7.1 μM·kg/μmol at 60 min, respectively.

Small amounts of Ole (Figure 1(c)) ($AUC_{0 \to 4\,h\text{-i.g.}}/Dose_{\text{i.g.}}$ 1.8 μM·min·kg/μmol) and HVA ($AUC_{0 \to 4\,h\text{-i.g.}}/Dose_{\text{i.g.}}$ 0.9 μM·min·kg/μmol) were detected after administration of Ole (300 mg/kg; 555.02 μmol/kg). Ole $T_{\max}$ was observed at 5 min. However, HVA peaked at 60 min.

3.2. Plasma Concentration of 3,4-DHPEA-EDA, HT, and Ole after Intravenous Administration.
The intact form of 3,4-DHPEA-EDA was not detected in the plasma and hematocytes after intravenous administration of 3,4-DHPEA-EDA (10 mg/kg; 31.22 μmol/kg) (Figure 2(a)). On the other hand, about 131 μM $C_{\max}$ of HT was detected at 5 min after intravenous administration of 3,4-DHPEA-EDA. HT was gradually metabolized in the systemic circulation, because

small amounts of HVA and HVAOH could be detected. The $AUC_{0 \to 4\,h\text{-i.v.}}/Dose_{\text{i.v.}}$ values for HT, HVA, and HVAOH were 259.2, 22.4, and 56.0 μM·min·kg/μmol, respectively. The higher amount of HVAOH observed was a typical characteristic of 3,4-DHPEA-EDA metabolism in plasma.

A maximum amount of intact HT was detected at 5 min (1744 μM) after intravenous administration of HT (10 mg/kg; 64.87 μmol/kg), which immediately decreased in the systemic venous plasma, and then almost disappeared about 15 min after dose (Figure 2(b)). The $AUC_{0 \to 4\,h\text{-i.v.}}/Dose_{\text{i.v.}}$ values for HVA and HVAOH were 15.4 and 15.0 μM·min·kg/μmol, respectively.

After administration of Ole (10 mg/kg; 18.50 μmol/kg) (Figure 2(c)), 502 μM Ole was detected in the plasma at 5 min and then disappeared from the systemic circulation within 60 min. Other metabolites could not be detected throughout the blood sampling.

3.3. Urinary Excretion of 3,4-DHPEA-EDA, HT, and Ole after Oral Administration.
After oral administration of 3,4-DHPEA-EDA, HT (0.51% of dose) and HVA (0.38% of dose) were detected in small amounts in the urine (Figure 3(a)), whereas 3,4-DHPEA-EDA and HVAOH were not detected in the urine.

HVAOH (0.17% of dose) and HVA (8.03% of dose) were the major metabolites of intact HT (1.14% of dose) in the urine

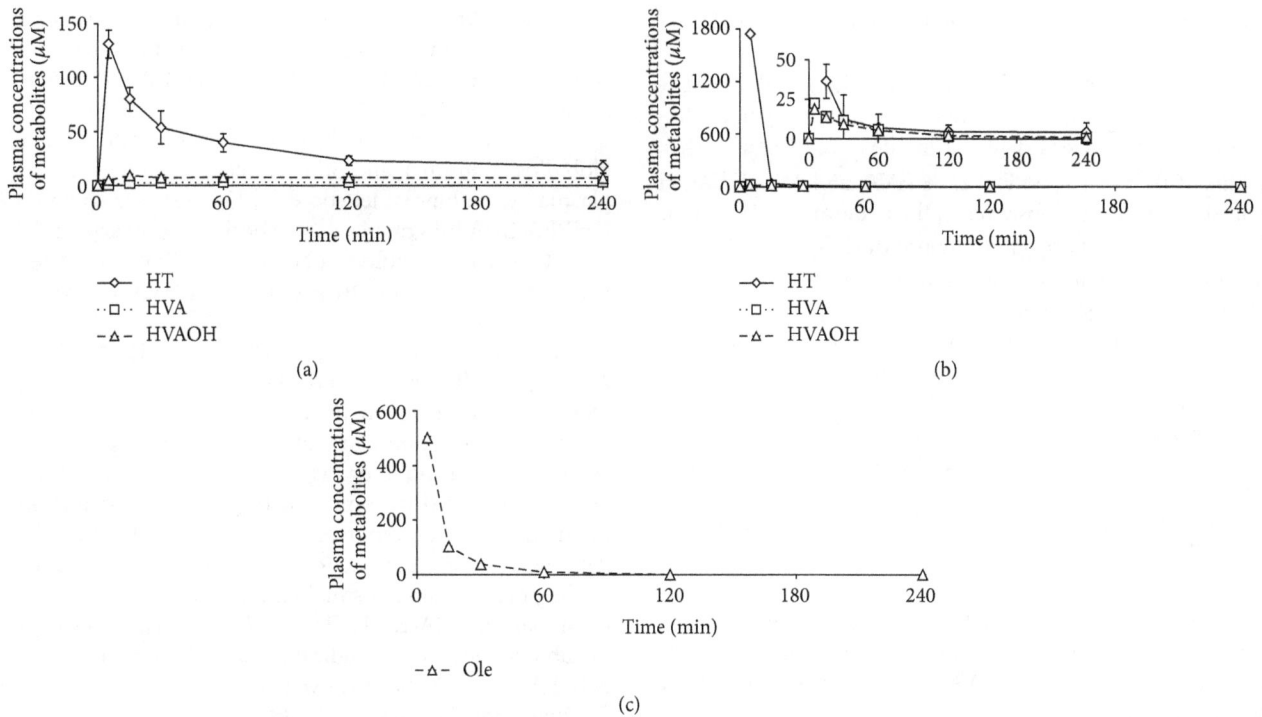

FIGURE 2: Metabolites from 3,4-DHPEA-EDA (a), hydroxytyrosol (b), and oleuropein (c) in plasma from the jugular vein over time after intravenous administration. Values of means and SEM are from measurements performed in triplicate. Analytical times are described in the text. HT: hydroxytyrosol, Ole: oleuropein, HVA: homovanillic acid, and HVAOH: homovanillyl alcohol.

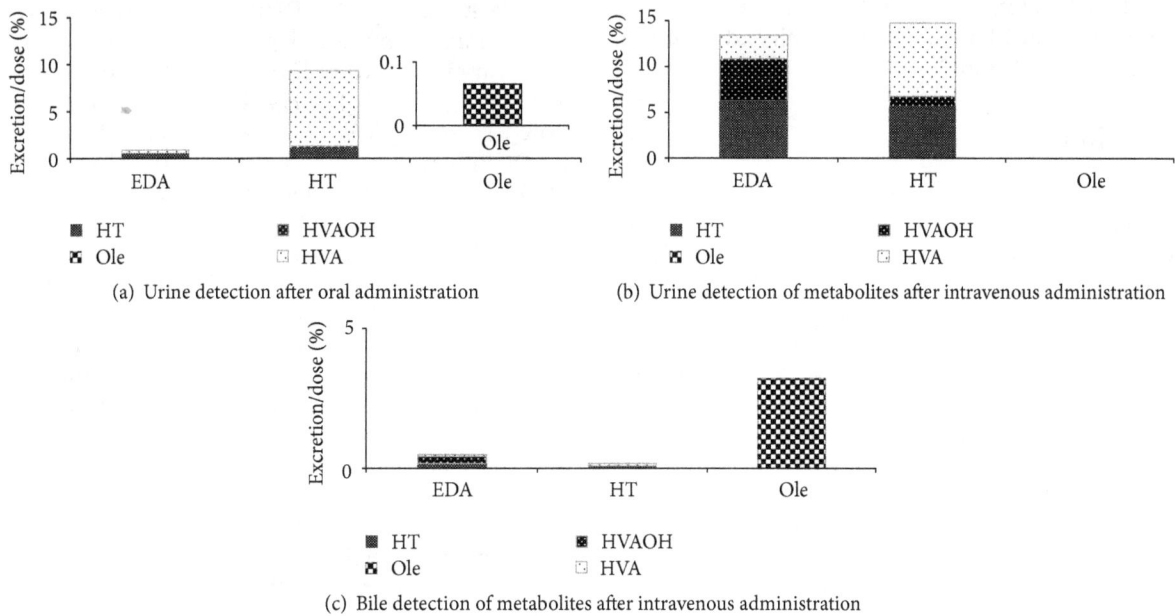

FIGURE 3: Monitoring of metabolites in urine and bile after oral and intravenous administration of 3,4-DHPEA-EDA, hydroxytyrosol, and oleuropein. Excretion rates are expressed as excretion/dose ratios. EDA: 3,4-DHPEA-EDA, HT: hydroxytyrosol, Ole: oleuropein, HVA: homovanillic acid, and HVAOH: homovanillyl alcohol.

of the rats after oral administration of HT (Figure 3(a)). Ole was detected in an extremely small amount (0.07% of dose) in the urine (Figure 3(a)) after oral administration of Ole.

3.4. Urinary Excretion of 3,4-DHPEA-EDA, HT, and Ole after Intravenous Administration. The intact form of 3,4-DHPEA-EDA was not detected in the urine (Figure 3(b)). On the other hand, 6.3% HT, 2.6% HVA, and 4.6% HVAOH against the dose were observed in the urine after intravenous administration. A fairly large amount of HVAOH detected is a characteristic of urinary excretion of 3,4-DHPEA-EDA after intravenous administration.

HT converted quickly into HVA (8.0%) in the urine after intravenous administration (Figure 3(b)), and then the transformation of HT into HVAOH after intravenous administration was 1.1% in the urine (Figure 3(b)). HT administrated intravenously leads to detecting a fair amount of intact HT.

The intact form of Ole and the metabolites were not detected in the urine.

3.5. Biliary Excretion of 3,4-DHPEA-EDA, HT, and Ole after Intravenous Administration. With the 3,4-DHPEA-EDA substrate, 0.174% HT, 0.048% HVA, and 0.255% HVAOH were detected in the bile (Figure 3(c)).

In contrast, 0.079% HT, 0.090% HVA, and 0.003% HVAOH were detected after HT was metabolized (Figure 3(c)). The chemical compositions of HT, HVA, and HVAOH in the bile from EDA and HT after intravenous administration were quite similar to those in the urine.

With the Ole substrate, the intact form of Ole was not detected in the urine, but 3.22% of the injected dose was observed in the bile (Figure 3(c)).

4. Discussion

4.1. Absorption of 3,4-DHPEA-EDA and Its Related Polyphenols. In comparing 3,4-DHPEA-EDA, HT, and Ole, the highest absorption was observed for HT, followed by 3,4-HPEA-EDA and Ole (Figure 1). Because the dose of 3,4-DHPEA-EDA (936.5 μmol/kg) was 1.44 times greater than that of HT (648.7 μmol/kg), the total absorption rate of 3,4-DHPEA-EDA (36.4 $AUC_{0 \rightarrow 4\,h\text{-}i.g.}/Dose_{i.g.}$), including its metabolites, was only 2.2 times less than that of HT (81.0 $AUC_{0 \rightarrow 4\,h\text{-}i.g.}/Dose_{i.g.}$) (Table 2). The difference in mutual absorption rates between 3,4-DHPEA-EDA and HT is quite small. On the other hand, the absorption rate of 3,4-DHPEA-EDA was 13.5 times greater than that of Ole (2.7 $AUC_{0 \rightarrow 4\,h\text{-}i.g.}/Dose_{i.g.}$). Therefore, total absorption of 3,4-DHPEA-EDA would be significant for the physiological function of olive fruit (ratio of AUC/Dose of 3,4-DHPEA-EDA : HT : Ole = 13.5 : 30 : 1). In addition, there is high specificity in the absorption rate among them.

After oral administration of 3,4-DHPEA-EDA, HT (27.7 $AUC_{0 \rightarrow 4\,h\text{-}i.g.}/Dose_{i.g.}$), HVAOH (4.2 $AUC_{0 \rightarrow 4\,h\text{-}i.g.}/Dose_{i.g.}$), and HVA (4.5 $AUC_{0 \rightarrow 4\,h\text{-}i.g.}/Dose_{i.g.}$) were detected in the portal plasma. Low absorption of 3,4-DHPEA-EDA is reported in some papers [16, 34]. However, we could not

detect any 3,4-DHPEA-EDA or 3,4-DHPEA-EDA-related compounds in a bound form, or an unbound form like 3,4-DHPEA-EDAH$_2$ [34], in the portal plasma. Plasma components such as serum albumin, serum lipoprotein, and glycoprotein may have bound 3,4-DHPEA-EDA and hindered its detection. 3,4-DHPEA-EDA as a dialdehyde form might quickly bind to proteins bearing primary amines such as globulins, albumins, and other functional proteins in blood plasma, as malondialdehyde does [35, 36]. We spiked 3,4-DHPEA-EDA into rat plasma to check the recovery yield, but 3,4-DHPEA-EDA could not be detected. To our knowledge, this is the first report of the formation of HVAOH from 3,4-DHPEA-EDA.

HT (51.1 $AUC_{0 \rightarrow 4\,h\text{-}i.g.}/Dose_{i.g.}$), HVAOH (7.1 $AUC_{0 \rightarrow 4\,h\text{-}i.g.}/Dose_{i.g.}$), and HVA (22.8 $AUC_{0 \rightarrow 4\,h\text{-}i.g.}/Dose_{i.g.}$) were observed in the portal plasma (Figure 1(b)) when HT was administered orally. Detecting HVA and HVAOH in plasma after ingestion of olive oil and HT has been documented in some papers [14, 15], but detection of them has not been reported so far from 3,4-DHPEA-EDA. Although a great amount of HVA was detected from HT in the portal plasma, a small amount of HVA was detected from 3,4-DHPEA-EDA (Figure 1(a)). If an intermediate metabolite from orally administered 3,4-DHPEA-EDA is HT, HVA should be observed in the portal plasma. Our findings probably indicated that 3,4-DHPEA-EDA or 3,4-DHPEA-EDA-related compounds in a bound form, or an unbound form like 3,4-DHPEA-EDAH$_2$ [34], would be gradually absorbed in the portal plasma.

The absorption of 3,4-DHPEA-EDA can be further supported by the metabolism of 3,4-DHPEA-EDA, HT, and Ole. Accordingly, 3,4-DHPEA-EDA was not observed in the plasma from the jugular vein after intravenous administration (Figure 2(a)). In addition, HT (259.2 $AUC_{0 \rightarrow 4\,h\text{-}i.g.}/Dose_{i.g.}$), HVAOH (56.0 $AUC_{0 \rightarrow 4\,h\text{-}i.g.}/Dose_{i.g.}$), and HVA (22.4 $AUC_{0 \rightarrow 4\,h\text{-}i.g.}/Dose_{i.g.}$) from 3,4-DHPEA-EDA were monitored in the plasma from the jugular vein after intravenous administration throughout the HPLC monitoring (240 min) (Figure 2(a)). Detection of high levels of HVAOH from 3,4-DHPEA-EDA was unique and it differs from the detection of HT, HVAOH, and HVA from HT (Figures 3(a) and 3(b)). Another interesting feature is the slower clearance rate of HT as a 3,4-DHPEA-EDA metabolite (Figure 2(a)) in the systemic circulation, compared with that of intact HT (Figure 2(b)). If orally administered 3,4-DHPEA-EDA can be converted to HT in the gastrointestinal tract, HT from 3,4-DHPEA-EDA and the intact HT should have rapid and similar clearance rates in the systemic venous plasma. These results might suggest that 3,4-DHPEA-EDA is present in the systemic venous blood (e.g., bound to plasma proteins) or some organs as the bound form, which would be gradually released. The slow release of 3,4-DHPEA-EDA and its related polyphenols to the free form and formation of the metabolites (HT, HVAOH, and HVA) maintain the biological function for a long period in the body. The formation of homovanillyl alcohol from 3,4-DHPEA-EDA administration might influence function, for example, as a neurotransmitter [37], which is also supporting evidence for the absorption of 3,4-DHPEA-EDA by rats.

4.2. Excretion of 3,4-DHPEA-EDA and Its Related Polyphenols after Oral and Intravenous Administration. After oral administration of 3,4-DHPEA-EDA, HT (0.51% of dose) and HVA (0.39% of dose) were detected in small amounts in the urine of the rats (Figure 3(a)), while HVAOH was not detected in the urine (Figure 3(a)). In total, 0.90% of the 3,4-DHPEA-EDA metabolites was excreted in the urine within 240 min after oral administration. Conversely, observation of the metabolites HT, HVAOH, and HVA from 3,4-DHPEA-EDA in the portal plasma revealed that HT (27.7 $AUC_{0 \rightarrow 4\,h\text{-i.g.}}/Dose_{\text{i.g.}}$), HVAOH (4.2 $AUC_{0 \rightarrow 4\,h\text{-i.g.}}/Dose_{\text{i.g.}}$), and HVA (4.5 $AUC_{0 \rightarrow 4\,h\text{-i.g.}}/Dose_{\text{i.g.}}$) were detected after oral administration (Figure 1(a)). These results indicate that 3,4-DHPEA-EDA lasts a long time in the systemic venous blood (e.g., bound to plasma proteins) or some organs as the bound form, which suggests that the physiological functions of 3,4-DHPEA-EDA, including the metabolites, will continue for a long period in the body.

On the other hand, after oral administration of HT, a large amount of HT (1.14% of dose), HVAOH (0.17% of dose), and HVA (8.03% of dose) were detected in the urine. In total, 9.34% of the HT metabolites including intact HT was rapidly excreted into the urine within 240 min after administration. The total excretion ratio of 3,4-DHPEA-EDA, HT, and Ole including their metabolites was 12.9 : 133.4 : 1 (3,4-DHPEA-EDA : HT : Ole). Comparing with the ratio of AUC/Dose of 3,4-DHPEA-EDA : HT : Ole (ratio of AUC/Dose of 3,4-DHPEA-EDA : HT : Ole = 13.5 : 30 : 1) in the portal plasma (Figure 1), it is clear that 3,4-DHPEA-EDA (bound form, its related compounds, and its metabolites), HT, and the metabolites in the body have high potential for contributing to physiological function as mentioned in Section 1.

With respect to 3,4-DHPEA-EDA, HT, and Ole excretion to urine and bile, monitoring and tracing these three chemicals and their metabolites after intravenous administration is simpler and easier than monitoring and tracing them after oral administration. Moreover, the first-pass hepatic metabolites after intravenous administration can be easily monitored and interpreted compared with metabolites detected in the systemic venous blood after oral administration. Therefore, oral administration results in more complicated and fewer metabolites than intravenous administration. Accordingly, a limited amount of chemicals could be detected when they were observed in the urine and bile because the absorption phase is subject to certain restrictions for selective transportation, conjugation, modification, and metabolism of the chemicals. Therefore, careful interpretation of the results is required.

Consequently, when 3,4-DHPEA-EDA was directly administrated to the jugular vein, 3,4-DHPEA-EDA was not observed, but HT (6.3% of dose), HVAOH (4.6% of dose), and HVA (2.6% of dose) were observed in the urine as excreted products (Figure 3(b)). The same chemicals, HT (0.17% of dose), HVAOH (0.26% of dose), and HVA (0.05% of dose), were observed in small amounts in the bile (Figure 3(c)). It can be concluded from the data above that mainly 3,4-DHPEA-EDA and its metabolites were excreted to the urine. This tendency was similarly observed when HT was directly administered to the jugular vein (Figures

3(b) and 3(c)). Thus, HT and its metabolites described above are mainly excreted to the urine. Conversely, Ole was not detected at all in the urine when Ole was administrated to the jugular vein. Instead, Ole (3.22% of dose) was excreted in the bile without any other metabolites (Figure 3(c)). Finally, excretion of Ole to the bile would be the most predominant pathway of Ole without any transformation of the chemical structure. The metabolic pathway of Ole is regulated and suppressed by inactivation of deglycosylation, hydrolysis, oxygenation, and methylation enzymes, which are related to metabolism after intravenous administration [26].

3,4-DHPEA-EDA, HT, and Ole could be easily excreted to urine or bile when these chemicals were intravenously administrated. When 3,4-DHPEA-EDA and HT in the urine after oral administration and intravenous administration were compared, 3,4-DHPEA-EDA showed a great difference in detection ratios of the metabolites against doses between i.g. and i.v. (Figures 3(a) and 3(b)). Thus, 13% metabolites are observed in the urine after intravenous administration but only 0.9% metabolites are observed in the urine after oral administration. HT had a similar tendency in the detection ratios against each dose. It is supposed that the ratios of metabolites from 3,4-DHPEA-EDA and HT in the urine after oral administration (Figure 3(a)) reflect those of 3,4-DHPEA-EDA, including its related substances, and HT absorbed in the body (Figures 2(a) and 2(b)) but do not reflect the metabolite ratios of 3,4-DHPEA-EDA and HT administered in the jugular vein, which means 3,4-DHPEA-EDA in the jugular plasma after oral administration is not the same as 3,4-DHPEA-EDA itself; it is modified like the bound form to certain proteins in the vein. Furthermore, 3,4-DHPEA-EDA or its related compounds with or without bound forms were retained in the body when 3,4-DHPEA-EDA was orally administrated, but HT was released easily to the urine as the excreted substance after oral administration in rats as reported in a previous paper [14]. In total, the physiological function of 3,4-DHPEA-EDA, its related polyphenols, and its metabolites last as active substances in the body.

4.3. Proposed Metabolic Pathway of 3,4-DHPEA-EDA. With the orally administrated 3,4-DHPEA-EDA, the late appearance of HVA ($T_{\max} = 60$ min), compared with HT ($T_{\max} = 30$ min) and HVAOH ($T_{\max} = 30$ min), supports the stepwise metabolism of 3,4-DHPEA-EDA, including 3,4-DHPEA-EDA-related polyphenols (bound form or unbound form), to HT, HVAOH, and HVA in this order. Furthermore, the ratios of the metabolites (HT, HVAOH, and HVA) from 3,4-DHPEA-EDA and HT were mainly influenced by enzyme activities in the gastrointestinal tract and the systemic venous blood. With oral administration of 3,4-DHPEA-EDA and HT, different ratios of metabolites were detected in the portal plasma. Thus, HT was observed as the main metabolite after oral administration of 3,4-DHPEA-EDA, but a large amount of HVA was observed as one of major metabolites of HT as expected from a previous paper [21] (Figures 2(a) and 2(b)). Enzymatic methylation and oxidation of HT by using

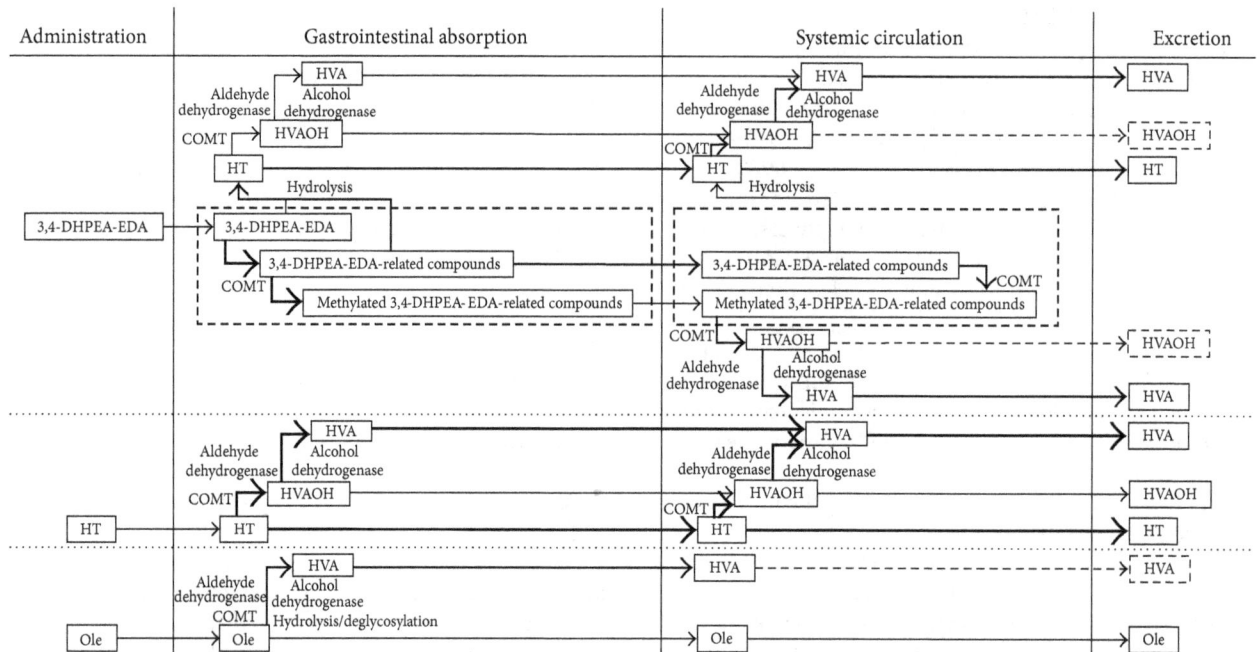

FIGURE 4: The possible flow of chemical conversion of 3,4-DHPEA-EDA, hydroxytyrosol, and oleuropein after administration in rats. HT: hydroxytyrosol, HVA: homovanillic acid, and HVAOH: homovanillyl alcohol.

catechol 3-O-methyltransferase (COMT), alcohol dehydrogenase, and aldehyde dehydrogenase have been reported in the gastrointestinal tract as the metabolization of HT using caco-2 cells [38]. However, 3,4-DHPEA-EDA metabolism was not clarified in previous studies. In fact, 3,4-DHPEA-EDA did not easily hydrolyze to HT in the gastrointestinal tract because HVA detection in the portal plasma from 3,4-DHPEA-EDA was not so high, when compared with HT metabolism under the same conditions.

Another point of view regarding 3,4-DHPEA-EDA metabolism is related to the large amount of HVAOH (4.6% of dose) from 3,4-DHPEA-EDA in the urine after intravenous administration, compared with that from HT. Systemic venous blood and the kidneys may easily metabolize 3,4-DHPEA-EDA or 3,4-DHPEA-EDA-related polyphenols (bound form or unbound form) to HVAOH but not to HVA (Figure 3(b)). The same tendency of HVAOH formation from 3,4-DHPEA-EDA and its related polyphenols was observed when it was detected in the jugular plasma after intravenous administration with a long period of detection of HT (Figure 2(a)). In contrast, HT was easily metabolized to HVA in the urine after intravenous administration (Figure 3(b)). Systemic venous blood and the kidneys may have different active enzymes for digesting 3,4-DHPEA-EDA, including 3,4-DHPEA-EDA-related polyphenols and HT. HVAOH formation from 3,4-DHPEA-EDA is much easier in the systemic venous blood than that of HT. COMT and aldehyde-alcohol dehydrogenase activities against 3,4-DHPEA-EDA and HT would be different in affinity. In considering the metabolism of 3,4-DHPEA-EDA and its related polyphenols from the data shown in the figures and tables, we have summarized the conversion of three typical substances in olive fruit (Figure 4).

5. Conclusion

The biokinetics of absorption, metabolism, and excretion of three olive phytochemicals (3,4-DHPEA-EDA, HT, and Ole) were studied in freely moving rats cannulated into the portal vein and jugular vein and restrained bile duct-cannulated rats. The ratio of AUC/Dose of 3,4-DHPEA-EDA : HT : Ole was 13.5 : 30 : 1 and these values are related to the rates of absorption of each substance. Furthermore, the total excretion ratio of 3,4-DHPEA-EDA, HT, and Ole including their metabolites was 12.9 : 133.4 : 1, which means that 3,4-DHPEA-EDA, including the metabolites, should be the most important source of biologically active substances in olive fruit overall when considering efficiency in the body. The amount of 3,4-DHPEA-EDA in olive fruit is also reported to be the highest [32].

In the circulation of metabolites in the blood, we found different enzymatic modification of 3,4-DHPEA-EDA and HT: 3,4-DHPEA-EDA mainly transforms to HVAOH, while HT transforms to HVA after intravenous administration. Conversely, transformation of Ole to other metabolites did not occur in the systemic venous blood. The low detection of Ole in the portal plasma after oral administration was due to low intestinal absorption and not fast metabolization by plasma enzymes. It has been supported in some papers [23, 26]. Finally, only minute amounts of Ole could be absorbed from the gastrointestinal tract after oral administration, and it was extensively excreted via the biliary route when administered intravenously, rendering as questionable its importance as a bioactive compound from olives.

In conclusion, the physiological function of 3,4-DHPEA-EDA and its related polyphenols and metabolites should make it an important functional food component in olive fruit.

Conflict of Interests

The authors declare that there is no conflict of interests regarding the publication of this paper.

Acknowledgment

The authors thank Ms. Marie Carroubi and Ms. Rika Matsuda for research assistance (both are from Kagawa University).

References

[1] M. Salami, C. Galli, L. De Angelis, and F. Visioli, "Formation of F_2-isoprostanes in oxidized low density lipoprotein: inhibitory effect of hydroxytyrosol," *Pharmacological Research*, vol. 31, no. 5, pp. 275–279, 1995.

[2] E. Tripoli, M. Giammanco, G. Tabacchi, D. Di Majo, S. Giammanco, and M. La Guardia, "The phenolic compounds of olive oil: structure, biological activity and beneficial effects on human health," *Nutrition Research Reviews*, vol. 18, no. 1, pp. 98–112, 2005.

[3] N. R. T. Damasceno, A. Pérez-Heras, M. Serra et al., "Crossover study of diets enriched with virgin olive oil, walnuts or almonds. Effects on lipids and other cardiovascular risk markers," *Nutrition, Metabolism and Cardiovascular Diseases*, vol. 21, no. 1, pp. S14–S20, 2011.

[4] Á. Hernáez, A. T. Remaley, M. Farràs et al., "Olive oil polyphenols decrease LDL concentrations and LDL atherogenicity in men in a randomized controlled trial," *Journal of Nutrition*, vol. 145, no. 8, pp. 1692–1697, 2015.

[5] M. A. Carluccio, L. Siculella, M. A. Ancora et al., "Olive oil and red wine antioxidant polyphenols inhibit endothelial activation: antiatherogenic properties of Mediterranean diet phytochemicals," *Arteriosclerosis, Thrombosis, and Vascular Biology*, vol. 23, no. 4, pp. 622–629, 2003.

[6] R. J. Widmer, M. A. Freund, A. J. Flammer et al., "Beneficial effects of polyphenol-rich olive oil in patients with early atherosclerosis," *European Journal of Nutrition*, vol. 52, no. 3, pp. 1223–1231, 2013.

[7] C. A. de la Lastra, M. D. Barranco, V. Motilva, and J. M. Herrerías, "Mediterranean diet and health: biological importance of olive oil," *Current Pharmaceutical Design*, vol. 7, no. 10, pp. 933–950, 2001.

[8] J. López-Miranda, F. Pérez-Jiménez, E. Ros et al., "Olive oil and health: summary of the II international conference on olive oil and health consensus report, Jaén and Córdoba (Spain) 2008," *Nutrition, Metabolism and Cardiovascular Diseases*, vol. 20, no. 4, pp. 284–294, 2010.

[9] S. Bulotta, M. Celano, S. M. Lepore, T. Montalcini, A. Pujia, and D. Russo, "Beneficial effects of the olive oil phenolic components oleuropein and hydroxytyrosol: focus on protection against cardiovascular and metabolic diseases," *Journal of Translational Medicine*, vol. 12, no. 1, article 219, 2014.

[10] G. K. Beauchamp, R. S. J. Keast, D. Morel et al., "Ibuprofen-like activity in extra-virgin olive oil. Enzymes in an inflammation pathway are inhibited by oleocanthal, a component of olive oil," *Nature*, vol. 437, pp. 45–46, 2005.

[11] A. Camargo, O. A. Rangel-Zuñiga, C. Haro et al., "Olive oil phenolic compounds decrease the postprandial inflammatory response by reducing postprandial plasma lipopolysaccharide levels," *Food Chemistry*, vol. 162, pp. 161–171, 2014.

[12] D. Caruso, F. Visioli, R. Patelli, C. Galli, and G. Galli, "Urinary excretion of olive oil phenols and their metabolites in humans," *Metabolism: Clinical and Experimental*, vol. 50, no. 12, pp. 1426–1428, 2001.

[13] M. N. Vissers, P. L. Zock, A. J. C. Roodenburg, R. Leenen, and M. B. Katan, "Olive oil phenols are absorbed in humans," *The Journal of Nutrition*, vol. 132, no. 3, pp. 409–417, 2002.

[14] F. Visioli, C. Galli, S. Grande et al., "Hydroxytyrosol excretion differs between rats and humans and depends on the vehicle of administration," *The Journal of Nutrition*, vol. 133, no. 8, pp. 2612–2615, 2003.

[15] E. Gallardo, R. Palma-Valdés, J. L. Espartero, and M. Santiago, "In vivo striatal measurement of hydroxytyrosol, and its metabolite (homovanillic alcohol), compared with its derivative nitrohydroxytyrosol," *Neuroscience Letters*, vol. 579, pp. 173–176, 2014.

[16] M. Suárez, M.-P. Romero, A. Macià et al., "Improved method for identifying and quantifying olive oil phenolic compounds and their metabolites in human plasma by microelution solid-phase extraction plate and liquid chromatography-tandem mass spectrometry," *Journal of Chromatography B: Analytical Technologies in the Biomedical and Life Sciences*, vol. 877, no. 32, pp. 4097–4106, 2009.

[17] M. González-Santiago, J. Fonollá, and E. Lopez-Huertas, "Human absorption of a supplement containing purified hydroxytyrosol, a natural antioxidant from olive oil, and evidence for its transient association with low-density lipoproteins," *Pharmacological Research*, vol. 61, no. 4, pp. 364–370, 2010.

[18] M. De Bock, E. B. Thorstensen, J. G. B. Derraik, H. V. Henderson, P. L. Hofman, and W. S. Cutfield, "Human absorption and metabolism of oleuropein and hydroxytyrosol ingested as olive (*Olea europaea* L.) leaf extract," *Molecular Nutrition & Food Research*, vol. 57, no. 11, pp. 2079–2085, 2013.

[19] P. Del Boccio, A. Di Deo, A. De Curtis, N. Celli, L. Iacoviello, and D. Rotilio, "Liquid chromatography—tandem mass spectrometry analysis of oleuropein and its metabolite hydroxytyrosol in rat plasma and urine after oral administration," *Journal of Chromatography B: Analytical Technologies in the Biomedical and Life Sciences*, vol. 785, no. 1, pp. 47–56, 2003.

[20] A. Serra, L. Rubió, X. Borràs, A. Macià, M.-P. Romero, and M.-J. Motilva, "Distribution of olive oil phenolic compounds in rat tissues after administration of a phenolic extract from olive cake," *Molecular Nutrition & Food Research*, vol. 56, no. 3, pp. 486–496, 2012.

[21] M. López de las Hazas, L. Rubió, A. Kotronoulas, R. De la Torre, R. Solà, and M. Motilva, "Dose effect on the uptake and accumulation of hydroxytyrosol and its metabolites in target tissues in rats," *Molecular Nutrition & Food Research*, vol. 59, no. 7, pp. 1395–1399, 2015.

[22] T. Ichiyanagi, Y. Kashiwada, Y. Shida et al., "Structural elucidation and biological fate of two glucuronyl metabolites of pelargonidin 3-*O*-β-D-glucopyranoside in rats," *Journal of Agricultural and Food Chemistry*, vol. 61, no. 3, pp. 569–578, 2013.

[23] S. C. Edgecombe, G. L. Stretch, and P. J. Hayball, "Oleuropein, an antioxidant polyphenol from olive oil, is poorly absorbed

from isolated perfused rat intestine," *The Journal of Nutrition*, vol. 130, no. 12, pp. 2996–3002, 2000.

[24] H.-W. Tan, K. L. Tuck, I. Stupans, and P. J. Hayball, "Simultaneous determination of oleuropein and hydroxytyrosol in rat plasma using liquid chromatography with fluorescence detection," *Journal of Chromatography B: Analytical Technologies in the Biomedical and Life Sciences*, vol. 785, no. 1, pp. 187–191, 2003.

[25] F. N. Bazoti, E. Gikas, and A. Tsarbopoulos, "Simultaneous quantification of oleuropein and its metabolites in rat plasma by liquid chromatography electrospray ionization tandem mass spectrometry," *Biomedical Chromatography*, vol. 24, no. 5, pp. 506–515, 2010.

[26] T. Zhou, T. Qian, X. Wang, X. Li, L. Cao, and S. Gui, "Application of LC-MS/MS method for the *in vivo* metabolite determination of oleuropein after intravenous administration to rat," *Biomedical Chromatography*, vol. 25, no. 12, pp. 1360–1363, 2011.

[27] P. Lin, W. Qian, X. Wang, L. Cao, S. Li, and T. Qian, "The biotransformation of oleuropein in rats," *Biomedical Chromatography*, vol. 27, no. 9, pp. 1162–1167, 2013.

[28] C. Bai, X. Yan, M. Takenaka, K. Sekiya, and T. Nagata, "Determination of synthetic hydroxytyrosol in rat plasma by GC-MS," *Journal of Agricultural and Food Chemistry*, vol. 46, no. 10, pp. 3998–4001, 1998.

[29] K. L. Tuck, M. P. Freeman, P. J. Hayball, G. L. Stretch, and I. Stupans, "The in vivo fate of hydroxytyrosol and tyrosol, antioxidant phenolic constituents of olive oil, after intravenous and oral dosing of labeled compounds to rats," *The Journal of Nutrition*, vol. 131, no. 7, pp. 1993–1996, 2001.

[30] S. D'Angelo, C. Manna, V. Migliardi et al., "Pharmacokinetics and metabolism of hydroxytyrosol, a natural antioxidant from olive oil," *Drug Metabolism and Disposition*, vol. 29, no. 11, pp. 1492–1498, 2001.

[31] L. Rubió, A. Serra, A. Macià, C. Piñol, M.-P. Romero, and M.-J. Motilva, "In vivo distribution and deconjugation of hydroxytyrosol phase II metabolites in red blood cells: a potential new target for hydroxytyrosol," *Journal of Functional Foods*, vol. 10, pp. 139–143, 2014.

[32] A. Sato, N. Shinozaki, and H. Tamura, "Secoiridoid type of antiallergic substances in olive waste materials of three Japanese varieties of *Olea europaea*," *Journal of Agricultural and Food Chemistry*, vol. 62, no. 31, pp. 7787–7795, 2014.

[33] M. Anastassiades, S. J. Lehotay, and D. Stajnbaher, "Quick, easy, cheap, effective, rugged, and safe (QuEChERS) approach for the determination of pesticide residues," in *Proceedings of the 18th Annual Waste Testing and Quality Assurance Symposium (WTQA '02)*, pp. 231–241, Arlington, Va, USA, August 2002.

[34] J. Pinto, F. Paiva-Martins, G. Corona et al., "Absorption and metabolism of olive oil secoiridoids in the small intestine," *British Journal of Nutrition*, vol. 105, no. 11, pp. 1607–1618, 2011.

[35] H. Tamura and T. Shibamoto, "Antioxidative activity measurement in lipid peroxidation systems with malonaldehyde and 4-hydroxy nonenal," *Journal of the American Oil Chemists Society*, vol. 68, no. 12, pp. 941–943, 1991.

[36] G. Lefèvre, C. Bonneau, S. Rahma et al., "Determination of plasma protein-bound malondialdehyde by derivative spectrophotometry," *European Journal of Clinical Chemistry and Clinical Biochemistry*, vol. 34, no. 8, pp. 631–636, 1996.

[37] K. T. Beggs and A. R. Mercer, "Dopamine receptor activation by honey bee queen pheromone," *Current Biology*, vol. 19, no. 14, pp. 1206–1209, 2009.

[38] C. Manna, P. Galletti, G. Maisto, V. Cucciolla, S. D'Angelo, and V. Zappia, "Transport mechanism and metabolism of olive oil hydroxytyrosol in Caco-2 cells," *FEBS Letters*, vol. 470, no. 3, pp. 341–344, 2000.

Alanine with the Precipitate of Tomato Juice Administered to Rats Enhances the Reduction in Blood Ethanol Levels

Shunji Oshima, Sachie Shiiya, Yoshimi Tokumaru, and Tomomasa Kanda

Research & Development Laboratories for Innovation, Asahi Group Holdings, Ltd., Ibaraki 302-0106, Japan

Correspondence should be addressed to Shunji Oshima; shunji.oshima@asahigroup-holdings.com

Academic Editor: Phillip B. Hylemon

Delay in gastric emptying (GE) lowers the blood ethanol concentration (BEC) after alcohol administration. We previously demonstrated that water-insoluble fractions, mainly comprising dietary fiber derived from many types of botanical foods, possessed the ability to absorb ethanol-containing aqueous solutions. Furthermore, there was a significant correlation between the absorption of ethanol and lowering of BEC because of delay in GE. Here we identified dietary nutrients that synergize with the water-insoluble fraction of tomatoes to lower BEC in rats. Consequently, unlike tomato juice without alanine, tomato juice with 5.0% alanine decreased BEC depending on the delay in GE and mediated the ethanol-induced decrease in the spontaneous motor activity (an indicator of drunkenness). Our findings indicate that the synergism between tomato juice and alanine to reduce the absorption of ethanol was attributable to the effect of alanine on precipitates such as the water-insoluble fraction of tomatoes.

1. Introduction

Alcoholic beverages comprise the most commonly consumed foods by adults worldwide, and their inappropriate and excessive use is associated with an increased risk of several diseases [1]. Because alcohol is usually consumed with meals, understanding the effects of the diet on the pharmacokinetics of alcohol is important to reduce the harmful effects of alcohol on human health. Several dietary components are often employed to increase alcohol metabolism. It is well known that the oral intake of fructose or sucrose stimulates the elimination of alcohol from the bloodstream of healthy or alcoholic subjects [2–4]. Moreover, the ability of alanine to lower the blood ethanol concentration (BEC) may be attributed to the formation of pyruvate by oxidative deamination. Metabolic pathways generate NAD+, which facilitates alcohol oxidation in the liver via the conversion of pyruvate to lactate [5]. However, the most significant effect that foods can have on the absorption of alcohol is delaying the gastric emptying (GE). GE delivers nutrients to the duodenum after meals [6], and numerous reports show that delayed GE decreases BEC [7–12]. Alcohol consumed with a meal is likely released into the duodenum in a gradual manner, regardless of the composition of nutritional constituents, thus reducing

BEC because of the reduction in the absorption rate [13]. Little is known about the effects of dietary components on the regulation of the rate of GE and its effect on the dynamics of alcohol metabolism.

We previously found that the water-insoluble fractions (WIFs) of several types of botanical foods besides tomatoes, mainly comprising water-insoluble dietary fibers, absorb ethanol-containing solutions. Moreover, the absorption of ethanol correlates with the inhibition of the blood ethanol elevation by delaying GE [14]. To the best of our knowledge, few studies have assessed the effects of dietary fibers on the dynamics of alcohol metabolism. In the current study, we analyzed BEC and the spontaneous motor activity as the indicators of drunkenness or intoxication. We were specifically interested in identifying nutrients that synergize with tomatoes to more effectively lower BEC.

2. Materials and Methods

2.1. Materials. Ethanol (99.5%) was purchased from Kanto Chemical Co., Inc. (Tokyo, Japan). Both the commercially available tomato paste and DL-alanine were kindly provided by Kagome Co., Ltd. (Tokyo, Japan) and Ajinomoto Healthy Supply, Inc. (Tokyo, Japan).

2.2. Sample Preparation. Tomato juice (5.8 Brix) was diluted 5-fold (w/v) using tomato paste and mineral water, and 200 mL of tomato juice was centrifuged at 3.000 rpm for 10 min. The supernatant, that is, the water-soluble fraction (WSF), and the precipitate, that is, the water-insoluble fraction (WIF), were diluted using mineral water to 200 mL. This yielded 200 mL of a WSF solution and 200 mL of a WIF suspension of tomato juice. Alanine (5.0 g) was added to 100 mL of each of the following: water, tomato juice, WSF, and WIF. Samples were stored at 5°C. The natural content of alanine in tomatoes is negligible (<0.01%) [15].

2.3. Preparation of Dry Powder of WIF Using Tomato Paste. Tomato paste (100 g) was added to 1 L of water, stirred thoroughly, and centrifuged at 3.000 rpm for 10 min. The supernatant was carefully removed. The washes were repeated until the Brix value of the supernatant solution reached 0. The precipitates were then frozen and freeze-dried to obtain WIF as a dry powder, which was stored at room temperature in a continuously evacuated desiccator until its ethanol-maintaining ability was assessed.

2.4. Determination of Ethanol-Maintaining Ability. Briefly, we prepared 15% (v/v) aqueous solutions of ethanol solution or suspensions containing 5.0% (w/v) alanine, 0.5% (w/v) dry WIF, or a mixture of 0.5% (w/v) dry WIF and 5.0% (w/v) alanine. A circular filter paper (110 mm diameter, Toyo Roshi Kaisha Ltd., Tokyo, Japan) was placed on a funnel, and 10 mL of each sample (or the 15% ethanol aqueous solution control) was poured on the filter paper. After 5 min, the filtered volume (mL) of ethanol solution was analyzed three times.

2.5. Animal Studies. All animal studies were conducted in accordance with the ethical guidelines for animal care, and the animal care committee of Asahi Group Holdings Ltd. (Ibaraki, Japan) approved the experimental protocol. F344 male rats (5 weeks old) were obtained from Japan SLC Inc. (Hamamatsu, Japan). The animals were housed in individual stainless steel cages in a room maintained at 25°C with 55% relative humidity and were allowed ad libitum access to laboratory chow pellets and tap water until use. For Experiment 1, the spontaneous motor activity of each rat weighing <360 g was measured continuously in a cage using the Supermex System (Muromachi Kikai, Japan), as described in previous studies [16, 17], for consecutive 4 h periods. Drunkenness induced by alcohol in rats decreases the spontaneous motor activity [18]. Four rats in each of the five groups (defined below) were fasted for 7 h. Before measuring the activity, the rats were orally given 20 mL/kg of tomato juice (tomato group), 5.0% aqueous alanine (alanine group), tomato juice with 5.0% alanine (tomato + alanine group), or water (water and control group). All groups were then given 8.45 mL/kg of 30% ethanol solution twice at 30 and 60 min intervals (total ethanol dose, 4.0 g/kg). The water group was given 8.45 mL/kg of water instead of ethanol solution. After the oral administration of ethanol, food and water were replenished and each rat was placed in a transparent plastic cage located in a sound-attenuating chamber. The sensor box, which was mounted in the center of the ceiling of the sound-attenuating chamber, detected the spontaneous motor activity; the signals were transmitted by an interface device to a computer and converted to the number of movements. Next, preliminary experiments were conducted to decide the time of blood collection after 2.0 g or 4.0 g/kg ethanol administrations to rats prior to Experiments 2 and 3. For Experiments 2A-2B, four rats weighing <270 g per group were fasted overnight and assigned to each group as follows: Before ethanol administration, the rats were orally given 20 mL/kg of tomato juice, 5.0% alanine aqueous solution, tomato juice with 5.0% alanine, or water (control) in Experiment 2A and an additional 20 mL/kg of tomato juice supernatant with or without 5.0% alanine or tomato juice precipitate with or without 5.0% alanine was given in Experiment 2B. The rats were then given 8.45 mL/kg of 30% ethanol solution twice, at 30 and 60 min intervals (total ethanol dose, 4.0 g/kg). Blood samples were then collected from the tail vein of each rat after 4 h because in preliminary experiments, using this method, we established that BEC peaked approximately 4 h after the administration of ethanol (4.0 g/kg) (see Figure 2). In Experiment 3, rats (n = 3 per group) weighing 240–340 g were fasted for 24 h and then orally given 20 mL/kg of tomato juice precipitate, 5.0% alanine aqueous solution, tomato juice precipitate with 5.0% alanine, or water (control). After 2.0 g/kg (8.45 mL/kg) of ethanol administration, blood samples were collected from the large vein of each rat after 2 h, when BEC reached maximal level (see Figure 2). After anesthetizing the rats with carbon dioxide, the stomach was ligated immediately at the cardiac and pyloric regions and harvested. The fluid in the stomach was diluted using distilled water to 50 mL. Residual ethanol in the stomach was calculated as a percentage of the total ethanol dose. Experiment 3 was performed twice to confirm repeatability of the result.

In Experiment 4, rats (n = 6 per group) weighing 370–480 g were fasted overnight and then orally given 20 mL/kg of tomato juice with 5.0% alanine or water (control). After 30 min, the rats were then orally or intraperitoneally given 16.9 mL/kg of 7.5% ethanol containing 0.9% NaCl solution (ethanol dose, 1.0 g/kg). After ethanol administration, blood samples were collected from the tail vein of each rat after 1 h.

2.6. Determination of BEC and Residual Ethanol in the Stomach. The amount of ethanol in the blood and residual ethanol in the stomach was determined using headspace gas chromatography, as previously described [19]. Briefly, blood or stomach-fluid samples (50 μL) were treated immediately with 2.5 mL of ice-cold 4% (w/v) perchloric acid (PCA) with ethanol-d6 (Merck, 99% D anhydrous) added as an internal standard to a capped tube. The supernatant (2.0 mL) collected after centrifugation at 3.000 rpm for 10 min at 4°C was transferred to a glass vial containing 0.50 g of solid NaCl, and the PCA-treated samples were applied to a GC-MS system. GC-MS analysis was performed using an Agilent 6890 N gas chromatograph coupled to an Agilent MS-5975C inert XL mass-selective detector and an Agilent headspace sampler G1888 (Agilent Technologies, Little Fall, NY, USA).

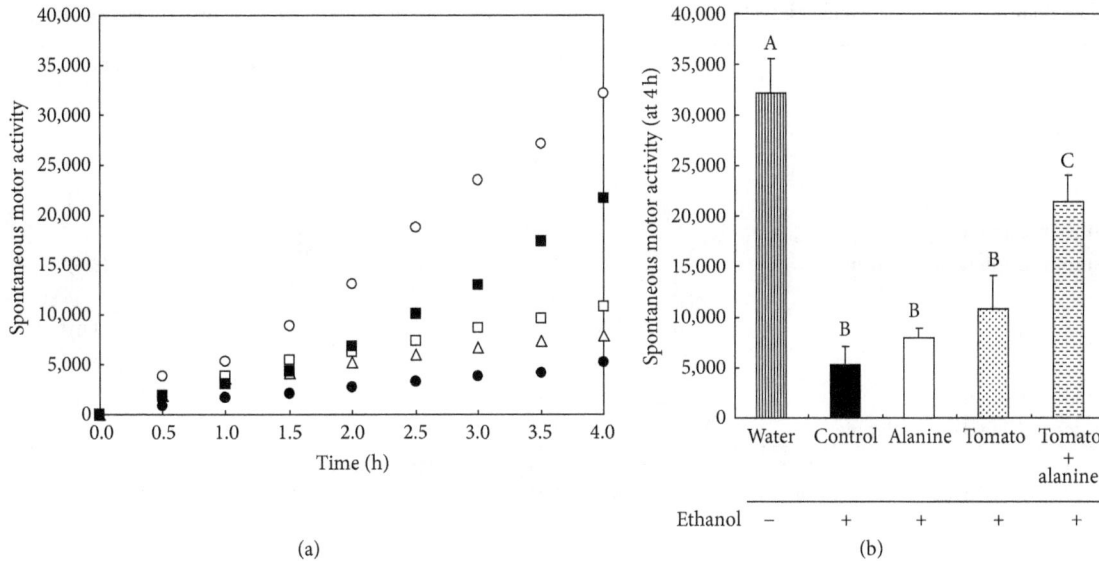

FIGURE 1: (a) The effects of a combination of tomato juice and alanine on the spontaneous motor activity 4 h following the ingestion of 4.0 g/kg of ethanol (Experiment 1). The rats were orally given 20 mL/kg of tomato juice (tomato group), 5.0% aqueous alanine (alanine group), tomato juice with 5.0% alanine (tomato + alanine group), or water (water and control group). Results are expressed as the mean values. O, water (ethanol-free); ●, control; △, alanine; □, tomato; ■, tomato + alanine. (b) The spontaneous motor activity at 4 h following the ingestion of 4.0 g/kg of ethanol. Results are expressed as the mean and standard deviations ($n = 4$). Mean values with different uppercase letters differ significantly (ANOVA, post hoc Tukey's HSD test, $p < 0.05$) among five groups.

A capillary DB-Wax column (60 m × 0.25 mm, Agilent J & W GC Columns) was used to separate the peak ethanol fraction. The initial temperature of 35°C was maintained for 3 min and then increased to 80°C at 4°C/min; subsequently, the temperature was increased to 240°C at 20°C/min and then maintained at 240°C for 4 min. The split ratio was 1 : 20, and helium was used as a carrier gas and delivered at 1.0 mL/min. The injector and detector temperatures were 210°C and 250°C, respectively. The peaks of ethanol and ethanol-d6 in the treated samples were identified by comparing with the retention times and mass spectra of standards.

2.7. Statistical Analyses. All statistical analyses were performed using Dr. SPSS II software (SPSS Inc.). The difference in the spontaneous motor activity, BEC, or residual ethanol in the stomach among the groups was analyzed using one-way analysis of variance followed by Tukey's honest significant difference test in Experiments 1–3. Mean values with the same superscript letter are significantly different ($p < 0.05$). Filtered volumes at different time intervals (control versus alanine and WIF versus WIF + alanine) were compared using Student's t-test. The acceptable level of significance was 5% for each analysis. Simple linear regression (Pearson's) analysis was performed to examine the correlation between BEC (mg/mL) and the residual amount of ethanol in the stomach as well as to assess the relationship between the lowering of blood alcohol and GE. The relationship is expressed as the correlation coefficient (r). In Experiment 4, BEC between the control and tomato + alanine groups was compared using Student's t-test.

3. Results

Spontaneous motor activity after the oral administration of a large amount of ethanol (4.0 g/kg) following the administration of each sample for 4 h periods is shown in Figures 1(a) and 1(b). The activity of the control group was significantly decreased compared with that of the water (ethanol-free) group. The activities of the tomato and alanine groups did not differ significantly from those of the control group. The activity of the tomato + alanine group was significantly greater than that of control and significantly lower than that of the water group.

We determined that BEC peaked approximately 4 h after the administration of 4.0 g/kg of ethanol in the preliminary experiment (Figure 2). BEC at 4 h after the oral administration of 4.0 g/kg of ethanol is shown in Figures 3(a) and 3(b). The BEC of the alanine group reached 3.26 mg/mL, similar to that of control group (3.54 mg/mL). The BEC (2.89 mg/mL) of the tomato group was significantly lower than that of the control group. Moreover, the BEC (2.04 mg/mL) of the tomato + alanine group was significantly lower than that of the three groups, including the tomato group. Next, we evaluated the supernatant or precipitate of tomato preparations with or without alanine. No difference in the BEC of the supernatant group with or without alanine (3.53 mg/mL versus 3.58 mg/mL) was detected; however, the BEC (2.89 mg/mL) of the precipitate group was significantly lower than that of the supernatant group. Furthermore, BEC after the administration of precipitate supplemented with alanine was significantly lower than that after the administration of precipitate without alanine (2.13 mg/mL versus 2.89 mg/mL).

TABLE 1: Blood ethanol concentration and residual ethanol in the stomach 2 h after the administration of 2.0 g/kg of ethanol in Experiment 3.

	BEC mg/mL	Residual amount of ethanol in stomach % dose
(First)		
Control	1.52 (0.25)[a]	16.9 (6.8)[a]
Alanine	1.47 (0.17)[a]	15.3 (7.4)[a]
Tomato juice precipitate	1.17 (0.12)[a]	22.4 (7.9)[a]
Tomato juice precipitate with alanine	0.57 (0.15)[b]	47.1 (3.7)[b]
(Second)		
Control	1.68 (0.10)[a]	6.5 (3.0)[a]
Alanine	1.66 (0.04)[a]	2.4 (0.7)[a]
Tomato juice precipitate	1.41 (0.18)[a]	13.9 (5.1)[a]
Tomato juice precipitate with alanine	0.95 (0.14)[b]	28.1 (7.2)[b]

Data are expressed as means and standard deviation (mean ± SD, $n = 3$). Means values with different lowercase letters are significantly different (ANOVA post hoc Tukey's HSD test, $p < 0.05$) among four groups. The secondary experiment was constructed to confirm repeatability of their effects.

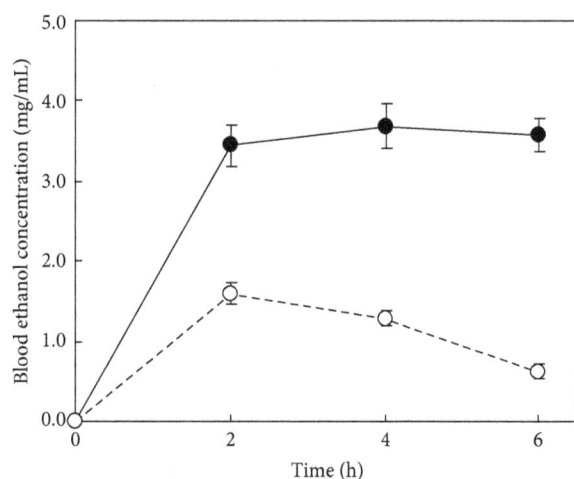

FIGURE 2: Changes in the blood ethanol concentration following the ingestion of 2.0 or 4.0 g/kg ethanol in preliminary experiments. Results are expressed as means and standard deviations. - - -O- - -, 2.0 g/kg ($n = 10$); —●—, 4.0 g/kg ($n = 8$).

The ability of each sample to maintain 15% (v/v) ethanol is shown in Figure 4. During filtration, the aqueous ethanol solution in the fluids dripped through the filter paper more slowly when the samples strongly absorbed ethanol. The total filtered volume of the ethanol solution after 20 min (control) was 7.3 mL. The volume of the ethanol solutions containing 5.0% (w/v) alanine was equivalent to that of the control solution. The volume of the ethanol solution containing WIF was significantly lower than that of the control or alanine solutions. Moreover, the volume of WIF with alanine was significantly lower than that of WIF at 10 min and 20 min.

BEC and residual ethanol concentration in the stomach 2 h after the administration of 2.0 g/kg ethanol are presented in Table 1. The BECs of the control and the alanine groups 2 h after the administration of 2.0 g/kg of ethanol were 1.52 or 1.68 mg/mL and 1.47 or 1.66 mg/mL, respectively. In contrast, the BEC of the tomato juice precipitate with alanine group (0.57 or 0.95 mg/mL) was significantly lower than that of

the tomato juice precipitate group (1.17 or 1.41 mg/mL). The residual gastric content of ethanol in the water control, alanine, tomato juice precipitate, and tomato juice precipitate with alanine groups was 16.9% or 6.5%, 15.3% or 2.4%, 22.4% or 13.9%, and 47.1% or 28.1%, respectively. There was no significant difference between the control and alanine groups; however, the amount of gastric ethanol was increased significantly in the tomato juice precipitate with alanine group compared with that of the tomato juice precipitate group. A simple linear regression (Pearson's) analysis was performed to examine the correlation between the residual amount of ethanol in the stomach and the BEC of each rat in the four groups. The correlation coefficient (r) for the difference between the residual amount and BEC was 0.931, and the two parameters correlated significantly ($p < 0.001$, Figure 5).

BEC at 1 h after the oral or intraperitoneal administration of 1.0 g/kg of ethanol is shown in Figure 6. When rats were orally given ethanol, the BEC of the tomato juice with 5.0% alanine group (0.26 mg/mL) was significantly lower than that of the control group (0.86 mg/mL). When rats were intraperitoneally given ethanol, BEC of the tomato juice with 5.0% alanine (0.95 mg/mL) group was significantly lower than that of the control group (1.03 mg/mL); however, the lowering was weaker than that observed after the oral administration of ethanol.

4. Discussion

Tomatoes contain various nutrients and functional components, including carotenoids, polyphenols, and phytosterols [20, 21]. In this study we found that the water-soluble fraction of tomato juice contains 4.5% sugar (glucose and fructose) and had no effect on BEC, whereas WIF lowered BEC. A 0.7% suspension of WIF solids in water corresponded to 0.14 g/kg, most of which was dietary fiber (70.7%) followed by small amounts of lipids or water-insoluble proteins. It seems unlikely that WIF accelerated alcohol metabolism; thus, we suspect that WIF inhibited alcohol absorption in the gastrointestinal tract. Ushida et al. previously reported

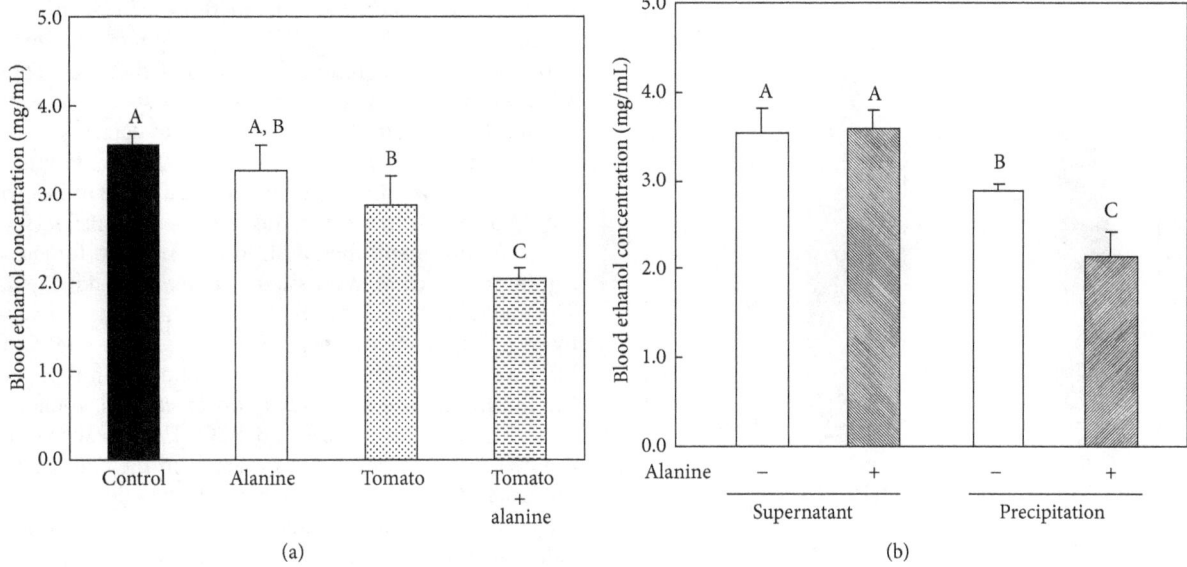

FIGURE 3: Blood ethanol concentrations 4 h following ingestion of 4.0 g/kg of ethanol after the administration of each sample in Experiments 2A and 2B. The rats were orally given 20 mL/kg of tomato juice, 5.0% alanine aqueous solution, tomato juice with 5.0% alanine, or water (control) in Experiment 2A and the tomato juice supernatant with or without 5.0% alanine or the tomato juice precipitate with or without 5.0% alanine in Experiment 2B. Results are expressed as mean and standard deviation ($n = 4$). Mean values with different uppercase letters differ significantly (ANOVA post hoc Tukey's HSD test, $p < 0.05$) among four groups.

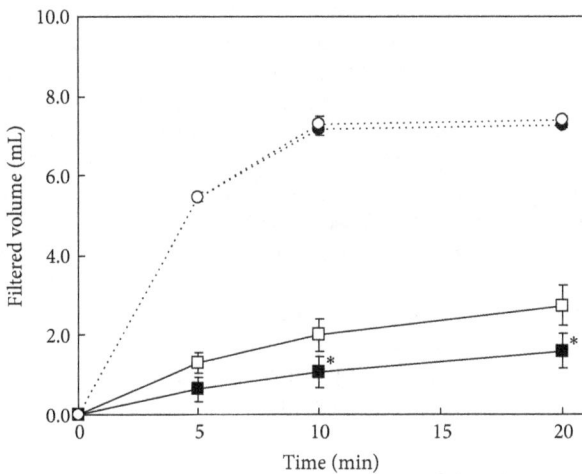

FIGURE 4: The ability of each sample to maintain a 15% (v/v) aqueous ethanol solution. Results are expressed as mean and standard deviation ($n = 3$). Asterisks indicate significant differences from the water-insoluble fraction (WIF) suspension group; $p < 0.05$, Student's t-test. ---●---, control group; ---○---, 5.0% aqueous alanine solution group; —□—, 0.5% WIF suspension group; —■—, 0.5% WIF + 5.0% alanine suspension group.

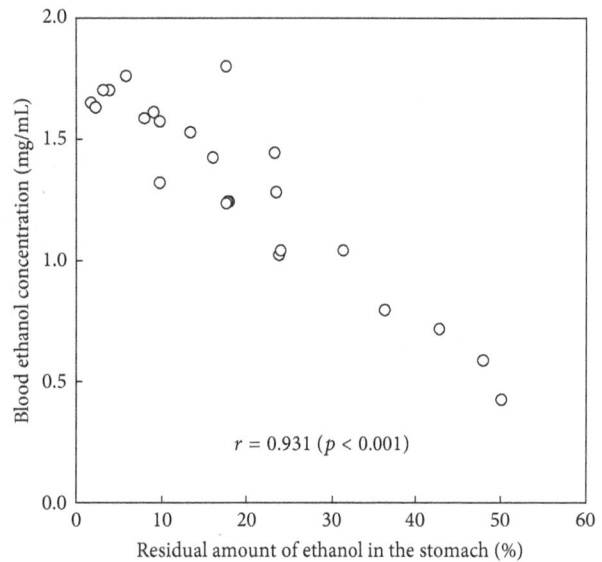

FIGURE 5: The correlation between the blood ethanol concentration and residual amount of ethanol in the stomach 2 h after ingestion of 2.0 g/kg ethanol after the administration of each sample in Experiment 3. Results represent the values of each rat ($n = 24$) and the correlation coefficient (r).

that the administration of a concentrated tomato supernatant attenuates elevated BEC [22]. However, the enriched water-soluble components were present in a concentrated tomato supernatant solution containing 11.3% glucose and 10.7% fructose. In contrast, tomatoes contain approximately 1%-2% glucose and fructose [23, 24]. Oral fructose (1.0 g/kg)

significantly lowers the peak BEC in humans [4]. Thus, higher quantities of sugars in the supernatant may contribute to the lowering of BEC. In contrast, Stice et al. found that whole tomato powder reduces the severity of alcohol-induced steatosis or hepatic inflammatory foci and that a lipid-soluble

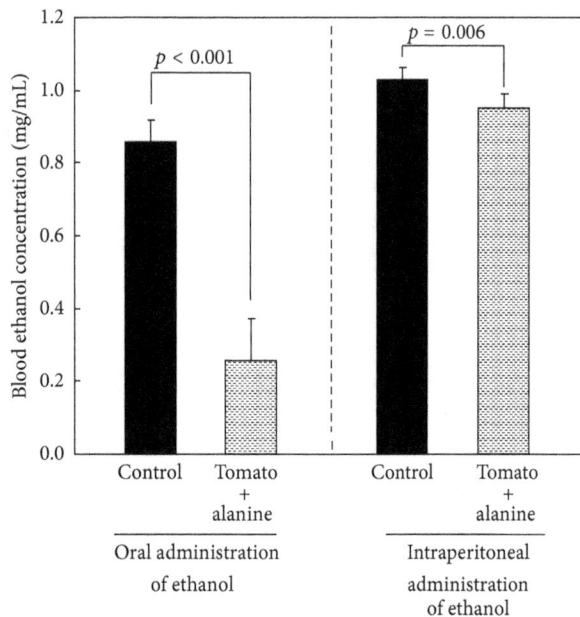

FIGURE 6: Blood ethanol concentrations 1 h after oral or intraperitoneal administration of 1.0 g/kg of ethanol (Experiment 4). Thirty minutes later, the rats received 20 mL/kg of either tomato juice with 5.0% alanine or water (control) orally. Results are expressed as mean and standard deviation ($n = 6$). Blood ethanol concentrations between control and tomato + alanine group were compared using the Student t-test.

tomato extract has no effect on these outcomes [25]. In this study, we did not identify the active components of whole tomatoes, and the lipid-soluble constituents were not measured. However, tomato varieties do contain dietary fiber [21]. Therefore, the results strongly suggest that the water-insoluble dietary fiber in whole tomatoes was responsible for reducing the adverse effects of alcohol by inhibiting alcohol absorption. To our knowledge, no previously published studies have addressed this question.

The initial goal of our study was to identify nutrients such as amino acids, sugars, and lipids that synergize with tomato WIF to produce effective functional foods for humans in the future. In an animal study, we evaluated the lowering of BEC because of supplementing tomato juice with one of the 22 amino acids. Supplementation of tomato juice with 3%–5% (w/v) of 7 amino acids (phenylalanine, histidine, ornithine, arginine, tryptophan, methionine, and alanine) significantly reduced BEC. Eighty percent of the free amino acids that naturally occur in tomatoes are glutamate, gamma-aminobutyrate, glutamine, and aspartate, and the alanine content of tomatoes is approximately < 0.01% [15], too low to influence BEC. We instead focused on the 7 amino acids that synergized with the components of tomatoes, in particular alanine, a superior tasting, sweet substance [26]. Oral administration of 4 g/kg alanine decreases the BEC of mice given 2 g/kg of ethanol [8]. However, an oral dose of 0.95 mmol (84.6 mg)/kg alanine does not affect rats given 1 g/kg ethanol [27]. In this study, the BEC of rats given 1.0 g/kg of alanine and 2.0 or 4.0 g/kg of ethanol

did not significantly differ from those of the controls. In contrast, the BECs of rats given precipitate supplemented with alanine were significantly lower than that of the controls. We were surprised that the supernatant supplemented with alanine had no effect on BEC, but our data suggests that alanine enhanced the ethanol-maintaining activity of WIF, whereas alanine alone had no effect. Further studies will be required to determine the mechanism responsible for this effect. In any event, the addition of alanine to tomato juice significantly increased the gastric content of residual ethanol. This augmentation of the ethanol-maintaining activity caused by WIF supplemented with alanine is likely the result of increasing the residual gastric content of ethanol. In case of the intraperitoneal administration of ethanol, tomato with alanine significantly decreased BEC. This effect may imply the metabolic acceleration of ethanol in the liver caused by the small dose of ethanol. However, the effect was weaker than that on oral ethanol administration. It is an undoubted fact that BEC lowering effect not involving absorptive process of ethanol is limited. We believe that our data clearly demonstrates that the delay in the GE of ethanol decreased BEC and that gastric ethanol was strongly correlated with BEC. Furthermore, the addition of alanine to tomato juice precipitate significantly increased the residual gastric content of ethanol to 47.1% or 28.1%, which is equivalent to 604 or 432 mg/kg ethanol compared with control (16.9% or 6.5%). In the body, ingested alcohol is mainly dissolved in aqueous fluids because alcohol is insoluble in body fat. Thus, alcohol is suitable for use as a solute to measure the volume of total body water [28]. In this study, we did not determine rats body-water volume, but the body-water volume was previously reported to be approximately 66% [29]. Therefore, the difference in BEC corresponds to approximately 0.92 or 0.65 mg/mL (604/660 or 432/660). The measured BEC was 0.95 or 0.73 mg/mL (1.52–0.57 or 1.68–0.95), approximately equivalent to the calculated values. The rough consistency suggests that the quantity of unabsorbed or residual ethanol in the stomach is sufficient to explain the decrease in BEC. Further studies will be required to determine an effective dose of tomato components and alanine for humans who consume alcohol for the purpose of ameliorating drunkenness.

5. Conclusions

Alanine enhanced the capacity of tomato juice to delay GE, thus lowering BEC. The administration of tomato juice supplemented with alanine ameliorated the spontaneous motor activity in rats given a high dose of ethanol. Alanine potentiated the effect of WIF, which comprises mainly dietary fiber present in tomato juice, although alanine alone had no effect. Hence, we conclude that tomato juice precipitate and alanine synergized to decrease ethanol absorption.

Abbreviations

BEC: Blood ethanol concentration
WSS: Water-soluble supernatant
WIP: Water-insoluble precipitation
GE: Gastric emptying.

Conflict of Interests

Shunji Oshima, Sachie Shiiya, Yoshimi Tokumaru, and Tomomasa Kanda are employed by Asahi Group Holdings, Ltd., Ibaraki, Japan. The authors declare no conflict of interests associated with this paper. There are no products in development or marketed products to declare.

Acknowledgments

The authors thank Akihiro Nemoto (Sankyo Labo Service Corporation, Inc.) for technical support and Enago (http://www.enago.jp/) for the English language review.

References

[1] M. Roerecke and J. Rehm, "Alcohol intake revisited: risks and benefits," *Current Atherosclerosis Reports*, vol. 14, no. 6, pp. 556–562, 2012.

[2] J. Soterakis and F. L. Iber, "Increased rate of alcohol removal from blood with oral fructose and sucrose," *The American Journal of Clinical Nutrition*, vol. 28, no. 3, pp. 254–257, 1975.

[3] A. K. Rawat, "Effects of fructose and other substances on ethanol and acetaldehyde metabolism in man," *Research Communications in Chemical Pathology and Pharmacology*, vol. 16, no. 2, pp. 281–290, 1977.

[4] U. E. Uzuegbu and I. Onyesom, "Fructose-induced increase in ethanol metabolism and the risk of Syndrome X in man," *Comptes Rendus Biologies*, vol. 332, no. 6, pp. 534–538, 2009.

[5] W. W. Westerfeld, E. Stotz, and R. L. Berg, "The role of pyruvate in the metabolism of ethyl alcohol," *The Journal of Biological Chemistry*, vol. 144, no. 3, pp. 657–665, 1942.

[6] J. N. Hunt and D. F. Stubbs, "The volume and energy content of meals as determinants of gastric emptying," *The Journal of Physiology*, vol. 245, no. 1, pp. 209–225, 1975.

[7] R. D. Johnson, M. Horowitz, A. F. Maddox, J. M. Wishart, and D. J. C. Shearman, "Cigarette smoking and rate of gastric emptying: effect on alcohol absorption," *British Medical Journal*, vol. 302, no. 6767, pp. 20–23, 1991.

[8] T. Akao and K. Kobashi, "Inhibitory effect of glycine on ethanol absorption from gastrointestinal tract," *Biological and Pharmaceutical Bulletin*, vol. 18, no. 12, pp. 1653–1656, 1995.

[9] S. Kechagias, K.-Å. Jönsson, and A. W. Jones, "Impact of gastric emptying on the pharmacokinetics of ethanol as influenced by cisapride," *British Journal of Clinical Pharmacology*, vol. 48, no. 5, pp. 728–732, 1999.

[10] A. Franke, S. Teyssen, H. Harder, and M. V. Singer, "Effect of ethanol and some alcoholic beverages on gastric emptying in humans," *Scandinavian Journal of Gastroenterology*, vol. 39, no. 7, pp. 638–644, 2004.

[11] R. Chaikomin, A. Russo, C. K. Rayner et al., "Effects of lipase inhibition on gastric emptying and alcohol absorption in healthy subjects," *British Journal of Nutrition*, vol. 96, no. 5, pp. 883–887, 2006.

[12] K.-L. Wu, R. Chaikomin, S. Doran, K. L. Jones, M. Horowitz, and C. K. Rayner, "Artificially sweetened versus regular mixers increase gastric emptying and alcohol absorption," *The American Journal of Medicine*, vol. 119, no. 9, pp. 802–804, 2006.

[13] A. W. Jones, K. Å. Jönsson, and S. Kechagias, "Effect of high-fat, high-protein, and high-carbohydrate meals on the pharmacokinetics of a small dose of ethanol," *British Journal of Clinical Pharmacology*, vol. 44, no. 6, pp. 521–526, 1997.

[14] S. Oshima, S. Shiiya, and T. Kanda, "Water-insoluble fractions of botanical foods lower blood ethanol levels in rats by physically maintaining the ethanol solution after ethanol administration," *The Journal of Functional Foods in Health and Disease*, vol. 5, no. 11, pp. 406–416, 2015.

[15] A. Kader, M. Stevens, M. Albright, and L. Morris, "Amino acid composition and flavor of fresh market tomatoes as influenced by fruit ripeness when harvested," *Journal of the American Society for Horticultural Science*, vol. 103, no. 4, pp. 541–544, 1978.

[16] Y. Masuo, J. Noguchi, S. Morita, and Y. Matsumoto, "Effects of intracerebroventricular administration of pituitary adenylate cyclase-activating polypeptide (PACAP) on the motor activity and reserpine-induced hypothermia in murines," *Brain Research*, vol. 700, no. 1-2, pp. 219–226, 1995.

[17] M. Ishido, Y. Masuo, S. Oka, M. Kunimoto, and M. Morita, "Application of supermex system to screen behavioral traits produced by tributyltin in the rat," *Journal of Health Science*, vol. 48, no. 5, pp. 451–454, 2002.

[18] E. P. Riley, E. X. Freed, and D. Lester, "Selective breeding of rats for differences in reactivity to alcohol. An approach to an animal model of alcoholism. I. General procedures," *Journal of Studies on Alcohol*, vol. 37, no. 11, pp. 1535–1547, 1976.

[19] T. Okada and Y. Mizoi, "Studies on the problem of blood acetaldehyde determination in man and level after alcohol intake," *Japanese Journal of Alcohol & Drug Dependence*, vol. 17, no. 2, pp. 141–159, 1982.

[20] N. Kalogeropoulos, A. Chiou, V. Pyriochou, A. Peristeraki, and V. T. Karathanos, "Bioactive phytochemicals in industrial tomatoes and their processing byproducts," *LWT—Food Science and Technology*, vol. 49, no. 2, pp. 213–216, 2012.

[21] J. L. Guil-Guerrero and M. M. Rebolloso-Fuentes, "Nutrient composition and antioxidant activity of eight tomato (Lycopersicon esculentum) varieties," *Journal of Food Composition and Analysis*, vol. 22, no. 2, pp. 123–129, 2009.

[22] Y. Ushida, S. Oshima, K. Aizawa et al., "Aqueous components of tomato accelerate alcohol metabolism by increasing pyruvate level," *Food and Nutrition Sciences*, vol. 5, no. 10, pp. 870–879, 2014.

[23] E. M. Widdowson and R. A. McCance, "The available carbohydrate of fruits: determination of glucose, fructose, sucrose and starch," *Biochemical Journal*, vol. 29, no. 1, pp. 151–156, 1935.

[24] I. Levin, N. Gilboa, E. Yeselson, S. Shen, and A. A. Schaffer, "Fgr, a major locus that modulates the fructose to glucose ratio in mature tomato fruits," *Theoretical and Applied Genetics*, vol. 100, no. 2, pp. 256–262, 2000.

[25] C. P. Stice, C. Liu, K. Aizawa, A. S. Greenberg, L. M. Ausman, and X.-D. Wang, "Dietary tomato powder inhibits alcohol-induced hepatic injury by suppressing cytochrome p450 2E1 induction in rodent models," *Archives of Biochemistry and Biophysics*, vol. 572, pp. 81–88, 2015.

[26] S. S. Schiffman and C. Dackis, "Taste of nutrients: amino acids, vitamins, and fatty acids," *Perception & Psychophysics*, vol. 17, no. 2, pp. 140–146, 1975.

[27] H. Murakami, H. Ito, Y. Furukawa, and M. Komai, "Leucine accelerates blood ethanol oxidation by enhancing the activity of ethanol metabolic enzymes in the livers of SHRSP rats," *Amino Acids*, vol. 43, no. 6, pp. 2545–2551, 2012.

[28] P. E. Watson, I. D. Watson, and R. D. Batt, "Prediction of blood alcohol concentrations in human subjects. Updating the Widmark equation," *Journal of Studies on Alcohol*, vol. 42, no. 7, pp. 547–556, 1981.

[29] S. C. Peckham, C. Entenman, and H. W. Carroll, "The influence of a hypercalric diet on gross body and adipose tissue composition in the rat," *The Journal of Nutrition*, vol. 77, no. 2, pp. 187–197, 1962.

21

Associations among Physical Activity, Diet, and Obesity Measures Change during Adolescence

Janne H. Maier[1,2,3] **and Ronald Barry**[1,4]

[1] University of Alaska Fairbanks, P.O. Box 757500, Fairbanks, AK 99775, USA
[2] Institute of Arctic Biology, University of Alaska Fairbanks, P.O. Box 757500, Fairbanks, AK 99775, USA
[3] Center for Alaska Native Health Research, University of Alaska Fairbanks, P.O. Box 757500, Fairbanks, AK 99775, USA
[4] Department of Mathematical Sciences, University of Alaska Fairbanks, P.O. Box 757500, Fairbanks, AK 99775, USA

Correspondence should be addressed to Janne H. Maier; jannehmaier@yahoo.com

Academic Editor: Christel Lamberg-Allardt

Background. Obesity in youth is highly prevalent. Physical activity and diet are influential in obesity development. However, there is a knowledge gap regarding links between activity and diet quality and their combined influence on obesity during adolescence. *Objectives.* We used five years of data from 2379 adolescent girls in the National Heart Lung and Blood Institute Growth and Health Study to evaluate the association between physical activity and diet quality during adolescence and to assess both as correlates of obesity. *Design.* Diet, activity, and body composition measures were evaluated pairwise for correlation. A canonical correlation analysis was used to evaluate relationships within and between variable groups. All statistics were examined for trends over time. *Results.* We found positive correlations between physical activity and diet quality that became stronger with age. Additionally we discovered an age-related decrease in association between obesity correlates and body composition. *Conclusion.* These results suggest that while health behaviors, like diet and activity, become more closely linked during growth, obesity becomes less influenced by health behaviors and other factors. This should motivate focus on juvenile obesity prevention capitalizing on the pliable framework for establishing healthy diet and physical activity patterns while impact on body composition is greatest.

1. Introduction

The 2009-2010 National Health and Nutrition Examination Survey (NHANES) found United States juvenile obesity (Body Mass Index (BMI) ≥ 95th percentile in BMI-for-age growth charts) at 16.9% and adult male/female obesity (BMI ≥ 30) at 35.5%/35.8%, respectively [1, 2]. While the high adult obesity prevalence is daunting, more concerning is the commonness of obesity in youth because juvenile obesity tracks into adulthood and is a precursor for obesity related diseases [3].

Current USDA physical activity guidelines for weight management advise a minimum of 60 minutes/day for ages 6–17 and 150 minutes/week of moderate activity or 75 minutes/week of moderate/vigorous activity for adults [4]. Accelerometer measures of physical activity, including occupational and transportation activity, from 2003-2004

NHANES data show that adherence to physical activity recommendation is 42% for 6–11-year-olds, 8% for 12–19-year-olds, and less than 5% for adults [5]. The suboptimal adherence to recommended physical activity, and the drastic decrease in physical activity during adolescence [5], particularly in girls [6], is directly correlated with the current rise in obesity [7–9].

Diet recommendations for weight management are limiting calorie intake and increasing diet quality by choosing fruits, vegetables, and foods high in fiber and low in sugar and saturated fat [4, 10]. In the industrialized world, diet quality may be assessed by using dietary energy density (kilocalories/gram) as a proxy because micronutrient intake, especially from fruits and vegetables, and diet quality are negatively correlated with dietary energy density [11, 12]. Most adults and children exceed suggested total and saturated fat intake [13, 14] in spite of a decline in intake of energy, fat, and saturated fat in the last decades of the 20th century [13], and adolescent girls are singled out as the only group in NHANES for whom energy intake has increased over this period [13].

While associations between physical activity and diet are well reported in adults and youths [15, 16], less is known about how they interact during adolescence. Diet and activity patterns change during adolescence [17], so it would be pertinent to examine potential changes in associations between these behaviors and their relationship to body composition during this time.

We report results from a secondary analysis on diet and physical activity associations using five years of data spanning seven years of development in adolescent girls from The National Heart Lung and Blood Institute Growth and Health Study (NGHS). We also investigated both diet and physical activity as correlates of body composition (BMI and body fat percent) during growth. Understanding the interactions and underpinnings of diet and physical activity is paramount to stemming the obesity epidemic, and identifying changes in associations during development may pinpoint an optimal intervention window for preventing adult obesity.

2. Subjects and Methods

2.1. The Growth and Health Study. The National Heart Lung and Blood Institute Growth and Health Study was conducted to evaluate racial differences in the development of obesity and CVD risk in girls. The study group recruited 1213 black girls and 1166 white girls from schools and clinics in the study areas, Berkeley, CA, Cincinnati, OH, and Rockville, MD, from January 1987 to May 1988 [18]. To be eligible for the study, the girls had to be white or black with no other mixed heritage. The girls were all 9-10 years old at recruitment (visit 1) and attended annual visits for 10 years with a follow-up rate of 89% [19]. The centers collected annual anthropometric measures (e.g., height, weight, skin fold thicknesses, and maturation stage indicators), as well as dietary and physical activity information. Extensive sampling design and study methods are described elsewhere [18, 20]. Our analysis to assess associations between diet and exercise and to identify correlates of obesity during adolescence in this cohort was approved by University of Alaska Institutional Review Board (ID# 231373-4). Subsequently, the National Heart Lung and Blood Institute's Biologic Specimen and Data Repository Information Coordinating Center (BioLINCC) approved the analysis objectives and fulfilled the NGHS data request. This paper was prepared using research materials obtained through BioLINCC, but it does not necessarily reflect the opinions or views of the NGHS or the National Heart Lung and Blood Institute.

2.2. Physical Activity and Diet Measures. In the NGHS, information about physical activity levels was collected using two questionnaires validated for use in children, a habitual activity questionnaire (HAQ) and an activity diary (AD) [18, 20]. The HAQ asked participants to report type and frequency of activities in school and outside of school throughout the year. The HAQ was interviewer administered in visit years 1, 3, and 5, and it was self-administered years 7-10.

The weekly scores were calculated by multiplying the weekly frequency of the activity by the fraction of year that the activity was engaged in and by the metabolic equivalent of task (MET) value for the activity. Weekly MET scores for reported activities for the previous year were added to yield the annual HAQ score as an estimate of physical activity throughout the year [20]. In visit years 1–5, 7, 8, and 10, the participant completed an AD on three consecutive days along with a 3-day food journal. The participants were instructed to record both active (e.g., jogging, kickball, and jumping rope) and sedentary (e.g., sitting to talk, watching TV, or reading) behaviors listed in their AD in designated timeslots. Along with the listed activity categories there was a blank "other activities" section to document activities not listed in the questionnaire. Daily AD scores were calculated by multiplying the MET value for each activity by its duration and summing the calculated MET scores for all the activities in one day. Final AD score was the average of the scores for all the usable days. The participant reviewed the AD and HAQ with staff at the centers using a common protocol for submission in each collection year [20].

Information about nutrient and calorie intake was collected with the 3-day food journal each year [18, 20]. The journals were completed by the participants on two consecutive weekdays and one weekend day, and the journal entries were reviewed and confirmed by the participant with staff following standardized protocol for all collection centers. Food journals were coded centrally and processed to yield information about average kilocalorie (kcal) intake, macronutrient distribution, and average intake of 50 different nutrients. During years 1 and 2 the data were processed at the Nutrition Coordinating Center in Minnesota, and the records from years 3 through 10 were processed at the NGHS Dietary Data Entry Center in Cincinnati [21].

Anthropometrics like height, weight, and skin fold measures were collected at annual visits [18]. The average of two measures was taken, and if the two measures deviated from each other excessively (more than 0.5 cm for height, 0.3 kg for weight, and 1.0 mm for skin folds), a third measure was included [18]. Maturation was assessed with a modified Tanner staging method to accurately gauge maturation at different body compositions [18].

2.3. Analysis. We used data from years 3, 5, 7, 8, and 10 because these years had physical activity information collected by both HAQ and AD ensuring a comprehensive physical activity assessment. We started at year three because data from this and subsequent years were processed at the Cincinnati Dietary Data Entry Center. These data include both total kcal and grams of intake which are needed for energy density calculations. Daily energy density for each journal period was calculated by dividing the total daily kcal by the daily sum of gram intake of all reported foods and drinks; energy density was then averaged for all valid journal days. To indicate diet quality we used average energy density, average kcal intake, average dietary saturated fat percent, and average (gram) fiber intake. To assess body composition we included BMI calculated from height and weight measurements and body fat percent estimated from skin fold measurements; in the NGHS body fat percent was calculated with a standard equation using measurements of triceps, subscapular, and suprailiac skin folds [18]. Household income, race, maturation stage, height, and weight were included to adjust for confounding effects. To account for racial differences in the recording of physical activity measures, we included variables for racial interaction with HAQ (race × HAQ) and racial interaction with AD (race × AD) in the analyses. All statistical analyses were conducted with SPSS 22 [22].

Initially we ran bivariate correlation analyses between all variables pairwise. To account for simultaneous evaluation of multiple variables we only considered $P < 0.001$ as significant. The NGHS dataset is large, so we used Spearman's rho correlation because it is less sensitive to potential outliers than Pearson's r.

To examine relationships between variable groups, we conducted a canonical correlation analysis (CCA). Canonical correlation is a multivariate approach where a program constructs linear combinations of two sets of variables:

Canonical variates (X):

$$a_1 X_{i1} + a_2 X_{i2} + \cdots + a_p X_{ip} = U_i. \qquad (1)$$

Canonical covariates (Y):

$$b_1 Y_{i1} + b_2 Y_{i2} + \cdots + b_q Y_{iq} = V_i. \qquad (2)$$

The linear indices are adjusted so that, for the ith set of observations, the correlation between the two resulting latent values, U and V, is maximized. The latent values represent underlying structure in the data. Statistics from CCA include several index sets, the first set of which explain the largest proportion of variance in the data and have the highest canonical correlation between variates and covariates. The canonical variates are evaluated for significance in the analysis with univariate F tests. Data from CCA can be interpreted by the canonical correlation (values ranging from 0 to 1), a measure of correlation between the latent values from the two linear indices. Additionally, the individual coefficients in each index tell us how the variables are related to each other within the group when the groups are most correlated with each other.

To evaluate the association between physical activity and diet quality we grouped variables for CCA as follows. Canonical variates (physical activity and confounders) were household income, race, height, weight, maturation stage, smoking status, race × HAQ, race × AD, HAQ, and AD, and canonical covariates (diet quality indicators) were average energy density, average kcal, average dietary saturated fat percent, and average fiber intake.

Using CCA to evaluate correlates of obesity during growth was done by grouping the variables by household income, race, maturation stage, smoking status, race × HAQ, race × AD, HAQ, AD, average energy density, average kcal, average dietary saturated fat percent, and average fiber intake (obesity correlates) as canonical variates and BMI and body fat percent (body composition) as canonical covariates; height and weight were not included as variates in the second

part of the analysis as they are direct predictors of BMI and including them gave an artificially high correlation. To allow evaluation of significance with regard to canonical correlation, the confounding variables were grouped as variates in this analysis. Statistics from all visit years were compared to identify trends during adolescence.

3. Results

3.1. Associations between Physical Activity and Diet. Figure 1(a) shows that the strength of the negative correlation, as indicated by Spearman's rho, between HAQ scores and dietary energy density and between HAQ scores and dietary saturated fat percent increased with age. The increase followed a linear trend with $R^2 = 0.89$ for energy density and $R^2 = 0.95$ saturated fat percent. Figure 1(b) shows that the strength of the positive correlation, as indicated by Spearman's rho, between HAQ scores and fiber intake also increased with age ($R^2 = 0.82$).

Examining association between physical activity and diet with CCA initially produced opposite coefficients for the two physical activity measures. Spearman's rho correlations between HAQ and 3d AD scores indicated that the measures were positively correlated with each other at each visit year ($P < 0.001$), but scores for HAQ had a positive coefficient and scores for 3d AD a negative coefficient for all years in the initial CCA results. Data stratification by race revealed racial differences in responses on physical activity questionnaire, so racial interaction variables were added as confounders in the CCA.

Table 1 presents standardized canonical correlation coefficients from the first index in the CCA examining association between physical activity and diet including the racial interaction variables with the confounders. The first index explained 66% to 75% of the variance in the data. Maturation stage, height, and weight were included as confounding variables in the analysis but their coefficients are not reported in Table 1. In visit years 3 and 5 HAQ scores and saturated fat percent had negative and positive coefficients, respectively, and positive and negative coefficients, respectively, for years 7–10, but for the rest of the variables, the coefficients are consistent from year to year. Correlation coefficients for visits 7, 8, and 10 indicate that higher household income, being white (white = 1, black = 2), and high physical activity scores yield the highest canonical correlation to low dietary energy density, low kcal intake, and low dietary saturated fat percent and high intake of fiber.

As illustrated in Figure 2, the canonical correlation of physical activity scores and confounding variables with diet quality indicators increases with time for the combined data and for white girls only. The correlation increase in Figure 2 follows a linear trend with $R^2 = 0.95$ for the combined data and $R^2 = 0.92$ for white girls. The canonical correlation for black girls does not mimic the trend for the combined data or for white girls. The correlation of physical activity scores and confounders with diet quality in black girls increases from years 3 to 5 but shows a sharp decrease in year 7, the year the HAQ went from interview to self-administered, which may

TABLE 1: Canonical correlation analysis evaluating physical activity and confounders with diet quality indicators. Percent variance explained by first indices and first indices' standardized canonical coefficients by visit year.

y	n	% var.	Physical activity, confounders						Diet quality			
			Inc	Race	RxHAQ	Rx3d AD	HAQ	3d AD	ED	Av kcal	Sat fat %	Fiber
3	1701	66.26	0.32**	−0.64**	0.14*	−0.66**	−0.03†	0.45*	−0.88	−0.55	0.35	0.23
5	1576	66.19	0.41**	−0.67**	0.35*	−0.54**	−0.09**	0.27**	−0.67	−0.79	0.10	0.76
7	1457	69.97	0.27**	−0.59**	−0.10**	−0.46**	0.37**	0.30**	−0.57	−0.89	−0.15	0.91
8	1423	70.60	0.36**	−0.41**	−0.56**	−0.31**	0.89**	0.29**	−0.54	−0.65	−0.08	0.86
10	1699	74.26	0.36**	−0.48**	−0.33**	−0.17**	0.72**	0.25**	−0.60	−0.49	−0.17	0.65

†Significance at $P < 0.1$; *significance at $P < 0.05$; **significance at $P < 0.001$; for dependent variates significance is evaluated with a univariate F test for association with first indices canonical correlation.
3-day activity diary score, 3d AD; average caloric intake, Av kcal; average dietary saturated fat percent, Sat fat %; average energy density, ED; average fiber intake, fiber; habitual physical activity questionnaire score, HAQ; income category, Inc; percent variance, % var.; race by 3-day AD, Rx3d AD; race by HAQ, RxHAQ; year, y.

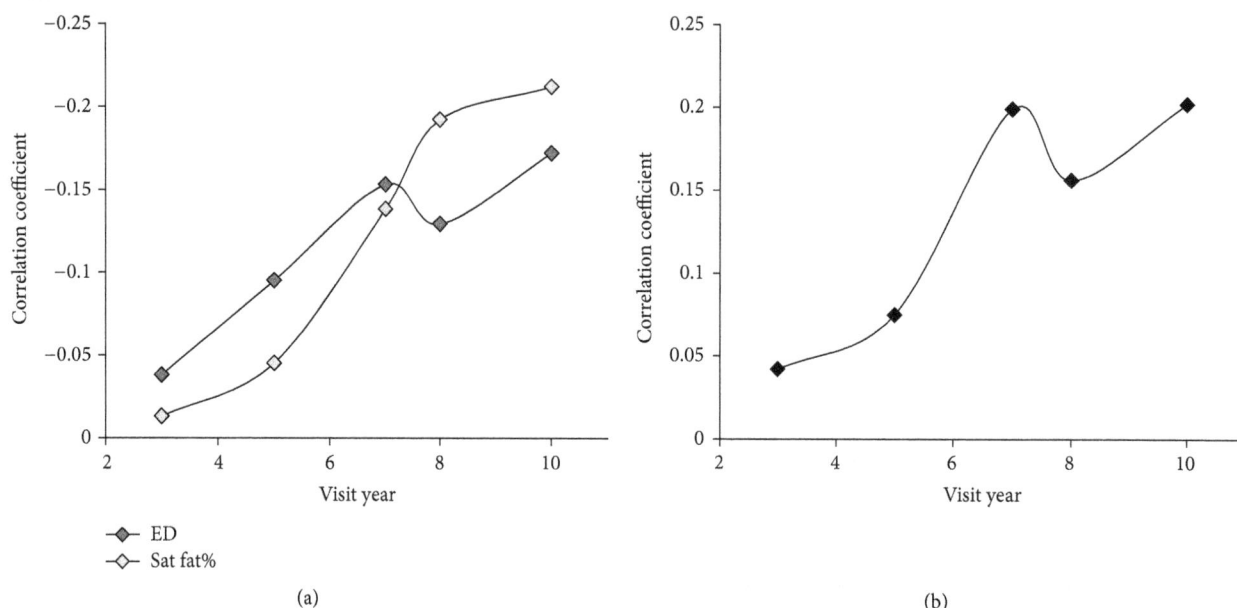

FIGURE 1: Spearman's rho correlation coefficients. (a) Correlation between habitual physical activity and average energy density (ED) and dietary saturated fat percent (Sat fat %) by visit year. (b) Correlation between habitual physical activity and fiber intake by visit year. Correlation is significant (two-tailed t, $P < 0.001$) for years 5, 7, 8, and 10 for correlation with ED and fiber, and for years 7, 8, and 10 for correlation with Sat fat %.

indicate reporting bias. From years 8 to 10 there is a weak increase in correlation for association of physical activity and confounders with diet quality in black girls indicating adherence to the general trend after adjusting to the change in recording methods.

3.2. Correlates of Obesity during Growth.

Bivariate analysis examining correlates of obesity shows that household income, HAQ scores, and fiber intake are negatively correlated with BMI and body fat percent (Spearman's rho correlation coefficients vary from −0.06 to −0.17), and race was positively correlated with both obesity measures (Spearman's rho correlation coefficients vary from 0.09 to 0.21) in every year ($P < 0.001$) indicating that being black was correlated with higher BMI and body fat percent. The strength of the correlation for BMI and body fat percent with household income, race, and HAQ increased a little during the study

years (increase range: 0.019 to 0.129), but the correlation of BMI and body fat percent with AD and diet quality variables did not increase and the sign, strength, and significance of the coefficients varied from year to year.

We conducted CCA between body composition (BMI and body fat percent) and obesity correlates (physical activity, diet quality indicators, and confounders) to identify important obesity correlates and to consider trends in correlation between the two during adolescence. Table 2 provides the standardized canonical correlation coefficients from the first index for canonical correlation between obesity correlates and body composition from this CCA. BMI is a measure of body fatness, but there are inverse coefficients for BMI and body fat percent. This is an effect of the two correlated variables simultaneously explaining variance in body composition. Correlation coefficients for obesity correlates indicate that higher income and lower race score (white)

TABLE 2: Canonical correlation analysis evaluating health behaviors and confounders (obesity correlates) with body composition. Percent variance explained by first indices and first indices' standardized canonical coefficients by visit year.

y	n	% var.	Obesity correlates										Body composition	
			Inc	Race	RxHAQ	Rx3d AD	HAQ	3d AD	ED	Av kcal	Sat fat %	Fiber	BMI	BF%
3	1693	81.43	0.21**	−0.50**	0.33**	0.11**	−0.36**	−0.29**	0.07*	0.05*	−0.21**	−0.05**	−1.80	1.12
5	1557	75.28	0.38**	−0.32**	0.01**	−0.45**	0.06**	0.18**	0.16*	0.30**	−0.12*	−0.26*	−1.61	0.75
7	1412	62.65	0.32**	−0.71**	0.24**	0.14*	−0.04**	−0.22	0.18	−0.02*	0.28†	0.26**	−1.19	0.22
8	1378	63.45	0.15**	−0.67**	0.37**	−0.64**	−0.19**	0.18**	0.18	0.07*	0.17*	0.03*	−1.73	0.90
10	1679	83.47	0.38**	−0.68**	−0.11**	0.11**	0.42**	−0.28*	0.32	0.14	0.08	0.06**	−1.14	0.16

†Significance at $P < 0.1$; *significance at $P < 0.05$; **significance at $P < 0.001$; for dependent variates significance is evaluated with a univariate F test for association with first indices canonical correlation.
3-day activity diary score, 3d AD; average caloric intake, Av kcal; average dietary saturated fat percent, Sat fat %; average energy density, ED; average fiber intake, fiber; habitual physical activity questionnaire score, HAQ; income category, Inc; body fat percent, BF%; percent variance, % var.; race by 3-day AD, Rx3d AD; race by HAQ, RxHAQ; year, y.

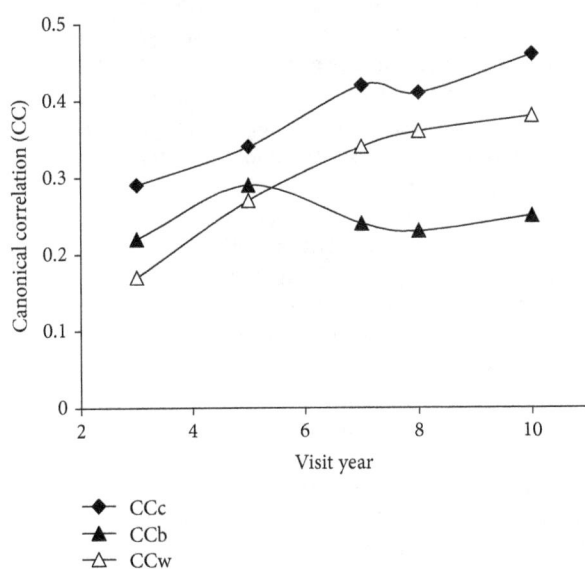

FIGURE 2: Canonical correlation of physical activity and confounders with diet quality indicators by visit year. Physical activity and confounders: income category, race, maturation stage, height, weight, habitual physical activity questionnaire score (HAQ), 3-day activity diary score (3d AD), racial interaction HAQ, and racial interaction 3d AD; diet quality indicators: average energy density, average caloric intake, average dietary saturated fat percent, and average fiber intake. Lines represent stratified data from black (CCb) and white (CCw) girls and combined (CCc) data.

consistently yielded the highest canonical correlation with body composition; however, the relationships between the other obesity correlates varied by year. For example, from years 5 to 7, when the groups are most correlated, the coefficients for activity measures, kcal, saturated fat percent, and fiber all changed signs (positive and negative). Nevertheless, some variables remained significant regardless of change in signs of the coefficients. In addition to income, race, and racial interaction variables, both activity measures and fiber intake were significant for canonical correlation with obesity measures for most (3d AD) or all (HAQ, fiber) visit years in the analysis. Other diet quality variables except energy

density were significant for at least three of the five study years. In this CCA, the first index explained 62% to 84% of the variance in the data (Table 2) indicating that income, race, activity, and diet are all important obesity correlates.

While the relationships between variables within the groups varied from year to year, the canonical correlation between the groups (obesity correlates and body composition) decreased with each subsequent visit as shown in Figure 3; the decreases in correlation followed linear trends with $R^2 = 0.72$ for the combined data, $R^2 = 0.93$ for black girls, and $R^2 = 0.75$ for white girls. This indicates an age-related reduction in association between obesity correlates and body composition in spite of an age-related increase in correlation for some of the variables in the bivariate results.

4. Discussion

In this study, physical activity was positively associated with diet quality indicators as early as 12 years of age (visit year 3) (Figures 1 and 2, Table 1). Furthermore, the increasing strength of correlation between diet quality indicators and physical activity during the study period indicates an increasingly tight association between these health behaviors with age (Figures 1 and 2). In evaluating correlates of obesity, white race and higher household income, habitual activity, and fiber intake were associated with having a lower BMI and body fat percent. The consistent correlation of race and income with obesity in the CCA (Table 2) supports that socioeconomic factors are a major consideration in obesity development which has been previously reported [18]. The age-related decline in canonical correlation of BMI and body fat percent with both health behaviors and socioeconomic factors (Figure 3) demonstrates a decreased connection between obesity correlates and body composition during transition into adulthood.

4.1. Physical Activity and Diet Interactions. The decreasing association between body composition and correlates of obesity during adolescence is contrasted by the increasingly tight relationship between physical activity and select predictors of diet quality with age. Similar to our findings of associations between diet quality indicators and physical

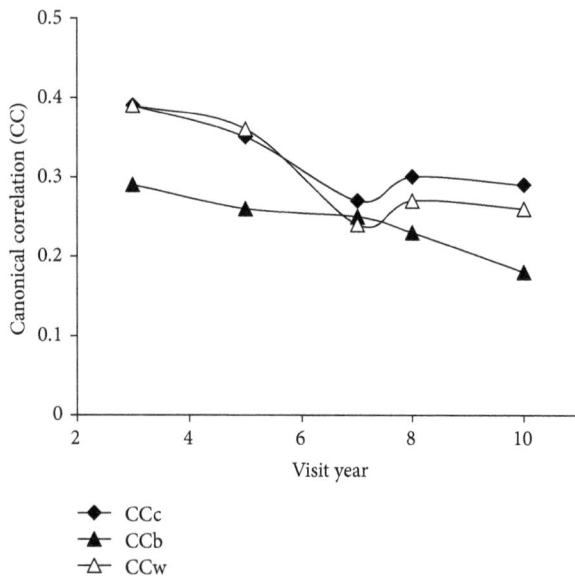

FIGURE 3: Canonical correlation between obesity correlates and body composition by visit year. Obesity correlates: income category, race, maturation stage, habitual physical activity questionnaire score (HAQ), 3-day activity diary score (3d AD), racial interaction HAQ, racial interaction 3d AD, average energy density, average caloric intake, average dietary saturated fat percent, and average fiber intake; body composition: BMI and body fat percent. Lines represent stratified data from black (CCb) and white (CCw) girls and combined (CCc) data.

activity, Gillman et al. [16] found that low fruit and vegetable intake was correlated with sedentary behaviors and more physical activity was related to lower saturated fat, trans fat, and cholesterol intake in a diverse cohort of 1322 men and women. Pate et al. [15] also found more physical activity to be positively associated with healthy diet choices as well as with other health behaviors like choosing not to smoke in the 1990 Youth Risk Behavior Survey of 11631 high school students suggesting that health behaviors have strong associations as early as high school. Our findings indicate that the association between physical activity and diet emerges already at 12 years of age (visit 3) and that the correlation between these health behaviors increases as adolescents mature.

While preference shapes health behaviors, physiological mechanisms like fatty acid oxidation may also influence the association between diet and activity because it responds to both dietary intake and physical activity [23]. High fat intake is associated with low carbohydrate intake and vice versa, and fatty acid oxidation increases with higher fat and lower carbohydrate intake [24–26]; exercise also increases fatty acid oxidation in muscle tissue performing or adapting to physical activity [23, 27]. Since high fat/low carbohydrate intake and physical activity have a similar physiological impact, accelerated fatty acid oxidation in response to physical activity may be reflexively counteracted by selection of a diet with low fat and energy density; low dietary fat intake suppresses fatty acid oxidation because it is generally higher in carbohydrate proportion [23, 26]. Conversely, decreasing physical activity decreases fatty acid oxidation [28]. As

higher fat/lower carbohydrate proportion in dietary intake accelerates fatty acid oxidation [26], oxidation decrease by decreasing physical activity level could be innately balanced by intake of a diet higher in fat and energy density. We found an increasing negative correlation between physical activity and dietary fat measures with age while physical activity declines (Figure 1(a)). This increase in negative correlation falls in line with potential physiological associations between these health behaviors as well as with the existing support for health behavior patterns.

4.2. Obesity during Growth. Along with increasing strength of association between health behaviors with age, the decrease in canonical correlation between obesity correlates and body composition indicates a potential for higher impact of behaviors on anthropometrics at younger ages. The age-related disconnect between obesity correlates and body composition that we saw is supported by a study demonstrating that various fat deposits are genetically distinct "miniorgans" subject to developmental processes and maturation [29]. Similarly, a four-year study of adipose tissue growth in 288 children found that fat tissue development in children played a role in developing enlarged fat depots in obese adults [30]. The decrease in canonical correlation between obesity correlates and body composition with age in our study may well reflect fat storage capacity maturing during adolescence.

In addition to, or as a consequence of, developmental effects on fat tissue, juvenile obesity increases adult obesity risk. A 35-year longitudinal British study ending in 2000 demonstrated that early weight gain predicted adult obesity [31], and a 1976 study found 2.4 odds' ratio for adult obesity associated with juvenile obesity and overweight [32]. The alarming increase in obesity in the juvenile population today consequently bodes ill for obesity and subsequent health odds for the generation to come [1, 8, 13, 33]. These bleak odds, in conjunction with our findings, stress the importance of attention to health habits in children when the framework of health behaviors is less settled and body composition is more responsive.

4.3. Limitations. Limitations of this analysis include self-report of physical activity which tends to overestimate actual activity levels particularly at lower levels of activity [5]. The fluctuation in correlations at years seven and eight (Figures 1, 2, and 3) may be an artifact of the change in questionnaire collection protocol at year seven [20]. Another limitation of this study is the inclusion of all beverages in the energy density calculation necessitated by the structure of the data set. Energy density depends on the water content in the food as water adds weight but no calories [11, 34, 35], and separate calculations for food only, and food and select beverages are typically used to adjust for energy density differences in liquid and solid intake [34, 36, 37]. The effect of these limitations is an underestimation of dietary energy density due to the low energy density in liquids and a reduced range for differences in activity. Future studies using data from an electronic physical activity monitor and dietary assessment coded for removal of beverages for multiple energy density calculations would better identify the extent of the trends uncovered here. Longitudinal data from a broader age group could additionally determine if our discovered trends continue with age guiding obesity prevention in the general population.

4.4. Conclusion. Not only did our study confirm positive associations between physical activity and diet quality indicators in adolescent girls, but we also documented an increasingly tight correlation between these two health behaviors with age. As a consequence, successful interventions to increase activity all the way through adolescence may have positive effects on diet quality as behavior patterns are established. While physical activity and diet become more tightly correlated with age, body composition becomes less associated with known obesity correlates. The age-related decrease in association between factors that affect obesity and measures of obesity is consistent with findings that predictions of adult overweight using adolescent weight status improve with age [3]. The increasing predictability of future weight with age and our finding of increasing disconnect between obesity correlates and body composition during adolescence indicate that maturation of fat tissues during growth may impact weight status throughout adult life. Fat tissue maturation, high adult obesity odds for overweight youth, and the increasing correlation between health behaviors with age should spur investments in interventions to promote healthy diet and physical activity behaviors continuously through adolescence when the potential for a return on the investment through long term impact on obesity odds and future health is greatest.

Conflict of Interests

Neither author reports any potential conflict of interests or financial support.

Acknowledgments

First the authors want to acknowledge the enormous contribution by the NGHS study group who conducted the research and the NGHS cohort who meticulously attended followup visits and completed questionnaires. For data requisition BioLINCC answered queries and fulfilled the authors' request professionally and expediently. The University of Alaska Fairbanks, Center for Alaska Native Health Research, assisted by making resources available to conduct this research. During the paper process, writing class instructor Syndonia BretHarte and fellow writing class students were instrumental in reviewing and clarifying the paper language and structure. Finally, the research and paper would not have been possible without the first author's committee and chairs.

References

[1] C. L. Ogden, M. D. Carroll, B. K. Kit, and K. M. Flegal, "Prevalence of obesity and trends in body mass index among US children and adolescents, 1999–2010," *The Journal of the American Medical Association*, vol. 307, no. 5, pp. 483–490, 2012.

[2] K. M. Flegal, M. D. Carroll, C. L. Ogden, and L. R. Curtin, "Prevalence and trends in obesity among US adults, 1999–2008," *The Journal of the American Medical Association*, vol. 303, no. 3, pp. 235–241, 2010.

[3] S. S. Guo and W. C. Chumlea, "Tracking of body mass index in children in relation to overweight in adulthood," *American Journal of Clinical Nutrition*, vol. 70, no. 1, part 2, pp. 145S–148S, 1999.

[4] US Department of Agriculture and US Department of Health and Human Services, *Dietary Guidelines for Americans 2010*, US Department of Agriculture, US Department of Health and Human Services, 2011.

[5] R. P. Troiano, D. Berrigan, K. W. Dodd, L. C. Mâsse, T. Tilert, and M. Mcdowell, "Physical activity in the United States measured by accelerometer," *Medicine and Science in Sports and Exercise*, vol. 40, no. 1, pp. 181–188, 2008.

[6] R. E. Andersen, C. J. Crespo, S. J. Bartlett, L. J. Cheskin, and M. Pratt, "Relationship of physical activity and television watching with body weight and level of fatness among children: results from the Third National Health and Nutrition Examination Survey," *The Journal of the American Medical Association*, vol. 279, no. 12, pp. 938–942, 1998.

[7] C. Simon, B. Schweitzer, M. Oujaa et al., "Successful overweight prevention in adolescents by increasing physical activity: a 4-year randomized controlled intervention," *International Journal of Obesity*, vol. 32, no. 10, pp. 1489–1498, 2008.

[8] L. B. Andersen, M. Harro, L. B. Sardinha et al., "Physical activity and clustered cardiovascular risk in children: a cross-sectional study (The European Youth Heart Study)," *The Lancet*, vol. 368, no. 9532, pp. 299–304, 2006.

[9] C. S. Berkey, H. R. H. Rockett, M. W. Gillman, and G. A. Colditz, "One-year changes in activity and in inactivity among 10- to 15-year-old boys and girls: relationship to change in body mass index," *Pediatrics*, vol. 111, no. 4, pp. 836–843, 2003.

[10] A. H. Lichtenstein, L. J. Appel, M. Brands et al., "Diet and lifestyle recommendations revision 2006: a scientific statement from the American Heart Association Nutrition Committee," *Circulation*, vol. 114, no. 1, pp. 82–96, 2006.

[11] J. H. Ledikwe, H. M. Blanck, L. K. Khan et al., "Low-energy-density diets are associated with high diet quality in adults in the United States," *Journal of the American Dietetic Association*, vol. 106, no. 8, pp. 1172–1180, 2006.

[12] A. K. Kant and B. I. Graubard, "Energy density of diets reported by American adults: association with food group intake, nutrient intake, and body weight," *International Journal of Obesity*, vol. 29, no. 8, pp. 950–956, 2005.

[13] R. P. Troiano, R. R. Briefel, M. D. Carroll, and K. Bialostosky, "Energy and fat intakes of children arid adolescents in the United States: Data from the National Health and Nutrition Examination Surveys," *American Journal of Clinical Nutrition*, vol. 72, no. 5, supplement, pp. 1343S–1353S, 2000.

[14] S. M. Krebs-Smith, P. M. Guenther, A. F. Subar, S. I. Kirkpatrick, and K. W. Dodd, "Americans do not meet federal dietary recommendations," *Journal of Nutrition*, vol. 140, no. 10, pp. 1832–1838, 2010.

[15] R. R. Pate, G. W. Heath, M. Dowda, and S. G. Trost, "Associations between physical activity and other health behaviors in a representative sample of US adolescents," *American Journal of Public Health*, vol. 86, no. 11, pp. 1577–1581, 1996.

[16] M. W. Gillman, B. M. Pinto, S. Tennstedt, K. Glanz, B. Marcus, and R. H. Friedman, "Relationships of physical activity with

dietary behaviors among adults," *Preventive Medicine*, vol. 32, no. 3, pp. 295–301, 2001.

[17] L. A. Lytle, S. Seifert, J. Greenstein, and P. McGovern, "How do children's eating patterns and food choices change over time? Results from a cohort study," *American Journal of Health Promotion*, vol. 14, no. 4, pp. 222–228, 2000.

[18] NHLBI Research Group, "Obesity and cardiovascular disease risk factors in black and white girls: the NHLBI Growth and Health Study," *American Journal of Public Health*, vol. 82, no. 12, pp. 1613–1620, 1992.

[19] S. Y. S. Kimm, N. W. Glynn, A. M. Kriska et al., "Decline in physical activity in black girls and white girls during adolescence," *The New England Journal of Medicine*, vol. 347, no. 10, pp. 709–715, 2002.

[20] S. Y. S. Kimm, N. W. Glynn, A. M. Kriska et al., "Longitudinal changes in physical activity in a biracial cohort during adolescence," *Medicine and Science in Sports and Exercise*, vol. 32, no. 8, pp. 1445–1454, 2000.

[21] NHLBI Growth and Health Study (NGHS) Master Documentation for Limited Data Sets, 2012.

[22] IBM Corporation, *IBM SPSS Statistics for Windows*, IBM Corporation, Armonk, NY, USA, 2013.

[23] S. R. Smith, L. De Jonge, J. J. Zachwieja et al., "Concurrent physical activity increases fat oxidation during the shift to a high-fat diet," *American Journal of Clinical Nutrition*, vol. 72, no. 1, pp. 131–138, 2000.

[24] B. J. Rolls, "Carbohydrates, fats, and satiety," *The American Journal of Clinical Nutrition*, vol. 61, no. 4, supplement, pp. 960S–967S, 1995.

[25] A. Tremblay, G. Plourde, J.-P. Despres, and C. Bouchard, "Impact of dietary fat content and fat oxidation on energy intake in humans," *American Journal of Clinical Nutrition*, vol. 49, no. 5, pp. 799–805, 1989.

[26] A. M. Prentice, "Manipulation of dietary fat and energy density and subsequent effects on substrate flux and food intake," *American Journal of Clinical Nutrition*, vol. 67, no. 3, supplement, pp. 535S–541S, 1998.

[27] E. C. Starritt, D. Angus, and M. Hargreaves, "Effect of short-term training on mitochondrial ATP production rate in human skeletal muscle," *Journal of Applied Physiology*, vol. 86, no. 2, pp. 450–454, 1999.

[28] J. O. Holloszy and E. F. Coyle, "Adaptations of skeletal muscle to endurance exercise and their metabolic consequences," *Journal of Applied Physiology Respiratory Environmental and Exercise Physiology*, vol. 56, no. 4, pp. 831–838, 1984.

[29] T. Tchkonia, M. Lenburg, T. Thomou et al., "Identification of depot-specific human fat cell progenitors through distinct expression profiles and developmental gene patterns," *American Journal of Physiology—Endocrinology and Metabolism*, vol. 292, no. 1, pp. E298–E307, 2007.

[30] J. L. Knittle, K. Timmers, F. Ginsberg-Fellner, R. E. Brown, and D. P. Katz, "The growth of adipose tissue in children and adolescents. Cross-sectional and longitudinal studies of adipose cell number and size," *Journal of Clinical Investigation*, vol. 63, no. 2, pp. 239–246, 1979.

[31] A. M. Toschke, S. Rückinger, T. Reinehr, and R. von Kries, "Growth around puberty as predictor of adult obesity," *European Journal of Clinical Nutrition*, vol. 62, no. 12, pp. 1405–1411, 2008.

[32] I. J. Rimm and A. A. Rimm, "Association between juvenile onset obesity and severe adult obesity in 73, 532 women," *American Journal of Public Health*, vol. 66, no. 5, pp. 479–481, 1976.

[33] N. Anderssen, D. R. Jacobs Jr., S. Sidney et al., "Change and secular trends in physical activity patterns in young adults: a seven-year longitudinal follow-up in the Coronary Artery Risk Development in Young Adults study (CARDIA)," *American Journal of Epidemiology*, vol. 143, no. 4, pp. 351–362, 1996.

[34] J. H. Ledikwe, H. M. Blanck, L. K. Khan et al., "Dietary energy density determined by eight calculation methods in a nationally representative United States population," *Journal of Nutrition*, vol. 135, no. 2, pp. 273–278, 2005.

[35] T. V. E. Kral, A. J. Stunkard, R. I. Berkowitz, V. A. Stallings, D. D. Brown, and M. S. Faith, "Daily food intake in relation to dietary energy density in the free-living environment: a prospective analysis of children born at different risk of obesity," *American Journal of Clinical Nutrition*, vol. 86, no. 1, pp. 41–47, 2007.

[36] T. V. E. Kral, R. I. Berkowitz, A. J. Stunkard, V. A. Stallings, D. D. Brown, and M. S. Faith, "Dietary energy density increases during early childhood irrespective of familial predisposition to obesity: results from a prospective cohort study," *International Journal of Obesity*, vol. 31, no. 7, pp. 1061–1067, 2007.

[37] J. A. Ello-Martin, L. S. Roe, J. H. Ledikwe, A. M. Beach, and B. J. Rolls, "Dietary energy density in the treatment of obesity: a year-long trial comparing 2 weight-loss diets," *American Journal of Clinical Nutrition*, vol. 85, no. 6, pp. 1465–1477, 2007.

22

Modulation of Metabolic Detoxification Pathways Using Foods and Food-Derived Components: A Scientific Review with Clinical Application

Romilly E. Hodges[1] and Deanna M. Minich[2,3]

[1]University of Bridgeport, 126 Park Avenue, Bridgeport, CT 07748, USA
[2]Institute for Functional Medicine, 505 S. 336th Street, Suite 500, Federal Way, WA 98003, USA
[3]University of Western States, 2900 NE 132nd Avenue, Portland, OR 97230, USA

Correspondence should be addressed to Deanna M. Minich; deannaminich@hotmail.com

Academic Editor: H. K. Biesalski

Research into human biotransformation and elimination systems continues to evolve. Various clinical and *in vivo* studies have been undertaken to evaluate the effects of foods and food-derived components on the activity of detoxification pathways, including phase I cytochrome P450 enzymes, phase II conjugation enzymes, Nrf2 signaling, and metallothionein. This review summarizes the research in this area to date, highlighting the potential for foods and nutrients to support and/or modulate detoxification functions. Clinical applications to alter detoxification pathway activity and improve patient outcomes are considered, drawing on the growing understanding of the relationship between detoxification functions and different disease states, genetic polymorphisms, and drug-nutrient interactions. Some caution is recommended, however, due to the limitations of current research as well as indications that many nutrients exert biphasic, dose-dependent effects and that genetic polymorphisms may alter outcomes. A whole-foods approach may, therefore, be prudent.

1. Introduction

Food-based nutrients have been and continue to be investigated for their role in the modulation of metabolic pathways involved in detoxification processes. Several publications to date have leveraged cell, animal, and clinical studies to demonstrate that food-derived components and nutrients can modulate processes of conversion and eventual excretion of toxins from the body [1]. In general, the nature of these findings indicates that specific foods may upregulate or favorably balance metabolic pathways to assist with toxin biotransformation and subsequent elimination [2, 3]. Various whole foods such as cruciferous vegetables [2, 4, 5], berries [6], soy [7], garlic [8, 9], and even spices like turmeric [10, 11] have been suggested to be beneficial and commonly prescribed as part of naturopathic-oriented and functional medicine-based therapies [12, 13].

While these foods are important to note, the science in this active area of inquiry continues to evolve to reveal new findings about food-based nutrients and their effect on health. Thus, the purpose of this review article is to summarize the science to date on the influence of whole foods, with a special focus directed towards phytonutrients and other food-based components, on influencing specific metabolic detoxification pathways, including phase I cytochrome enzymes, phase II conjugation enzymes, antioxidant support systems, and metallothionein upregulation for heavy metal metabolism. Based on this current science, the paper will conclude with clinical recommendations that may be applied in a personalized manner for patients via the discretion of a qualified health professional.

2. The Metabolic Pathways of Detoxification

Discussion of physiological pathways for detoxification has been mainly centered around phase I and phase II enzyme systems. This review will cover phase I cytochrome P450 enzymes as well as phase II enzymes, specifically UDP-glucuronosyl transferases, glutathione S-transferases, amino acid transferases, N-acetyl transferases, and methyltransferases. Note that there are other important classes of phase I enzymes, namely, hydroxylation and reduction, which are not covered in this review. While these important enzymes are pivotal to consider, this review of the effect of food on detoxification will also extend into other pathways, including ways to promote gene expression of antioxidant-related enzymes and of metallothionein, an endogenous protein carrier for heavy metals. Each of these four classes of detoxification-related pathways will be discussed within the context of nutrients.

2.1. Phase I Cytochrome P450 Enzymes. Initially, the "phases" of detoxification were described as functionalization (or phase I), or the addition of oxygen to form a reactive site on the toxic compound, and conjugation (phase II), or the process of adding a water-soluble group to this now reactive site [14, 15]. The "Phase I" cytochrome P450 superfamily of enzymes (CYP450) is generally the first defense employed by the body to biotransform xenobiotics, steroid hormones, and pharmaceuticals. These microsomal membrane-bound, heme-thiolate proteins, located mainly in the liver, but also in enterocytes, kidneys, lung, and even the brain, are responsible for the oxidation, peroxidation, and reduction of several endogenous and exogenous substrates [13, 15, 16]. Specifically, the function of CYP450 enzymes is to add a reactive group such as a hydroxyl, carboxyl, or an amino group through oxidation, reduction, and/or hydrolysis reactions [15]. These initial reactions have the potential to create oxidative damage within cell systems because of the resulting formation of reactive electrophilic species.

It is accepted that any variability in the number of CYP450 enzymes could have benefit(s) and/or consequence(s) for how an individual responds to the effect(s) of (a) toxin(s). Clinical application of the knowledge of these phase I CYP450 enzymes has been primarily addressed within pharmacology to understand the nature of drug interactions, side effects, and interindividual variability in drug metabolism [15]. The ability of an individual to metabolize 90% of currently used drugs will largely depend on the genetic expression of these enzymes [17]. It is established that many of these CYP450 genes are subject to genetic polymorphisms, resulting in an altered expression and function of individual enzymes. Currently, there exist some laboratory tests to identify the presence of these genetic variants. It is conceivable that having knowledge about foods and their individual (phyto)nutrients, especially in the case of dietary supplements and functional foods, could be worthwhile for clinicians to consider for patients who are taking a polypharmacy approach. Furthermore, as nutritional strategies become more personalized, it would seem that this information could be interfaced with a patient's known CYP450 polymorphisms to determine how to best optimize health outcomes.

2.1.1. CYP1 Enzymes. The CYP1A family is involved in metabolizing procarcinogens, hormones, and pharmaceuticals.

It is well-known for its role in the carcinogenic bioactivation of polycyclic aromatic hydrocarbons (PAHs), heterocyclic aromatic amines/amides, polychlorinated biphenyls (PCBs), and other environmental toxins [18, 19]. Low CYP1A2 activity, for example, has been linked to higher risk of testicular cancer [20]. However, due to their rapid conversion to highly reactive intermediates, excessive activity of CYP1A enzymes without adequate phase II support may enhance the destructive effects of environmental procarcinogens [21]. Indeed, genetic polymorphisms in this cytochrome family have been suggested as useful markers for predisposition to certain cancers [15]. CYP1 enzymes are also involved in the formation of clinically relevant estrogen metabolites: CYP1A1/1A2 and CYP1B1 catalyze the 2-hydroxylation and 4-hydroxylation of estrogens, respectively [22]. The potential role of 4-hydroxyestradiol in estrogen-related carcinogenesis, via the production of free radicals and related cellular damage [22], has prompted investigation into factors that modulate CYP1 enzymes.

Various foods and phytonutrients alter CYP1 activity (Tables 1(a) and 1(b)). Cruciferous vegetables have been shown, in humans, to act as inducers of CYP1A1 and 1A2, and animal studies also suggest an upregulation of CYP1B1 [4, 23–27]. The inductory effect of crucifers on CYP1A2 seems especially well established. Clinical studies also indicate that resveratrol and resveratrol-containing foods are CYP1A1 enhancers [28]. Conversely, berries and their constituent polyphenol, ellagic acid, may reduce CYP1A1 overactivity [6], and apiaceous vegetables and quercetin may attenuate excessive CYP1A2 action [24, 29]. Cruciferous vegetables and berries have been suggested as possible modulators of estrogen metabolites: berries for their reducing effect on CYP1A1 [6] and cruciferous vegetables for their stronger induction of CYP1A versus 1B1 enzymes [25–27, 30]. Chrysoeriol, present in rooibos tea and celery, acts selectively to inhibit CYP1B1 *in vitro* [31] and may be especially relevant to patients with CYP1B1 overactivity. However, further research is needed to confirm this finding.

Many foods appear to act as both inducers and inhibitors of CYP1 enzymes, an effect which may be dose dependent or altered by the isolation of bioactive compounds derived from food. Curcumin at 0.1% of the diet has been shown, in animals, to induce CYP1A1, for example, [35], yet a diet of 1% turmeric was inhibitory [46]. Black tea at 54 mL/d induced both CYP1A1 and 1A2 [33], yet 20 mg/kg of theaflavins was inhibitory to CYP1A1 [45]. Soybean intake at 100 mg/kg upregulated CYP1A1 activity [7], yet at 1 g/kg black soybean extract [44] and 200 mg daidzein twice daily [49], its effect was inhibitory. Further research is needed to confirm different dose effects and impact in humans.

Varied effects may also occur from different members of the same food group. Seemingly contradictory to research showing that cruciferous vegetables activate CYP1 enzymes, kale (another member of the cruciferous family) appears to inhibit CYP1A2 (as well as 2C19, 2D6, and 3A4) in animals [51]. The dose used, at 2 g/kg per day, is 15-fold higher than the typical level of human consumption [51], and more research would be required to determine whether lower intake levels would also have a similar effect. The same authors also tested

TABLE 1: (a) Human and *in vivo* example nutrient inducers of CYP1 enzymes. (b) Human and *in vivo* example nutrient inhibitors of CYP1 enzymes.

(a)

Enzyme	Food, beverage, or bioactive compounds *Food sources in italics*	Type of study	Dosages used and references
CYP1A1	Cruciferous vegetables	Clinical	500 mg/d indole-3-carbinol [23]
	Resveratrol *Grapes, wine, peanuts, soy, and itadori tea* [32]	Clinical	1 g/d resveratrol [28]: *note high dose used*
	Green tea	*In vivo*	45 mL/d/rat (avg. 150 g animal weight) green tea [33]
	Black tea	*In vivo*	54 mL/d/rat (avg. 150 g animal weight) black tea [33]
	Curcumin *Turmeric, curry powder* [34]	*In vivo*	1,000 mg/kg/d/rat curcumin [35], or about 150 mg per rat per day
	Soybean	*In vivo*	100 mg/kg soybean extract [7]
	Garlic	*In vivo*	30 to 200 mg/kg garlic oil [36]
	Fish oil	*In vivo*	20.5 g/kg fish oil [36]: *note high dose used*
	Rosemary	*In vivo*	Diet of 0.5% rosemary extract [37]
	Astaxanthin *Algae, yeast, salmon, trout, krill, shrimp, and crayfish* [38]	*In vivo*	Diets of 0.001–0.03% astaxanthin for 15 days [39]
CYP1A2	Cruciferous vegetables	Clinical	7–14 g/kg cruciferous vegetables including frozen broccoli and cauliflower, fresh daikon radish sprouts and raw shredded cabbage, and red and green [24] 500 g/d broccoli [4] 250 g/d each of Brussel sprouts and broccoli [25] 500 g/d broccoli [26]
	Green tea	*In vivo*	45 mL/d/rat (avg. 150 g animal weight) green tea [33] Green tea (2.5% w/v) as sole beverage [40]
	Black tea	*In vivo*	54 mL/d/rat (avg. 150 g animal weight) black tea [33]
	Chicory root	*In vivo*	Diet of 10% dried chicory root [41]
	Astaxanthin *Algae, yeast, salmon, trout, krill, shrimp, and crayfish* [38]	*In vivo*	Diets of 0.001–0.03% astaxanthin for 15 days [39]
CYP1B1	Curcumin *Turmeric, curry powder* [34]	*In vivo*	Diet of 0.1% curcumin [35]
	Cruciferous vegtables	*In vivo*	25–250 mg/kg indole-3-carbinol [27]

(b)

Enzyme	Food, beverage, or bioactive compounds *Food sources in italics*	Type of study	Dosages used and references
CYP1A1	Black raspberry	*In vivo*	Diet of 2.5% black raspberry [6]
	Blueberry	*In vivo*	Diet of 2.5% blueberry [6]
	Ellagic acid *Berries, pomegranate, grapes, walnuts, and blackcurrants* [42]	*In vivo*	30 mg/kg/d ellagic acid [43] 400 ppm ellagic acid [6]
	Black soybean	*In vivo*	1 g/kg black soybean seed coat extract [44]: *note high dose used*
	Black tea	*In vivo*	20 mg/kg theaflavins [45]
	Turmeric	*In vivo*	Diet of 1% turmeric [46]

(b) Continued.

Enzyme	Food, beverage, or bioactive compounds *Food sources in italics*	Type of study	Dosages used and references
CYP1A2	Apiaceous vegetables	Clinical	4 g/kg apiaceous vegetables, including frozen carrots and fresh celery, dill, parsley, and parsnips [24]
	Quercetin *Apple, apricot, blueberries, yellow onion, kale, alfalfa sprouts, green beans, broccoli, black tea, and chili powder* [47, 48]	Clinical	500 mg/d quercetin [29]
	Daidzein *Soybean* [49]	Clinical	200 mg twice daily dosing of daidzein [49]
	Grapefruit	Clinical	300 mL grapefruit juice [50]
	Kale	*In vivo*	2 g/kg/d kale, as freeze-dried kale drink [51]
	Garlic	*In vivo*	100 mg/kg garlic oil [52]
	Chamomile	*In vivo*	Free access to 2% chamomile tea solution [53]
	Peppermint	*In vivo*	Free access to 2% peppermint tea solution [53]
	Dandelion	*In vivo*	Free access to 2% dandelion tea solution [53]
	Turmeric	*In vivo*	Diet of 1% turmeric [46]

the effects of an equivalent volume of cabbage consumption and found no such inhibitory effect, pointing to the possibility that different cruciferous vegetables may have distinct effects on cytochrome activity.

2.1.2. CYP2A-E Enzymes. The large CYP2 family of enzymes is involved in the metabolism of drugs, xenobiotics, hormones, and other endogenous compounds such as ketones, glycerol, and fatty acids [15, 54]. Some notable polymorphisms occur in the CYP2C and CYP2D subgroups, leading to the classification of patients as "poor metabolizers" of various pharmaceuticals: warfarin and CYP2C9, antiarrhythmia agents, metoprolol and propafenone, and CYP2D6, phenytoin, cyclobarbital, omeprazole, and CYP2C19, for example, [15, 17]. CYP2D polymorphisms may be associated with Parkinson's disease and lung cancer [15]. Clinical evidence exists for the induction of CYP2A6 by quercetin and broccoli [4, 29] (Table 2(a)). In animals, chicory appears to induce CYP2A enzymes [41] and rosemary and garlic may upregulate CYP2B activity [9, 37]. Clinical studies using resveratrol and garden cress indicate CYP2D6 inhibition [28, 55] (Table 2(b)). Ellagic acid, green tea, black tea, and cruciferous vegetables also appear to inhibit various CYP2 enzymes.

CYP2E1 enzymes have also attracted particular interest for their role in various diseases. 2E1 metabolizes nervous system agents such as halothane, isoflurane, chlorzoxazone, and ethanol and bioactivates procarcinogenic nitrosamines and aflatoxin B1 [15, 65]. It produces free radicals regardless of substrate [15], and CYP2E1 polymorphisms have been associated with altered risk for coronary artery disease [66]

and gastric cancer [67]. CYP2E1-induced oxidative stress has also been shown to lead to impaired insulin action via the suppression of GLUT4 expression [68]. Attenuation of 2E1 overactivity may therefore be an important consideration in high-risk patients.

Watercress and garlic are CYP2E1 inhibitors in humans [59, 60]. *In vivo* evidence also suggests that N-acetyl cysteine, ellagic acid, green tea, black tea, dandelion, chrysin, and medium chain triglycerides (MCTs) may downregulate CYP2E1 [33, 43, 54, 61, 63, 64]. MCT oil may specifically attenuate the ethanol-induced upregulation of CYP2E1 and production of mitochondrial 4-hydroxynonenal, a marker of oxidative stress [64].

2.1.3. CYP3A Enzymes. The occurrence of the different CYP3A isoforms is tissue-specific [15]. Rooibos tea, garlic, and fish oil appear to induce the activity of CYP3A, 3A1, and 3A2 [8, 36, 69, 70] (Table 3(a)). Possible inhibitory foods include green tea, black tea, and quercetin [33, 56, 71, 72] (Table 3(b)). The most clinically relevant of the enzymes is CYP3A4, which is expressed mainly in the liver and to a lesser extent in the kidney [13]. Caffeine, testosterone, progesterone, and androstenedione are substrates of the CYP3A4 enzyme system, as are various procarcinogens including PAHs and aflatoxin B1 [15]. To date, however, the principal driver for research on CYP3A4 has been due to its role in the metabolism of over 50 percent of all pharmaceuticals [73]. The potential for drug interaction with this single enzyme, coupled with the wide interindividual differences in enzymatic activity, generates some level of risk in administration of high doses and multiple drugs as well as food-drug

TABLE 2: (a) Human and *in vivo* example nutrient inducers of selected CYP2 enzymes. (b) Human and *in vivo* example nutrient inhibitors of selected CYP2 enzymes.

(a)

Enzyme	Food, beverage, or bioactive compounds *Food sources in italics*	Type of study	Dosages used and references
CYP2A	Chicory root	*In vivo*	Diet of 10% dried chicory root [41]
CYP2A6	Quercetin *Apple, apricot, blueberries, yellow onion, kale, alfalfa sprouts, green beans, broccoli, black tea,* and *chili powder* [47, 48]	Clinical	500 mg/d quercetin [29]
	Broccoli	Clinical	500 g/d broccoli [4]
CYP2B1	Rosemary	*In vivo*	Diet of 0.5% rosemary extract [37]
	Garlic	*In vivo*	0.5 and 2.0 mmol/kg diallyl sulfide, or about 75 and 300 mg, respectively [9]
CYP2B2	Rosemary	*In vivo*	Diet of 0.5% rosemary extract [37]
CYP2E1	Fish oil	*In vivo*	20.5 g/kg fish oil [36]: *note high dose used*
	Chicory root	*In vivo*	Diet of 10% dried chicory root [41]

(b)

Enzyme	Food, beverage, or bioactive compounds *Food sources in italics*	Type of study	Dosages used and references
CYP2B	Ellagic acid *Berries, pomegranate, grapes, walnuts,* and *blackcurrants* [42]	*In vivo*	10 and 30 mg/kg/d ellagic acid [43]
	Green tea	*In vivo*	100 mg/kg/d green tea extract [56]
	Cruciferous vegetables	*In vivo*	3 and 12 mg/kg/d sulforaphane [57]
CYP2B1	Turmeric	*In vivo*	Diet of 1% turmeric [46]
CYP2C	Green tea	*In vivo*	45 mL/d/rat (avg. 150 g animal weight) green tea [33]
	Black tea	*In vivo*	54 mL/d/rat (avg. 150 g animal weight) black tea [33]
	Ellagic acid *Berries, pomegranate, grapes, walnuts,* and *blackcurrants* [42]	*In vivo*	30 mg/kg/d ellagic acid [43]
CYP2C6	Ellagic acid *Berries, pomegranate, grapes, walnuts,* and *blackcurrants* [42]	*In vivo*	30 mg/kg/d ellagic acid [43]
CYP2C9	Resveratrol *Grapes, wine, peanuts, soy,* and *itadori tea* [32]	Clinical	1 g/d resveratrol [28]: *note high dose used*
	Myricetin *Onions, berries, grapes,* and *red wine* [58]	*In vivo*	2 and 8 mg/kg myricetin [58]
CYP2C19	Kale	*In vivo*	2 g/kg/d kale, as freeze-dried kale drink [51]
CYP2D6	Resveratrol *Grapes, wine, peanuts, soy,* and *itadori tea* [32]	*Clinical*	1 g/d resveratrol [28]: *note high dose used*
	Garden cress	Clinical	7.5 g twice daily intake of garden cress seed powder [55]
	Kale	*In vivo*	2 g/kg/d kale, as freeze-dried kale drink [51]

Enzyme	Food, beverage, or bioactive compounds *Food sources in italics*	Type of study	Dosages used and references
	Watercress	Clinical	50 g watercress homogenate [59]
	Garlic	Clinical and *in vivo*	0.2 mg/kg diallyl sulfide, equivalent to high human garlic consumption [60] 100 mg/kg garlic oil [52] 200 mg/kg diallyl sulfide [8] 30 to 200 mg/kg garlic oil [36] Diet of 2% and 5% garlic powder [61]
	N-acetyl cysteine *Allium vegetables* [54]	*In vivo*	25 mg/kg and 50 mg/kg N-acetyl cysteine [54]
CYP2E1	Ellagic acid *Berries, pomegranate, grapes, walnuts,* and *blackcurrants* [42]	*In vivo*	10 and 30 mg/kg/d ellagic acid [43]
	Green tea	*In vivo*	45 mL/d/rat (avg. 150 g animal weight) green tea [33]
	Black tea	*In vivo*	54 mL/d/rat (avg. 150 g animal weight) black tea [33]
	Dandelion	*In vivo*	0.5 and 2 g/kg dandelion leaf water extract [62]
	Chrysin *Honey, honeycomb* [63]	*In vivo*	20 and 40 mg/kg/d chrysin [63]
	Medium-chain triglycerides (MCTs) *Coconut and coconut oil*	*In vivo*	32% calories as MCTs [64]

and herb-drug interactions. Grapefruit juice is perhaps the most well-known food inhibitor of this enzyme [74], though resveratrol and garden cress, a member of the cruciferous vegetable family, appear to have similar effects in humans, albeit at intakes above what would be expected without high-dose supplementation [28, 55]. Curcumin may upregulate 3A4 activity [11].

Once again, there are indications that a biphasic effect may be seen from dietary bioactive compounds; Davenport and Wargovich (2005) found that shorter-term or lower dosing with garlic organosulfur compounds produced potentially anticarcinogenic effects but that longer-term higher doses (200 mg/kg) of allyl sulfides led to minor hepatic toxicity [8]. One garlic clove contains only 2,500–4,500 μg of the allyl sulfide precursor, allicin [76], so the higher dose is much more than would be consumed in a typical human diet. In another example, two components of cruciferous vegetables, sulforaphanes and indole-3-carbinol, inhibited and increased activity, respectively [57, 75], highlighting the potential for human studies using whole foods to clarify the outcome of consumption.

2.1.4. CYP4 Enzymes. Less is known about this family of enzymes, since it is thought to play a smaller role in drug metabolism. It is, however, understood to be a primarily extrahepatic family of cytochromes, inducible by clofibrate and ciprofibrate (hypolipidemic drugs), NSAIDs, prostaglandins, and toxicants such as phthalate esters [15, 77]. The CYP4B1 isoform is involved in the metabolism of MCTs

(medium chain triglycerides), as well as the bioactivation of pneumotoxic and carcinogenic compounds [78].

Polymorphisms and overexpression of this subgroup may be associated with bladder cancer [15] and colitis [79]. A report by Ye et al. (2009) which examined the link between colitis and CYP4B1 activity found that the promotion of CYP4B1 activity by caffeic acid (found in caffeine-containing foods) (Table 4) correlated with reduced inflammation and disease activity [79]. Green tea may act to induce CYP4A1, as suggested by animal studies [40]. More research is needed to clearly identify food influences on this enzyme family.

2.2. Phase II Conjugation Enzymes. After a xenobiotic has gone through the process of becoming hydrophilic through reactions overseen by CYP450 enzymes, its reactive site can be conjugated with an endogenous hydrophilic substance. This reaction is often referred to as "phase II detoxification." Conjugation involves the transfer of a number of hydrophilic compounds (via their corresponding enzymes), including glucuronic acid (glucuronyl transferases), sulfate (sulfotransferases), glutathione (glutathione transferases), amino acids (amino acid transferases), an acetyl group (N-acetyl transferases), and a methyl group (N- and O-methyltransferases) [81]. The result of the collective activity of these enzymes is an increase in the hydrophilicity of the metabolite, theoretically leading to enhanced excretion in the bile and/or urine [81]. Similar to the CYP450 enzymes, genetic polymorphisms can have profound influence on the function of these conjugating

TABLE 3: (a) Human and *in vivo* example nutrient inducers of selected CYP3 enzymes. (b) Human and *in vivo* example nutrient inhibitors of selected CYP3 enzymes.

(a)

Enzyme	Food, beverage, or bioactive compounds *Food sources in italics*	Type of study	Dosages used and references
CYP3A	Rooibos tea	*In vivo*	Rooibos tea, 4 g/L simmered for 5 minutes, as sole beverage [69]
CYP3A1	Garlic	*In vivo*	30 to 200 mg/kg garlic oil [36] 80 and 200 mg/kg garlic oil 3 times weekly [70]
	Fish oil	*In vivo*	20.5 g/kg fish oil [36]: *note high dose used*
CYP3A2	Garlic	*In vivo*	200 mg/kg diallyl sulfide [8]
	Cruciferous vegetables	*In vivo*	50 mg/kg/d indole-3-carbinol [75]
CYP3A4	Curcumin *Turmeric, curry powder* [34]	*In vivo*	50 and 100 mg/kg curcumin [11]

(b)

Enzyme	Food, beverage, or bioactive compounds *Food sources in italics*	Type of study	Dosages used and references
CYP3A	Green tea	*In vivo*	45 mL/d/rat (avg. 150 g animal weight) green tea [33] 400 mg/kg green tea extract [71] 100 mg/kg/d green tea extract [56]
	Black tea	*In vivo*	54 mL/d/rat (avg. 150 g animal weight) black tea [33]
	Quercetin *Apple, apricot, blueberries, yellow onion, kale, alfalfa sprouts, green beans, broccoli, black tea,* and *chili powder* [47, 48]	*In vivo*	10 and 20 mg/kg [72]
CYP3A2	Cruciferous vegetables	*In vivo*	12 mg/kg/d sulforaphane [57]
CYP3A4	Grapefruit	Clinical	200 mL grapefruit juice 3 times daily [74]
	Resveratrol *Grapes, wine, peanuts, soy,* and *itadori tea* [32]	Clinical	1 g/d resveratrol [28]: *note high dose used*
	Garden cress	Clinical	7.5 g twice daily dose of garden cress seed powder [55]
	Soybean	*In vivo*	100 mg/kg soybean extract [7]
	Kale	*In vivo*	2 g/kg/d kale, as freeze-dried kale drink [51]
	Myricetin *Onions, berries, grapes,* and *red wine* [58]	*In vivo*	0.4, 2, and 8 mg/kg myricetin [58]

TABLE 4: Human and *in vivo* example nutrient inducers of selected CYP4 enzymes.

Enzyme	Food, beverage, or bioactive compounds *Food sources in italics*	Type of study	Dosages used and references
CYP4A1	Green tea	*In vivo*	Green tea (2.5% w/v) as sole beverage [40]
CYP4B1	Caffeic acid *Coffee* [80]	*In vivo*	179 mg/kg caffeic acid [79]

enzymes [82], with potential implication in the development of several forms of cancer [83].

It is conceivable that modulation of phase II enzymes by food-based bioactive compounds may be advantageous in patients who have altered enzyme activity due to genetic polymorphisms or who have a high toxic burden due to chronic exposure to environmental pollutants, overactive phase I activity, or hormonal imbalance. For example, James et al. (2008) suggest that upregulation of glucuronidation and sulfonation by certain bioactive compounds may be a useful consideration for the elimination of environmental PCBs [19].

2.2.1. UDP-Glucuronosyltransferases.

This class of enzymes, comprising multiple proteins and even subfamilies, plays an essential role in enhancing the elimination of biotransformed toxins in urine and feces, as well as metabolizing steroid hormones and bilirubin [84, 85]. Their function is to catalyze the covalent linkage of glucuronic acid from UDP-glucuronic acid to an accepting functional group on the molecule, a process referred to as glucuronidation [86]. Glucuronidation occurs primarily in the liver but can occur in other tissues, such as the small intestine [86, 87]. Bilirubin, specifically, is principally conjugated by UGT1A1 in hepatocytes [88] and then excreted with bile into the intestinal tract. It has been estimated that 40–70% of all medications are subject to glucuronidation reactions in humans, thereby suggesting the significance of this conjugation enzyme family [88]. Since UDP-glucuronosyltransferases (UGTs) also metabolize phytochemicals, alterations in their effects may be seen with genetically downregulated enzyme activity; flavonoids are conjugated with glucuronide and sulfate; therefore, UGT or sulfotransferase (SULT) polymorphisms may produce variability in phytochemical clearance and efficacy [89].

Clinical and observational studies point to cruciferous vegetables, resveratrol, and citrus as foods and bioactive compounds that induce UGT enzymes [25, 28, 90–92] (Table 5(a)). Animal studies also suggest the potential for other foods and nutrients, including dandelion, rooibos tea, honeybush tea, rosemary, soy, ellagic acid, ferulic acid, curcumin, and astaxanthin, to enhance UGT activity [37, 39, 53, 93–95]. Interestingly, the effect of resveratrol was seen only in individuals with low baseline enzyme levels/activity, suggesting that some phytochemicals may modulate, rather than outright induce, enzymatic activity [28]. In addition, many studies note that effects are variable depending on gender and genotype [85, 90, 92]; for example, women with the UGT1A1 *28 polymorphism (7/7) were responsive to citrus intervention, whereas those with other genetic variants were not [92].

Meaningful interpretations of these studies may still be elusive, however: in one combined dietary trial, the consumption of 10 servings per day of a combination of cruciferous vegetables, soy foods, and citrus fruits did not have a significant effect on UGT enzyme activity compared with a diet devoid of fruits and vegetables [85]. The authors hypothesize that these results may be due to their choice of specific foods within those groups or due to Nrf2 activation (discussed in subsequent sections) when fruits and vegetables were avoided.

The effects of UGT activity may also be enhanced by D-glucaric acid by theoretical inhibition of beta-glucuronidase enzymes [100]. Beta-glucuronidase enzymes act to reverse UGT conjugation reactions. D-glucaric acid is found in many fruits, vegetables and legumes (Table 5(b)). When tested in humans, however, a diet supplemented with cruciferous vegetables (2/3 cup broccoli, 1/2 cup cabbage, and 1/2 cup radish sprouts), citrus fruits (1 cup grapefruit juice, 1/2 cup orange juice, 1 cup orange/grapefruit segments, and 1 orange peel), and soy foods was found to have no effect on beta-glucuronidase activity [101] (amounts standardized for 55 kg body weight), indicating that the clinical effects of D-glucaric acid consumption still need further clarification.

In vivo research suggests that polyphenol extracts of certain berries, specifically strawberries and blackcurrant, may inhibit beta-glucuronidase activity in the intestinal lumen; Kosmala et al. (2014) observed this effect using both strawberry pomace water extract and water-alcohol extract containing 5.1% and 17.1% ellagic acid, and 0.2% and 10.9% proanthocyanidins, respectively [100]. Jurgoński et al. (2014) found a similar inhibitory effect using a diet of 1.5% blackcurrant extract (total polyphenolic content 66.8 g/100 g extract) [102]. Interestingly, the highest levels of beta-glucuronidase activity were seen in rabbits fed a high fat diet (32% calories from fat, including 10% from lard), without blackcurrant extract supplementation, suggesting that dietary fat may also alter enzyme activity [102].

Inhibition of UGT enzymatic activity may be a consideration for modulation of hormone levels and the risk of certain cancers, such as prostate cancer [84]. *In vitro* studies suggest that various foods and food-based components may inhibit UGT activity, including green and black tea, quercetin, rutin, naringenin, allspice, peppermint oil, cacao, and silymarin [84], although further research is needed to evaluate their *in vivo* and clinical effects.

2.2.2. Sulfotransferases.

As the name of this superfamily of enzymes might suggest, SULTs are responsible for the transfer of a sulfuryl group donated by $3'$-phosphoadenosine-$5'$-phosphosulfate (PAPS) to hydroxyl or amine groups, particularly in the areas of liver, intestine, adrenal gland, brain, and skin tissues [103]. This process is often referred to as sulfation but is more accurately termed sulfonation or sulfurylation. Decreased function of these enzymes, through genetic variability or presence of environmental chemicals, can lead to eventual interference with thyroid hormone, estrogen, and androgen levels [104, 105], as well as variable polyphenol effects [106], since the active forms of these compounds can be degraded via sulfonation. Typically, once compounds have been conjugated with sulfate, there is less reactivity and toxicity incurred from the precursor molecule [105].

Few *in vivo* studies have examined the effects of dietary components on SULT activity, although caffeine and retinoic acid are possible SULT inducers according to animal studies [107, 108] (Table 6(a)). Although it is uncertain how their outcomes will translate *in vivo*, various *in vitro* studies have indicated the possibility of sulfotransferase inhibition (including competitive inhibition) by wine anthocyanins and flavonols, synthetic food colors (especially red colors), apple and grape juice, catechins including epigallocatechin gallate, quercetin, curcumin, resveratrol, flavonoids (apigenin, chrysin, fisetin, galangin, kaempferol, quercetin, myricetin, naringenin, and naringin), and certain phytoestrogens (daidzein, genistein) [3, 105]. Pyridoxal-6-phosphate, the active form of vitamin B6 (which is widely distributed in foods), may also be a competitive SULT inhibitor, according to one *in vitro* study [109], although human tissue concentrations and clinical effects

TABLE 5: (a) Human and *in vivo* example nutrient inducers of UGT enzymes. (b) Selected dietary sources of D-glucaric acid.

(a)

Enzyme	Food, beverage, or bioactive compounds *Food sources in italics*	Type of study	Dosages used and references
UGTs	Cruciferous vegetables	Clinical	Approximately 5 and 10 servings/d of cruciferous vegetables including frozen broccoli, cauliflower, fresh cabbage (red and green), and fresh radish sprouts [90] 250 g/d each of Brussel sprouts and broccoli [25] 2 oz (56.8 g) watercress three times daily [91]
	Resveratrol *Grapes, wine, peanuts, soy,* and *itadori tea* [32]	Clinical	1 g/d resveratrol [28]: *note high dose used*
	Citrus	Observational	0.5+ servings/day of citrus fruits or foods [92]
	Dandelion	*In vivo*	Free access to 2% dandelion tea solution [53]
	Rooibos tea	*In vivo*	Rooibos tea as sole beverage; concentration 2 g tea leaves/100 mL water steeped for 30 minutes [93]
	Honeybush tea	*In vivo*	Honeybush tea as sole beverage; concentration 4 g tea leaves/100 mL water steeped for 30 minutes [93]
	Rosemary	*In vivo*	Diet of 0.5% rosemary extract [37]
	Soy	*In vivo*	150 and 500 mg/kg soy extract [94]
	Ellagic acid *Berries, pomegranate, grapes, walnuts,* and *blackcurrants* [42]	*In vivo*	Diet of 1% ellagic acid [95]
	Ferulic acid *Whole grains, roasted coffee, tomatoes, asparagus, olives, berries, peas, vegetables,* and *citrus* [96]	*In vivo*	Diet of 1% ferulic acid [95]
	Curcumin *Turmeric, curry powder* [34]	*In vivo*	Diet of 1% curcumin [95]
	Astaxanthin *Algae, yeast, salmon, trout, krill, shrimp,* and *crayfish* [38]	*In vivo*	Diets of 0.001–0.03% astaxanthin for 15 days [39]

(b)

Legumes	Mung bean seeds, adzuki bean sprouts [97]
Vegetables and fruits	Oranges, spinach, apples, carrots, alfalfa sprouts, cabbage, Brussel sprouts, cauliflower, broccoli, grapefruit, grapes, peaches, plums, lemons, apricots, sweet cherries, corn, cucumber, lettuce, celery, green pepper, tomato, and potatoes [97–99]

may be vastly different. Of note, caffeic acid demonstrates *in vitro* SULT-inhibitory properties [105]. This finding conflicts with its *in vivo* ability to induce SULT enzymes, as described by Zhou et al. (2012) [107], highlighting the difficulty of extrapolating meaningful conclusions from *in vitro* data.

SULT enzyme activity is dependent on a depletable reserve of inorganic sulfate [112]. Dietary sources of sulfur-containing compounds may therefore play an essential role in SULT function, by providing the substrate for enzyme action (Table 6(b)).

2.2.3. Glutathione S-Transferases. Similar to the aforementioned categories of conjugating enzymes, glutathione S-transferases (GSTs) include a complex of enzymes, whose main function is to attach a glutathione group to a biotransformed metabolite. The production of these enzymes can be induced through the production of reactive oxygen species and via gene transcription involving the antioxidant-responsive element (ARE) and the xenobiotic-responsive element (XRE), which will be subsequently discussed in this paper [113].

Cruciferous and allium vegetables and resveratrol demonstrate ability to induce GSTs in humans [28, 114–117] (Table 7(a)). Observational research also associates citrus consumption with increased GST activity [115]. *In vivo* data also suggest many foods and food constituents to be upregulators of these enzymes, including garlic, fish oil, black soybean, purple sweet potato, curcumin, green tea, rooibos tea, honeybush tea, ellagic acid, rosemary, ghee, and genistein [36, 43, 44, 70, 93, 118–123]. Conjugated linoleic acid has been shown to be at least partly responsible for the effect of ghee [122]. It is possible that the effects of at least some of these foods and bioactive compounds may be due to their upregulation of the Nrf2 signaling pathway.

TABLE 6: (a) *In vivo* example nutrient inducers of sulfotransferases (SULTs). (b) Selected dietary sources of sulfur-containing compounds (adapted from [110]).

(a)

Enzyme	Food, beverage, or bioactive compounds *Food sources in italics*	Type of study	Dosages used and references
SULTs	Caffeine *Coffee, cocoa, black tea,* and *green tea* [111]	*In vivo*	2, 10, and 50 mg/kg caffeine [107]
	Retinoic acid (bioactive form of vitamin A) *Meat (especially liver), fish, egg,* and *dairy products* contain retinol; *apple, apricot, artichokes, arugula, asparagus,* and other plant foods contain provitamin A carotenes [111]	*In vivo*	2, 10, and 50 mg/kg/d retinoic acid suspension in corn oil [108]

(b)

Animal products	Fish, shellfish, lamb, beef, chicken, pork, duck, goose, turkey, egg, and cheese
Legumes	Lentils, peas, and butter beans
Grains	Barley, oatmeal
Vegetables and fruits	Cabbage, horseradish, Brussel sprouts, leeks, cress, haricot beans, apricots, peaches, spinach, and watercress
Nuts and seeds	Brazil nuts, almonds, peanuts, and walnuts
Herbs and spices	Mustard, ginger

Genetic variances, gender, and even possibly body weight appear to play a role in the effects of dietary factors on GST enzymes [114–116]. Clinical investigation of cruciferous and allium vegetables by Lampe et al. (2000) found that an upregulated effect was most marked in women, indicating gender variability, and that the effect was also genotype-dependent, occurring only in GSTM1-null individuals [116]. The same investigators also found that apiaceous vegetables inhibited GST activity, but only in GSTM1+ men [116] (Table 7(b)). High doses of quercetin and genistein have also shown inhibitory effects [123, 126].

There is evidence that at least some of these foods and phytonutrients may exert modulatory rather than absolute inductive/inhibitory effects; Chow et al. (2010) found that resveratrol increased GST only in those with low baseline enzyme levels or activity [28]. It is also noteworthy that bioactive components of crucifers, including isothiocyanates, are substrates for GST enzymes and that GST genotype may therefore alter the response to cruciferous vegetables consumption on other mechanisms such as glutathione peroxidase and superoxide dismutase [134, 135]. GSTM1-null genotype is associated with a more rapid excretion of isothiocyanates, leading some researchers to conclude that the benefits of cruciferous vegetable consumption may be lessened in individuals with this genetic variation [89].

Support for glutathione conjugation also involves enhancing reduced glutathione (GSH) status. Glutathione is a low-molecular weight tripeptide containing residues of cysteine, glutamate, and glycine [136]. Most glutathione from foods and supplements is poorly absorbed, so liposomal delivery has been used [137]. The sulfur-containing amino acids methionine and cystine are important precursors to glutathione formation; their depletion leads to depressed GSH levels [138]. N-acetyl cysteine has also been used to restore depleted GSH levels in a clinical setting [139].

Various nutrients may also enhance endogenous glutathione synthesis, including vitamin B6, magnesium, and selenium [140, 141]. Curcuminoids (from turmeric), silymarin (from milk thistle), folic acid, and alpha-lipoic acid have been shown, in humans, to restore depleted GSH [129, 130, 142, 143]. In animal studies, cruciferous vegetables and artichoke have also demonstrated a GSH-protective effect [131–133]. There is therefore the potential to improve glutathione status via diet or supplementation (Table 7(c)).

2.2.4. Amino Acid Transferases. Amino acids of various types (e.g., taurine, glycine), whether endogenous or exogenous (from dietary sources) in origin, can be utilized for attaching to molecules for their excretion. For the benefit of providing a substrate to these enzymes, it is generally thought that dietary protein is required for an effective detoxification protocol. Table 8 lists amino acids used in phase II conjugation reactions and selected food sources.

2.2.5. N-Acetyl Transferases (NAT). This class of enzymes is responsible for the transfer of an acetyl group to convert aromatic amines or hydrazines to aromatic amides and hydrazides, which is significant for those taking pharmaceuticals such as isoniazid, hydralazine, and sulphonamides [83]. Polymorphisms in genes for this category of enzymes, leading to slow metabolism, have been shown to be associated with hepatoxicity during drug treatment [146]. One small human study found that 500 mg quercetin daily enhanced NAT activity [29]. However, more research is needed to understand the relationship between dietary nutrients and NAT function.

TABLE 7: (a) *In vivo* example nutrient inducers of glutathione S-transferases (GSTs). (b) *In vivo* example nutrient inhibitors of glutathione S-transferases (GSTs). (c) Selected dietary sources of nutrients for glutathione support ([111] unless otherwise noted).

(a)

Enzyme	Food, beverage, or bioactive compounds *Food sources in italics*	Type of study	Dosages used and references
GSTs	Cruciferous vegetables	Clinical, observational	Approximately 5 and 10 servings/d of cruciferous vegetables including frozen broccoli, cauliflower, fresh cabbage (red and green), and fresh radish sprouts [114] >31.2 g/d cruciferous vegetables [115] 4.5 cups of cruciferous vegetables/d, including 0.5 cups of radish sprouts, 1 cup of frozen cauliflower, 2 cups of frozen broccoli, and 1 cup of fresh cabbage [116] 300 g/d cooked Brussels sprouts [117]
	Allium vegetables	Clinical	3 tbsp fresh chives, 1.33 cups of fresh leeks, 1 tsp garlic, and 0.5 cups of fresh onion [116]
	Resveratrol *Grapes, wine, peanuts, soy,* and *itadori tea* [32]	Clinical	1 g/d resveratrol [28]: *note high dose used*
	Citrus	Observational, *in vivo*	>76 g/d citrus [115] 20 mg limonoid mixture every 2 days [124]
	Garlic	*In vivo*	30 to 200 mg/kg garlic oil [36] 80 and 200 mg/kg garlic oil 3 times weekly [70]
	Fish oil	*In vivo*	20.5 g/kg fish oil [36]: *note high dose used*
	Black soybean	*In vivo*	1 g/kg black soybean seed coat extract [44]
	Purple sweet potato	*In vivo*	100 and 200 mg/kg anthocyanin extract from purple sweet potato [118]
	Curcumin	*In vivo*	Diet of 2% curcumin [119]
	Green tea	*In vivo*	Equivalent of 4 cups/d (200 mL each) of green tea [120]
	Rooibos tea	*In vivo*	Rooibos tea as sole beverage; concentration 2 g tea leaves/100 mL water steeped for 30 minutes [93]
	Honeybush tea	*In vivo*	Honeybush tea as sole beverage; concentration 4 g tea leaves/100 mL water steeped for 30 minutes [93]
	Ellagic acid *Berries, pomegranate, grapes, walnuts,* and *blackcurrants* [42]	*In vivo*	30 mg/kg/d ellagic acid [43]
	Rosemary	*In vivo*	20 mg/kg carnosic acid 3 times weekly [121]
	Ghee (clarified butter)	*In vivo*	19.5 mg CLA (conjugated linoleic acid)/g fat [122]
	Genistein (kidney GSTs) *Fermented soy* (e.g., *miso, tempeh*) contains up to 40% bioavailable genistein versus 1% or less in other soy products [125]	*In vivo*	1.5 g/kg genistein [123]: *note high dose used*

(b)

Enzyme	Food, beverage, or bioactive compounds *Food sources in italics*	Type of study	Dosages used and references
	Apiaceous vegetables	Clinical	1 tsp fresh dill weed, 0.5 cups of fresh celery, 3 tbsp. fresh parsley, 1.25 cups of grated parsnips, and 0.75 cups of frozen carrots [116]

(b) Continued.

Enzyme	Food, beverage, or bioactive compounds *Food sources in italics*	Type of study	Dosages used and references
GSTs	Quercetin *Apple, apricot, blueberries, yellow onion, kale,* and *alfalfa sprouts, green beans, broccoli, black tea,* and *chili powder* [47, 48]	*In vivo*	2 g/kg quercetin [126]: *note high dose used*
	Genistein (liver GSTs) *Fermented soy* (e.g., *miso, tempeh*) containsup to 40% bioavailable genistein, versus 1% or less in other soy products [125]	*In vivo*	1.5 g/kg genistein [123]: *note high dose used*

(c)

Vitamin B6	Turkey, pork, chicken, beef, amaranth, lentils, pistachio nuts, sunflower seeds, garlic, and prunes
Magnesium	Nuts, seeds, beans, and whole grains
Selenium	Brazil nuts, pork, turkey, lamb, chicken, and egg
Methionine	Turkey, pork, chicken, beef, egg, Brazil nuts, soybean, sesame seeds, and spirulina
Cystine	Pork, turkey, chicken, egg, soybean, spirulina, sesame seeds, and oats
Glycine	Turkey, pork, chicken, amaranth, soybean, peanuts, pumpkin seed, and beef
Folate (dietary form of folic acid)	Mung bean, adzuki bean, and other legumes, liver, sunflower seeds, quinoa, spinach, asparagus, avocados, mustard greens, and artichokes
Alpha-lipoic acid	Spinach, broccoli, tomato, peas, Brussels sprouts, and visceral meats [127, 128]
Functional foods	Turmeric, milk thistle, cruciferous vegetables, and artichoke [129–133]

TABLE 8: Amino acids used in phase II conjugation and selected food sources.

Glycine	Turkey, pork, chicken, soybean, seaweed, eggs, amaranth, beef, mollusks, peanuts, pumpkin seeds, almonds, duck, goose, mung beans, sunflower seeds, lentils, lamb, bison, lobster, and fish [111]
Taurine	Many cooked meats and fish supply taurine. Taurine is also synthesized in the body from cystine (requiring niacin and vitamin B6) and homocysteine (requiring additionally betaine and serine) [144]
Glutamine	Plant and animal proteins such as beef, pork, chicken, dairy products, spinach, parsley, and cabbage [145]
Ornithine	Ornithine is synthesized endogenously via the urea cycle, requiring arginine and magnesium [144]
Arginine	Turkey and pork are especially rich sources; also chicken, pumpkin seeds, soybean, butternuts, egg, peanuts, walnuts, split peas, mollusks, almonds, sesame seeds, lentils, fava beans, mung beans, pine nuts, beef, sunflower seeds, and white beans [111]

2.2.6. Methyltransferases. Relatively significant attention has been given in various medical communities to this class of phase II enzymes due to the increasing importance of methylation for reducing disease risk. The conjugating donor compound in methyltransferase reactions is a methionine group from S-adenosyl-L-methionine (SAMe) [147]. Catechol O-methyltransferase (COMT) is one of the prominent methyltransferases that has received wide attention due to its role in estrogen detoxification [148].

Support for methylation consists of nutrient cofactors and methyl donors, such as methionine, vitamin B12, vitamin B6, betaine, folate, and magnesium [144]. Various foods can provide these nutrients (Table 9). Conversely, a high sucrose diet may inhibit methylation enzymes such as COMT [149].

3. Gene Induction of Phase II Detoxification and Antioxidant Enzymes through Nrf2

The transcription factor, Nrf2 [nuclear factor erythroid 2 (NF-E2) p45-related factor 2], is key to regulating the body's detoxification and antioxidant system. When activated, Nrf2 dissociates from the cytosolic protein, Keap1 (Kelch-like ECH associated protein 1), and translocates to the nucleus to bind to AREs in the promoter/enhancer portion of genes associated with phase II detoxification and antioxidant enzyme genes [150] (Figure 1). Nrf2-deficient animals experience increased toxicity from drugs [151], carcinogens, allergens, and environmental pollutants [152] and do not respond as well to the anti-inflammatory effects of phytochemicals [153], indicating the essentiality of these enzymes. Conversely, Nrf2

TABLE 9: Selected dietary sources of nutrients for methylation support (adapted from [111]).

Methionine	Meats, poultry, fish, shellfish, egg, nuts (especially Brazil nuts), seeds (especially sesame seeds and pumpkin seeds), spirulina, teff, soybeans Lower amounts found in other legumes and whole grains (especially teff and oats)
Vitamin B12	Meats and meat products (especially liver and kidney), poultry, fish, shellfish, and eggs
Vitamin B6	Meats, nuts (especially pistachio), garlic, whole grains, seeds (especially sesame and sunflower seeds), legumes (especially chickpeas and lentils), and prunes
Betaine	Quinoa, beets, spinach, whole grains (especially rye, kamut, bulgur, amaranth, barley, and oats) sweet potato, meats, and poultry
Folate	Beans and legumes (especially mung beans, adzuki beans, chickpeas, and lentils), liver, nuts (especially peanuts), seeds (especially sunflower seeds), spinach, asparagus, mustard greens, and avocado
Magnesium	Seeds (especially pumpkin seeds and sesame seeds), beans (especially soybeans), nuts (especially Brazil nuts and almonds), and whole grains (especially amaranth)

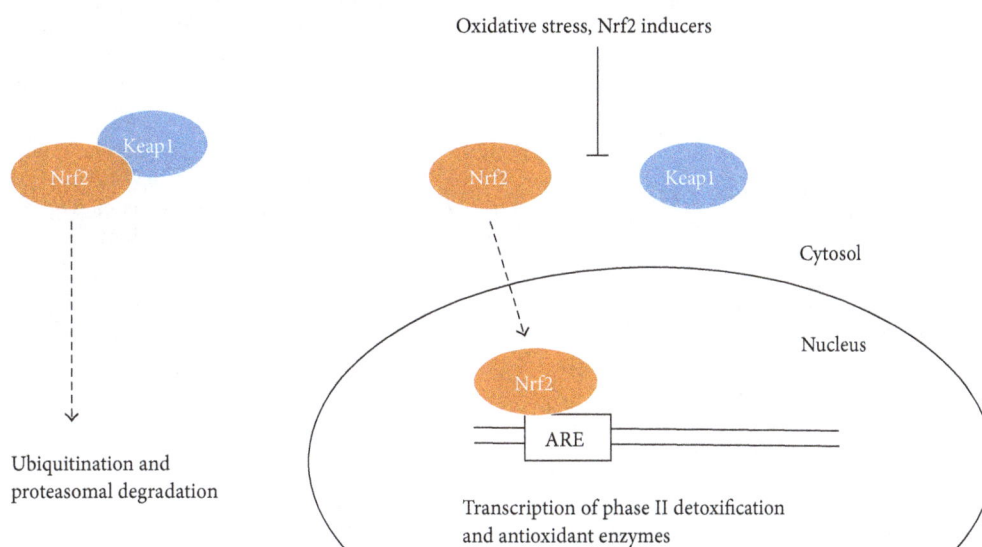

FIGURE 1: Nrf2/Keap1 signaling (created from text in [154]).

induction is considered protective against various oxidative stress-related conditions such as cancer, kidney dysfunction, pulmonary disorders, arthritis, neurological disease, and cardiovascular disease [154].

Research demonstrates that dietary components, especially phytochemicals, not only scavenge reactive oxygen species, thereby acting as direct antioxidants, but also regulate Nrf2 activity [150]. *In vivo* evidence exists for Nrf2-modulation by curcumin [155–158], broccoli constituents [159, 160], garlic [161–163], epicatechins [164–167], resveratrol [168, 169], ginger [170, 171], purple sweet potato [118], isoflavones [172, 173], coffee [174], rosemary [175, 176], blueberry [166, 177], pomegranate [178], naringenin [179], ellagic acid [166], astaxanthin [166], and γ-tocopherol [180] (Table 10(a)). A clinical trial by Magbanua et al. (2011), investigating the Nrf2 modulation effects of fish oil and lycopene in the context of prostate cancer risk, also demonstrated that these dietary compounds can upregulate Nrf2 signaling and response to oxidative stress in humans [181]. Direct comparison of the magnitude of effect between these compounds can be difficult to gauge. Some information on

their relative effects is provided by Kavitha et al. (2013), who ranked the order of potency of the compounds they tested (from highest to lowest) as chlorophyllin (a semisynthetic compound derived from chlorophyll), blueberry, ellagic acid, astaxanthin, and EGCG [166].

Various studies point to the advantageous effects of whole foods, and food combinations, versus specific bioactive compounds. Zhou et al. (2014), for example, illustrate how organosulfur compounds are not the only Nrf2-enhancing bioactive compounds in garlic; garlic carbohydrate derivatives also show Nrf2-modulatory activity [186]. Balstad et al. (2011), in testing the effects of a combination of food extracts on Nrf2 activity in mice, found that the combination produced a larger-than-expected effect, indicating an additive or synergistic effect [176]. By their calculations, the food extract they used equated to a human (70 kg) dose of 14–23 g each of turmeric, rosemary, and thyme, which is clearly not practical for clinical application, as well as 140–233 g each of coffee, red onion, and broccoli. Calabrese et al. (2010) and Houghton et al. (2013) have also argued that Nrf2 inducers exhibit biphasic effects, with lower doses demonstrating stimulatory effects

TABLE 10: (a) *In vivo* example nutrient inducers of the Nrf2 pathway. (b) *In vivo* example nutrient inhibitors of the Nrf2 pathway.

(a)

Enzyme	Food, beverage, or bioactive compounds *Food sources in italics*	Type of study	Dosages used and references
	Fish oil	Clinical	3×1 g/d fish oil containing 1098 mg EPA and 549 mg DHA [181]
	Lycopene *Tomatoes, rose hips, guava, watermelon,* and *papaya* [111]	Clinical	2×15 mg/d lycopene [181]
	Curcumin *Turmeric, curry powder* [34]	*In vivo*	200 mg/kg/d curcumin [155] 75 mg/kg/d curcumin [156] 50 mg/kg/d curcumin [157] 200 mg/kg/d curcumin [158]
	Cruciferous vegetables	*In vivo*	0.5 mg/kg/d sulforaphane [159] Diet of 15% crushed broccoli seed [160]
	Garlic	*In vivo*	50 and 100 mg/kg/d diallyl disulfide [161] 250 mg/kg/d raw garlic [162] 25 mg/kg S-allyl cysteine [163]
	Catechins *Tea* (especially *green tea*), *cocoa, legumes,* and *grapes* [182]	*In vivo*	5, 15, and 45 mg/kg epicatechin [164] 15 mg/kg epicatechin [165] 20 mg/kg Theaphenon E (95% EGCG) [166] 5, 15, and 30 mg/kg epicatechin [167]
	Resveratrol *Grapes, wine, peanuts, soy,* and *itadori tea* [32]	*In vivo*	10 mg/kg/d [168] 20 mg/kg/d [169]
Nrf2	Ginger	*In vivo*	100 mg/kg/d [6]-shogaol [170] 10 and 100 mg/kg dried ginger extract [171]
	Purple sweet potato	*In vivo*	100 and 200 mg/kg anthocyanin extract from purple sweet potato [118]
	Isoflavones *Soy, kudzu root,* and *red clover* [183]	*In vivo*	80 mg/kg/d soy isoflavones [172] 60 and 120 mg/kg puerarin from kudzu root [173]
	Coffee	*In vivo*	2.0 mL/d coffee to an average animal weight of 200 g $\pm$ 10 g [174]
	Rosemary	*In vivo*	50 and 100 mg/kg carnosic acid [175] 5 mg/animal carnosol extract [176]
	Blueberry	*In vivo*	200 mg/kg blueberry [166] 0.6 and 10 g/day [177]
	Pomegranate	*In vivo*	1 and 10 g/kg pomegranate extract [178]: *note high doses used*
	Naringenin *Citrus* [179]	*In vivo*	50 mg/kg/d naringenin [179]
	Ellagic acid *Berries, pomegranate, grapes, walnuts,* and *blackcurrants* [42]	*In vivo*	Diet of 0.4% ellagic acid [166]
	Asthaxanthin *Algae, yeast, salmon, trout, krill, shrimp,* and *crayfish* [38]	*In vivo*	15 mg/kg astaxanthin [166]
	γ-tocopherol *Nuts, seeds, whole grains, vegetable oils,* and *legumes* [111]	*In vivo*	20.8 mg/kg γ-tocopherol [180]

(b)

Enzyme	Food, beverage, or bioactive compounds	Type of study	Dosages used and references
Nrf2	Luteolin	*In vivo*	40 mg/kg luteolin three times per week [184]
	Quercetin	*In vivo*	50 mg/kg/d quercetin [185]

and higher doses exhibiting Nrf2-interference [187, 188]. These data suggest that the doses found in whole foods may be more beneficial than supplements at supraphysiological doses. In fact, it may well be their weak prooxidant effects that stimulate Nrf2 inducers' favorable antioxidant responses [188].

Nonuniform activities of different foods within the same food group should, once again, be considered; in their recent review of the effects of plant-derived compounds on Nrf2 activation, Stefanson and Bakovic (2014) noted that pak choi, via presumed Nrf2 activation, was more effective at reducing inflammation in the colon than broccoli and that broccoli upregulated some additional Nrf2-related antioxidant enzymes compared with pak choi [189]. Interestingly, this effect was only apparent when steamed, rather than cooked, broccoli was used [189], indicating that food preparation may be an important consideration.

Conversely to its role in cancer prevention, overexpression of Nrf2 is found in many cancer cells and has been shown to promote tumor growth and resistance to anticancer therapy [154]. Consequently, the inhibition of Nrf2 signaling may be clinically relevant for patients receiving cancer chemotherapy [184, 185]. Overexpression of Nrf2 and CYP2E1 has also been associated with impaired GLUT4 activity and insulin resistance [68]. As noted above, supplementation (above levels normally consumed through diet) with certain phytochemicals may have inhibitory effects on Nrf2 activation, including luteolin [184] and quercetin [185] (Table 10(b)). Vitamins A, C, and E and N-acetyl cysteine have also been implicated as Nrf2 inhibitors at high doses [188]. These findings point to the need for further research to clarify outcomes as they relate to specific disease states as well as potential biphasic dose effects.

4. Metallothionein

Metallothionein, a cysteine-rich protein with the ability to bind divalent cations, including toxic metals such as mercury, cadmium, lead, and arsenic, is gaining recognition as an important component in heavy metal detoxification [190–192]. Similar to the upregulation of phase II and antioxidant enzymes, metallothionein can be induced at specific promoter regions of genes by stimuli such as heavy metals, oxidative stress, glucocorticoids, and even zinc [192]. In addition to sequestering heavy metals, it is capable of scavenging free radicals and reducing injury from oxidative stress [192], as well as inhibiting NF-κB signaling [193].

Dietary patterns and nutrients may result in changes in metallothionein production. Lamb et al. (2011) reported a 54% increase in metallothionein mRNA production in a small clinical trial in women with fibromyalgia following an elimination diet in conjunction with a phytonutrient-rich medical food consisting of hops, pomegranate, prune skin, and watercress [194]. Zinc supplementation (15 mg/day) to healthy men over 10 days led to significantly increased metallothionein mRNA, up to 2-fold in leukocytes and up to 4-fold from dried blood spots [195]. Metallothionein has been shown to be decreased in the intestinal mucosa of patients with inflammatory bowel disease (IBD); however, zinc supplementation (300 mg zinc aspartate, equal to 60 mg elemental zinc per day for 4 weeks) in 14 zinc-deficient patients with IBD resulted in slightly higher metallothionein concentration in the intestinal mucosa [196]. Cruciferous phytonutrients may also modulate metallothionein expression, as suggested by a 10-fold increase following a single oral dose of 50 μmol sulforaphane to rats [197]. Chromium may *inhibit* zinc-induced metallothionein expression, according to animal studies by Kimura et al. (2011) [198]. Early-stage, *in vitro* studies also suggest that quercetin and *Cordyceps sinensis*, a mushroom native to the Himalayan region, may upregulate metallothionein expression [199, 200].

5. Clinical Applications

With the continued emergence of data supporting the role of toxins in chronic disease processes, it is becoming increasingly necessary for clinicians to understand how to provide therapeutic modalities to reduce toxin load in patients. In this paper, several studies regarding the influence of foods and food-based nutrients on the systems of detoxification were presented. From the current information presented, listed below are some key concepts for translation into the clinical setting.

5.1. Nonclinical versus Clinical Studies. One of the limitations that comes to the forefront in this collection of studies is how the information, in many cases, is constrained primarily to studies in cells or animals. It remains questionable as to whether similar effects would be seen in humans at moderate, reasonable doses. In the cell studies, it is difficult to anticipate findings due to the lack of pleiotropic activity that occurs in a complex, living system with multiple detoxification systems working simultaneously. Along similar lines, animal studies are often difficult to extrapolate to individuals due to the degree of variability in genotype and environmental phenotype seen in the diverse human population. Therefore, at this time, it is best to take precaution in firmly advocating foods or food-based nutrients that only have cell or animal data as support. It is best to rely on the clinical studies that have been published to date in making more firm recommendations.

5.2. Single Agent versus Lifestyle. While this paper focuses on isolated nutrients and foods that contain those nutrients, it might be optimal from a clinical perspective to consider how an entire lifestyle might induce or inhibit the array of detoxification enzymes. For example, this paper has not addressed behaviors like smoking, physical activity, or stress. The modern clinician needs to weigh all these variables against each other. Yet, science has not fully demonstrated the individual impacts of these factors, along with all of them together. Therefore, at this time, a dietary pattern favoring whole, unprocessed, plant-based foods and the removal or reduction of toxic substances in one's environment is a two-prong approach that would seem to have the best overarching scientific underpinning.

5.3. Modulating versus Inhibiting/Inducing Effects. In several instances, certain foods exhibited a particular activity on an enzyme, while, at higher doses, they had another, opposite effect. Essentially, many foods serve as what is commonly referred to as being "bifunctional modulators," possessing the ability to effectively induce or inhibit detoxification enzyme activity based on the dose response. Therefore, the resulting clinical takeaway might be to encourage patients to follow a mixed, varied diet, full of different plant-based, whole foods. Smaller amounts of many compounds might be more therapeutic and supportive for biochemical pathways rather than overriding signals derived from high concentrations of nutrients through high-dose supplementation or the repeat, daily ingestion of large quantities of the same food.

5.4. Polypharmacy. For patients who are taking multiple pharmaceuticals, it is important to know which detoxification systems will be influenced by nutrients and foods so that side effects are minimized or avoided.

5.5. Dietary Supplements versus Foods. Since there can be potent effects of food-based nutrients on detoxification pathways, it would be best for the average patient to follow, as indicated above, a mixed, complex, and whole-foods diet. Additionally, dietary supplements may be a helpful adjunct in patients in which the practitioner has information about the patient's genetic variability, so that nutrients can be tailored accordingly. Without a full understanding of a patient's SNPs (single nucleotide polymorphisms), it becomes difficult to make accurate assessments about nutrients and dosing.

5.6. Duration of Dosing. Another factor to consider in therapeutic intervention is the timing and duration of the dose of nutrient or the food. In some of the research presented here, effects on detoxification enzymes were seen after several days of food intake or supplementation, while, in other cases, induction of an enzyme might be fairly rapid, followed by efficient adaptability. This variable needs to be considered in further clinical research and requires close monitoring in clinical practice.

5.7. Foods Known to Impact Detoxification. Based on the four systems examined in this paper, there are several foods which seem to have demonstrated an influence on detoxification systems. Many of them have been acknowledged as part of naturopathic medicine. Hence, it would be useful to have a knowledge base of this cumulative set of foods as patients embark upon detoxification protocols. This recent scientific update notes clinical evidence of effects from cruciferous vegetables (in combination, and specifically watercress, garden cress, and broccoli), allium vegetables, apiaceous vegetables, grapefruit, resveratrol, fish oil, quercetin, daidzein, and lycopene. Many other foods, beverages, and nutrient bioactive compounds, based on this review of scientific literature, are also suggested as modulators of detoxification enzymes *in vivo* (Table 11).

TABLE 11: Food, beverages, and bioactive compounds with demonstrated, or potential, clinical impact on detoxification systems.

Food or beverage	Nutrient bioactive compounds
Allium vegetables	Astaxanthin
Apiaceous vegetables	Caffeic acid
Black raspberry	Catechins (*including EGCG*)
Black tea	Chrysin
Blueberry	Curcumin
Chamomile tea	Daidzein
Chicory root	Ellagic acid
Citrus	Ferulic acid
Coffee	Fish oil
Cruciferous vegetables (*with potential for distinct effects of different crucifers*)	Genistein
	Luteolin
	Lycopene
Dandelion tea	MCTs
Garlic	Myricetin
Ghee	N-acetyl cysteine
Ginger	Naringenin
Grapefruit	Quercetin
Green tea	Resveratrol
Honeybush tea	Retinoic acid (*vitamin A*)
Peppermint tea	
Pomegranate	
Purple sweet potato	
Rooibos tea	
Rosemary	
Soybean/black soybean	
Turmeric	

6. Conclusions

Over the past decade, there has been investigation into nutrigenomic and epigenetic influences of food constituents on chronic diseases [201, 202]. Similarly, studies have revealed that exposure to and accumulation of toxins play a significant role in cardiovascular disease, type 2 diabetes, and obesity [203–207]. Thus, one's dietary intake and environmental influences may have large bearing on the incidence of chronic disease. In fact, these influences may be significant not just for the individual, but for several generations due to the transgenerational inheritance of epigenetic changes [208, 209]. Therefore, it would seem that designing clinical recommendations to maximize the effects of food and reduce the impact of toxins is essential. However, it is not without caution and critical thinking that a detoxification protocol should be assembled for patients by trained clinicians. There remain many unresolved issues regarding knowing how and what foods modulate detoxification pathways.

Conflict of Interests

The authors declare that there is no conflict of interests regarding the publication of this paper.

Authors' Contribution

All authors read and approved the final version of the paper.

References

[1] W. Baer-Dubowska and H. Szaefer, "Modulation of carcinogen-metabolizing cytochromes P450 by phytochemicals in humans," *Expert Opinion on Drug Metabolism and Toxicology*, vol. 9, no. 8, pp. 927–941, 2013.

[2] H. Steinkellner, S. Rabot, C. Freywald et al., "Effects of cruciferous vegetables and their constituents on drug metabolizing enzymes involved in the bioactivation of DNA-reactive dietary carcinogens," *Mutation Research*, vol. 480-481, pp. 285–297, 2001.

[3] Y. J. Moon, X. Wang, and M. E. Morris, "Dietary flavonoids: effects on xenobiotic and carcinogen metabolism," *Toxicology in Vitro*, vol. 20, no. 2, pp. 187–210, 2006.

[4] N. Hakooz and I. Hamdan, "Effects of dietary broccoli on human in vivo caffeine metabolism: a pilot study on a group of Jordanian volunteers," *Current Drug Metabolism*, vol. 8, no. 1, pp. 9–15, 2007.

[5] D. James, S. Devaraj, P. Bellur, S. Lakkanna, J. Vicini, and S. Boddupalli, "Novel concepts of broccoli sulforaphanes and disease: induction of phase II antioxidant and detoxification enzymes by enhanced-glucoraphanin broccoli," *Nutrition Reviews*, vol. 70, no. 11, pp. 654–665, 2012.

[6] H. S. Aiyer and R. C. Gupta, "Berries and ellagic acid prevent estrogen-induced mammary tumorigenesis by modulating enzymes of estrogen metabolism," *Cancer Prevention Research*, vol. 3, no. 6, pp. 727–737, 2010.

[7] Bogacz A, P. Ł. Mikołajczak, P. Ł. Mikołajczak et al., "The influence of soybean extract on the expression level of selected drug transporters, transcription factors and cytochrome P450 genes encoding phase I drug-metabolizing enzymes," *Ginekologia Polska*, vol. 85, no. 5, pp. 348–353, 2014.

[8] D. M. Davenport and M. J. Wargovich, "Modulation of cytochrome P450 enzymes by organosulfur compounds from garlic," *Food and Chemical Toxicology*, vol. 43, no. 12, pp. 1753–1762, 2005.

[9] C. K. Lii, C. W. Tsai, and C. C. Wu, "Garlic allyl sulfides display differential modulation of rat cytochrome P450 2B1 and the placental form glutathione S-transferase in various organs," *Journal of Agricultural and Food Chemistry*, vol. 54, no. 14, pp. 5191–5196, 2006.

[10] C. M. Kaefer and J. A. Milner, "The role of herbs and spices in cancer prevention," *Journal of Nutritional Biochemistry*, vol. 19, no. 6, pp. 347–361, 2008.

[11] Y. W. Hsieh, C. Y. Huang, S. Y. Yang et al., "Oral intake of curcumin markedly activated CYP 3A4: in vivo and ex-vivo studies," *Scientific Reports*, vol. 4, article 6587, 2014.

[12] M. Murray and J. Pizzorno, *Encyclopedia of Natural Medicine*, Prima Publishing, Rocklin, Calif, USA, 2nd edition, 1998.

[13] Institute for Functional Medicine, *Textbook of Functional Medicine*, Johnston Printing, Boulder, Colo, USA, 2006.

[14] V. Ullrich, "Cytochrome P450 and biological hydroxylation reactions," *Topics in Current Chemistry*, vol. 83, pp. 67–104, 1979.

[15] P. B. Danielson, "The cytochrome P450 superfamily: biochemistry, evolution and drug metabolism in humans," *Current Drug Metabolism*, vol. 3, no. 6, pp. 561–597, 2002.

[16] A. J. Paine, "Hepatic cytochrome P-450," *Essays in Biochemistry*, vol. 17, pp. 85–126, 1981.

[17] Q. Chen, T. Zhang, J. F. Wang, and D. Q. Wei, "Advances in human cytochrome P450 and personalized medicine," *Current Drug Metabolism*, vol. 12, no. 5, pp. 436–444, 2011.

[18] Q. Ma and A. Y. H. Lu, "CYP1A induction and human risk assessment: an evolving tale of in vitro and in vivo studies," *Drug Metabolism and Disposition*, vol. 35, no. 7, pp. 1009–1016, 2007.

[19] M. O. James, J. C. Sacco, and L. R. Faux, "Effects of food natural products on the biotransformation of PCBs," *Environmental Toxicology and Pharmacology*, vol. 25, no. 2, pp. 211–217, 2008.

[20] K. Vistisen, S. Loft, J. H. Olsen et al., "Low CYP1A2 activity associated with testicular cancer," *Carcinogenesis*, vol. 25, no. 6, pp. 923–929, 2004.

[21] N. Božina, V. Bradamante, and M. Lovrić, "Genetic polymorphism of metabolic enzymes P450 (CYP) as a susceptibility factor for drug response, toxicity, and cancer risk," *Arhiv za Higijenu Rada i Toksikologiju*, vol. 60, no. 2, pp. 217–242, 2009.

[22] Y. Tsuchiya, M. Nakajima, and T. Yokoi, "Cytochrome P450-mediated metabolism of estrogens and its regulation in human," *Cancer Letters*, vol. 227, no. 2, pp. 115–124, 2005.

[23] J. J. Michnovicz and H. L. Bradlow, "Induction of estradiol metabolism by dietary indole-3-carbinol in humans," *Journal of the National Cancer Institute*, vol. 82, no. 11, pp. 947–949, 1990.

[24] S. Peterson, Y. Schwarz, S. S. Li et al., "CYP1A2, GSTM1, and GSTT1 polymorphisms and diet effects on CYP1A2 activity in a crossover feeding trial," *Cancer Epidemiology Biomarkers and Prevention*, vol. 18, no. 11, pp. 3118–3125, 2009.

[25] D. G. Walters, P. J. Young, C. Agus et al., "Cruciferous vegetable consumption alters the metabolism of the dietary carcinogen 2-amino-1-methyl-6-phenylimidazo[4,5-b]pyridine (PhIP) in humans," *Carcinogenesis*, vol. 25, no. 9, pp. 1659–1669, 2004.

[26] M. A. Kall, O. Vang, and J. Clausen, "Effects of dietary broccoli on human in vivo drug metabolizing enzymes: evaluation of caffeine, oestrone and chlorzoxazone metabolism," *Carcinogenesis*, vol. 17, no. 4, pp. 793–799, 1996.

[27] T. L. Horn, M. A. Reichert, R. L. Bliss, and D. Malejka-Giganti, "Modulations of P450 mRNA in liver and mammary gland and P450 activities and metabolism of estrogen in liver by treatment of rats with indole-3-carbinol," *Biochemical Pharmacology*, vol. 64, no. 3, pp. 393–404, 2002.

[28] H. H. S. Chow, L. L. Garland, C. H. Hsu et al., "Resveratrol modulates drug- and carcinogen-metabolizing enzymes in a healthy volunteer study," *Cancer Prevention Research*, vol. 3, no. 9, pp. 1168–1175, 2010.

[29] Y. Chen, P. Xiao, D. S. Ou-Yang et al., "Simultaneous action of the flavonoid quercetin on cytochrome p450 (cyp) 1a2, cyp2a6, n-acetyltransferase and xanthine oxidase activity in healthy volunteers," *Clinical and Experimental Pharmacology and Physiology*, vol. 36, no. 8, pp. 828–833, 2009.

[30] R. S. Lord, B. Bongiovanni, and J. A. Bralley, "Estrogen metabolism and the diet-cancer connection: rationale for assessing the ratio of urinary hydroxylated estrogen metabolites," *Alternative Medicine Review*, vol. 7, no. 2, pp. 112–129, 2002.

[31] H. Takemura, H. Sakakibara, S. Yamazaki, and K. Shimoi, "Breast cancer and flavonoids—a role in prevention," *Current Pharmaceutical Design*, vol. 19, no. 34, pp. 6125–6132, 2013.

[32] J. Burns, T. Yokota, H. Ashihara, M. E. J. Lean, and A. Crozier, "Plant foods and herbal sources of resveratrol," *Journal of Agricultural and Food Chemistry*, vol. 50, no. 11, pp. 3337–3340, 2002.

[33] H. T. Yao, Y. R. Hsu, C. K. Lii, A. H. Lin, K. H. Chang, and H. T. Yang, "Effect of commercially available green and black tea beverages on drug-metabolizing enzymes and oxidative stress in Wistar rats," *Food and Chemical Toxicology*, vol. 70, pp. 120–127, 2014.

[34] R. F. Tayyem, D. D. Heath, W. K. Al-Delaimy, and C. L. Rock, "Curcumin content of turmeric and curry powders," *Nutrition and Cancer*, vol. 55, no. 2, pp. 126–131, 2006.

[35] S. S. Bansal, H. Kausar, M. V. Vadhanam et al., "Curcumin implants, not curcumin diet, inhibit estrogen-induced mammary carcinogenesis in ACI rats," *Cancer Prevention Research*, vol. 7, no. 4, pp. 456–465, 2014.

[36] H. W. Chen, C. W. Tsai, J. J. Yang, C. T. Liu, W. W. Kuo, and C. K. Lii, "The combined effects of garlic oil and fish oil on the hepatic antioxidant and drug-metabolizing enzymes of rats," *British Journal of Nutrition*, vol. 89, no. 2, pp. 189–200, 2003.

[37] P. Debersac, J. M. Heydel, M. J. Amiot et al., "Induction of cytochrome P450 and/or detoxication enzymes by various extracts of rosemary: Description of specific patterns," *Food and Chemical Toxicology*, vol. 39, no. 9, pp. 907–918, 2001.

[38] R. R. Ambati, P. S. Moi, S. Ravi, and R. G. Aswathanarayana, "Astaxanthin: sources, extraction, stability, biological activities and its commercial applications—a review," *Marine Drugs*, vol. 12, no. 1, pp. 128–152, 2014.

[39] S. Gradelet, P. Astorg, J. Leclerc, J. Chevalier, M.-F. Vernevaut, and M.-H. Siess, "Effects of canthaxanthin, astaxanthin, lycopene and lutein on liver xenobiotic-metabolizing enzymes in the rat," *Xenobiotica*, vol. 26, no. 1, pp. 49–63, 1996.

[40] A. Bu-Abbas, M. N. Clifford, R. Walker, and C. Ioannides, "Selective induction of rat hepatic CYP1 and CYP4 proteins and of peroxisomal proliferation by green tea," *Carcinogenesis*, vol. 15, no. 11, pp. 2575–2579, 1994.

[41] M. K. Rasmussen, C. Brunius, G. Zamaratskaia, and B. Ekstrand, "Feeding dried chicory root to pigs decrease androstenone accumulation in fat by increasing hepatic 3β hydroxysteroid dehydrogenase expression," *Journal of Steroid Biochemistry and Molecular Biology*, vol. 130, no. 1-2, pp. 90–95, 2012.

[42] C. Usta, S. Ozdemir, M. Schiariti, and P. E. Puddu, "The pharmacological use of ellagic acid-rich pomegranate fruit," *International Journal of Food Sciences and Nutrition*, vol. 64, no. 7, pp. 907–913, 2013.

[43] G. Celik, A. Semiz, S. Karakurt, S. Arslan, O. Adali, and A. Sen, "A comparative study for the evaluation of two doses of ellagic acid on hepatic drug metabolizing and antioxidant enzymes in the rat," *BioMed Research International*, vol. 2013, Article ID 358945, 9 pages, 2013.

[44] T. Zhang, S. Jiang, C. He, Y. Kimura, Y. Yamashita, and H. Ashida, "Black soybean seed coat polyphenols prevent B(a)P-induced DNA damage through modulating drug-metabolizing enzymes in HepG2 cells and ICR mice," *Mutation Research*, vol. 752, no. 1-2, pp. 34–41, 2013.

[45] F. Catterall, N. J. McArdle, L. Mitchell, A. Papayanni, M. N. Clifford, and C. Ioannides, "Hepatic and intestinal cytochrome P450 and conjugase activities in rats treated with black tea theafulvins and theaflavins," *Food and Chemical Toxicology*, vol. 41, no. 8, pp. 1141–1147, 2003.

[46] R. Thapliyal and G. B. Maru, "Inhibition of cytochrome P450 isozymes by curcumins in vitro and in vivo," *Food and Chemical Toxicology*, vol. 39, no. 6, pp. 541–547, 2001.

[47] L. Sampson, E. Rimm, P. C. H. Hollman, J. H. M. de Vries, and M. B. Katan, "Flavonol and flavone intakes in US health professionals," *Journal of the American Dietetic Association*, vol. 102, no. 10, pp. 1414–1420, 2002.

[48] M. G. L. Hertog, E. J. M. Feskens, P. C. H. Hollman et al., "Content of potentially anticarcinogenic flavonoids of 28 vegetables and 9 fruits commonly consumed in the Netherlands," *Journal of Agricultural and Food Chemistry*, vol. 40, no. 12, pp. 2379–2383, 1992.

[49] W. X. Peng, H. D. Li, and H. H. Zhou, "Effect of daidzein on CYP1A2 activity and pharmacokinetics of theophylline in healthy volunteers," *European Journal of Clinical Pharmacology*, vol. 59, no. 3, pp. 237–241, 2003.

[50] U. Fuhr, K. Klittich, and A. H. Staib, "Inhibitory effect of grapefruit juice and its bitter principal, naringenin, on CYP1A2 dependent metabolism of caffeine in man," *British Journal of Clinical Pharmacology*, vol. 35, no. 4, pp. 431–436, 1993.

[51] I. Yamasaki, M. Yamada, N. Uotsu, S. Teramoto, R. Takayanagi, and Y. Yamada, "Inhibitory effects of kale ingestion on metabolism by cytochrome P450 enzymes in rats," *Biomedical Research*, vol. 33, no. 4, pp. 235–242, 2012.

[52] T. Zeng, C. L. Zhang, F. Y. Song, X. Y. Han, and K. Q. Xie, "The modulatory effects of garlic oil on hepatic cytochrome P450s in mice," *Human and Experimental Toxicology*, vol. 28, no. 12, pp. 777–783, 2009.

[53] P. P. Maliakal and S. Wanwimolruk, "Effect of herbal teas on hepatic drug metabolizing enzymes in rats," *Journal of Pharmacy and Pharmacology*, vol. 53, no. 10, pp. 1323–1329, 2001.

[54] A. U. Nissar, M. R. Farrukh, P. J. Kaiser et al., "Effect of N-acetyl cysteine (NAC), an organosulfur compound from Allium plants, on experimentally induced hepatic prefibrogenic events in wistar rat," *Phytomedicine*, vol. 20, no. 10, pp. 828–833, 2013.

[55] F. I. Al-Jenoobi, A. A. Al-Thukair, M. A. Alam et al., "Effect of garden cress seeds powder and its alcoholic extract on the metabolic activity of CYP2D6 and CYP3A4," *Evidence-Based Complementary and Alternative Medicine*, vol. 2014, Article ID 634592, 6 pages, 2014.

[56] D. Park, J. H. Jeon, S. Shin et al., "Green tea extract increases cyclophosphamide-induced teratogenesis by modulating the expression of cytochrome P-450 mRNA," *Reproductive Toxicology*, vol. 27, no. 1, pp. 79–84, 2009.

[57] V. Yoxall, P. Kentish, N. Coldham, N. Kuhnert, M. J. Sauer, and C. Ioannides, "Modulation of hepatic cytochromes P450 and phase II enzymes by dietary doses of sulforaphane in rats: implications for its chemopreventive activity," *International Journal of Cancer*, vol. 117, no. 3, pp. 356–362, 2005.

[58] C. Li, S. C. Lim, J. Kim, and J. S. Choi, "Effects of myricetin, an anticancer compound, on the bioavailability and pharmacokinetics of tamoxifen and its main metabolite, 4-hydroxytamoxifen, in rats," *European Journal of Drug Metabolism and Pharmacokinetics*, vol. 36, no. 3, pp. 175–182, 2011.

[59] I. Leclercq, J. P. Desager, and Y. Horsmans, "Inhibition of chlorzoxazone metabolism, a clinical probe for CYP2E1, by a single ingestion of watercress," *Clinical Pharmacology and Therapeutics*, vol. 64, no. 2, pp. 144–149, 1998.

[60] G. D. Loizou and J. Cocker, "The effects of alcohol and diallyl sulphide on CYP2E1 activity in humans: a phenotyping study

using chlorzoxazone," *Human and Experimental Toxicology*, vol. 20, no. 7, pp. 321–327, 2001.

[61] K. A. Park, S. Kweon, and H. Choi, "Anticarcinogenic effect and modification of cytochrome P450 2E1 by dietary garlic powder in diethylnitrosamine-initiated rat hepatocarcinogenesis," *Journal of Biochemistry and Molecular Biology*, vol. 35, no. 6, pp. 615–622, 2002.

[62] C. M. Park, Y. S. Cha, H. J. Youn, C. W. Cho, and Y. S. Song, "Amelioration of oxidative stress by dandelion extract through CYP2E1 suppression against acute liver injury induced by carbon tetrachloride in sprague-dawley rats," *Phytotherapy Research*, vol. 24, no. 9, pp. 1347–1353, 2010.

[63] M. Tahir and S. Sultana, "Chrysin modulates ethanol metabolism in Wistar rats: a promising role against organ toxicities," *Alcohol and Alcoholism*, vol. 46, no. 4, Article ID agr038, pp. 383–392, 2011.

[64] C. S. Lieber, Q. Cao, L. M. Decarli et al., "Role of medium-chain triglycerides in the alcohol-mediated cytochrome P450 2E1 induction of mitochondria," *Alcoholism: Clinical and Experimental Research*, vol. 31, no. 10, pp. 1660–1668, 2007.

[65] S. A. Sheweita, "Drug-metabolizing enzymes: mechanisms and functions," *Current Drug Metabolism*, vol. 1, no. 2, pp. 107–132, 2000.

[66] N. K. Zgheib, Z. Mitri, E. Geryess, and P. Noutsi, "Cytochrome P4502E1 (CYP2E1) genetic polymorphisms in a Lebanese population: frequency distribution and association with morbid diseases," *Genetic Testing and Molecular Biomarkers*, vol. 14, no. 3, pp. 393–397, 2010.

[67] C. A. González, N. Sala, and G. Capellá, "Genetic susceptibility and gastric cancer risk," *International Journal of Cancer*, vol. 100, no. 3, pp. 249–260, 2002.

[68] M. Armoni, C. Harel, M. Ramdas, and E. Karnieli, "CYP2E1 impairs GLUT4 gene expression and function: NRF2 as a possible mediator," *Hormone and Metabolic Research*, vol. 46, no. 7, pp. 477–483, 2014.

[69] K. Matsuda, Y. Nishimura, N. Kurata, M. Iwase, and H. Yasuhara, "Effects of continuous ingestion of herbal teas on intestinal CYP3A in the rat," *Journal of Pharmacological Sciences*, vol. 103, no. 2, pp. 214–221, 2007.

[70] C. C. Wu, L. Y. Sheen, H. W. Chen, W. W. Kuo, S. J. Tsai, and C. K. Lii, "Differential effects of garlic oil and its three major organosulfur components on the hepatic detoxification system in rats," *Journal of Agricultural and Food Chemistry*, vol. 50, no. 2, pp. 378–383, 2002.

[71] S. Misaka, K. Kawabe, S. Onoue et al., "Green tea extract affects the cytochrome P450 3A activity and pharmacokinetics of simvastatin in rats," *Drug Metabolism and Pharmacokinetics*, vol. 28, no. 6, pp. 514–518, 2013.

[72] S. N. Umathe, P. V. Dixit, V. Kumar, K. U. Bansod, and M. M. Wanjari, "Quercetin pretreatment increases the bioavailability of pioglitazone in rats: involvement of CYP3A inhibition," *Biochemical Pharmacology*, vol. 75, no. 8, pp. 1670–1676, 2008.

[73] J. Liu, G. J. Tawa, and A. Wallqvist, "Identifying cytochrome P450 functional networks and their allosteric regulatory elements," *PLoS ONE*, vol. 8, no. 12, Article ID e81980, 2013.

[74] S. Tanaka, S. Uchida, S. Miyakawa et al., "Comparison of inhibitory duration of grapefruit juice on organic anion-transporting polypeptide and cytochrome P450 3A4," *Biological and Pharmaceutical Bulletin*, vol. 36, no. 12, pp. 1936–1941, 2013.

[75] D. A. Leibelt, O. R. Hedstrom, K. A. Fisher, C. B. Pereira, and D. E. Williams, "Evaluation of chronic dietary exposure to indole-3-carbinol and absorption-enhanced 3,3′-diidolylmethane in Sprague-Dawley rats," *Toxicological Sciences*, vol. 74, no. 1, pp. 10–21, 2003.

[76] Linus Pauling Institute, *Garlic and Organosulfur Compounds*, Micronutrient Information Center, Corvallis, Ore, USA, 2008, http://lpi.oregonstate.edu/infocenter/phytochemicals/garlic/.

[77] C. Ioannides, "Effect of diet and nutrition on the expression of cytochromes P450," *Xenobiotica*, vol. 29, no. 2, pp. 109–154, 1999.

[78] B. Baer and A. Rettie, "CYP4B1: an enigmatic P450 at the interface between xenobiotic and endobiotic metabolism," *Drug Metabolism Reviews*, vol. 38, no. 3, pp. 451–476, 2006.

[79] Z. Ye, Z. Liu, A. Henderson et al., "Increased CYP4B1 mRNA is associated with the inhibition of dextran sulfate sodium-induced colitis by caffeic acid in mice," *Experimental Biology and Medicine (Maywood)*, vol. 234, no. 6, pp. 606–616, 2009.

[80] S. Lafay, C. Morand, C. Manach, C. Besson, and A. Scalbert, "Absorption and metabolism of caffeic acid and chlorogenic acid in the small intestine of rats," *British Journal of Nutrition*, vol. 96, no. 1, pp. 39–46, 2006.

[81] C. Xu, C. Y. Li, and A. T. Kong, "Induction of phase I, II and III drug metabolism/transport by xenobiotics," *Archives of Pharmacal Research*, vol. 28, no. 3, pp. 249–268, 2005.

[82] G. Ginsberg, K. Guyton, D. Johns, J. Schimek, K. Angle, and B. Sonawane, "Genetic polymorphism in metabolism and host defense enzymes: implications for human health risk assessment," *Critical Reviews in Toxicology*, vol. 40, no. 7, pp. 575–619, 2010.

[83] P. Jancova, P. Anzenbacher, and E. Anzenbacherova, "Phase II drug metabolizing enzymes," *Biomedical Papers*, vol. 154, no. 2, pp. 103–116, 2010.

[84] C. Jenkinson, A. Petroczi, and D. P. Naughton, "Effects of dietary components on testosterone metabolism via UDP-glucuronosyltransferase," *Frontiers in Endocrinology*, vol. 4, article 80, 2013.

[85] J. L. Chang, J. Bigler, Y. Schwarz et al., "UGT1A1 polymorphism is associated with serum bilirubin concentrations in a randomized, controlled, fruit and vegetable feeding trial," *Journal of Nutrition*, vol. 137, no. 4, pp. 890–897, 2007.

[86] A. Rowland, J. O. Miners, and P. I. Mackenzie, "The UDP-glucuronosyltransferases: their role in drug metabolism and detoxification," *International Journal of Biochemistry and Cell Biology*, vol. 45, no. 6, pp. 1121–1132, 2013.

[87] C. P. Strassburg, S. Kneip, J. Topp et al., "Polymorphic gene regulation and interindividual variation of UDP-glucuronosyltransferase activity in human small intestine," *The Journal of Biological Chemistry*, vol. 275, no. 46, pp. 36164–36171, 2000.

[88] P. G. Wells, P. I. Mackenzie, J. R. Chowdhury et al., "Glucuronidation and the UDP-glucuronosyltransferases in health and disease," *Drug Metabolism and Disposition*, vol. 32, no. 3, pp. 281–290, 2004.

[89] J. W. Lampe, "Interindividual differences in response to plant-based diets: Implications for cancer risk," *The American Journal of Clinical Nutrition*, vol. 89, no. 5, pp. 1553S–1557S, 2009.

[90] S. L. Navarro, S. Peterson, C. Chen et al., "Cruciferous vegetable feeding alters UGT1A1 activity: diet- and genotype-dependent changes in serum bilirubin in a controlled feeding trial," *Cancer Prevention Research (Phila)*, vol. 2, no. 4, pp. 345–352, 2009.

[91] S. S. Hecht, S. G. Carmella, and S. E. Murphy, "Effects of watercress consumption on urinary metabolites of nicotine in smokers," *Cancer Epidemiology Biomarkers and Prevention*, vol. 8, no. 10, pp. 907–913, 1999.

[92] M. R. Saracino, J. Bigler, Y. Schwarz et al., "Citrus fruit intake is associated with lower serum bilirubin concentration among women with the UGT1A1*28 polymorphism," *Journal of Nutrition*, vol. 139, no. 3, pp. 555–560, 2009.

[93] J. L. Marnewick, E. Joubert, P. Swart, F. van der Westhuizen, and W. C. Gelderblom, "Modulation of hepatic drug metabolizing enzymes and oxidative status by rooibos (*Aspalathus linearis*) and Honeybush (*Cyclopia intermedia*), green and black (*Camellia sinensis*) teas in rats," *Journal of Agricultural and Food Chemistry*, vol. 51, no. 27, pp. 8113–8119, 2003.

[94] A. Marahatta, B. Bhandary, S.-K. Jeong, H.-R. Kim, and H.-J. Chae, "Soybean greatly reduces valproic acid plasma concentrations: a food-drug interaction study," *Scientific Reports*, vol. 4, article 4362, 2014.

[95] E. M. J. van der Logt, H. M. J. Roelofs, F. M. Nagengast, and W. H. M. Peters, "Induction of rat hepatic and intestinal UDP-glucuronosyltransferases by naturally occurring dietary anticarcinogens," *Carcinogenesis*, vol. 24, no. 10, pp. 1651–1656, 2003.

[96] E. Graf, "Antioxidant potential of ferulic acid," *Free Radical Biology and Medicine*, vol. 13, no. 4, pp. 435–448, 1992.

[97] C. B. Simone II, N. L. Simone, M. Pallante, and C. B. Simone, "Cancer, lifestyle modification and glucarate," *Journal of Orthomolecular Medicine*, vol. 16, no. 2, pp. 83–90, 2001.

[98] R. Zółtaszek, M. Hanausek, Z. M. Kiliańska, and Z. Walaszek, "The biological role of D-glucaric acid and its derivatives: potential use in medicine," *Postpy Higieny i Medycyny Doświadczalnej*, vol. 62, pp. 451–462, 2008.

[99] C. Dwivedi, W. J. Heck, A. A. Downie, S. Larroya, and T. E. Webb, "Effect of calcium glucarate on beta-glucuronidase activity and glucarate content of certain vegetable and fruits," *Biochemical Medicine and Metabolic Biology*, vol. 43, no. 2, pp. 83–92, 1990.

[100] M. Kosmala, Z. Zduńczyk, K. Kołodziejczyk, E. Klimczak, J. Jukiewicz, and P. Zduńczyk, "Chemical composition of polyphenols extracted from strawberry pomace and their effect on physiological properties of diets supplemented with different types of dietary fibre in rats," *European Journal of Nutrition*, vol. 53, no. 2, pp. 521–532, 2014.

[101] S. S. Maruti, J. L. Chang, J. A. Prunty et al., "Serum β-glucuronidase activity in response to fruit and vegetable supplementation: a controlled feeding study," *Cancer Epidemiology Biomarkers and Prevention*, vol. 17, no. 7, pp. 1808–1812, 2008.

[102] A. Jurgoński, J. Juśkiewicz, Z. Zduńczyk, P. Matusevicius, and K. Kołodziejczyk, "Polyphenol-rich extract from blackcurrant pomace attenuates the intestinal tract and serum lipid changes induced by a high-fat diet in rabbits," *European Journal of Nutrition*, vol. 53, no. 8, pp. 1603–1613, 2014.

[103] M. O. James and S. Ambadapadi, "Interactions of cytosolic sulfotransferases with xenobiotics," *Drug Metabolism Reviews*, vol. 45, no. 4, pp. 401–414, 2013.

[104] S. Kodama and M. Negishi, "Sulfotransferase genes: regulation by nuclear receptors in response to xeno/endo-biotics," *Drug Metabolism Reviews*, vol. 45, no. 4, pp. 441–449, 2013.

[105] L.-Q. Wang and M. O. James, "Inhibition of sulfotransferases by xenobiotics," *Current Drug Metabolism*, vol. 7, no. 1, pp. 83–104, 2006.

[106] D. Ung and S. Nagar, "Variable sulfation of dietary polyphenols by recombinant human sulfotransferase (SULT) 1A1 genetic variants and SULT1E1," *Drug Metabolism and Disposition*, vol. 35, no. 5, pp. 740–746, 2007.

[107] T. Zhou, Y. Chen, C. Huang, and G. Chen, "Caffeine induction of sulfotransferases in rat liver and intestine," *Journal of Applied Toxicology*, vol. 32, no. 10, pp. 804–809, 2012.

[108] S. Maiti, X. Chen, and G. Chen, "All-trans retinoic acid induction of sulfotransferases," *Basic and Clinical Pharmacology and Toxicology*, vol. 96, no. 1, pp. 44–53, 2005.

[109] K. Kamio, K. Honke, and A. Makita, "Pyridoxal 5'-phosphate binds to a lysine residue in the adenosine 3'-phosphate 5'-phosphosulfate recognition site of glycolipid sulfotransferase from human renal cancer cells," *Glycoconjugate Journal*, vol. 12, no. 6, pp. 762–766, 1995.

[110] M. Masters and R. A. McCance, "The sulfur content of foods," *Biochemical Journal*, vol. 33, no. 8, pp. 1304–1312, 1939.

[111] USDA National Nutrient Database for Standard Reference, *Nutrient Data Laboratory. Release 27*, Agriculture Research Service, Washington, DC, USA, 2011, http://ndb.nal.usda.gov/ndb/.

[112] S. A. McFadden, "Phenotypic variation in xenobiotic metabolism and adverse environmental response: focus on sulfur-dependent detoxification pathways," *Toxicology*, vol. 111, no. 1–3, pp. 43–65, 1996.

[113] J. D. Hayes and D. J. Pulford, "The glutathione S-transferase supergene family: regulation of GST and the contribution of the isoenzymes to cancer chemoprotection and drug resistance," *Critical Reviews in Biochemistry and Molecular Biology*, vol. 30, no. 6, pp. 445–600, 1995.

[114] S. L. Navarro, J. L. Chang, S. Peterson et al., "Modulation of human serum glutathione *S*-transferase A1/2 concentration by cruciferous vegetables in a controlled feeding study is influenced by *GSTM1* and *GSTT1* genotypes," *Cancer Epidemiology Biomarkers and Prevention*, vol. 18, no. 11, pp. 2974–2978, 2009.

[115] P. A. Wark, M. J. A. L. Grubben, W. H. M. Peters et al., "Habitual consumption of fruits and vegetables: associations with human rectal glutathione S-transferase," *Carcinogenesis*, vol. 25, no. 11, pp. 2135–2142, 2004.

[116] J. W. Lampe, C. Chen, S. Li et al., "Modulation of human glutathione S-transferases by botanically defined vegetable diets," *Cancer Epidemiology Biomarkers and Prevention*, vol. 9, no. 8, pp. 787–793, 2000.

[117] W. A. Nijhoff, T. P. J. Mulder, H. Verhagen, G. van Poppel, and W. H. M. Peters, "Effects of consumption of brussels sprouts on plasma and urinary glutathione S-transferase class-alpha and -pi in humans," *Carcinogenesis*, vol. 16, no. 4, pp. 955–957, 1995.

[118] Y. P. Hwang, J. H. Choi, H. J. Yun et al., "Anthocyanins from purple sweet potato attenuate dimethylnitrosamine-induced liver injury in rats by inducing Nrf2-mediated antioxidant enzymes and reducing COX-2 and iNOS expression," *Food and Chemical Toxicology*, vol. 49, no. 1, pp. 93–99, 2011.

[119] M. Iqbal, S. D. Sharma, Y. Okazaki, M. Fujisawa, and S. Okada, "Dietary supplementation of curcumin enhances antioxidant and phase II metabolizing enzymes in ddY male mice: possible role in protection against chemical carcinogenesis and toxicity," *Pharmacology and Toxicology*, vol. 92, no. 1, pp. 33–38, 2003.

[120] B. J. Newsome, M. C. Petriello, S. G. Han et al., "Green tea diet decreases PCB 126-induced oxidative stress in mice by up-regulating antioxidant enzymes," *Journal of Nutritional Biochemistry*, vol. 25, no. 2, pp. 126–135, 2014.

[121] C. Y. Lin, J. H. Chen, R. H. Fu, and C. W. Tsai, "Induction of Pi form of glutathione S-transferase by carnosic acid is mediated through PI3K/Akt/NF-κB pathway and protects against neurotoxicity," *Chemical Research in Toxicology*, vol. 27, no. 11, pp. 1958–1966, 2014.

[122] K. Chinnadurai, H. K. Kanwal, A. K. Tyagi, C. Stanton, and P. Ross, "High conjugated linoleic acid enriched ghee (clarified butter) increases the antioxidant and antiatherogenic potency in female Wistar rats," *Lipids in Health and Disease*, vol. 12, no. 1, article 121, 2013.

[123] E. B. Froyen, J. L. R. Reeves, A. E. Mitchell, and F. M. Steinberg, "Regulation of phase II enzymes by Genistein and daidzein in male and female Swiss Webster mice," *Journal of Medicinal Food*, vol. 12, no. 6, pp. 1227–1237, 2009.

[124] J. L. Perez, G. K. Jayaprakasha, A. Cadena, E. Martinez, H. Ahmad, and B. S. Patil, "In vivo induction of phase II detoxifying enzymes, glutathione transferase and quinone reductase by citrus triterpenoids," *BMC Complementary and Alternative Medicine*, vol. 10, article 51, 2010.

[125] J. R. Barrett, "The science of soy: what do we really know?" *Environmental Health Perspectives*, vol. 114, no. 6, pp. A352–A358, 2006.

[126] H. Wiegand, C. Boesch-Saadatmandi, I. Regos, D. Treutter, S. Wolffram, and G. Rimbach, "Effects of quercetin and catechin on hepatic glutathione-s transferase (GST), NAD(P)H quinone oxidoreductase 1 (NQO1), and antioxidant enzyme activity levels in rats," *Nutrition and Cancer*, vol. 61, no. 5, pp. 717–722, 2009.

[127] M. B. Gomes and C. A. Negrato, "Alpha-lipoic acid as a pleiotropic compound with potential therapeutic use in diabetes and other chronic diseases," *Diabetology & Metabolic Syndrome*, vol. 6, no. 1, article 80, 2014.

[128] Linus Pauling Institute, *Lipoic Acid*, Micronutrient Information Center, Corvalis, Ore, USA, 2012, http://lpi.oregonstate.edu/infocenter/othernuts/la/.

[129] R. W. Kalpravidh, N. Siritanaratkul, P. Insain et al., "Improvement in oxidative stress and antioxidant parameters in beta-thalassemia/Hb E patients treated with curcuminoids," *Clinical Biochemistry*, vol. 43, no. 4-5, pp. 424–429, 2010.

[130] M. I. Lucena, R. J. Andrade, J. P. de la Cruz, M. Rodriguez-Mendizabal, E. Blanco, and F. Sánchez de la Cuesta, "Effects of silymarin MZ-80 on oxidative stress in patients with alcoholic cirrhosis," *International Journal of Clinical Pharmacology and Therapeutics*, vol. 40, no. 1, pp. 2–8, 2002.

[131] R. A. Santana-Martínez, S. Galván-Arzáte, R. Hernández-Pando et al., "Sulforaphane reduces the alterations induced by quinolinic acid: modulation of glutathione levels," *Neuroscience*, vol. 272, pp. 188–198, 2014.

[132] M. F. Chen, L. T. Chen, and H. W. Boyce Jr., "Cruciferous vegetables and glutathione: their effects on colon mucosal glutathione level and colon tumor development in rats induced by DMH," *Nutrition and Cancer*, vol. 23, no. 1, pp. 77–83, 1995.

[133] E. M. El Morsy and R. Kamel, "Protective effect of artichoke leaf extract against paracetamol-induced hepatotoxicity in rats," *Pharmaceutical Biology*, vol. 53, no. 2, pp. 167–173, 2015.

[134] H. A. Brauer, T. E. Libby, B. L. Mitchell et al., "Cruciferous vegetable supplementation in a controlled diet study alters the serum peptidome in a GSTM1-genotype dependent manner," *Nutrition Journal*, vol. 10, no. 1, article 11, 2011.

[135] T. Hofmann, A. Kuhnert, A. Schubert et al., "Modulation of detoxification enzymes by watercress: *in vitro* and *in vivo* investigations in human peripheral blood cells," *European Journal of Nutrition*, vol. 48, no. 8, pp. 483–491, 2009.

[136] H. J. Forman, H. Zhang, and A. Rinna, "Glutathione: overview of its protective roles, measurement, and biosynthesis," *Molecular Aspects of Medicine*, vol. 30, no. 1-2, pp. 1–12, 2009.

[137] J. K. Kern, D. A. Geier, J. B. Adams, C. R. Garver, T. Audhya, and M. R. Geier, "A clinical trial of glutathione supplementation in autism spectrum disorders," *Medical Science Monitor*, vol. 17, no. 12, pp. CR677–CR682, 2011.

[138] P. G. Paterson, A. W. Lyon, H. Kamencic, L. B. Andersen, and B. H. J. Juurlink, "Sulfur amino acid deficiency depresses brain glutathione concentration," *Nutritional Neuroscience*, vol. 4, no. 3, pp. 213–222, 2001.

[139] A. T. Treweeke, T. J. Winterburn, I. Mackenzie et al., "N-Acetylcysteine inhibits platelet-monocyte conjugation in patients with type 2 diabetes with depleted intraplatelet glutathione: a randomised controlled trial," *Diabetologia*, vol. 55, no. 11, pp. 2920–2928, 2012.

[140] L. Galluzzi, I. Vitale, L. Senovilla et al., "Prognostic impact of vitamin B6 metabolism in lung cancer," *Cell Reports*, vol. 2, no. 2, pp. 257–269, 2012.

[141] J. M. Howard, S. Davies, and A. Hunnisett, "Red cell magnesium and glutathione peroxidase in infertile women—effects of oral supplementation with magnesium and selenium," *Magnesium Research*, vol. 7, no. 1, pp. 49–57, 1994.

[142] D. F. Child, P. R. Hudson, H. Jones et al., "The effect of oral folic acid on glutathione, glycaemia and lipids in type 2 diabetes," *Diabetes, Nutrition and Metabolism—Clinical and Experimental*, vol. 17, no. 2, pp. 95–102, 2004.

[143] H. Ansar, Z. Mazloom, F. Kazemi, and N. Hejazi, "Effect of alpha-lipoic acid on blood glucose, insulin resistance, and glutathione peroxidase of type 2 diabetic patients," *Saudi Medical Journal*, vol. 32, no. 6, pp. 584–588, 2011.

[144] R. S. Lord and J. A. Bralley, Eds., *Laboratory Evaluations for Integrative and Functional Medicine*, Genova Diagnostics, Duluth, Ga, USA, 2nd edition, 2012.

[145] University of Maryland Medical Center, Glutamine, University of Maryland Medical Center, Baltimore, Md, USA, 2014, http://umm.edu/health/medical/altmed/supplement/glutamine.

[146] S. I. Makarova, "Human N-acetyltransferases and drug-induced hepatotoxicity," *Current Drug Metabolism*, vol. 9, no. 6, pp. 538–545, 2008.

[147] K. Kohalmy and R. Vrzal, "Regulation of phase II biotransformation enzymes by steroid hormones," *Current Drug Metabolism*, vol. 12, no. 2, pp. 104–123, 2011.

[148] J. D. Yager, "Mechanisms of estrogen carcinogenesis: the role of E2/E1-quinone metabolites suggests new approaches to preventive intervention—a review," *Steroids*, 2014.

[149] J. Busserolles, W. Zimowska, E. Rock, Y. Rayssiguier, and A. Mazur, "Rats fed a high sucrose diet have altered heart antioxidant enzyme activity and gene expression," *Life Sciences*, vol. 71, no. 11, pp. 1303–1312, 2002.

[150] Z. Y. Su, L. Shu, T. O. Khor, J. H. Lee, F. Fuentes, and A. N. T. Kong, "A perspective on dietary phytochemicals and cancer chemoprevention: oxidative stress, Nrf2, and epigenomics," *Topics in Current Chemistry*, vol. 329, pp. 133–162, 2013.

[151] K. Chan, X. D. Han, and Y. W. Kan, "An important function of Nrf2 in combating oxidative stress: detoxification of acetaminophen," *Proceedings of the National Academy of Sciences of the United States of America*, vol. 98, no. 8, pp. 4611–4616, 2001.

[152] V. Calabrese, C. Cornelius, C. Mancuso et al., "Cellular stress response: A novel target for chemoprevention and nutritional neuroprotection in aging, neurodegenerative disorders and longevity," *Neurochemical Research*, vol. 33, no. 12, pp. 2444–2471, 2008.

[153] S. S. Boyanapalli, X. Paredes-Gonzalez, F. Fuentes et al., "Nrf2 knockout attenuates the anti-inflammatory effects of phenethyl isothiocyanate and curcumin," *Chemical Research in Toxicology*, vol. 27, no. 12, pp. 2036–2043, 2014.

[154] S. K. Niture, R. Khatri, and A. K. Jaiswal, "Regulation of Nrf2—an update," *Free Radical Biology and Medicine*, vol. 66, pp. 36–44, 2014.

[155] Y. Xie, Q. Y. Zhao, H. Y. Li, X. Zhou, Y. Liu, and H. Zhang, "Curcumin ameliorates cognitive deficits heavy ion irradiation-induced learning and memory deficits through enhancing of Nrf2 antioxidant signaling pathways," *Pharmacology Biochemistry and Behavior*, 2014.

[156] V. Soetikno, F. R. Sari, A. P. Lakshmanan et al., "Curcumin alleviates oxidative stress, inflammation, and renal fibrosis in remnant kidney through the Nrf2-keap1 pathway," *Molecular Nutrition & Food Research*, vol. 57, no. 9, pp. 1649–1659, 2013.

[157] H. J. He, G. Y. Wang, Y. Gao, W. H. Ling, Z. W. Yu, and T. R. Jin, "Curcumin attenuates Nrf2 signaling defect, oxidative stress in muscle and glucose intolerance in high fat diet-fed mice," *World Journal of Diabetes*, vol. 3, no. 5, pp. 94–104, 2012.

[158] E. O. Farombi, S. Shrotriya, H. K. Na, S. H. Kim, and Y. J. Surh, "Curcumin attenuates dimethylnitrosamine-induced liver injury in rats through Nrf2-mediated induction of heme oxygenase-1," *Food and Chemical Toxicology*, vol. 46, no. 4, pp. 1279–1287, 2008.

[159] Z. Zhang, S. Wang, S. Zhou et al., "Sulforaphane prevents the development of cardiomyopathy in type 2 diabetic mice probably by reversing oxidative stress-induced inhibition of LKB1/AMPK pathway," *Journal of Molecular and Cellular Cardiology*, vol. 77, pp. 42–52, 2014.

[160] G. K. McWalter, L. G. Higgins, L. I. McLellan et al., "Transcription factor Nrf2 is essential for induction of NAD(P)H:quinone oxidoreductase 1, glutathione S-transferases, and glutamate cysteine ligase by broccoli seeds and isothiocyanates," *Journal of Nutrition*, vol. 134, no. 12, supplement, pp. 3499S–3506S, 2004.

[161] I. C. Lee, S. H. Kim, H. S. Baek et al., "The involvement of Nrf2 in the protective effects of diallyl disulfide on carbon tetrachloride-induced hepatic oxidative damage and inflammatory response in rats," *Food and Chemical Toxicology*, vol. 63, pp. 174–185, 2014.

[162] R. Padiya, D. Chowdhury, R. Borkar, R. Srinivas, M. Pal Bhadra, and S. K. Banerjee, "Garlic attenuates cardiac oxidative stress via activation of PI3K/AKT/Nrf2-Keap1 pathway in fructose-fed diabetic rat," *PLoS ONE*, vol. 9, no. 5, Article ID e94228, 2014.

[163] T. Gómez-Sierra, E. Molina-Jijón, E. Tapia et al., "S-allylcysteine prevents cisplatin-induced nephrotoxicity and oxidative stress," *Journal of Pharmacy and Pharmacology*, vol. 66, no. 9, pp. 1271–1281, 2014.

[164] C. F. Chang, S. Cho, and J. Wang, "(-)-Epicatechin protects hemorrhagic brain via synergistic Nrf2 pathways," *Annals of Clinical and Translational Neurology*, vol. 1, no. 4, pp. 258–271, 2014.

[165] C. C. Leonardo, M. Agrawal, N. Singh, J. R. Moore, S. Biswal, and S. Doré, "Oral administration of the flavanol (-)-epicatechin bolsters endogenous protection against focal ischemia through the Nrf2 cytoprotective pathway," *European Journal of Neuroscience*, vol. 38, no. 11, pp. 3659–3668, 2013.

[166] K. Kavitha, P. Thiyagarajan, J. Rathna, R. Mishra, and S. Nagini, "Chemopreventive effects of diverse dietary phytochemicals against DMBA-induced hamster buccal pouch carcinogenesis via the induction of Nrf2-mediated cytoprotective antioxidant, detoxification, and DNA repair enzymes," *Biochimie*, vol. 95, no. 8, pp. 1629–1639, 2013.

[167] Z. A. Shah, R.-C. Li, A. S. Ahmad et al., "The flavanol (−)-epicatechin prevents stroke damage through the Nrf2/HO1 pathway," *Journal of Cerebral Blood Flow and Metabolism*, vol. 30, no. 12, pp. 1951–1961, 2010.

[168] N. Tamaki, R. Cristina Orihuela-Campos, Y. Inagaki, M. Fukui, T. Nagata, and H. Ito, "Resveratrol improves oxidative stress and prevents the progression of periodontitis via the activation of the Sirt1/AMPK and the Nrf2/antioxidant defense pathways in a rat periodontitis model," *Free Radical Biology and Medicine*, vol. 75, pp. 222–229, 2014.

[169] G. Sadi, D. Bozan, and H. B. Yildiz, "Redox regulation of antioxidant enzymes: post-translational modulation of catalase and glutathione peroxidase activity by resveratrol in diabetic rat liver," *Molecular and Cellular Biochemistry*, vol. 393, no. 1-2, pp. 111–122, 2014.

[170] H. Chen, J. Fu, Y. Hu et al., "Ginger compound [6]-shogaol and its cysteine-conjugated metabolite (M2) activate Nrf2 in colon epithelial cells *in vitro* and *in vivo*," *Chemical Research in Toxicology*, vol. 27, no. 9, pp. 1575–1585, 2014.

[171] M. J. Bak, S. Ok, M. Jun, and W. S. Jeong, "6-shogaol-rich extract from ginger up-regulates the antioxidant defense systems in cells and mice," *Molecules*, vol. 17, no. 7, pp. 8037–8055, 2012.

[172] Y. D. Xi, X. Y. Li, H. L. Yu et al., "Soy isoflavone antagonizes the oxidative cerebrovascular injury induced by β-Amyloid Peptides 1–42 in Rats," *Neurochemical Research*, vol. 39, no. 7, pp. 1374–1381, 2014.

[173] R. Li, T. Liang, L. Xu, N. Zheng, K. Zhang, and X. Duan, "Puerarin attenuates neuronal degeneration in the substantia nigra of 6-OHDA-lesioned rats through regulating BDNF expression and activating the Nrf2/ARE signaling pathway," *Brain Research*, vol. 1523, pp. 1–9, 2013.

[174] S. J. V. Vicente, E. Y. Ishimoto, and E. A. F. S. Torres, "Coffee modulates transcription factor Nrf2 and highly increases the activity of antioxidant enzymes in rats," *Journal of Agricultural and Food Chemistry*, vol. 62, no. 1, pp. 116–122, 2014.

[175] B. D. Sahu, U. K. Putcha, M. Kuncha, S. S. Rachamalla, and R. Sistla, "Carnosic acid promotes myocardial antioxidant response and prevents isoproterenol-induced myocardial oxidative stress and apoptosis in mice," *Molecular and Cellular Biochemistry*, vol. 394, no. 1-2, pp. 163–176, 2014.

[176] T. R. Balstad, H. Carlsen, M. C. W. Myhrstad et al., "Coffee, broccoli and spices are strong inducers of electrophile response element-dependent transcription in vitro and in vivo—studies in electrophile response element transgenic mice," *Molecular Nutrition and Food Research*, vol. 55, no. 2, pp. 185–197, 2011.

[177] Y. P. Wang, M. L. Cheng, B. F. Zhang et al., "Effect of blueberry on hepatic and immunological functions in mice," *Hepatobiliary and Pancreatic Diseases International*, vol. 9, no. 2, pp. 164–168, 2010.

[178] A. Bishayee, D. Bhatia, R. J. Thoppil, A. S. Darvesh, E. Nevo, and E. P. Lansky, "Pomegranate-mediated chemoprevention of experimental hepatocarcinogenesis involves Nrf2-regulated antioxidant mechanisms," *Carcinogenesis*, vol. 32, no. 6, pp. 888–896, 2011.

[179] M. A. Esmaeili and M. Alilou, "Naringenin attenuates CCl4-induced hepatic inflammation by the activation of an Nrf2-mediated pathway in rats," *Clinical and Experimental Pharmacology and Physiology*, vol. 41, no. 6, pp. 416–422, 2014.

[180] C. K. Singh, M. A. Ndiaye, I. A. Siddiqui et al., "Methaneseleninic acid and γ-tocopherol combination inhibits prostate tumor growth in vivo in a xenograft mouse model," *Oncotarget*, vol. 5, no. 11, pp. 3651–3661, 2014.

[181] M. J. M. Magbanua, R. Roy, E. V. Sosa et al., "Gene expression and biological pathways in tissue of men with prostate cancer in a randomized clinical trial of lycopene and fish oil supplementation," *PLoS ONE*, vol. 6, no. 9, Article ID e24004, 2011.

[182] C. Manach, A. Scalbert, C. Morand, C. Rémésy, and L. Jiménez, "Polyphenols: food sources and bioavailability," *The American Journal of Clinical Nutrition*, vol. 79, no. 5, pp. 727–747, 2004.

[183] P. Delmonte and J. I. Rader, "Analysis of isoflavones in foods and dietary supplements," *Journal of AOAC International*, vol. 89, no. 4, pp. 1138–1146, 2006.

[184] S. Chian, R. Thapa, Z. Chi, X. J. Wang, and X. Tang, "Luteolin inhibits the Nrf2 signaling pathway and tumor growth in vivo," *Biochemical and Biophysical Research Communications*, vol. 447, no. 4, pp. 602–608, 2014.

[185] R. Marina, P. González, M. C. Ferreras, S. Costilla, and J. P. Barrio, "Hepatic Nrf2 expression is altered by quercetin supplementation in Xirradiated rats," *Molecular Medicine Reports*, vol. 11, no. 1, pp. 539–546, 2015.

[186] H. Zhou, Z. Qu, V. V. Mossine et al., "Proteomic analysis of the effects of aged garlic extract and its fruarg component on lipopolysaccharide-induced neuroinflammatory response in microglial cells," *PLoS ONE*, vol. 9, no. 11, Article ID e113531, 2014.

[187] V. Calabrese, C. Cornelius, A. T. Dinkova-Kostova, E. J. Calabrese, and M. P. Mattson, "Cellular stress responses, the hormesis paradigm, and vitagenes: novel targets for therapeutic intervention in neurodegenerative disorders," *Antioxidants & Redox Signaling*, vol. 13, no. 11, pp. 1763–1811, 2010.

[188] C. A. Houghton, R. G. Fassett, and J. S. Coombes, "Sulforaphane: translational research from laboratory bench to clinic," *Nutrition Reviews*, vol. 71, no. 11, pp. 709–726, 2013.

[189] A. L. Stefanson and M. Bakovic, "Dietary regulation of Keap1/Nrf2/ARE pathway: focus on plant-derived compounds and trace minerals," *Nutrients*, vol. 6, no. 9, pp. 3777–3801, 2014.

[190] G. K. Andrews, "Regulation of metallothionein gene expression by oxidative stress and metal ions," *Biochemical Pharmacology*, vol. 59, no. 1, pp. 95–104, 2000.

[191] P. Lichtlen and W. Schaffner, "Putting its fingers on stressful situations: the heavy metal-regulatory transcription factor MTF-1," *BioEssays*, vol. 23, no. 11, pp. 1010–1017, 2001.

[192] M. Sato and M. Kondoh, "Recent studies on metallothionein: protection against toxicity of heavy metals and oxygen free radicals," *Tohoku Journal of Experimental Medicine*, vol. 196, no. 1, pp. 9–22, 2002.

[193] Y. Pan, J. Huang, R. Xing et al., "Metallothionein 2A inhibits NF-κB pathway activation and predicts clinical outcome segregated with TNM stage in gastric cancer patients following radical resection," *Journal of Translational Medicine*, vol. 11, no. 1, article 173, 2013.

[194] J. J. Lamb, V. R. Konda, D. W. Quig et al., "A program consisting of a phytonutrient-rich medical food and an elimination diet ameliorated fibromyalgia symptoms and promoted toxic-element detoxification in a pilot trial," *Alternative Therapies in Health and Medicine*, vol. 17, no. 2, pp. 36–44, 2011.

[195] T. B. Aydemir, R. K. Blanchard, and R. J. Cousins, "Zinc supplementation of young men alters metallothionein, zinc transporter, and cytokine gene expression in leukocyte populations," *Proceedings of the National Academy of Sciences of the United States of America*, vol. 103, no. 6, pp. 1699–1704, 2006.

[196] T. P. J. Mulder, A. van der Sluys Veer, H. W. Verspaget et al., "Effect of oral zinc supplementation on metallothionein and superoxide dismutase concentrations in patients with inflammatory bowel disease," *Journal of Gastroenterology and Hepatology*, vol. 9, no. 5, pp. 472–477, 1994.

[197] R. Hu, V. Hebbar, B. R. Kim et al., "In vivo pharmacokinetics and regulation of gene expression profiles by isothiocyanate sulforaphane in the rat," *Journal of Pharmacology and Experimental Therapeutics*, vol. 310, no. 1, pp. 263–271, 2004.

[198] T. Kimura, F. Okumura, A. Onodera, T. Nakanishi, N. Itoh, and M. Isobe, "Chromium (VI) inhibits mouse *metallothionein-I* gene transcription by modifying the transcription potential of the co-activator p300," *The Journal of Toxicological Sciences*, vol. 36, no. 2, pp. 173–180, 2011.

[199] C. J. Weng, M. J. Chen, C. T. Yeh, and G. C. Yen, "Hepatoprotection of quercetin against oxidative stress by induction of metallothionein expression through activating MAPK and PI3K pathways and enhancing Nrf2 DNA-binding activity," *New Biotechnology*, vol. 28, no. 6, pp. 767–777, 2011.

[200] M. Singh, R. Tulsawani, P. Koganti, A. Chauhan, M. Manickam, and K. Misra, "Cordyceps sinensis increases hypoxia tolerance by inducing heme oxygenase-1 and metallothionein via Nrf2 activation in human lung epithelial cells," *BioMed Research International*, vol. 2013, Article ID 569206, 13 pages, 2013.

[201] N. M. R. Sales, P. B. Pelegrini, and M. C. Goersch, "Nutrigenomics: definitions and advances of this new science," *Journal of Nutrition and Metabolism*, vol. 2014, Article ID 202759, 6 pages, 2014.

[202] U. Lim and M. A. Song, "Dietary and lifestyle factors of DNA methylation," *Methods in Molecular Biology*, vol. 863, pp. 359–376, 2012.

[203] I. A. Lang, T. S. Galloway, A. Scarlett et al., "Association of urinary bisphenol A concentration with medical disorders and laboratory abnormalities in adults," *The Journal of the American Medical Association*, vol. 300, no. 11, pp. 1303–1310, 2008.

[204] R. Rezg, S. El-Fazaa, N. Gharbi, and B. Mornagui, "Bisphenol A and human chronic diseases: current evidences, possible mechanisms, and future perspectives," *Environment International*, vol. 64, pp. 83–90, 2014.

[205] S. Mostafalou and M. Abdollahi, "Pesticides and human chronic diseases: evidences, mechanisms, and perspectives," *Toxicology and Applied Pharmacology*, vol. 268, no. 2, pp. 157–177, 2013.

[206] D. J. Magliano, V. H. Y. Loh, J. L. Harding, J. Botton, and J. E. Shaw, "Persistent organic pollutants and diabetes: a review of the epidemiological evidence," *Diabetes & Metabolism*, vol. 40, no. 1, pp. 1–14, 2014.

[207] S. Agarwal, T. Zaman, E. M. Tuzcu, and S. R. Kapadia, "Heavy metals and cardiovascular disease: results from the National Health and Nutrition Examination Survey (NHANES) 1999–2006," *Angiology*, vol. 62, no. 5, pp. 422–429, 2011.

[208] E. F. Rissman and M. Adli, "Minireview: transgenerational epigenetic inheritance: focus on endocrine disrupting compounds," *Endocrinology*, vol. 155, no. 8, pp. 2770–2780, 2014.

[209] D. M. Walker and A. C. Gore, "Transgenerational neuroendocrine disruption of reproduction," *Nature Reviews Endocrinology*, vol. 7, no. 4, pp. 197–207, 2011.

Treatment Outcome of Severe Acute Malnutrition Cases at the Tamale Teaching Hospital

Mahama Saaka,[1] **Shaibu Mohammed Osman,**[2] **Anthony Amponsem,**[1,2]
Juventus B. Ziem,[1,2] **Alhassan Abdul-Mumin,**[1,2] **Prosper Akanbong,**[1,2]
Ernestina Yirkyio,[2] **Eliasu Yakubu,**[2] **and Sean Ervin**[3]

[1] *School of Medicine and Health Sciences, University for Development Studies, P.O. Box 1883, Tamale, Ghana*
[2] *Tamale Teaching Hospital, P.O. Box 16, Tamale, Ghana*
[3] *School of Medicine, Wake Forest University, Winston-Salem, NC, USA*

Correspondence should be addressed to Mahama Saaka; mmsaaka@gmail.com

Academic Editor: Maurizio Muscaritoli

Objective. This study investigated the treatment outcomes and determinant factors likely to be associated with recovery rate. *Methods.* A retrospective chart review (RCR) was performed on 348 patients who were enrolled in the outpatient care (OPC) during the study period. *Results.* Of the 348 cases, 33.6% recovered (having MUAC ≥125 mm), 49.1% defaulted, and 11.5% transferred to other OPC units to continue with treatment. There were 187 (53.7%) males and 161 (46.3%) females with severe malnutrition. The average weight gain rate was 28 g/kg/day. Controlling for other factors, patients who completed the treatment plan had 3.2 times higher probability of recovery from severe acute malnutrition (SAM) as compared to patients who defaulted (adjusted odds ratio (AOR) = 3.2, 95% CI = 1.9, 5.3, and $p < 0.001$). The children aged 24–59 months had 5.8 times higher probability of recovery from SAM as compared to children aged 6–11 months (AOR = 5.8, 95% CI = 2.5, 10.6, and $p < 0.001$). *Conclusions.* Cure rate was low and the default rate was quite high. Children who were diagnosed as having marasmus on admission stayed longer before recovery than their kwashiorkor counterparts. Younger children were of greater risk of nonrecovery.

1. Introduction

Severe acute malnutrition (SAM) affects nearly 20 million children under five and contributes to one million child deaths yearly [1, 2]. SAM is an important cofactor in the development of severe infections. SAM as defined by WHO-UNICEF includes severe wasting and nutritional oedema. Severe wasting (marasmus) is defined as weight-for-height (WH) below −3 standard deviations (SD or Z-scores) or mid-upper arm circumference (MUAC) <115 mm [2–4].

A child with SAM has a limited ability to respond to stressors (infection and environmental), is highly vulnerable, and has a high mortality risk [5, 6]. Severely underweight and wasted children had an approximately eight- to nine-fold increased risk of mortality as shown in, respectively, [7] and [1]. Stunting, severe wasting, and intrauterine growth restriction together are responsible for 2.2 million deaths per year and 21% of disability-adjusted life-years for children younger than 5 years [1]. It is critical that such children are treated proactively with intensive treatment regimes of short duration, aiming to rehabilitate the child in a few weeks.

SAM is a common indication for hospital admission among pediatric patients in sub-Saharan Africa. In Ethiopia, severe acute malnutrition is the primary diagnosis in 20% of pediatric hospital admissions [8], while 41.4% of preschool-aged children are affected by malnutrition of any degree [9].

Relatively little has been published on treatment outcomes for SAM in outpatient care settings (OPC). One study in rural Malawi demonstrative of children enrolled in an outpatient treatment programme for moderate acute malnutrition demonstrated that 80% recovered, 4% defaulted, and 0.4% died [10]. In the same setting, 30% of children who completed treatment for moderate acute malnutrition either relapsed or died within one-year following treatment [11]. In northwest Ethiopia, patients hospitalized for severe malnutrition had a case fatality rate of 18%, and 9% abandoned treatment [12]. However, it is unknown how these results translate to in other regions of Africa.

UNICEF supports the rehabilitation of SAM cases in Ghana with Plumpy'Nut which is a ready-to-use therapeutic food (RUTF). Plumpy'Nut is energy, mineral, and vitamin enriched paste-food designed to treat SAM. A sachet has a serving size of 92 gm and gives energy of 2,100 kJ (500 kcal). Severely malnourished children are also provided with routine medications such as deworming tabs, antibiotics, vitamin A, folic acid, and measles vaccine. Only children who fulfil the criteria for SAM, who do not have medical complications, and who have passed the appetite test with Plumpy'Nut are managed at the OPC. Once admitted, children get a weekly Plumpy'Nut ration. They receive different amount of Plumpy'Nut sachets according to their body weight and appetite. They are also given routine medications during the course of the treatment such as vitamin A, folic acid tabs, antibiotics, deworming tabs, and measles vaccine.

On each visit, children receive medical review. This includes temperature, respiratory rates, and pulse measurements. Medical history is also taken every week and that includes history of vomiting, diarrhoea, cough, and temperature in the last 7 days. Nutrition education is also given to mothers or caretakers. They are counseled on how to provide adequate diets for their children at home in order to prevent reoccurrence of malnutrition or other siblings from becoming malnourished.

The treatment of SAM cases in outpatient care (OPC) has been going on since 2011 in the TTH; however little is known about the treatment outcomes and factors determining the recovery rate of the children presenting with SAM in the OPC. This study sought to assess the performance of the programme and determinants of recovery rate in an Outpatient Paediatric Nutrition Clinic in Northern Ghana.

2. Materials and Methods

2.1. Study Site. All study patients were enrolled from the OPC of the Tamale Teaching Hospital in Tamale, Ghana. This hospital has a 452-bed capacity and serves as a major referral center for Northern Ghana, with an estimated catchment population of 2.1 million.

2.2. Study Design. We performed a retrospective chart review (RCR) of patients enrolled in the outpatient care (OPC) clinic 2011–2013 at the Tamale Teaching Hospital.

2.3. Study Population Selection (Exclusion Criteria). No sampling was done but all malnourished children aged 6–59 months who sought outpatient care between 2011 and 2013 were included in the study. However, those children with some variables not recorded were excluded from the study. A total of 353 outpatient cases were reviewed but 348 patients had a complete set of data and were included in the analysis.

2.4. Outcome Measures. The main outcome indicators were average length of stay, average rate of weight gain, cure rate, death rate, default rate, and transfer rate.

2.5. Data Collection. The source of data for the study was individual OPC record documents including registers and monitoring cards. Information extracted were patient age, sex, residence, admission criteria, the number of admissions, death, defaulters, date of admission and discharge, length of stay, diagnosis, and discharge condition (resolved malnutrition, death, and lost to follow-up). Additionally, anthropometric measurements, including weight, presence of bilateral oedema, and mid-upper arm circumference, were collected at the time of enrollment, at discharge, and during treatment.

The average length of stay was calculated by adding the total number of days that each child discharged as cured stayed in the OPC and dividing this by the number of children cured for a specific month.

The length of stay (LOS) in the OPC and the rate of weight gain were computed only for those children admitted with marasmus and who recovered from SAM. This is methodology within the Guidelines of the International SPHERE standard [13, 14].

Rate of weight gain was calculated using the formula

$$\frac{\text{Discharge Weight (g)} - \text{Minimum Weight (g)}}{\text{Minimum Weight} \times \text{no. of days between minimum weight and discharged weight}}. \tag{1}$$

2.6. Data Processing and Analysis. Data was checked for correctness and consistency and were analyzed using SPSS for Windows (version 21; SPSS Inc., Chicago, IL, USA). First, frequency tables were produced for different variables and cross tables were produced accordingly. Mean values were produced for continuous variables. Comparison between groups was done using chi-square tests for proportions and t-tests or ANOVA procedures for continuous variables. Regression analysis was performed to identify independent outcome predictors.

A trend analysis was carried out to help assess if severe malnutrition cases are increasing or declining over time.

Seasonal changes in severe malnutrition cases were also investigated.

2.7. Ethics Consideration. Ethical approval was obtained from the Tamale Teaching Hospital with reference number TTH/R&M/SR/13/91. In this study there was no direct contact with patients and secondary data was used anonymously by using identity numbers instead of names in order to protect patient identity. As this was a retrospective chart review, there was minimal risk involved to participants. All protected health information (PHI) was deidentified prior to data analysis and publication; subject identities were known only

TABLE 1: Sociodemographic characteristics of study sample (N = 348).

Characteristics	N	%
Age (months)		
6–11	92	26.4
12–23	128	36.8
24–59	128	36.8
Total	**348**	**100.0**
MUAC (mm) on admission		
Severe (<115)	234	67.2
Moderate (115 to <125)	95	27.3
Normal (≥125)	19	5.5
Total	**348**	**100.0**
MUAC (mm) on discharge		
Severe (<115)	119	34.2
Moderate (115 to <125)	112	32.2
Normal (≥125)	117	33.6
Total	**348**	**100.0**
Duration in OPC (days)		
1–14	323	92.8
15–28	21	6.0
29–42	3	0.9
>42 days	1	0.3
Total	**348**	**100.0**
Admission		
Direct from community	79	22.7
Referred from health centre	34	9.8
Referred from inpatient care	235	67.5
Total	**348**	**100.0**

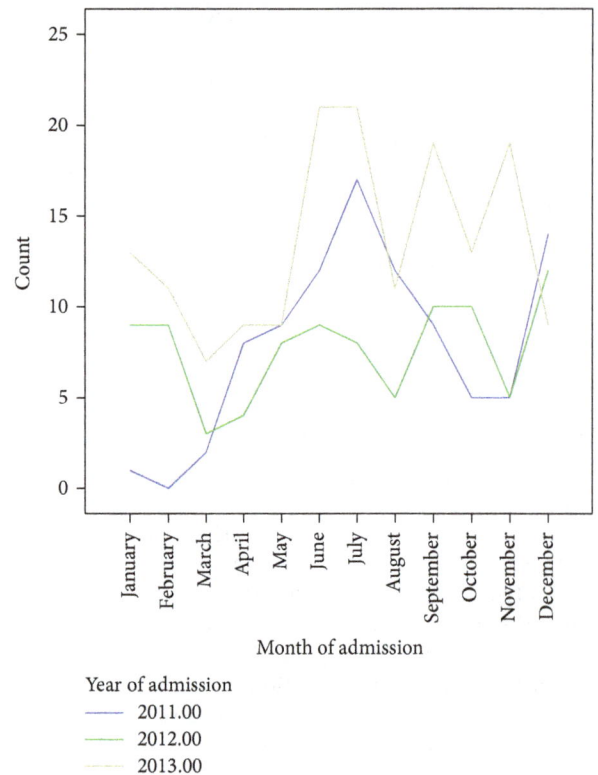

FIGURE 1: Trends in admissions in the OPC.

to the study staff. No reference to any individual participant is made in the study reports.

To help ensure subject privacy and confidentiality, only a unique study identifier appeared on the data collection form for each subject. Any collected subject identifying information corresponding to the unique study identifier was maintained on a linkage file, stored separately from the data. Data access was limited to study staff.

3. Results

3.1. Sample Characteristics. The mean ± SD age of the children was 19.2 ± 10.9 months, with 36.8% being in the age groups of 12–23 and 24–59 months. There were 187 (53.7%) males and 161 (46.3%) females with severe malnutrition. Children below the age of 24 months accounted for 63.2% of all the admissions. Overall, 67.2% of cases met the criteria for SAM (MUAC <115 mm), 27.3% of cases met the criterion for moderate acute malnutrition (115 to <125), and 5.5% were normal (MUAC ≥125 mm). This latter category of children with normal MUAC could have had some form of oedema suggesting SAM cases. The mean MUAC of children on admission and discharge was 11.1 ± 0.9 and 11.9 ± 1.2, respectively. The average

number of days spent in the facility was 7.0 ± 5.4 with the minimum and maximum being 1 and 47 days, respectively. Well over 67% of cases admitted to the OPC were referred from the inpatient care (IPC) of the same hospital (Table 1).

Based on MUAC, 25.5% SAM cases on admission progressed to a normal nutritional status whereas 24.8% improved to moderate acute malnutrition. Similarly, the proportion of moderate acute malnutrition that converted to normal status was 40.0% on discharge (Table 2).

3.2. Trend in Admissions in the OPC. The trend analysis was carried out to help assess if severe malnutrition cases are increasing or declining over time. Seasonal changes in severe malnutrition cases were also investigated as shown in Figure 1. The peak months for admissions over the three-year period were June and July and the highest number of admissions was recorded in 2013. The least number of cases was admitted in March.

There was no significant difference in the trend of admissions of malnutrition cases in terms of sex over the three-year period (Table 3).

3.3. Programme Effectiveness as Gauged by the Global SPHERE Standards. The main outcome indicators were cure rate, death rate, default rate, and transfer rate. The treatment outcomes of severely malnourished children from 2011 to 2013 at the OPC are shown in Table 4. In all, 348 children aged 6–59 months with SAM were admitted to the OPC during the period of study. Of these, 33.6% recovered (having MUAC ≥125), 49.1% defaulted, and 11.5% transferred to other OPC

TABLE 2: Cross-tabulation between mid-upper arm circumference (MUAC) on admission and MUAC at discharge.

| | | MUAC at discharge | | | Total |
		Severe	Moderate	Normal	
MUAC on admission					
Severe (<11.5 cm)	Count	116	58	60	234
	%	49.6	24.8	25.6	100.0
Moderate (11.5 to <12.5 cm)	Count	3	54	38	95
	%	3.2	56.8	40.0	100.0
Normal (at least 12.5 cm)	Count	0	0	19	19
	%	0.0	0.0	100.0	100.0
Total	Count	**119**	**112**	**117**	**348**
	%	**34.2**	**32.2**	**33.6**	**100.0**

TABLE 3: Annual and seasonal trends in admissions in the OPC (bivariate analysis).

| Factor | N | Sex of child | | Test statistic |
		Male n (%)	Female n (%)	
Year of admission				
2011	94	48 (51.1)	46 (48.9)	
2012	92	52 (56.5)	40 (43.5)	Chi-square $(\chi^2) = 0.6, p = 0.8$
2013	162	87 (53.7)	75 (46.3)	
Season of admission				
Dry	111	60 (54.1)	51 (45.9)	$\chi^2 = 0.0, p = 0.94$
Rainy	237	127 (53.6)	110 (46.4)	

TABLE 4: Comparison of the treatment outcomes of severely malnourished children at the outpatient care (OPC) with international SPHERE standards ($N = 348$).

| Outcome indicator | Outcome (%) | SPHERE standards | |
		Acceptable	Alarming
Recovery rate	33.6	>75%	<50%
Nonresponse rate	0.3	15	
Defaulter rate	49.1	<15%	>25%
Case fatality rate (CFR)	0.0	<10%	>15%
Average rate of weight gain (g/kg/day)	28.2	≥8	<8
Average length of stay (weeks)	1.0	<8 weeks	>6 weeks
Referred to IPC	0.3	—	—
Transferred to other OPC	11.5	—	—

units to continue with treatment. The case fatality rate was 0.0%.

Among SAM cases who were cured, the average length of stay was 8.0 ± 5.34 days and the maximum was 33 days (4 weeks). The children who recovered from SAM had an average weight gain of 28.3 ± 23.9 gm/kg/day.

Children that were diagnosed with kwashiorkor (presence of bilateral pitting oedema) on admission seemed to have greater weight and MUAC on discharge than their marasmic counterparts but the average length of stay was not significantly different among children who were discharged as cured (Table 5).

3.4. Factors Associated with the Type of Malnutrition on Admission. Marasmic children came more from an urban

setting whilst the kwashiorkor came more from rural settings. Marasmus was more predominant in dry season compared to the rainy season (96.4% versus 86.1) (Chi = 8.5, p = 0.004). Kwashiorkor was more prevalent in the rainy season than the dry season at the OPC (Table 6). However, no significant difference was observed in the occurrence of type of malnutrition between males and females.

3.5. Factors Associated with Recovery from Severe Acute Malnutrition. Bivariate and logistic regression analysis was performed to identify factors that independently predict recovery from SAM. In the bivariate analysis, the factors associated with recovery from SAM are shown in Table 7. No significant difference in recovery rate was observed with respect to where child was admitted from, place of residence,

TABLE 5: Length of stay and treatment outcomes among cured children under different admission criteria.

Outcome/admission criteria	N	Mean	Std. deviation	95% confidence interval for mean		Test statistic
				Lower bound	Upper bound	
MUAC on discharge						
MUAC < 115 mm	98	13.2	0.66	13.04	13.31	$F(1, 116) = 4.6$, $p = 0.04$
Bilateral pitting oedema	19	13.6	1.15	13.03	14.14	
Total	**117**	**13.2**	**0.77**	**13.10**	**13.38**	
Discharge weight						
MUAC < 115 mm	98	8.8	1.83	8.47	9.21	$F(1, 116) = 6.2$, $p = 0.01$
Bilateral pitting oedema	19	9.9	1.31	9.30	10.57	
Total	**117**	**9.0**	**1.80**	**8.69**	**9.35**	
Number of days on treatment						
MUAC < 115 mm	98	8.4	5.13	7.35	9.41	$F(1, 116) = 0.005$, $p = 0.9$
Bilateral pitting oedema	19	8.5	5.91	5.63	11.32	
Total	**117**	**8.4**	**5.24**	**7.43**	**9.35**	

TABLE 6: Factors associated with the type of malnutrition on admission.

Factor	N	Type of malnutrition		Test statistic
		Marasmus n (%)	Kwashiorkor n (%)	
Age (months)				
6–11	92	90 (97.8)	2 (2.2)	Chi-square $(\chi^2) = 24.3, p < 0.001$
12–23	128	120 (93.8)	8 (6.3)	
24–59	128	101 (78.9)	27 (21.1)	
Total	**348**	**311 (89.4)**	**37 (10.6)**	
Season of admission				
Dry	111	107 (96.4)	4 (3.6)	$\chi^2 = 8.5, p = 0.004$
Rainy	237	204 (86.1)	33 (13.9)	
Total	**348**	**311 (89.4)**	**37 (10.6)**	
Place of residence				
Urban	201	187 (93.0)	14 (7.0)	$\chi^2 = 6.7, p = 0.009$
Rural	147	124 (84.4)	23 (15.6)	
Total	**348**	**311 (89.4)**	**37 (10.6)**	
Gender				
Male	187	169 (90.4)	18 (9.6)	$\chi^2 = 0.4, p = 0.5$
Female	161	142 (88.2)	19 (11.8)	
Total	**348**	**311 (89.4)**	**37 (10.6)**	

sex of the child, and duration of stay in the programme. Children aged 24–59 months were more likely to recover compared to children 6–11 months (Chi = 23.4, $p < 0.001$).

Controlling for other factors, the patients who did not default during treatment had 3.2 times higher probability of recovery from SAM as compared to the patients who defaulted (AOR = 3.2, 95% CI = 1.9, 5.3, and $p < 0.001$). Likewise, the children aged 24–59 months had 5.8 times higher probability of recovery from SAM as compared to children aged 6–11 months (AOR = 5.8, 95% CI = 2.5, 10.6, and $p < 0.001$) (Table 8).

The set of factors accounted for 23.1% of the variance in cure or recovery rate (Nagelkerke R square = 0.231).

4. Discussion

A MUAC of less than 115 mm and/or bipedal oedema are currently the only admission criteria for the programme. This study analyzed the treatment outcomes of infants and children aged 6–59 months and who have a mid-upper arm circumference <115 mm and/or the presence of bilateral oedema (swelling of both feet). These children were admitted for the management of severe acute malnutrition.

Most of the children admitted were marasmic (MUAC <115 mm but without oedema). This is in agreement with what other studies have found that marasmus is more prevalent than kwashiorkor in Northern Ghana [15].

TABLE 7: Factors associated with recovery from severe acute malnutrition (bivariate analysis).

Factor	N	Recovery from SAM?		Test statistic
		No n (%)	Yes n (%)	
Age group (months)				
6–11	92	77 (83.7)	15 (16.3)	
12–23	128	81 (63.3)	47 (36.7)	Chi-square (χ^2) = 17.9, $p < 0.001$
24–59	128	73 (57.0)	55 (43.0)	
Admission criteria				
MUAC < 115 mm	311	213 (68.5)	98 (31.5)	$\chi^2 = 5.8$, $p = 0.02$
Bilateral pitting oedema	37	18 (48.6)	19 (51.4)	
Classification of % change in weight				
Up to 20%	256	187 (73.0)	69 (27.0)	$\chi^2 = 19.2$, $p < 0.001$
More than 20%	92	44 (47.8)	48 (52.2)	
Defaulted				
No	163	83 (50.9)	80 (49.1)	$\chi^2 = 32.8$, $p < 0.001$
Yes	185	148 (80.0)	37 (20.0)	

TABLE 8: Factors associated with recovery from severe acute malnutrition (multivariate analysis).

	Wald	Sig.	Exp(β)	95% CI for EXP(β)	
				Lower	Upper
Not defaulted during treatment	18.99	<0.001	3.2	1.9	5.3
>20% change in weight	10.84	0.001	2.6	1.5	4.6
Age group (reference: 6–11 months)	20.12	<0.001			
12–23 months	10.35	0.001	3.2	1.6	6.5
24–59 months	20.12	<0.001	5.8	2.5	10.6
Constant	54.91	<0.001	0.07		

The aetiology of severe wasting (marasmus) is linked to the situation where the child consumes much less food than required for his or her energy needs and so energy is mobilized from both body fat and muscle. Gluconeogenesis in the liver is enhanced, and there is loss of subcutaneous fat and wasting of muscles.

4.1. Treatment Outcomes. The effectiveness of an OPC programme is gauged by the Global SPHERE standards. The defaulter and recovery rates in this study were outside the acceptable range of Global SPHERE standards [13]. However, the average rate of weight gain was substantially more than the International SPHERE standard.

The cure rate was quite low whereas the defaulter rate was three times higher than the acceptable SPHERE standard. A defaulter is a patient that is absent for two consecutive weeks and it is confirmed that the patient is not dead. It is unclear what factors contribute to the very high defaulter rate observed in this study (49.1%). The less than 40% recovery rate may be directly attributable to the 50% defaulter rate and the significant number of cases transferred to other outpatients' clinics. The outcomes for these patients are unknown and limit a complete interpretation of the data for this programme. No significant age and sex difference was found in terms of default rate.

If malnourished children access nutritional care early in the onset of their condition and comply with treatment until they have recovered, one might expect improved medical and nutritional outcomes. Conversely, if patients access care late and/or they are deterred from staying in the programme for as long as necessary, then success rates will be low. To achieve the Millenium Development Goal 4, reduction in child mortality [16], the management of SAM needs to be implemented correctly. This means that outpatient programmes should decrease barriers to access, encourage early identification of malnutrition, reduce inpatient caseloads and so decrease the risks of cross infection, reduce costs associated with treatment, encourage compliance by patients, and increase the time available to staff to help the sickest children [17, 18].

The rate of weight gain observed among marasmic children who recovered was 28.3 gm/kg/day. This is relatively high when compared to the goals set by the SPHERE standards (Table 4). Usually, a weight gain of 10–15 g/kg/day is considered satisfactory; and conversely if the weight gain is less than 5 g/kg for three consecutive days, it shows that the child is not responding to the treatment. The discharge weight is usually seen as 90% of expected weight for age and most children reach this target weight within two to four weeks of therapy [19].

4.2. Patient Factors That Might Have Contributed to Patient Outcome. Among SAM cases who were cured, the average length of stay was 8.0 ± 5.34 days and the maximum was 33 days (4 weeks). SAM cases who were admitted on the basis of having oedema (i.e., kwashiorkor) recovered faster than cases that were admitted based on low MUAC. Those children who were diagnosed with marasmus on admission stayed longer (LOS) before recovery than their kwashiorkor counterparts (LOS). This finding is consistent with the findings of earlier studies [20, 21].

Children with severe acute malnutrition should only be discharged from treatment when their weight-for-height/length is ≥ -2 Z-score and they have had no oedema for at least 2 weeks, or mid-upper arm circumference is ≥ 125 mm and they have had no oedema for at least 2 weeks [22].

5. Conclusion

This study demonstrates a high defaulter rate but a recovery rate of 34% from severe malnutrition in our Outpatient Clinic. The mean weight gain was 28 gm/kg/day with a mean LOS of 8 days. These results are in concordance with the goals of the SPHERE standards and WHO criteria for discharge from nutritional programmes. The study was limited by a high default rate of 49% of patients. Reasons for default from the programme were not identified. The findings reinforce the need to curb default during management of severe malnutrition. It is thus recommended that a study be done which may include interviews with key informants such as programme managers and focus group discussions with caretakers to come up with strategies that may help to reduce the default rate and improve on recovery rate. Overall mortality rate of those completing the programme is 0.0%.

Conflict of Interests

The authors declare that they have no competing interests.

Authors' Contribution

Sean Ervin, Shaibu Mohammed Osman, Anthony Amponsem, and Alhassan Abdul-Mumin conceived and designed the study. Shaibu Mohammed Osman, Ernestina Yirkyio, and Eliasu Yakubu acquired and compiled data. Prosper Akanbong and Alhassan Abdul-Mumin performed experiments. Mahama Saaka analyzed data and drafted the paper. Sean Ervin and Juventus B. Ziem revised it critically for important intellectual content. All authors read and approved the final paper.

Acknowledgment

The authors thank Dr. Elizabeth Halvorson for helpful discussions on the topic.

References

[1] R. E. Black, L. H. Allen, Z. A. Bhutta et al., "Maternal and child undernutrition: global and regional exposures and health consequences," *The Lancet*, vol. 371, no. 9608, pp. 243–260, 2008.

[2] UNICEF, *Joint Statement on Community-Based Management of Severe Acute Malnutrition*, UNICEF Publications, 2007, http://www.unicef.org/publications/index_39468.html.

[3] WHO and UNICEF, *WHO Child Growth Standards and the Identification of Severe Acute Malnutrition in Infants and Children. A Joint WHO-UNICEF Statement*, WHO, Geneva, Switzerland; UNICEF, New York, NY, USA, 2009, http://www.who.int/nutrition/publications/severemalnutrition/9789241598163_eng.pdf.

[4] C. Prudhon, Z. W. Prinzo, A. Briend, B. M. E. G. Daelmans, and J. B. Mason, "Proceedings of the WHO, UNICEF, and SCN informal consultation on community-based management of severe malnutrition in children," *Food and Nutrition Bulletin*, vol. 27, no. 3, pp. S99–S104, 2006.

[5] A. Briend, M. Garenne, O. Fontaine, and K. Dieng, "Nutritional status, age and survival: the muscle mass hypothesis," *European Journal of Clinical Nutrition*, vol. 43, no. 10, pp. 715–726, 1989.

[6] R. Bairagi, "On validity of some anthropometric indicators as predictors of mortality," *American Journal of Clinical Nutrition*, vol. 34, no. 11, pp. 2592–2594, 1981.

[7] D. L. Pelletier, E. A. Frongillo Jr., D. G. Schroeder, and J.-P. Habicht, "A methodology for estimating the contribution of malnutrition to child mortality in developing countries," *Journal of Nutrition*, vol. 124, no. 10, pp. 2047S–2081S, 1994.

[8] D. M. Gordon, S. Frenning, H. R. Draper, and M. Kokeb, "Prevalence and burden of diseases presenting to a general pediatrics ward in gondar, ethiopia," *Journal of Tropical Pediatrics*, vol. 59, no. 5, pp. 350–357, 2013.

[9] G. Asres and A. I. Eidelman, "Nutritional assessment of ethiopian beta-israel children: a cross-sectional survey," *Breastfeeding Medicine*, vol. 6, no. 4, pp. 171–176, 2011.

[10] L. Lagrone, S. Cole, A. Schondelmeyer, K. Maleta, and M. J. Manary, "Locally produced ready-to-use supplementary food is an effective treatment of moderate acute malnutrition in an operational setting," *Annals of Tropical Paediatrics*, vol. 30, no. 2, pp. 103–108, 2010.

[11] C. Y. Chang, I. Trehan, R. J. Wang et al., "Children successfully treated for moderate acute malnutrition remain at risk for malnutrition and death in the subsequent year after recovery," *Journal of Nutrition*, vol. 143, no. 2, pp. 215–220, 2013.

[12] S. Amsalu and G. Asnakew, "The outcome of severe malnutrition in northwest Ethiopia: retrospective analysis of admissions," *Ethiopian Medical Journal*, vol. 44, no. 2, pp. 151–157, 2006.

[13] SPHERE Project Team, *The SPHERE Humanitarian Charter and Minimum Standards in Disaster Response*, The SPHERE Project, Geneva, Switzerland, 2003.

[14] S. Collins and K. Sadler, "The outpatient treatment of severe malnutrition during humanitarian relief programs," *The Lancet*, vol. 360, pp. 1824–1830, 2002.

[15] B. Minkah, *Pro-Inflammatory Cytokines as Markers for the Diagnosis of Protein Energy Malnutrition*, Kwame Nkrumah University of Science & Technology, Kumasi, Ghana, 2010.

[16] UNICEF, *Management of Severe Acute Malnutrition in Children: Programme and Supply Components of Scaling-Up an Integrated Approach*, UNICEF, New York, NY, USA, 2008.

[17] S. Collins, N. Dent, P. Binns, P. Bahwere, K. Sadler, and A. Hallam, "Management of severe acute malnutrition in children," *The Lancet*, vol. 368, no. 9551, pp. 1992–2000, 2006.

[18] G. Fuchs, T. Ahmed, M. Araya, S. Baker, N. Croft, and L. Weaver, "Malnutrition: working group report of the second world congress of pediatric gastroenterology, hepatology, and nutrition," *Journal of Pediatric Gastroenterology and Nutrition*, vol. 39, pp. S670–677, 2004.

[19] WHO, *Management of Severe Malnutrition: A Manual for Physicians and Other Senior Health Workers*, WHO, Geneva, Switzerland, 1999.

[20] N. Lapidus, A. Minetti, A. Djibo et al., "Mortality risk among children admitted in a large-scale nutritional program in Niger," *PLoS ONE*, vol. 4, no. 1, pp. 13–50, 2009.

[21] A. U. Ahmed, T. U. Ahmed, M. S. Uddin, M. H. Chowdhury, M. H. Rahman, and M. I. Hossain, "Outcome of standardized case management of under-5 children with severe acute malnutrition in three hospitals of Dhaka city in Bangladesh," *Bangladesh Journal of Child Health*, vol. 37, no. 1, pp. 5–13, 2013.

[22] WHO, *Guideline: Updates on the Management of Severe Acute Malnutrition in Infants and Children*, World Health Organization, Geneva, Switzerland, 2013.

Permissions

List of Contributors

Marlieke Visser
Department of Cardiothoracic Surgery, Academic Medical Center, University of Amsterdam, P.O. Box 22700, 1100 DE Amsterdam, Netherlands
Department of Surgery, VU University Medical Center, P.O. Box 7057, 1007 MB Amsterdam, Netherlands

Hans W. M. Niessen
Department of Pathology and Cardiac Surgery, ICaR-VU, VU University Medical Center, P.O. Box 7057, 1007 MB Amsterdam, Netherlands

Wouter E. M. Kok
Department of Cardiology, Academic Medical Center, University of Amsterdam, P.O. Box 22700, 1100 DE Amsterdam, Netherlands

Riccardo Cocchieri and Bas A. J. M. de Mol
Department of Cardiothoracic Surgery, Academic Medical Center, University of Amsterdam, P.O. Box 22700, 1100 DE Amsterdam, Netherlands

Paul A. M. van Leeuwen and Willem Wisselink
Department of Surgery, VU University Medical Center, P.O. Box 7057, 1007 MB Amsterdam, Netherlands

Abel Gebre
Department of Public Health, College of Health Sciences, Samara University, Samara, Ethiopia

Afework Mulugeta
Department of Public Health, College of Health Sciences, Mekelle University, Mekelle, Ethiopia

W. Aekplakorn
Department of Community Medicine, Faculty of Medicine, Ramathibodi Hospital, Mahidol University, Rama VI Road, Ratchathewi, Bangkok 10400, Thailand
National Health Examination Survey Office, Nonthaburi 11000, Thailand

W. Satheannoppakao
Department of Nutrition, Faculty of Public Health, Mahidol University, Bangkok 10400, Thailand

P. Putwatana
Ramathibodi School of Nursing, Faculty of Medicine, Ramathibodi Hospital, Mahidol University, Bangkok 10400, Thailand

S. Taneepanichskul
College of Public Health Sciences, Chulalongkorn University, Bangkok 10330, Thailand

P. Kessomboon
Faculty of Medicine, Khon Kaen University, Khon Kaen 40002, Thailand

V. Chongsuvivatwong
Epidemiology Unit, Faculty of Medicine, Prince of Songkla University, Songkhla 90110, Thailand

S. Chariyalertsak
Faculty of Medicine, Chiang Mai University, Chiang Mai 50002, Thailand

Lourdes Rodríguez, Paola Otero, María I. Panadero, Silvia Rodrigo and Carlos Bocos
Facultad de Farmacia, Universidad CEU San Pablo, Urbanización Montepríncipe, Boadilla del Monte, 28668 Madrid, Spain

Juan J. Álvarez-Millán
Facultad de Farmacia, Universidad CEU San Pablo, Urbanización Montepríncipe, Boadilla del Monte, 28668 Madrid, Spain
CQS Laboratory, C/Artistas 1, 28020 Madrid, Spain

Paul J. Arciero and Emery Ward
Human Nutrition and Metabolism Laboratory, Health and Exercise Sciences Department, Skidmore College, Saratoga Springs, NY 12866, USA

Vincent J. Miller
College of Graduate Health Studies, A. T. StillUniversity, Mesa, AZ 85206, USA

Gabriela Ruth Mendeluk
Laboratory of Male Fertility, Hospital de Clínicas "José de SanMartín", INFIBIOC, Faculty of Pharmacy and Biochemistry, University of Buenos Aires, 5950-800 Buenos Aires, Argentina

Mariano Isaac Cohen
Urology Division, Hospital de Clínicas "José de SanMartín", University of Buenos Aires, 5950-800 Buenos Aires, Argentina

Carla Ferreri
Consiglio Nazionale delle Ricerche (CNR), Istituto per la Sintesi Organica e la Fotoreattivitá (ISOF), 40129 Bologna, Italy

Chryssostomos Chatgilialoglu
Institute of Nanoscience and Nanotechnology, National Center of Scientific Research "Demokritos", Agia Paraskevi, 15310 Athens, Greece

Christopher K. Nyirenda
Ndola Central Hospital, School of Medicine, 10101 Ndola, Zambia
School of Medicine, Copperbelt University, 10101 Ndola, Zambia
Vanderbilt Institute for Global Health, Vanderbilt University, Nashville, TN 37203, USA

Edmond K. Kabagambe
Vanderbilt Institute for Global Health, Vanderbilt University, Nashville, TN 37203, USA
Division of Epidemiology, Department of Medicine, Vanderbilt University Medical Center, Nashville, TN 37203, USA

John R. Koethe
Vanderbilt Institute for Global Health, Vanderbilt University, Nashville, TN 37203, USA
Division of Infectious Diseases, Department of Medicine, Vanderbilt University Medical Center, Nashville, TN 37232, USA

James N. Kiage
Division of Epidemiology, Department of Medicine, Vanderbilt University Medical Center, Nashville, TN 37203, USA

Benjamin H. Chi
Centre for Infectious Disease Research in Zambia, 10101 Lusaka, Zambia
Department of Obstetrics and Gynecology, University of North Carolina at Chapel Hill, Chapel Hill, NC 27599, USA

PatrickMusonda
Centre for Infectious Disease Research in Zambia, 10101 Lusaka, Zambia

Meridith Blevins
Vanderbilt Institute for Global Health, Vanderbilt University, Nashville, TN 37203, USA
Department of Biostatistics, Vanderbilt University Medical Center, Nashville, TN 37203, USA

Claire N. Bosire
Division of Cancer Epidemiology and Genetics, National Cancer Institute, Nutritional Epidemiology Branch, Bethesda, MD 20850, USA

Michael Y. Tsai
Department of Laboratory Medicine and Pathology, University of Minnesota Medical School, Minneapolis, MN 55455, USA

Douglas C. Heimburger
Vanderbilt Institute for Global Health, Vanderbilt University, Nashville, TN 37203, USA
Division of Epidemiology, Department of Medicine, Vanderbilt University Medical Center, Nashville, TN 37203, USA

D. B. Kumah, K. O. Akuffo, J. E. Abaka-Cann, D. E. Affram and E. A. Osae
Department of Optometry and Visual Science, College of Science, Kwame Nkrumah University of Science and Technology, Kumasi, Ghana

Elizabeth Rendina-Ruedy, Kelsey D. Hembree, Angela Sasaki, McKale R. Davis, Stephen L. Clarke, Edralin A. Lucas and Brenda J. Smith
Department of Nutritional Sciences, Oklahoma State University, Stillwater, OK 74078, USA

Stan A. Lightfoot
Center for Cancer Prevention and Drug Development, University of Oklahoma Health Sciences Center, Oklahoma City, OK 73104, USA

Imane ElMenchawy, Asmaa El Hamdouchi, Khalid El Kari, Naima Saeid, Fatima Ezzahra Zahrou, Nada Benajiba, Imane El Harchaoui, Mohamed ElMzibri, Noureddine El Haloui and Hassan Aguenaou
Joint Unit of Nutrition and Food Research (URAC39), Ibn Tofaïl University-CNESTEN, Regional Designated Center for Nutrition (AFRA/IAEA), Kenitra, 14000 Rabat, Morocco

I. O. Oluwayemi
Department of Paediatrics, College of Medicine, Ekiti State University, Ado-Ekiti, Ekiti State, Nigeria

S. J. Brink
New England Diabetes and Endocrine Center, USA

E. E. Oyenusi and O. A. Oduwole
Paediatric Endocrinology Training Centre forWest Africa, Lagos University Teaching Hospital, Idi-Araba, Lagos, Nigeria

M. A. Oluwayemi
Clinical Nursing Services, Ekiti State University Teaching Hospital, Ado-Ekiti, Ekiti State, Nigeria

Daniel B. Hoffmann, Markus H. Griesel, Bastian Brockhusen, Mohammad Tezval, Marina Komrakova, Bjoern Menger, Klaus Michael Stuermer and Stephan Sehmisch
Department of Trauma and Reconstructive Surgery, University of Goettingen, 37075 Goettingen, Germany

Marco Wassmann
Medical Institute of General Hygiene and Environmental Health, University of Goettingen, 37075 Goettingen, Germany

Haile Woldie and Amare Tariku
Department of Human Nutrition, Institute of Public Health, College of Medicine and Health Science, University of Gondar, Gondar, Ethiopia

Yigzaw Kebede
Department of Epidemiology and Biostatistics, Institute of Public Health, College of Medicine and Health Science, University of Gondar, Gondar, Ethiopia

Astrid Kolderup
Faculty of Public Health, Hedmark University College, P.O. Box 400, 2418 Elverum, Norway

Birger Svihus
Norwegian University of Life Sciences, P.O. Box 5003, 1432 Aas, Norway

Christina M. Crowder, Brianna L. Neumann and Jamie I. Baum
Department of Food Science, University of Arkansas, 2650 North Young Avenue, Fayetteville, AR 72704, USA

Catherine Crinigan, Matthew Calhoun and Karen L. Sweazea
School of Nutrition and Health Promotion, School of Life Sciences, Arizona State University, Tempe, AZ 85287, USA

Vincent Lee
Diabetes Education Centre, Southlake Regional Health Centre, Newmarket, ON, Canada L3Y 2B1

Taylor McKay
Department of Human Health and Nutritional Science, University of Guelph, Guelph, ON, Canada N1G 2W1

Chris I. Ardern
School of Kinesiology and Health Science, York University, Toronto, ON, Canada M3J 1P3

Diego Rocha de Lucena Herrera Mascato, Michele M. Passarinho, Denise Morais Lopes Galeno and Rosany Piccolotto Carvalho
1Programa Multi-Institucional de Pós-Graduação em Biotecnologia, Universidade Federal do Amazonas, 69077-000 Manaus, AM, Brazil

Janice B. Monteiro
Division of Biochemistry, PonceHealth Sciences University, School ofMedicine, Ponce Research Institute, Ponce, PR 00732-7004, USA

Rubén J. Cruz
Biology Department, University of Puerto Rico at Ponce, Ponce, PR 00734, USA

Carmen Ortiz
Divisions of Pharmacology & Toxicology & Cancer Biology, Ponce Health Sciences University, School of Medicine, Ponce Research Institute, Ponce, PR 00732-7004, USA

LuisaMorales
Public Health Program, Ponce Health Sciences University, Ponce, PR 00732-7004, USA

Emerson Silva Lima
Faculdade de Ciências Farmacêuticas, Universidade Federal do Amazonas, 69010-300 Manaus, AM, Brazil

Shunsuke Kano
The United Graduate School of Agricultural Sciences, Ehime University, 3-5-7 Tarumi, Matsuyama 790-8566, Japan

Haruna Komada and Akihiko Sato
The Graduate School of Agriculture, Kagawa University, 2393 Ikenobe, Miki-cho, Kagawa 761-0795, Japan

Lina Yonekura and Hirotoshi Tamura
The United Graduate School of Agricultural Sciences, Ehime University, 3-5-7 Tarumi, Matsuyama 790-8566, Japan
The Graduate School of Agriculture, Kagawa University, 2393 Ikenobe, Miki-cho, Kagawa 761-0795, Japan

Hisashi Nishiwaki
Faculty of Agriculture, Ehime University, 3-5-7 Tarumi, Matsuyama 790-8566, Japan

Shunji Oshima, Sachie Shiiya, Yoshimi Tokumaru and Tomomasa Kanda
Research & Development Laboratories for Innovation, Asahi Group Holdings, Ltd., Ibaraki 302-0106, Japan

Janne H. Maier
University of Alaska Fairbanks, P.O. Box 757500, Fairbanks, AK 99775, USA
Institute of Arctic Biology, University of Alaska Fairbanks, P.O. Box 757500, Fairbanks, AK 99775, USA
Center for Alaska Native Health Research, University of Alaska Fairbanks, P.O. Box 757500, Fairbanks, AK 99775, USA

Ronald Barry
University of Alaska Fairbanks, P.O. Box 757500, Fairbanks, AK 99775, USA
Department of Mathematical Sciences, University of Alaska Fairbanks, P.O. Box 757500, Fairbanks, AK 99775, USA

Romilly E. Hodges and
University of Bridgeport, 126 Park Avenue, Bridgeport, CT 07748, USA

Deanna M. Minich
Institute for Functional Medicine, 505 S. 336th Street, Suite 500, Federal Way, WA 98003, USA
University of Western States, 2900 NE 132nd Avenue, Portland, OR 97230, USA

Mahama Saaka
School of Medicine and Health Sciences, University for Development Studies, P.O. Box 1883, Tamale, Ghana

Shaibu Mohammed Osman, Ernestina Yirkyio and Eliasu Yakubu
Tamale Teaching Hospital, P.O. Box 16, Tamale, Ghana

Anthony Amponsem, Juventus B. Ziem, Alhassan Abdul-Mumin and Prosper Akanbong
School of Medicine and Health Sciences, University for Development Studies, P.O. Box 1883, Tamale, Ghana
Tamale Teaching Hospital, P.O. Box 16, Tamale, Ghana

Sean Ervin
School of Medicine, Wake Forest University, Winston-Salem, NC, USA